# Descriptive Geometry:

## An Integrated Approach Using AutoCAD

# Descriptive Geometry:

## An Integrated Approach Using AutoCAD

**Kevin Standiford**
**Debrah Standiford**

Delmar
Thomson Learning™

Africa • Australia • Canada • Denmark • Japan • Mexico • New Zealand
Philippines • Puerto Rico • Singapore • Spain • United Kingdom • United States

## NOTICE TO THE READER



---

**Delmar Staff**

Business Unit Director: Alar Elken
Executive Editor: Sandy Clark
Acquisitions Editor: Michael Kopf
Developmental Editor: John Fisher
Editorial Assistant: Jasmine Hartman
Executive Marketing Manager: Maura Theriault
Channel Manager: Mona Caron
Marketing Coordinator: Paula Collins
Executive Production Manager: Mary Ellen Black
Production Coordinator: Jennifer Gaines
Art Director: Mary Beth Vought
Technology Project Manager: Tom Smith



Printed in the United States of America
3 4 5 6 7 8 9 10 XXX 05 04 03

For more information, contact Delmar at 3 Columbia Circle, PO Box 15015, Albany, New York 12212-5015; or find us on the World Wide Web at http://www.delmar.com

**Asia**
Thomson Learning
60 Albert Street, #15-01
Albert Complex
Singapore 189969

**Australia/New Zealand**
Nelson/Thomson Learning
102 Dodds Street
South Melbourne, Victoria 3205
Australia

**Canada**
Nelson/Thomson Learning
1120 Birchmont Road
Scarborough, Ontario
Canada M1K 5G4

**International Headquarters**
Thomson Learning
International Division
290 Harbor Drive, 2nd Floor
Stamford, CT 06902-7477
USA

**Japan**
Thomson Learning
Palaceside Building 5F
1-1-1 Hitotsubashi, Chiyoda-ku
Tokyo 100 0003
Japan

**Latin America**
Thomson Learning
Seneca, 53
Colonia Polanco
11560 Mexico D. F. Mexico

**Spain**
Thomson Learning
Calle Magallanes, 25
28015-Madrid
Espana

**UK/Europe/Middle East**
Thomson Learning
Berkshire House
168-173 High Holborn
London
WC1V 7AA United Kingdom

**Thomas Nelson & Sons Ltd.**
Nelson House
Mayfield Road
Walton-on-Thames
KT 12 5PL United Kingdom



**Library of Congress Cataloging-in-Publication Data**

Standiford, Kevin.
Descriptive geometry: An Integrated Approach Using AutoCAD / Kevin Standiford.
p. cm.
ISBN 0-7668-1123-9
1. AutoCAD. 2. Computer graphics. I. Title
T385 .S69 2000
620'.001' 00-026926

## Quick Contents

# contents

## Chapter 5 Revolutions 235



## Chapter 6 Developments 269

# preface

A technician is a specialist in the technical details of any given discipline. Their responsibilities include the creation and execution of plans for a project. However, their work would be meaningless without the ability to convey their findings to the outside world through effective written, oral, and graphic communication. Of these three forms of transmitting knowledge, graphics is often the most important. A picture is worth a thousand words. How much more difficult would it be to write out the directions for the construction and assembly of each piece of a building rather than simply draw those pieces in their proper location on a sheet of paper? For this reason, graphic communication is positively crucial to preparing and presenting a project for construction. This textbook is designed to help the student gain a better understanding of the principles involved in graphically describing a technical project through the use of descriptive geometry.

*Descriptive Geometry: an Integrated Approach using AutoCAD* will give the student an edge over using traditional descriptive geometry textbooks. Using the current technology as a building block will enable the student to gain more accuracy and efficiency in solving problems. This book covers the basic theories involved in solving descriptive geometry problems, but it also incorporates the practical use of computer aided drafting and design software into 75 percent of the theories and solutions of problems. AutoCAD was selected as the CADD package to be covered in this text for several reasons. First, AutoCAD currently holds 70+ percent of the world's PC market, making it the most popular CADD software available today. AutoCAD's releases 13, 14, and 2000 are the most common versions used in the world market, and the authors have structured the text to be compatible with all three, making this text more suitable for students entering the work force. The second reason is AutoCAD's user-friendly approach to the generation and manipulation of two- and three-dimensional drawings, which will facilitate the student's application of theoretical solutions. Third, AutoCAD can load and execute four major programming languages: AutoLISP, Visual LISP, ObjectARX, and Visual Basic. This makes it a valuable tool for solving descriptive geometry problems because these programming languages allow the technician to customize AutoCAD for specific applications. A section has been added to the end of each chapter in this book utilizing AutoLISP programming to solve a specific spacial problem. Finally, AutoCAD's capability to construct solid models using the AME package in release 12, and the ability to construct parametric solid models using Mechanical Desktop in releases 13, 14, and 2000, puts AutoCAD ahead of the competition. These features and their usefulness are incorporated into *Descriptive Geometry: an Integrated Approach using AutoCAD.*

Along with the basic theory and the enhanced sections dealing with AutoCAD, this textbook also utilizes Microsoft Excel in the solution of vector problems as well as introducing the student to Microsoft Access and its potential applications in the area of developments. By introducing the student to the principles of using spreadsheets, databases, graphic-based design packages, and current programming languages, the student becomes more valuable to a future employer. At the end of each chapter a series of questions and problems have been provided to help the student in applying the theories presented in that chapter. There is also a set of advanced problems at the end of each chapter so that students who would like to be challenged beyond the limits of the normal chapter practice problems may do so.

## features of this edition

- Extensive coverage of the concepts and theories of descriptive geometry using AutoCAD
- Step-by-step coverage of solving descriptive geometry problems using AutoCAD
- Inclusion of Microsoft Excel and Access for solving vector analyst and developments
- Integration of AutoLISP and Visual LISP programming for solving descriptive geometry problems
- Windows 95/98- and NT-based help files allowing students instant access to additional review questions, practice problems, animations, definitions, key topics, and concepts
- Chapter review questions and exercises designed for all levels
- Basic concepts of geometric construction using AutoCAD

## styles and conventions

The material covered in this textbook, exercises, review questions, and practice problems are designed to work with AutoCAD 12, 13, 14, and 2000. However, all drawing files contained on the student CD-ROM and included with the instructor resource material are saved in AutoCAD 2000 format. All drawings and illustrations associated with the material have been created at full scale; however, because of the dimensions of the pages contained within this textbook, many of these drawing had to be reduced in size and therefore may not be actually shown at full scale. Finally, the student CD-ROM contains many Windows-based help files that are designed to aid the student with the material. These help files contain links that automatically call up several of the examples provided in the text. These links are hard coded and many not work if AutoCAD is installed in any location other than C:\Program Files\ACAD2000 (this is the default installation setting for AutoCAD 2000).

## how to use this book

This book is designed to be used in a course ranging in length from eleven to sixteen weeks. It is not intended to teach students the basics of manual drafting or even AutoCAD, but instead is designed to enhance their understanding of the theory and execution of descriptive geometry problems. Therefore a basic understanding of manual drafting and AutoCAD is necessary before this text can be effectively used. The same holds true for spreadsheets. Because Excel is used in this book, it is preferable that the student have some understanding or exposure to Excel or spreadsheets in general before undertaking this course. Finally, it is not the intent of this text to teach the student Visual Basic, Visual LISP, or even AutoLISP. It is, however, the purpose of this text to demonstrate to the student how these programming languages can be incorporated into AutoCAD to enhance the productivity of the software. Consequently, a basic understanding of programming languages is also needed to better use the programming section at the end of each chapter. Additional information on customizing AutoCAD using Visual LISP, Visual Basic, and ObjectARX can be found in Autodesk Press Programmer Series. The success of this textbook is largely dependent upon the instructor's dedication to lab time, since the hands-on application of the theories presented is the main focus of this book.

A CD-ROM has been included to help the student set up the descriptive geometry problems using AutoCAD, Microsoft Access, and Microsoft Excel spreadsheets. On this disk are several descriptive geometry problems for the student to complete. Some of these problems can be found in the practice problem sections at the end of each chapter. Several AutoLISP programs also have been included to help the student when setting up some of the featured programs provided in the text. Modifying these programs will allow the student a better understanding of the mechanics of AutoLISP programming. Also provided on the student CD-ROM are nine Windows 95/98–based help files. These files are intended to be a valuable teaching aid, providing the student with quick access to definitions, examples, animations, video captures, Web links, and practice tests.

## how to use the help files

Each chapter, including the appendix, has its own help file. These files can be launched with AutoCAD, or independently, or accessed using the main menu (descriptive_geometry_today.hlp), as shown in Figure P–1. Each chapter contains the following sections: Glossary, How to, Links to Example Drawings, Links to Delmar's Web Page, and a Practice Exam (see Figure P–2). Many of the help files contain Segmented HyperGraphics or "hotspots," as they are also called. A hotspot allows the user to click on a portion of a graphic to request additional information regarding that section. To activate a Segmented HyperGraphic, the user positions the cursor over the section for which they would like to request additional information. If a Segmented HyperGraphic exists, the cursor will change from an arrow to a hand. If the cursor does not change, then a hotspot does not exist in the graphic, (see Figure P–3).

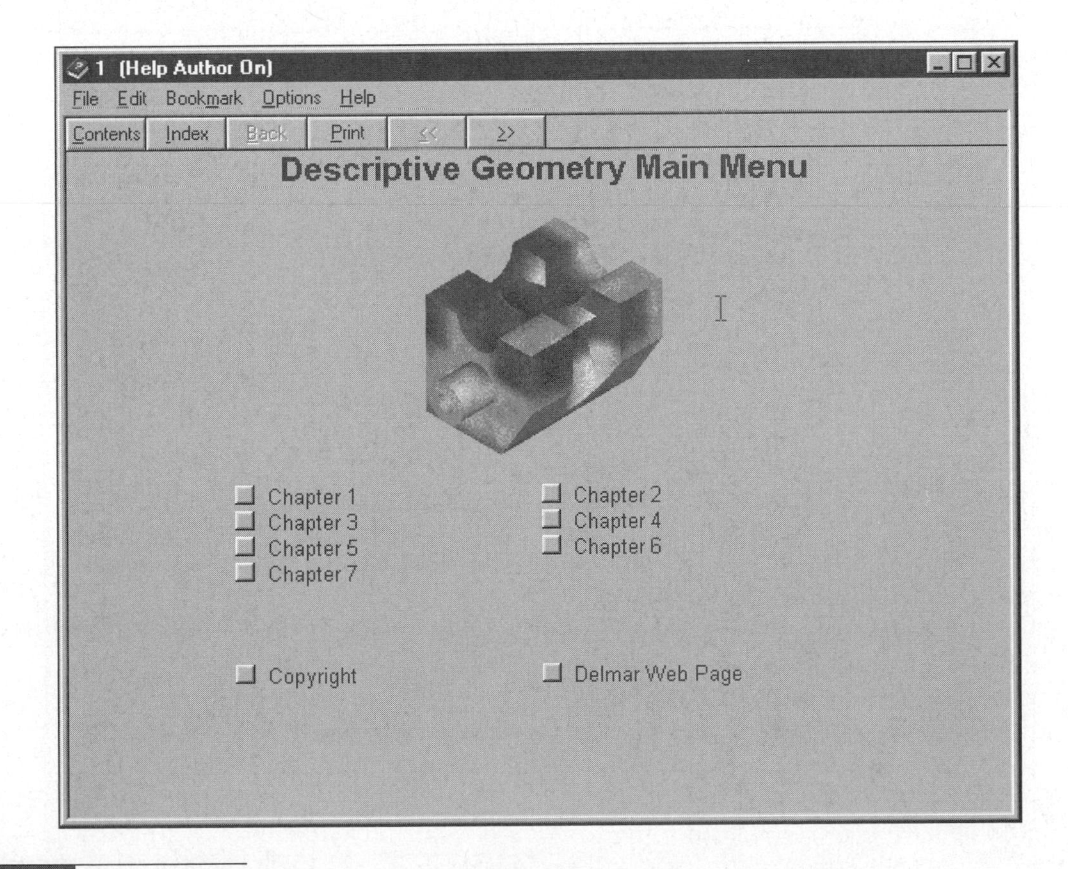


**Figure P–1** *Main screen.*

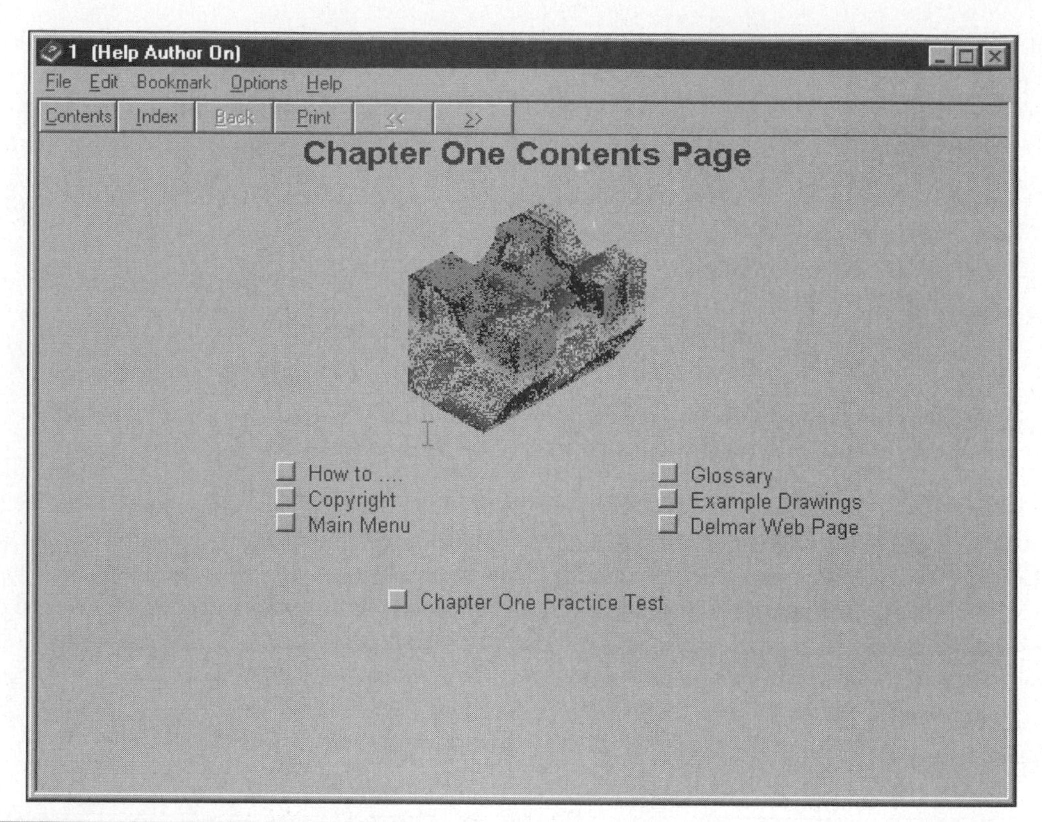


**Figure P-2** *Main screen for Chapter one*

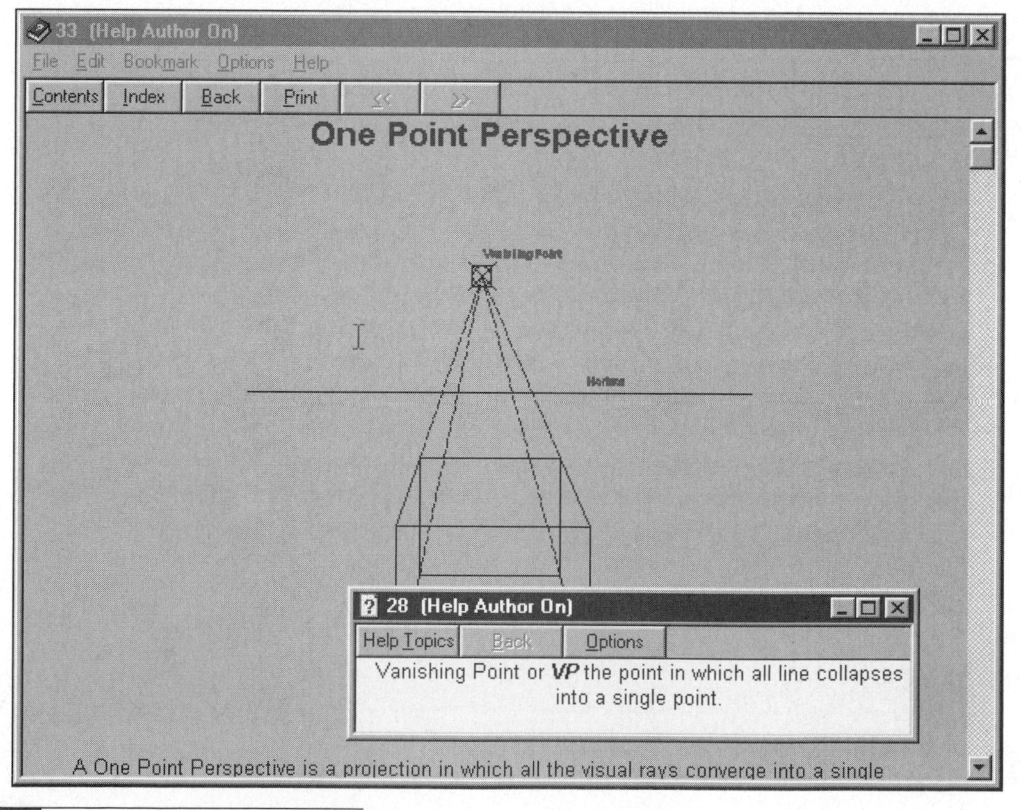


**Figure P-3** *Segmented HyperGraphic.*

## online companion

The online companion is a link to *Descriptive Geometry: an Integrated Approach Using AutoCAD* on the Internet. The site contains additional information regarding descriptive geometry in the form of links to different descriptive geometry reference materials, additional illustrations and exercises not contained with either the student or instructor supplemental material, and additional AutoLISP, Visual LISP, and Visual Basic applications designed to solve descriptive geometry problems. You can find the online companion at:

**http://www.autodeskpress.com/resources/olc.html**

## we want to hear from you

If you have any questions or comments about this text, please contact:

The CADD-Drafting Team
c/o Delmar, Thomson Learning
3 Columbia Circle
P.O. Box 15015
Albany, NY 12212-5015

www.autodeskpress.com

## supplements

**e.resource**—this is an educational resource that creates a truly electronic classroom. It is a CD-ROM containing tools and instructional resources that enriches your classroom and makes your preparation time shorter. The elements of e.resource link directly to the text and tie together to provide a unified instructional system. Spend your time teaching, not preparing to teach.

Features contained in e.resource include:

- Syllabus: Lesson that lists goals and discussion topics. You have the option of using these lesson plans with your own course information.
- Chapter Hints: Objectives and teaching hints that provide direction on how to present the material and coordinate the subject matter with student projects.
- Answers to Review Questions: These solutions enable you to grade and evaluate end of chapter tests and exercises.
- PowerPoint® Presentations: These slides provide the basis for a lecture outline that helps you to present concepts and material. Key points and concepts can be graphically highlighted for student retention.
- Test Questions: Over 800 questions of varying levels of difficulty are provided in true/false and multiple-choice formats. These questions can be used to assess student comprehension or can be made available to the student for self-evaluation.

## about the authors

Kevin Standiford received his Bachelor of Science in Mechanical Engineering Technology from the University of Arkansas at Little Rock. He has over seven years combined experience consulting in the fields of mechanical, electrical, and civil engineering. He has developed several short courses, seminars, and college classes, including Introduction to AutoCAD, Mechanical Desktop, AutoLISP Programming, Computer Numeric Control, Descriptive Geometry, Mechanical Drafting and Design, (Mechanical Desktop 3.0 & CNC), Electrical/Electronics Drafting, Algebra and Trigonometry, and Professional Project Management. In addition, he was been developing instructional material for the past five years for ITT Educational Services Inc.

Debrah Standiford received her Bachelor of Arts in Art from the University of Arkansas at Little Rock. While in college she received various scholarships and awards in the fields of art and advertising, in addition to receiving the Dean's, Chancellor's, and President's awards for outstanding academic achievement. She has over six years experience in the graphic art field, working with both technical and non-technical illustrations and designs using various software packages. Among these are Adobe Photoshop, Illustrator, PageMaker, QuarkXPress, TrapWise, PressWise, Infini-D, and AutoCAD Releases 14 and 2000. Debrah has worked with various groups and committees to help promote education in the community. She was a key advisor and contributor to the first and second annual Technology Fair, an event designed to help high school instructors gain knowledge about the technology currently available and how to effectively implement this technology into the classroom.

## acknowledgments

The authors would like to thank and acknowledge the many professionals who reviewed the manuscript, helping us publish this textbook. The technical edit and development of exam questions was performed by Dorothy Kent. Chris Mikesell, Education Supervisor, ITT Technical Institute of Little Rock, created the instructor's resource material.

A special acknowledgment is due to the following instructors, who reviewed the chapters in detail:

Bruce Bainbridge, Clinton Community College, Clinton, IA

Tom Bledsaw, ITT Educational Services, Indianapolis, IN

Gregg Cress, University of Arkansas, Little Rock, AR

Kenneth E. Critten, Texas State Technical College, Harlingen, TX

Larry Lamont, Moraine Park Technical College, Fond du Lac, WI

Robert H. Walder, Clark State Community College, Springfield, OH

The authors would also like to acknowledge and thank the staff members of Delmar Publishers for their help and support. Thanks also goes to John Shanley of Phoenix Creative Graphics for his help in the production of this book.

## dedication

This textbook is dedicated to my loving and caring wife, Debrah, who has not only made a major contribution to the content of the material but has also supported me through all the long hours. Without her contribution, support, and understanding, this project would not have been possible.

Kevin Standiford

# Viewing an Object

## Objectives

**Upon completion of this chapter the student will be able to do the following:**

- *Define descriptive geometry.*
- *Define the following terms: projection planes, projection lines, and line-of-sight.*
- *Describe the two types of orthographic projections and where they are used.*
- *Name the six views produced by a multiview drawing.*
- *Define folding lines.*
- *Explain the importance of folding lines.*
- *Describe how measurements are transferred from view to view.*
- *Construct an orthographic (multiview) projection using manual methods and AutoCAD.*
- *Define and list the three categories of axonometric projections.*
- *Describe how an isometric scale is constructed.*
- *Construct an isometric projection using manual methods and AutoCAD.*
- *Define and describe the three categories of oblique projections.*
- *Construct an oblique projection using manual methods and AutoCAD.*
- *Define ground plane, station plane, vanishing point, picture plane, ground line, and visual rays.*
- *Define one-point perspective, two-point perspective, and multiview perspective.*
- *Describe and construct a multiview perspective.*
- *List and describe the three types of three-dimensional models produced using AutoCAD.*
- *Describe the difference between nonparametric and parametric solid models.*
- *Describe the difference between paper space and model space.*

- *Know how to toggle between paper space and model space.*
- *Use* MVIEW *and* VPOINT *to produce an orthographic and isometric projection.*
- *Use* MVSETUP *to generate orthographic and isometric projections.*

## KEY WORDS AND TERMS

**Adjacent view**
**Axonometric projection**
**Bottom view**
**Cabinet oblique**
**Cavalier oblique**
**Descriptive geometry**
**Dimetric projection**
**First angle projection**
**Floating viewport**
**Folding lines**
**Front view**
**Frontal plane**
**Gaspard Monge**
**General oblique**
**Ground line**
**Ground plane**
**Horizontal plane**
**Isometric projection**
**Isometric scale**
**Layout**
**Left side view**
**Line-of-sight**
**Miter line**
**Multiview drawings**
**Non-parametric**
**Object lines**
**Oblique projection**
**Orthographic projection**
**Parametric**
**Perspective**
**Picture plane**
**Plan view**
**Plane**
**Profile plane**
**Projection plane**
**Rear view**
**Right side view**
**Solid model**
**Station point**
**Surface model**
**Third angle projection**
**Top view**
**Trimetric projection**
**Vanishing point**
**Visual ray**
**Wire frame model**

## Introduction to Descriptive Geometry

When taking a course like descriptive geometry for the first time, there are two questions that a student will often ask: "What is this course about?" and "How will it benefit me?" The first question can be answered by examining the course title one word at a time. The word descriptive means to tell about. Geometry is the branch of mathematics that studies the relationships of three-dimensional lines, angles, surfaces, and solids by a means of comparing and measuring. So by deduction it can be reasoned that descriptive geometry involves describing a problem by the use of pictures. When these two words are combined a comprehensive explanation of descriptive geometry is derived. ***Descriptive geometry*** is the branch of mathematics that precisely describes three-dimensional objects by projecting their three-dimensional reality onto a two-dimensional surface such as paper or a computer screen. It incorporates techniques and procedures developed for both manual and computer-aided drafting. To answer the second question, by studying descriptive geometry the student gains the necessary tools and principles to graphically visualize, manipulate, and solve engineering and architectural design problems.

Traditionally these problems were solved by doing mathematical calculations. It was not until the late 1700s that a young mathematician named ***Gaspard Monge*** proved that he could solve three-dimensional spatial problems graphically, and by doing so, fathered the science of ***descriptive geometry***. Monge, whose developments were kept a military secret for a number of years afterwards, originally developed orthographic projection and the revolution method to solve complex military fortress design problems. Using the techniques he pioneered, Monge could often solve these problems more quickly than the traditional methods of mathematics, and with equal accuracy.

Until recently, the basic tools used in descriptive geometry have remained the same as those employed 200 years ago by Monge for the development of his procedures. The instruments and procedures have been refined throughout time, as better materials became available for the production of manual drafting equipment. Even a few techniques have been refined to utilize the better drafting equipment, but the basic methods are still Monge's. Although the improvements to the manual drafting equipment created more reliable and accurate tools, some solutions remained less reliable than those secured by mathematics. For example, when solving vector problems using manual drafting procedures and equipment, the results are not as precise as those derived from trigonometry or calculus. This is often because these problems have magnitudes so large that to illustrate them on paper would require the use of a scale so small that the width of the pencil lead could cause a tremendous variation in the results. Depending upon the application of the problem being solved, this variation could become too great and render the result unacceptable. With the introduction of the computer and computer-aided drafting and design software (CADD), this is no longer a problem. Objects can now be drawn to actual size, alleviating the problems associated with manual scales.

Even though the computer has enhanced many aspects of drafting, especially the accuracy, the technician should still have a good understanding of the basics of manual drafting. By studying the principles of manual drafting, the technician is more capable of executing the operations on the computer, and understanding the mechanics of setting up the problem and interpreting the results. This is why the mechanics of solving these problems is still an important aspect, even if the computer is used as the main drafting tool.

## Types of Two-Dimensional Views

Of all the techniques covered in descriptive geometry, the way in which an object is presented or viewed is the most important. If a part is to be manufactured and the angle or view shown does not provide the necessary information, then the part could never be manufactured to design expectations. The following section examines the four major classes of views: orthographic projections, axonometric projections, oblique projections, and perspectives. It is from these four classes of views that the results of many of the problems solved in descriptive geometry are obtained. It is therefore important to fully understand these concepts—how they are generated and where they are used.

### Orthographic Projection (multiview)

When designing a part to be manufactured, supplying all the necessary information regarding the part's sizes and the location of its features is vital. All this can easily be accomplished when the part is relatively simple, and all the essential data can be provided in one view. For example, suppose that a two-inch square piece of steel plating that is 1/2" thick is to be fabricated with a 1/2" diameter hole drilled at its center. All the required information about this part can be contained on one drawing consisting of a single view (see Figure 1–1). When a more complex part is produced, a single view is not sufficient to clearly show all its features. In other words, a series of grooves on the opposite side of this same part would be shown as hidden features, resulting in a drawing that is difficult to interpret (see Figure 1–2). To solve this problem, additional views must be created that will reveal all the hidden attributes of the part. In this instance, an additional view of the side is required (see Figure 1–3). Very complex parts might contain views showing the front, sides, back, bottom, and top. A drawing containing two or more of these views is called a ***multiview drawing*** or ***orthographic projection.***

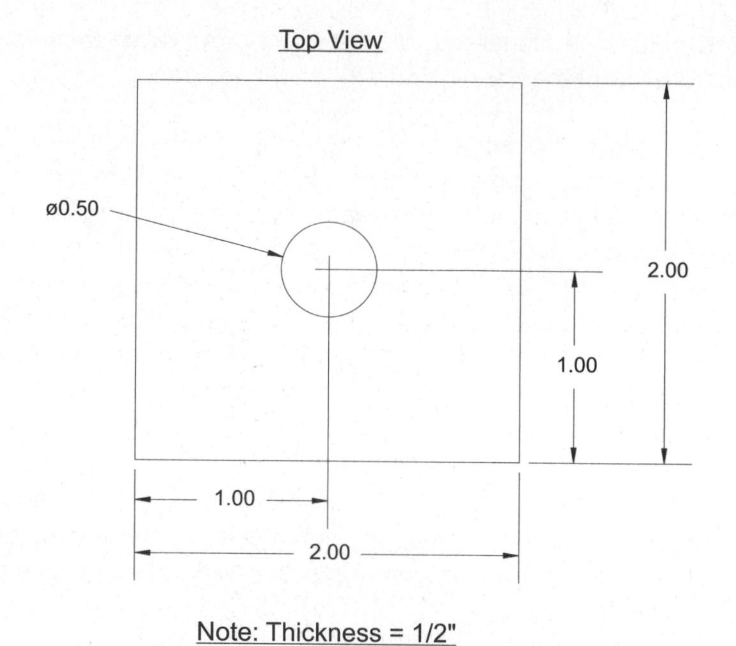


**Figure 1–1** *Single view of a 2" x 2" x 1/2" steel plate with a 1/2" hole drilled in the center.*

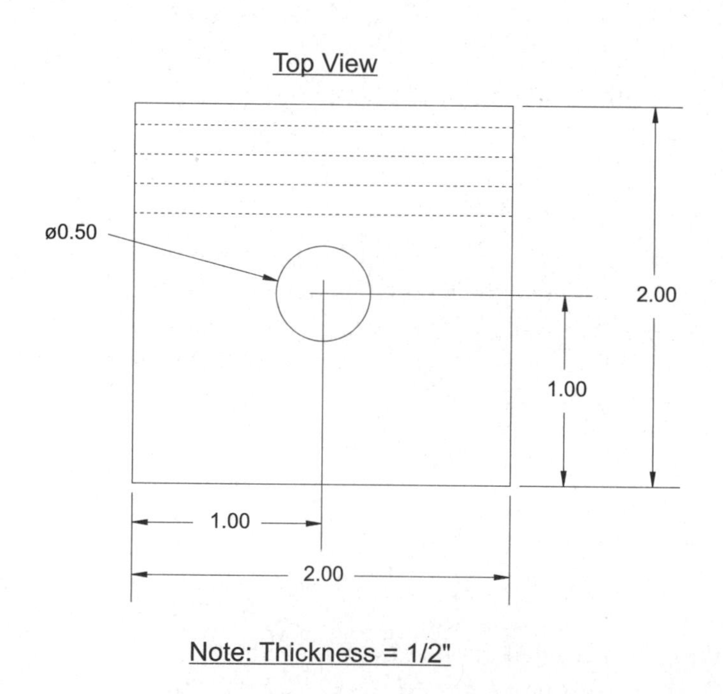


**Figure 1–2** *Single view of a 2" x 2" x 1/2" steel plate with a 1/2" hole drilled in the center and containing a series of grooves as hidden features.*

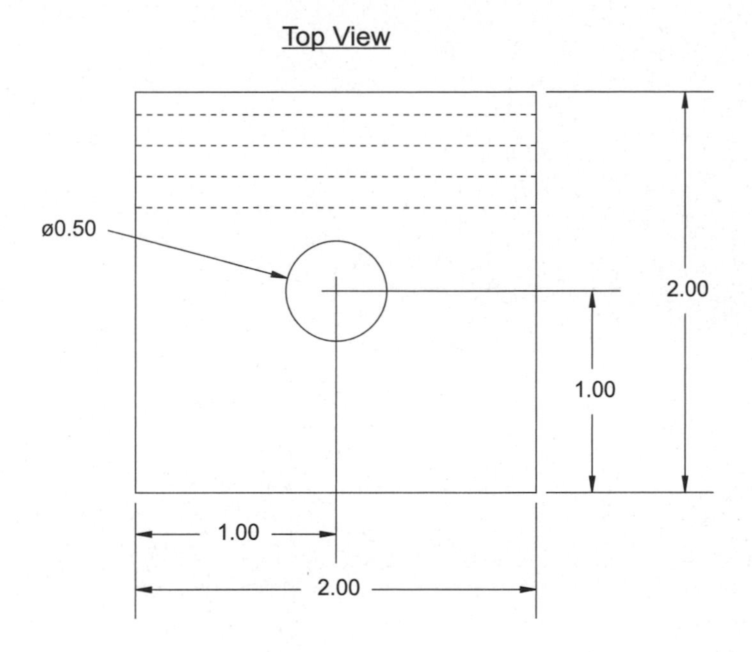


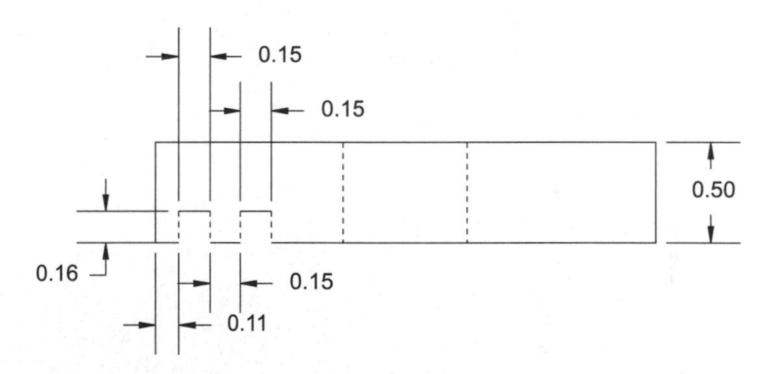


**Figure 1-3** *Orthographic projection of a 2" x 2" x 1/2" steel plate with a 1/2" hole drilled in the center and containing a series of grooves.*

There are some basic terms associated with orthographic projections to become familiar with before attempting to create one. A ***plane*** is a noncurved surface; it can be thought of as a sheet of paper or glass. In descriptive geometry there are three principal planes on which an object is projected: ***horizontal plane***, ***profile plane***, and ***frontal plane*** (see Figure 1–4). The current plane in which the projection is taking place is called the ***projection plane***; therefore, the projection plane can be any of the three principal planes. ***Line-of-sight*** is the angle from which an observer is viewing an object. The ***object lines*** are the contours defining the part. ***Adjacent views*** are two orthographic views that share a common boundary line.

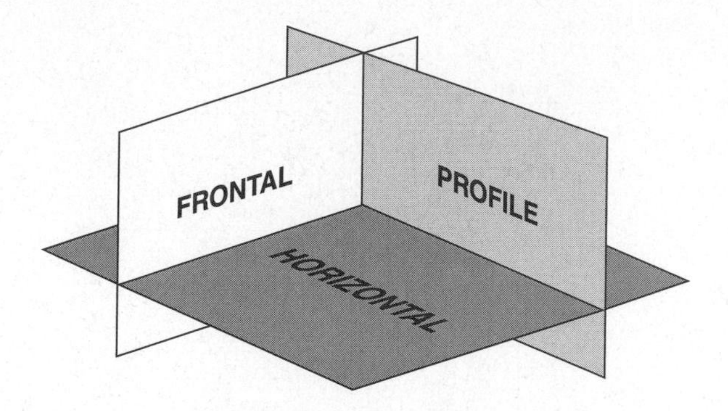


**Figure 1–4** *Three principal planes of projection: Horizontal, Profile, and Frontal. In this figure the Frontal and Profile Planes are adjacent, the Frontal and Horizontal Planes are adjacent, and the Profile and Horizontal Planes are adjacent.*

In a multiview drawing, the object can be visualized as being encased in a glass cube (see Figure 1–5), with the object lines projected onto the surface of the glass (see Figure 1–6). The glass represents the projection plane, with the object's contours transferred via the projection lines. Once the object's contours have been transferred (projected) to the projection plane, the cube is then unfolded to reveal six different views of the object (see Figure 1–7).

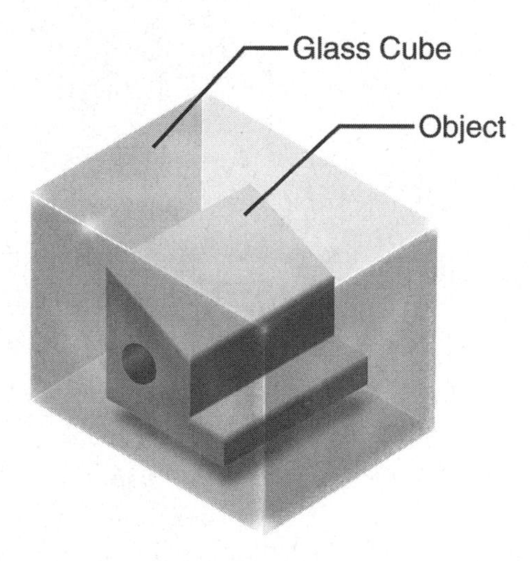


**Figure 1–5** *Object encased in a glass cube.*

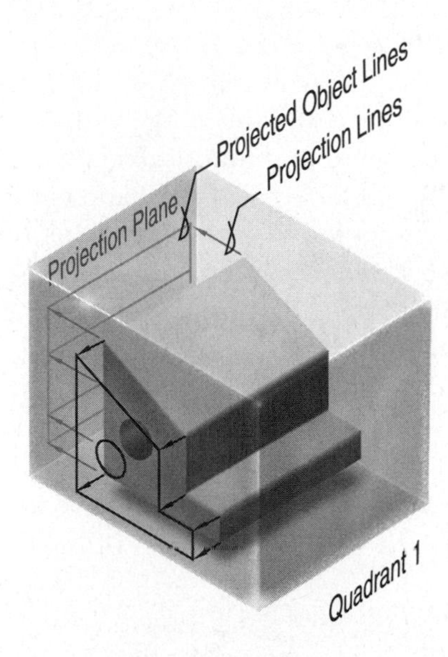


**Figure 1-6** *Object encased in glass cube with object lines projected onto surface of glass cube.*

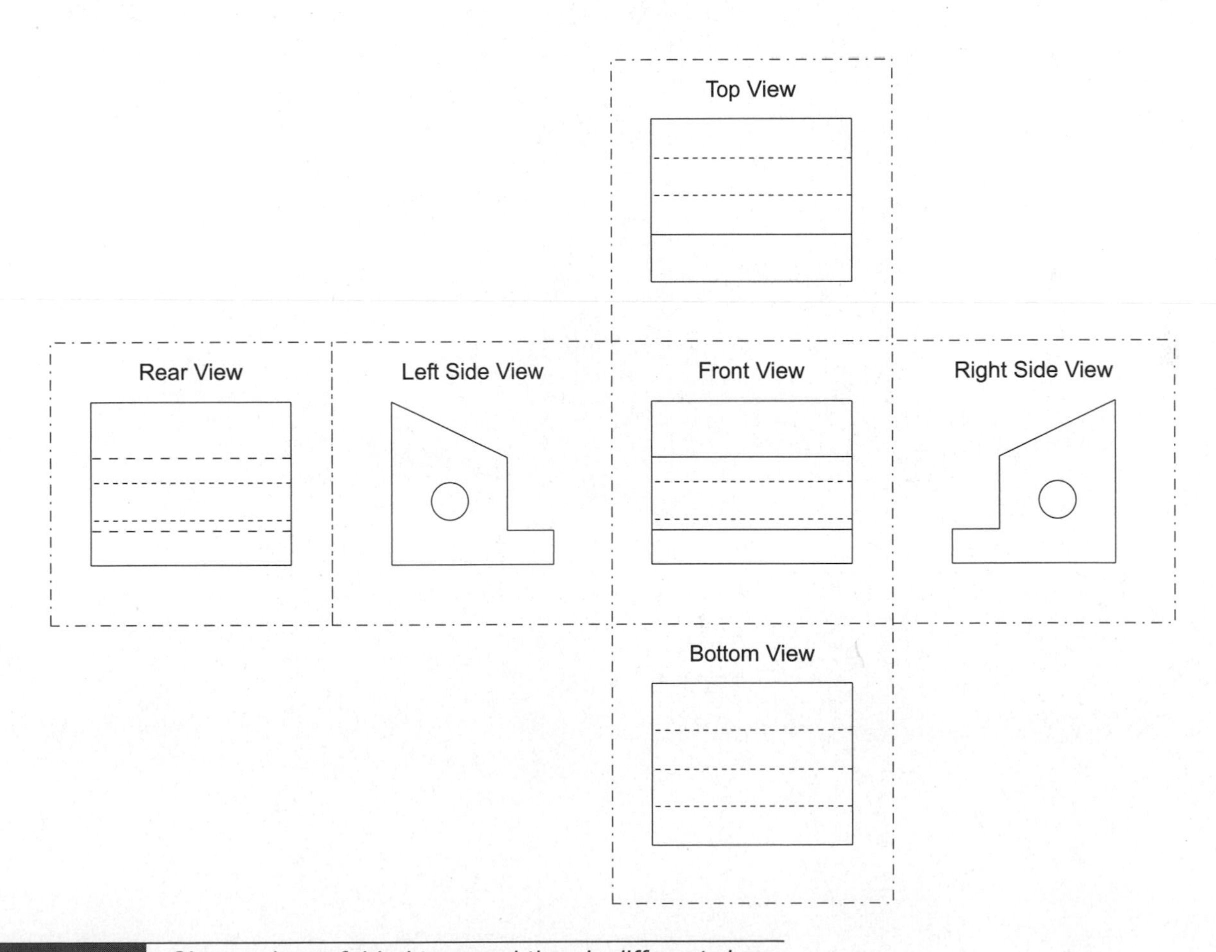


**Figure 1-7** *Glass cube unfolded to reveal the six different views.*

### First Angle Projection (Standard in Europe)

When constructing a multiview drawing, the arrangement of the views is determined by the quadrant in which the object is placed. A quadrant is an area formed by the three principal planes intersecting at right angles to one another. These intersections produce four distinct quadrants (see Figure 1–8). The quadrants are numbered one through four starting in the upper right-hand corner and continuing counterclockwise. This formation of quadrants allows the object to be placed in one of four possible positions, but only two of these positions are standard. These two standard positions are the first and third quadrants. When the object is placed in the first quadrant, it is called a first angle projection (see Figure 1–9). In a first angle projection, the top view is generated by projecting the object's contours *down* onto the horizontal plane. The right side view is generated by projecting the right side of the object *through* the object to the profile plane. This results in an arrangement in which the top view is positioned below the front view and the right side view is located to the left of the front view. The rest of the views are arranged as shown in Figure 1–10. This method of projection is used extensively in foreign countries, but is rarely used in the United States.

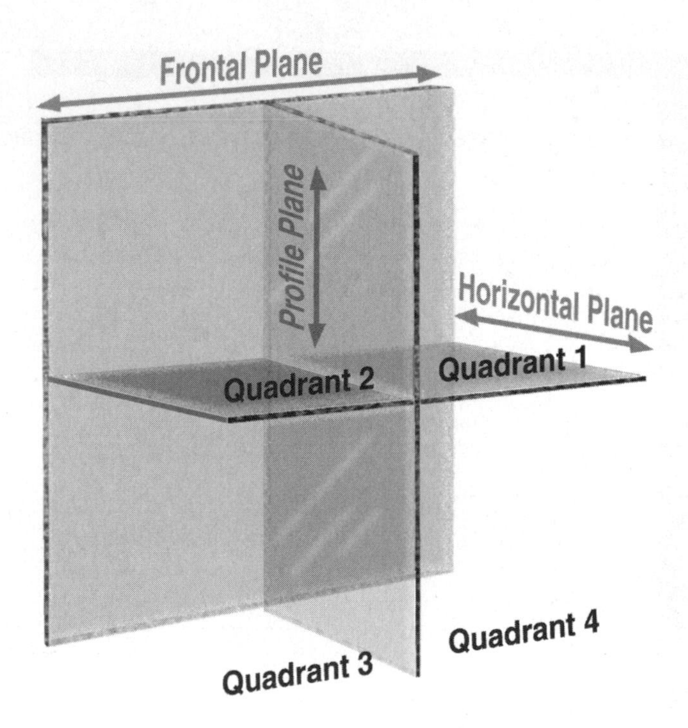


**Figure 1–8** *Four quadrants produced by the intersection of the three principal planes.*

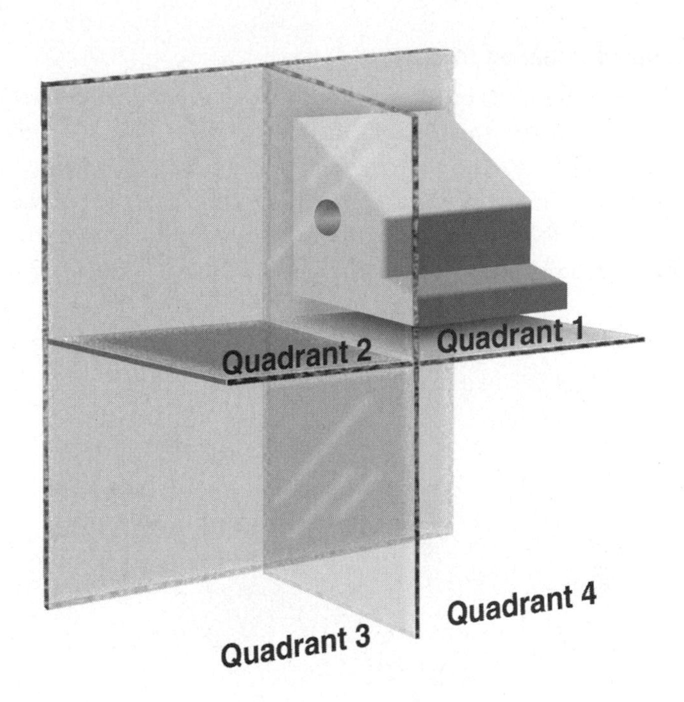


**Figure 1–9** *Object is positioned in the first quadrant. This arrangement is known as first angle projection.*

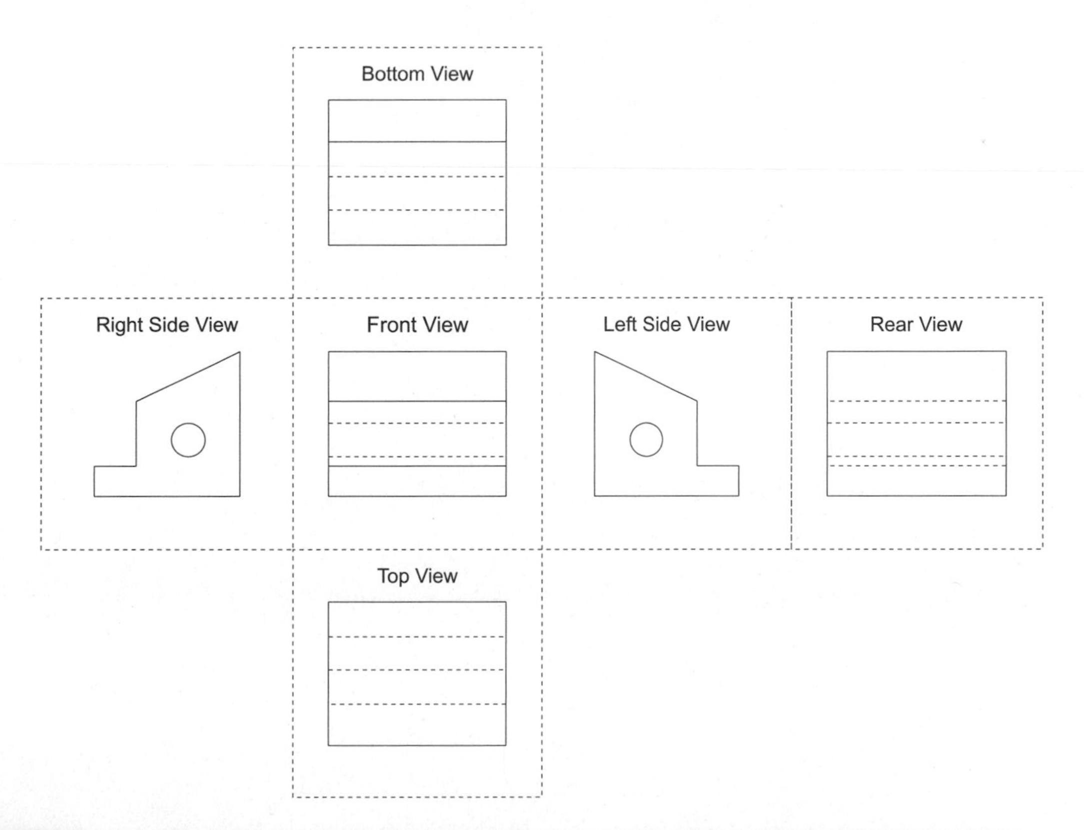


**Figure 1–10** *The arrangement of the views as a result of the object being positioned in the first quadrant.*

#### Third Angle Projection (Standard in United States)

The second method of arranging the views of a multiview drawing, ***third angle projection*** receives its name from the fact that the object is placed in the third quadrant (see Figure 1–11). In this method, the top view is projected *up* onto the horizontal plane. The right side view is generated by projecting the object's contours onto the profile plane. This results in an arrangement of views in which the top view is positioned above the front view and the right side view is located to the right of the front view. The rest of the views are arranged as shown in Figure 1–12. This method of projection is used extensively in the United States.

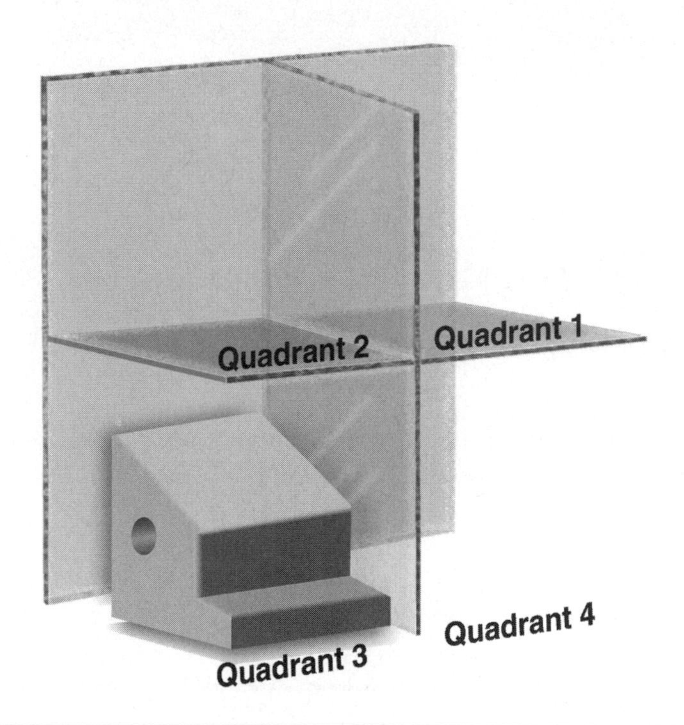


**Figure 1–11** *Object is positioned in the third quadrant. This arrangement of views is known as third angle projection.*

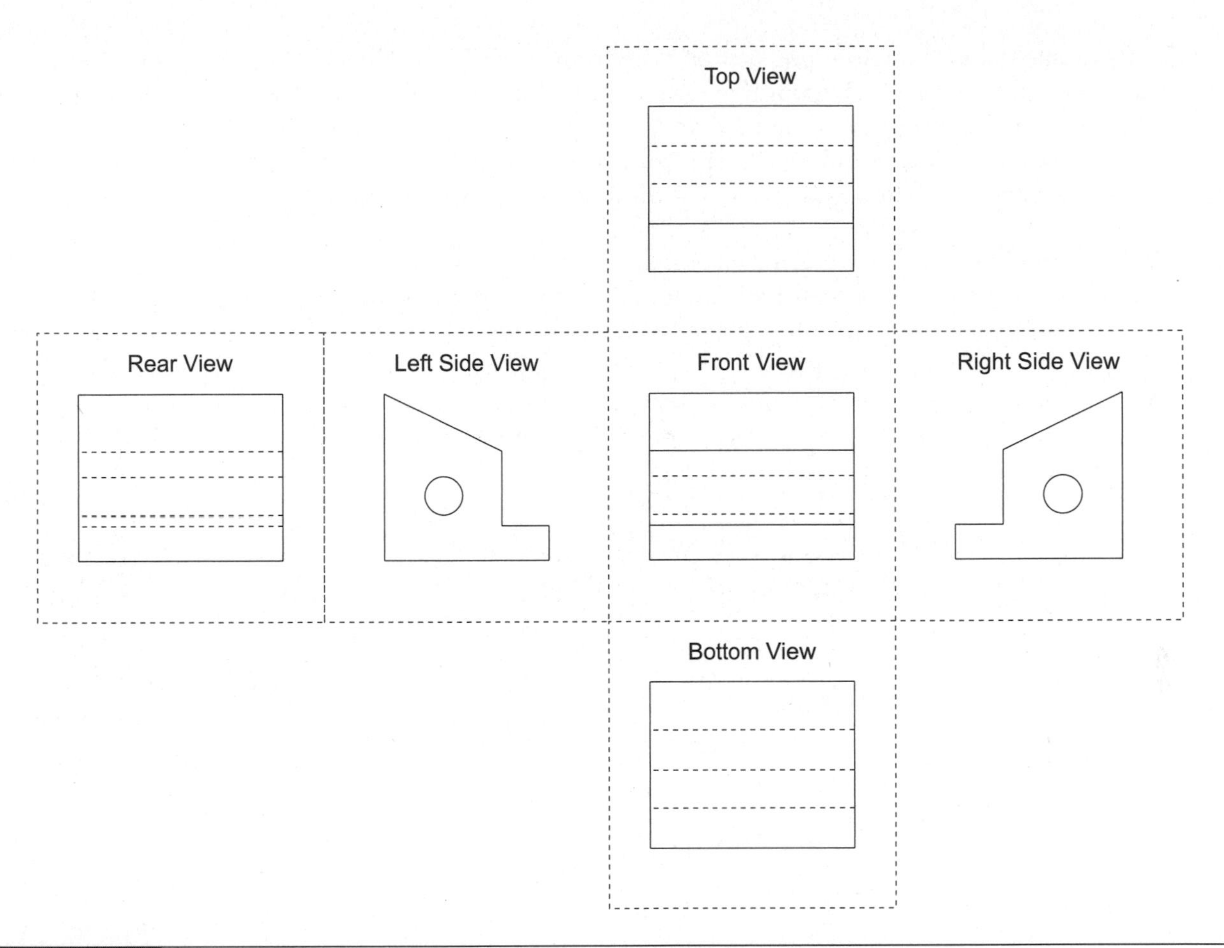


**Figure 1–12** *The arrangement of views as a result of the object being positioned in the third quadrant.*

**Note: Notice the difference in the location of views between the first angle projection and the third angle projection. The views are mirrored in the different projections.**

### Constructing a Multiview Drawing

The views produced from the unfolding of the glass cube are ***right side view, left side view, plan*** *or* ***top view, front view, bottom view,*** and ***rear view.*** Recall that in the previous example of a multiview drawing the object was encased in a glass cube and its contour lines were then projected onto the surface of the cube. Thereafter the cube was unfolded and the six views were produced. The edges of the glass cube play an important role in the construction of a multiview drawing. These edges are called ***folding lines*** (see Figure 1–13) and they show where the planes begin and end. The folding lines are used as a basis for locating points, lines, and surfaces in other views. When creating a multiview drawing, the features of the object are measured from the folding lines and the measurements are then transferred to other views. For example, to generate a multiview drawing of the part illustrated in Figure 1–13, the bottom folding line of the top view is used to measure the horizontal placement of the features in this view, which are labeled TD1-TD5 (top view dimension). These measurements are then rotated 90° and placed along the left folding line of the side view. Any remaining features from the top view are located in their proper position in the side view by following the same procedure. The top folding line of the front view is used to measure the horizontal features in this view labeled FD1-FD3, which are then transferred straight across into the side view. TD1-TD5 will intersect with FD1-FD3 to create the side views of the object with all the features in their proper location, as shown in Figure 1–13.

Frequently in manual drafting, the top view and the front view are created simultaneously. Recall that in a third angle projection the top view is positioned above the front view, which allows the features in the top view to be transferred to the front view by merely extending a line from the feature in the top view down into the front view. As a result, portions of the top and front view can be made without making repetitive measurements. This saves time and cuts down on the possibility of mistakes from multiple interpretations of the scale.

A common method used in manual drafting to transfer dimensions from the top view to the side view is the ***miter line***. A ***miter line*** is a line drawn at a 45° angle from the top view's bottom folding line upward to the side view's right folding line. Points are then extended from the top view across to the miter line; at the point where they intersect with the miter line, they are turned 90° and extended down into the side view. The actual location of the points in the side view is determined by extending lines across from the front view to the point at which they intersect with the lines extended down from the miter line. These intersections indicate the position of points in the side view, as shown in Figure 1–14.

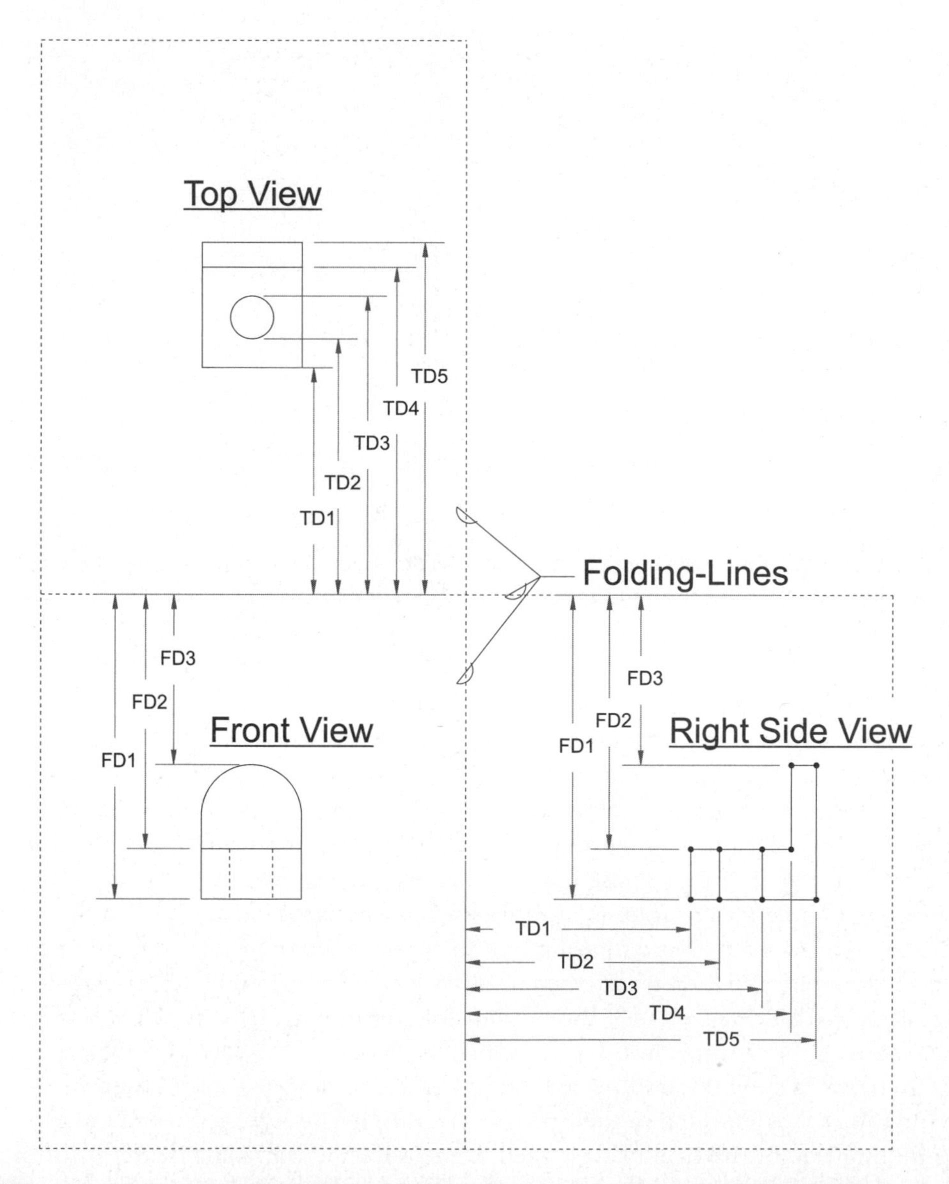


**Figure 1–13** *The transfer of measurements using the folding lines as a working edge.*

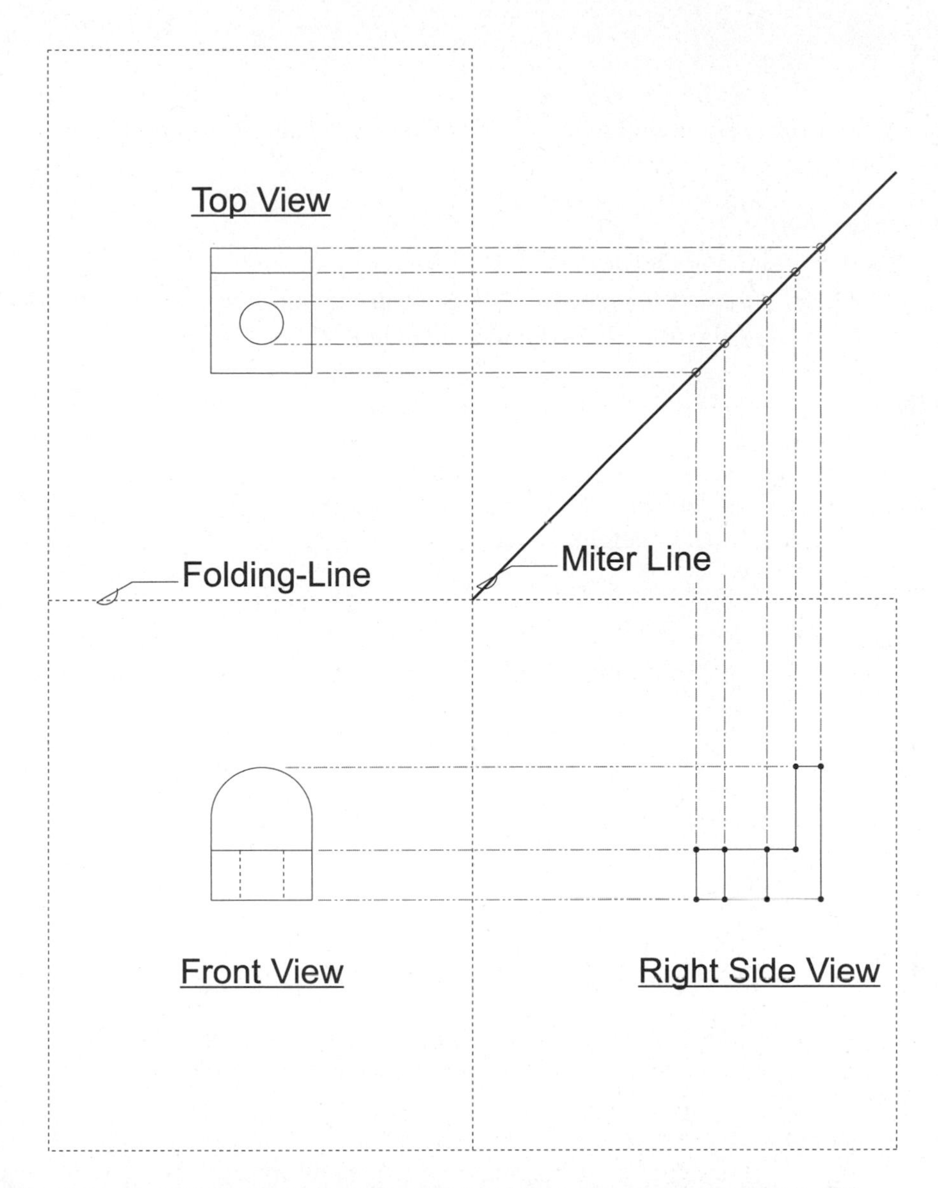


**Figure 1–14** *The transfer of measurements using the miter line as a turning point.*

#### Using AutoCAD to Create Two-Dimensional Multiview Drawings

When using AutoCAD to create a two-dimensional multiview drawing, the same techniques used in manual drafting are used to begin the computer-generated drawing. The top and front views are again created simultaneously, allowing the transfer of information from one view to the other. For example, in creating a multiview drawing of Figure 1–13, the technician could start by drawing a 2" square in the top view, then extending a line from the lower left-hand corner of the square down into the front view. The corner of the 2" square is now partially established in the front view. The other side of this corner can be found by choosing an arbitrary point along this line and extending a line out perpendicular to the first. This creates the left-hand corner and bottom edge of the square in the front view. Now a line can be extended from the lower right-hand side of the square into the front view. The point at which it intersects with the bottom edge of the object line (in the front view) forms the lower right-hand corner of the square. To establish the height of the square the OFFSET command is used. In the following example, the techniques for transferring measurements are illustrated by drawing the top and front view of a 2" x 1/2" square.

### Step #1

Start by using the RECTANG command to draw a 2" square. The coordinates for the location of the square are not important at this time.

Command: RECTANG (ENTER)

First corner: *(Select an arbitrary point.)*

Other corner: @2,2 *(Press* ENTER *to create a rectangle starting from the first corner selected, 2 units in length along the positive X-axis, and 2 units in length along the positive Y-axis.)*

### Step #2

Now start transferring measurements to the front view by first drawing a line from the lower left-hand corner of the object to an undetermined point below the object. Then, draw a line from the lower right-hand corner of the object to an undetermined point below the object, thus establishing the sides of the object in the front view.

Command: LINE (ENTER)

From point: end of *(Select the lower left-hand corner of the object in the top view.)*

To point: *(Select an arbitrary point below the top view.)*

To point: *(Press* ENTER *to terminate the* LINE *command.)*

Command: LINE (ENTER)

From point: *(Select the lower right-hand corner of the object in the top view.)*

To point: *(Select an arbitrary point below the top view.)*

To point: *(Press* ENTER *to terminate the* LINE *command.)*

### Step #3

The bottom edge of the object is created by moving down several inches below the top view of the part and drawing a line that crosses both of the lines that extended downward.

Command: LINE (ENTER)

From point: end of *(Select an arbitrary point along the left-hand line below the object. This establishes the lower left-hand corner of the front view.)*

To point: per of *(Select the line extended from the lower right-hand corner. This completes the bottom edge of the object in the front view.)*

To point: *(Press* ENTER *to terminate the* LINE *command.)*

### Step #4

The top edge of the object (in the front view) is created by using the OFFSET command. Finally, the FILLET command with a zero radius can be used to trim off excess line lengths, revealing a plan and front view of the object.

Command: OFFSET (ENTER)

Offset distance or Through <Through>: 0.5 *(Press* ENTER *to set the offset distance to 0.5 inches.)*

Select object to offset: *(Select the bottom edge of the object in the front view.)*

Side to offset? *(Select a point above the bottom edge.)*

Select object to offset: *(Press* ENTER *to terminate the* OFFSET *command.)*

Command: FILLET *(Press* ENTER *to use the* FILLET *command with a zero radius to clean up the object by trimming the excess from the lines.)*
(TRIM mode) Current fillet radius = 0.0000
Polyline/Radius/Trim/<Select first object>: *(Select first corner)*
Select second object:
Command: FILLET (ENTER)
(TRIM mode) Current fillet radius = 0.0000
Polyline/Radius/Trim/<Select first object>: *(Select second corner)*
Select second object:
Command: FILLET (ENTER)
(TRIM mode) Current fillet radius = 0.0000
Polyline/Radius/Trim/<Select first object>: *(Select third corner)*
Select second object:
Command: FILLET (ENTER)
(TRIM mode) Current fillet radius = 0.0000
Polyline/Radius/Trim/<Select first object>: *(Select fourth corner)*
Select second object:

The advantage of using AutoCAD to produce a multiview drawing is that each view does not have to be drawn separately. By using the MIRROR command a left-side view can be created from a right-side view, a bottom view from a top view, and a rear view from a front view, with only minor changes.

## Axonometric Projections

***Axonometric projections*** are pictorial drawings in which the object is rotated so that all three dimensions are seen at once (length, width, and depth). This type of projection is especially helpful when a non-technical person is trying to visualize what a part is going to look like. Axonometric projections can be divided into three categories. These categories are ***isometric***, ***dimetric***, and ***trimetric*** projections.

### Isometric Projection

An ***isometric projection*** is an object that has been projected so that the three principle axes are inclined equally 120° (see Figure 1–15). In an isometric projection, an imaginary axis is placed at the lower left-hand corner of the object. The object is then rotated 45° counterclockwise from the X-axis. Finally, the object is tilted forward 35°16' counter clockwise from the Y-axis along the YZ plane (see Figure 1–16). This produces a view of the object revealing all three dimensions: length, width, and depth.

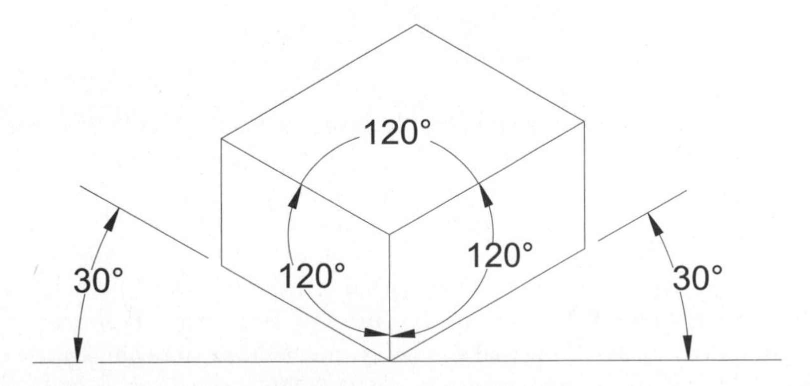


**Figure 1–15** *The three principle axes are inclined equally 120°.*

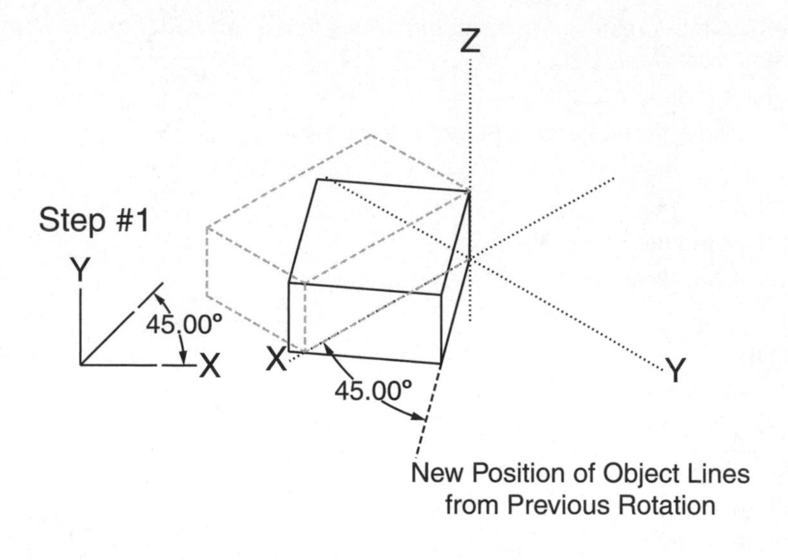


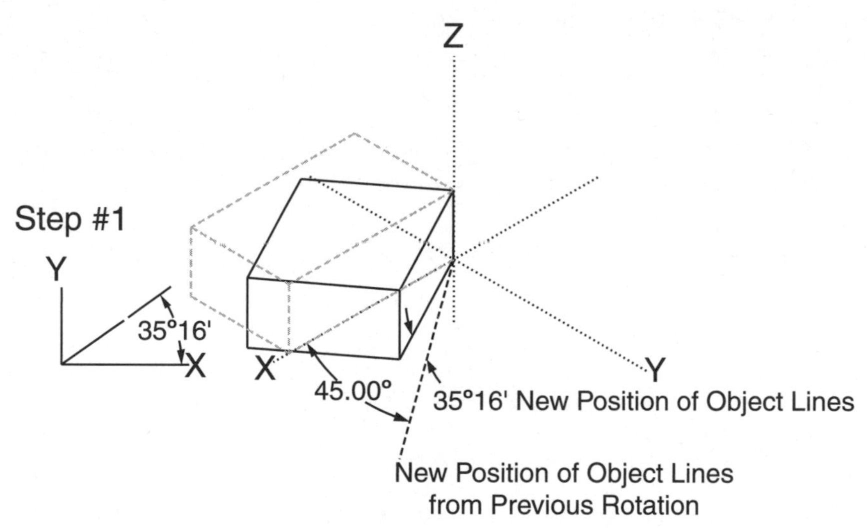


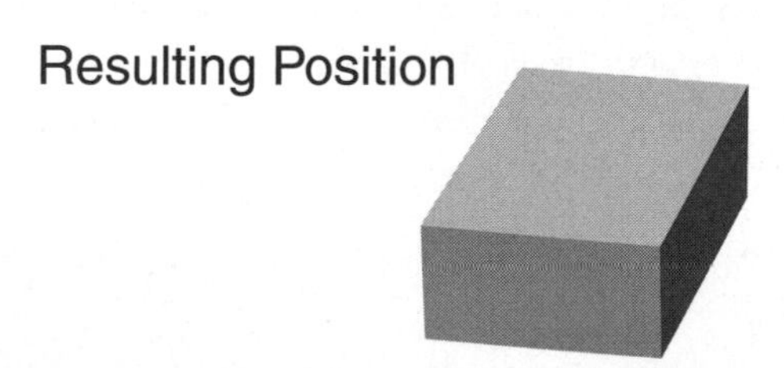


**Figure 1–16** *In the construction of an isometric, the object is first rotated 45° in the XY plane and then 35°16' in the YZ plane.*

#### Isometric Scale

Because the object has been rotated in such a way that the three principal planes are inclined 120°, the isometric projection is foreshortened. This means that the true isometric will appear to be smaller than its actual size, approximately 80% or /(2/3) actual size. This scale factor can be illustrated by first constructing an XY-axis (see Figure 1–17A). Next, starting at the origin (X0, Y0), draw a line that is at a 45° angle counter-clockwise from the X-axis (this line can be any length), as shown in Figure 1–17B. Placing a scale along the line just drawn with zero set at the origin, make a tick mark every 1/8 inch (see Figure 1–17C). Construct a line starting at the origin that is at a 30°-angle counterclockwise from the X-axis (this line can be any length), as shown in Figure 1–17D. Draw a series of lines from the tick marks on the 45° line perpendicular to the X-axis (see Figure 1–17E). These lines should cross the 30° line. Finally, placing a scale along the line drawn at 30° with zero set at the origin, you will find that the distance is now foreshortened, and the eight tick marks that equaled one inch on the 45° line now equal 0.8165 on the 30° line (see Figure 1–17F). In an isometric projection, all three axes are drawn at this scale of 0.82.

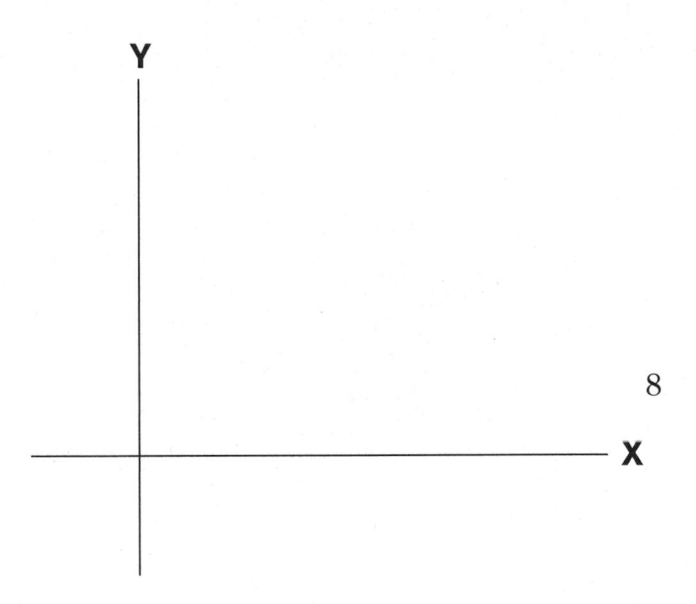


**Figure 1–17A** *The X- and Y-axis are constructed.*

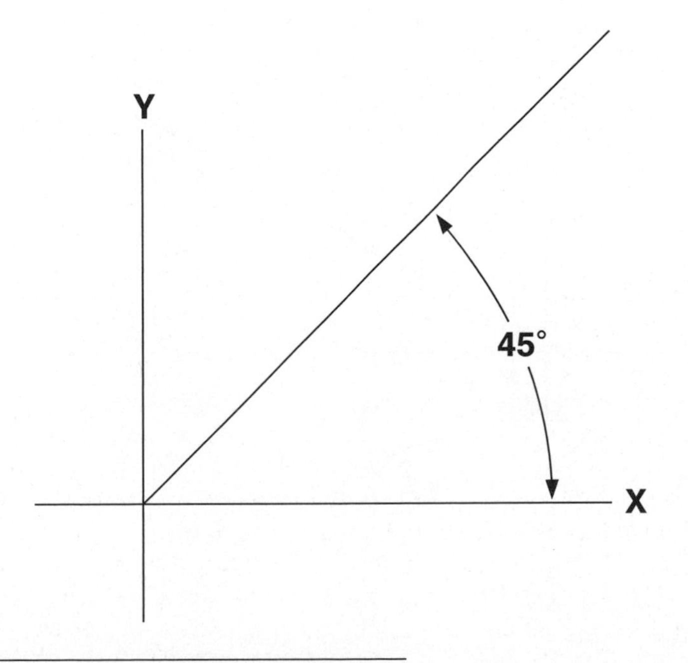


**Figure 1–17B** *A line drawn 45° from the positive X-axis.*

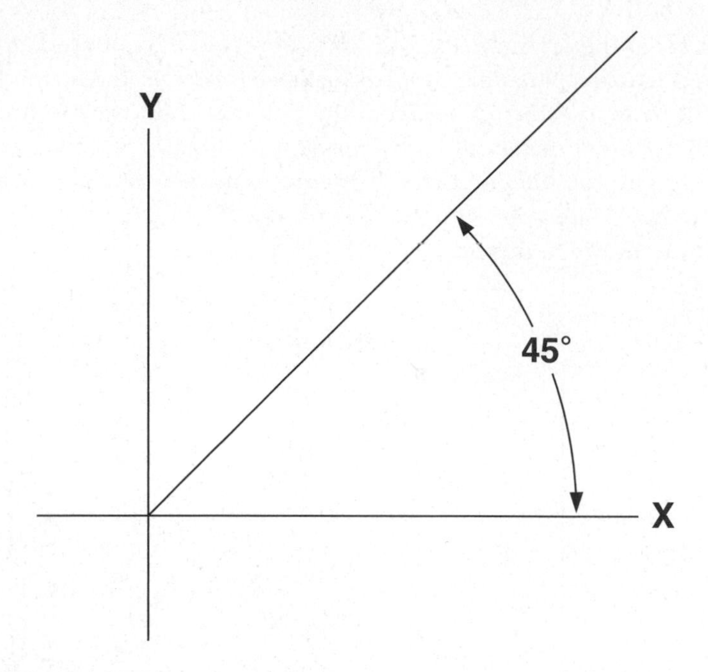


**Figure 1–17C** *A scale is positioned along the line just drawn and the line is divided into equal segments.*

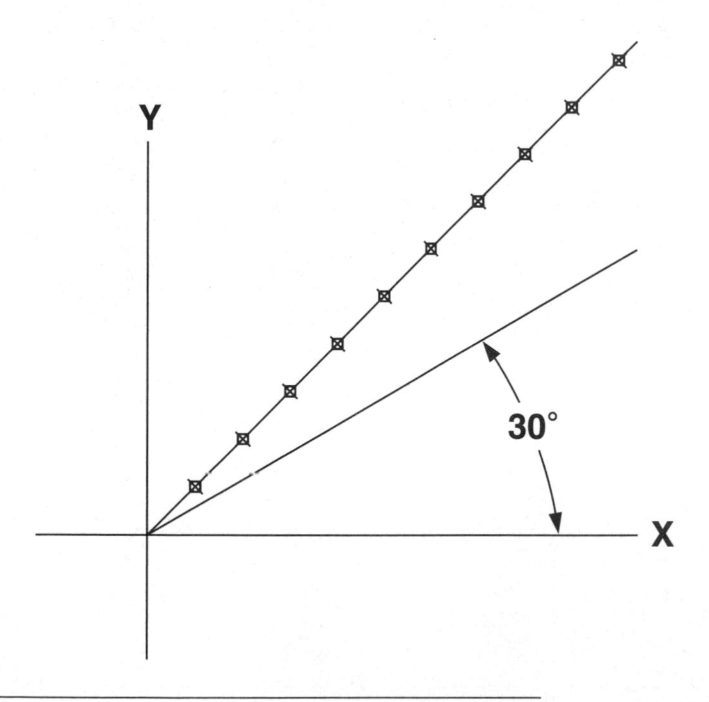


**Figure 1–17D** *A second line is drawn 30° from the positive X-axis.*

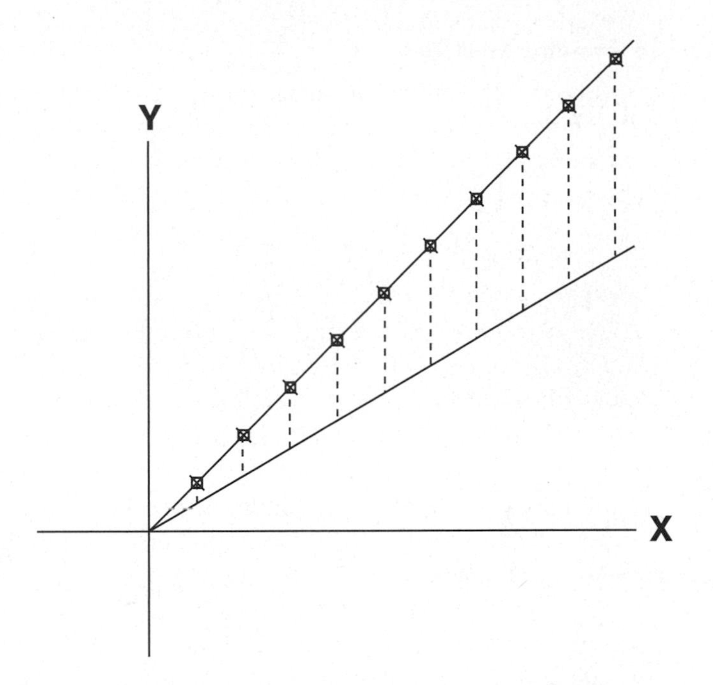


**Figure 1-17E** *Lines are extended from the division marks on the 45° degree line perpendicular to the positive X-axis.*

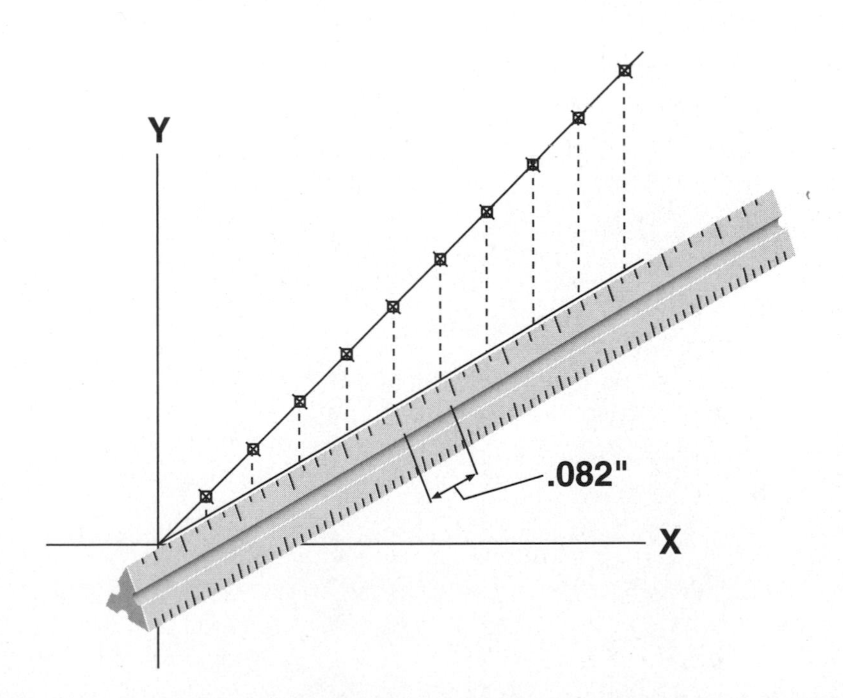


**Figure 1-17F** *Measuring the distance between the intersections of the extended lines reveals that the isometric scale is equal to 0.82 or 82% of the actual scale.*

### Using AutoCAD to Produce an Isometric Projection

AutoCAD has several different ways of handling an isometric drawing. The object can be drawn as a two-dimensional drawing using techniques similar to the traditional manual drafting procedures, or it can be drawn as a three-dimensional solid object using a totally different approach. The following section focuses on AutoCAD's two-dimensional ability. However, later in this chapter AutoCAD's three-dimensional capabilities will be explored.

A two-dimensional isometric projection can be drawn in AutoCAD using the *I*sometric option of the SNAP command. This option rotates the cross hairs into one of three different isometric planes (also know as an isoplane. An isoplane is a term used in AutoCAD to denote the projection planes while in isometric mode. When the isometric mode is enabled, the technician can draw lines using AutoCAD's ORTHOgraphic command (F8), that snaps to 30°, 90°, 150° 210°, 250°, and 330° angles from the X-axis. Once the SNAP command has been executed, the cross hairs will rotate to match one of the configurations shown in Figure 1–18, 1–19, and 1–20. The technician can then toggle between these planes by pressing CTRL and E simultaneously (In AutoCAD 2000 this can also be achieved by pressing F5). Now an isometric projection can be produced with the aid of most of the standard AutoCAD commands.

The only feature that requires a special command is a circle. Circles are created using the Isocircle option of the ELLIPSE command. This command option is only visible if the snap mode is set to isometric. The following examples illustrate how to set up an isometric drawing and construct an isometric circle.

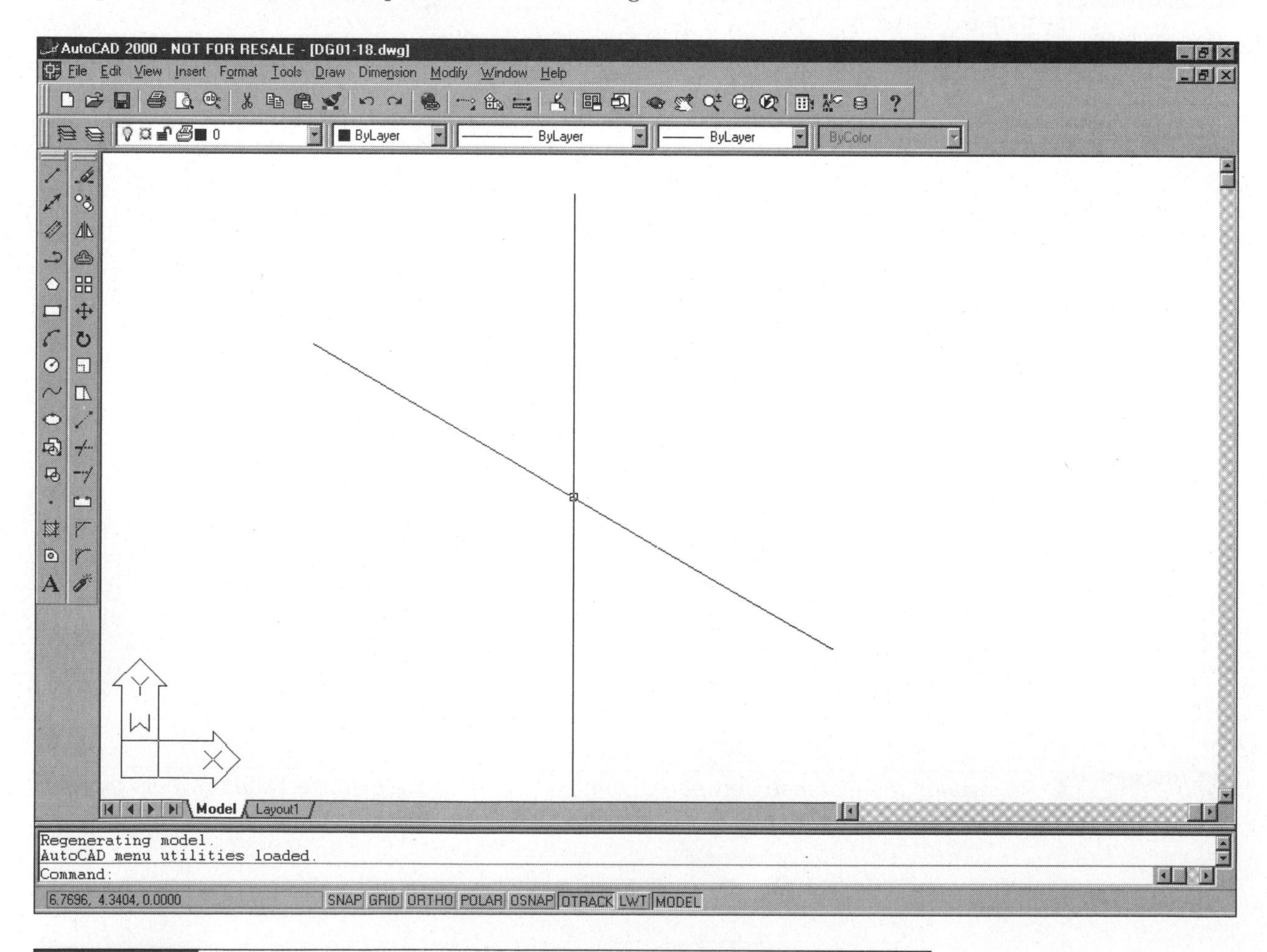


**Figure 1–18** *AutoCAD with cross-hairs shifted into an isometric left position.*

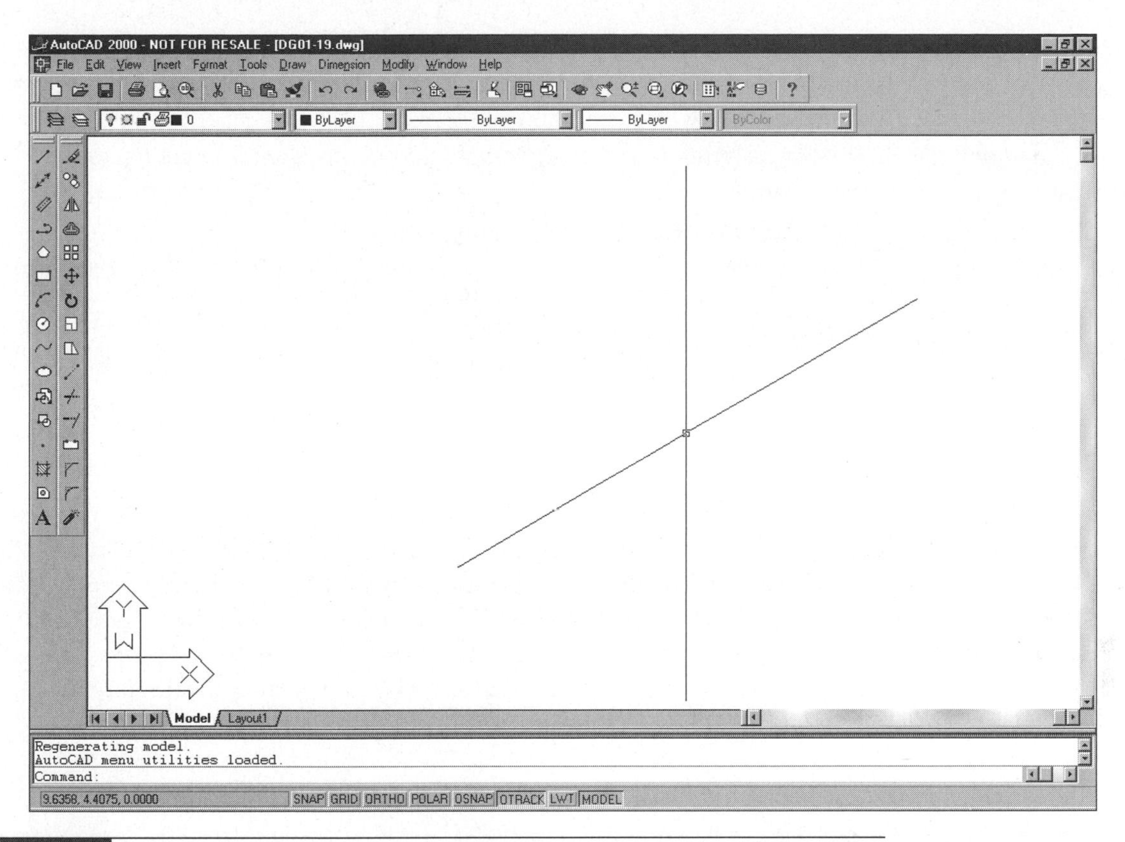


**Figure 1–19** *AutoCAD with cross-hairs shifted into an isometric right position.*

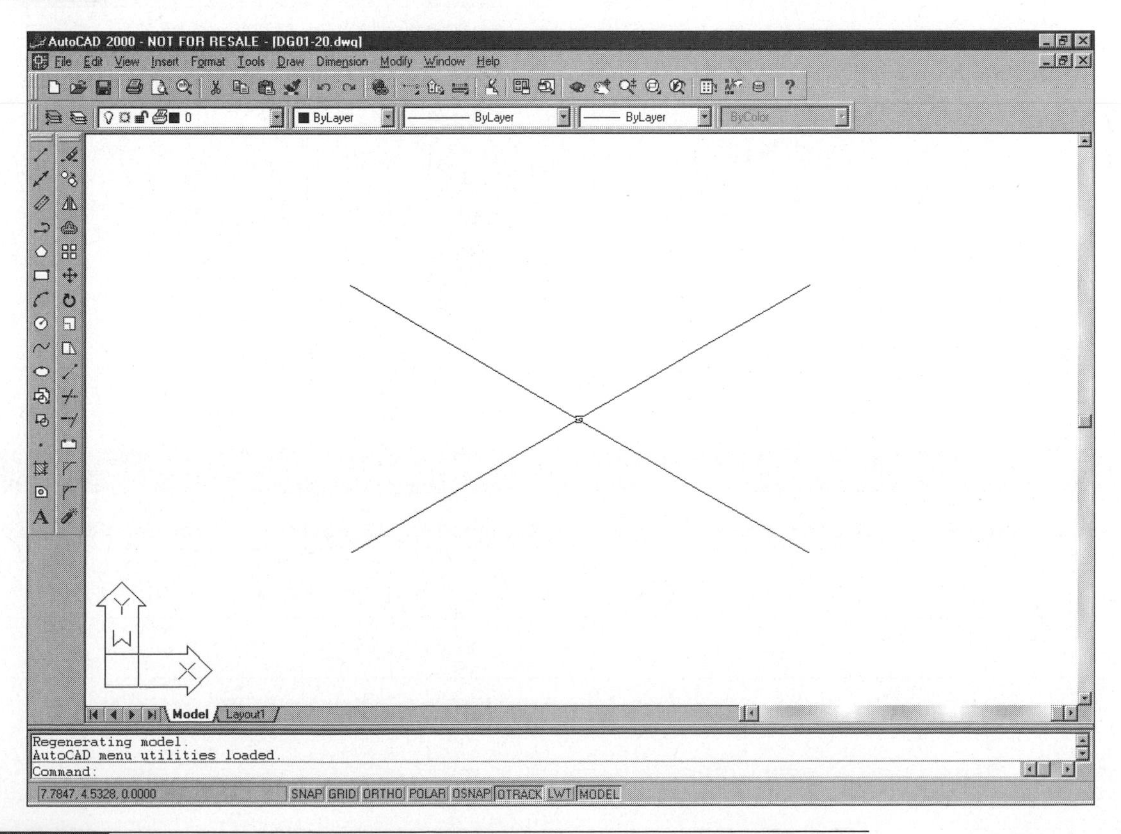


**Figure 1–20** *AutoCAD with cross-hairs shifted into an isometric top position.*

Command: SNAP (ENTER)

Snap spacing or ON/OFF/Aspect/Rotate/Style <1.0000>: S *(Press* ENTER *to set the snap style to isometric.)*

Standard/Isometric <S>: I *(Press* ENTER *to rotate the cross hairs to an isometric configuration.)*

Vertical spacing <1.0000>: (ENTER)

Command: <Snap off> ELLIPSE *(Press* ENTER *then* F9 *to turn snap off.)*

Arc/Center/Isocircle/<Axis endpoint 1>: I *(Press* ENTER *to create an isometric circle on the current isometric plane.)*

Center of circle: *(Select location of isometric circle.)*

<Circle radius>/Diameter: *(Enter radius of circle.)*

Command:

### Dimetric Projections

***Dimetric*** projections have only two axes that form equal angles instead of all three like the isometric projection. These angles will be greater than 90° and less than 180°, but not equal to 120°. An angle of 120° is not used because it would produce an isometric projection. The third angle will be less than or greater than the angle chosen for the first two. This produces a view in which two of the axes are foreshortened equally while the third is at a different scale. Common angles and scale factors used for this type of projection are illustrated in Figure 1–21 and Table 1–1. The information listed in the table under the columns labeled isoplanes are the angles required to construct an ellipse in that plane.

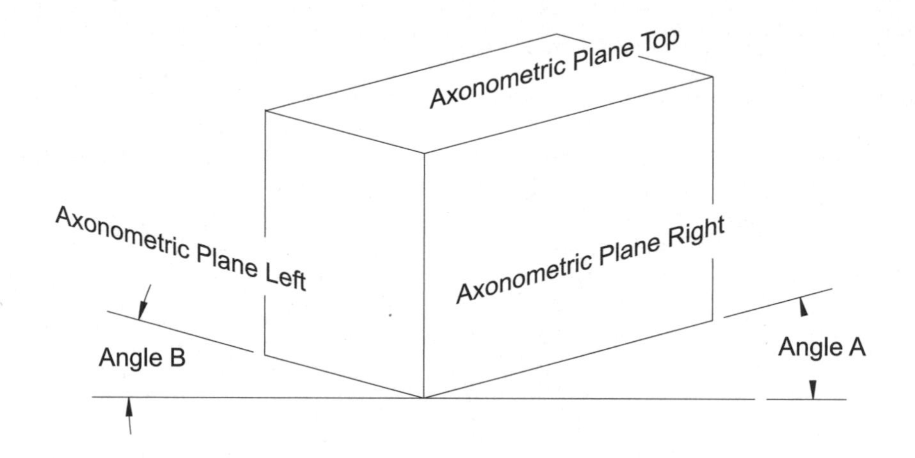


**Figure 1–21** *This figure shows the different isoplanes in their proper position.*

**Table 1.1 Common Angles and Scales Used for Dimetric Projections**

| Angle A | Angle B | Axonometric Right/Scale | Axonometric Left/Scale | Axonometric Top/Scale |
|---|---|---|---|---|
| 15° | 15° | 45° .73 | 45° .73 | 15° .96 |
| 20° | 20° | 40° .75 | 40° .75 | 20° .93 |
| 25° | 25° | 40° .78 | 40° .78 | 30° .88 |
| 35° | 35° | 30° .88 | 30° .71 | 45° .71 |
| 40° | 40° | 25° .92 | 25° .64 | 55° .64 |

**Table 1.1 (continued)**

| Angle A | Angle B | Axonometric Right/Scale | Axonometric Left/Scale | Axonometric Top/Scale |
|---|---|---|---|---|
| 60° | 15° | 15° .73 | 45° .93 | 45° .73 |
| 40° | 25° | 30° .75 | 40° .88 | 40° .75 |
| 60° | 20° | 20° .75 | 40° .93 | 40° .75 |
| 40° | 10° | 25° .54 | 55° .93 | 25° .93 |

**Trimetric Projections**

None of the three principle axes form equal angles in a ***trimetric*** projection. The result is a view that contains three different scale factors, making this type of projection difficult to generate. However, an infinite number of possible projections can be achieved from a trimetric projection. Common angles and scale factors used for this type of projection are illustrated in Figure 1–21 and Table 1–2. The information listed under the columns labeled isoplanes are the angles required to construct an ellipse in that plane.

**Table 1.2 Common Angles and Scales Used for Trimetric Projections**

| Angle A | Angle B | Axonometric Right/Scale | Axonometric Left/Scale | Axonometric Top/Scale |
|---|---|---|---|---|
| 20° | 10° | 35° .62 | 50° .83 | 15° .95 |
| 30° | 10° | 25° .57 | 50° .90 | 20° .95 |
| 50° | 10° | 20° .56 | 50° .94 | 25° .9 |
| 60° | 10° | 15° .5 | 50° .98 | 35° .86 |
| 20° | 15° | 40° .70 | 45° .78 | 20° .95 |
| 30° | 15° | 30° .6 | 50° .66 | 25° .96 |
| 40° | 15° | 30° .65 | 50° .92 | 25° .86 |
| 45° | 15° | 25° .70 | 50° .61 | 35° .95 |
| 50° | 15° | 20° .65 | 50° .94 | 35° .62 |
| 30° | 20° | 35° .71 | 45° .83 | 25 .90 |
| 35° | 25° | 30° .75 | 40° .86 | 35 .82 |
| 40° | 25° | 25° .65 | 35° .90 | 45 .73 |

## Oblique Projection

***Oblique projections*** are another form of pictorial drawing. In this form of drawing, the projection plane is set parallel with the front surface of the object. The line-of-sight is at an angle to the projection plane, producing a view that reveals all three axes (length, width, and depth). This is similar to an isometric because all three axes are shown, but the front of the object is not rotated. As a result the front is drawn at full scale. There are three categories of oblique drawings, ***cavalier***, ***cabinet***, and ***general***.

#### Cavalier Oblique

***Cavalier obliques*** are drawn with the projection plane set parallel to the front of the object and the line-of-sight drawn at a 45° angle from the X-axis. The protruding lines (representing the depth) are drawn full scale. This produces an object that appears distorted if the depth is greater than the width (see Figure 1–22). For this reason, cavalier oblique projections are better suited to depicting an object that is wider than it is deep.

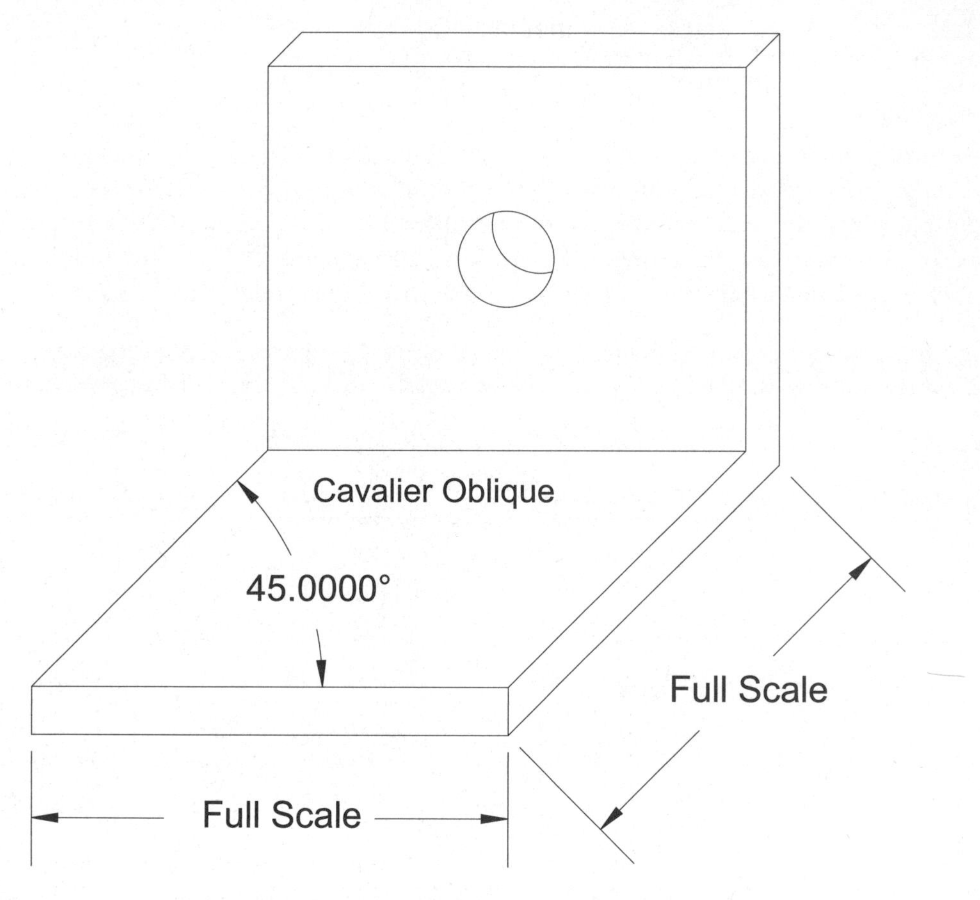


**Figure 1–22** *This figure illustrates a cavalier oblique projection with the line-of-sight set at a 45° angle to the X-axis. The protruding lines are drawn at full scale.*

#### Cabinet Oblique

A ***cabinet oblique*** projection is similar to the cavalier oblique drawing because the line-of-sight is drawn at an angle of 45° from the X-axis, but the protruding lines of the cabinet oblique are drawn at a scale of one-half their actual size. This type of oblique projection is well suited for objects that are deeper than they are wide. It produces an object that appears less distorted than one created using the cabinet oblique method (see Figure 1–23).

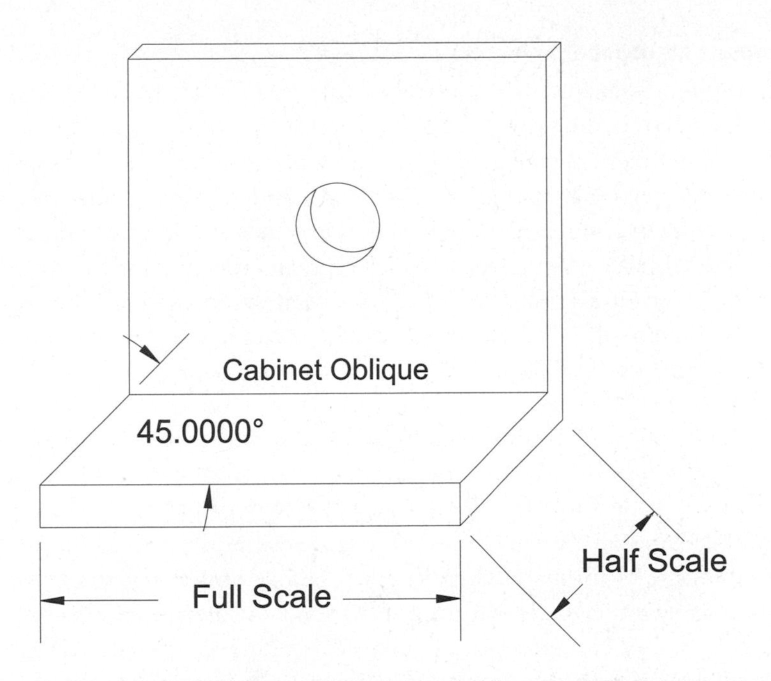


**Figure 1–23** *Cabinet oblique projection with the line-of-sight set at a 45° angle to the X-axis. The protruding lines are drawn at half scale.*

**General Oblique**

For a ***general oblique*** projection, the line-of-sight is drawn at an angle other than 45° from the X-axis. Any angle can be used when creating this form of oblique, but the most common angles are 30° and 60°. Any scale can be used to draw the protruding lines as long as it is greater than one-half and less than full. The most common scale used for a general oblique is the three-quarter scale (see Figure 1–24).

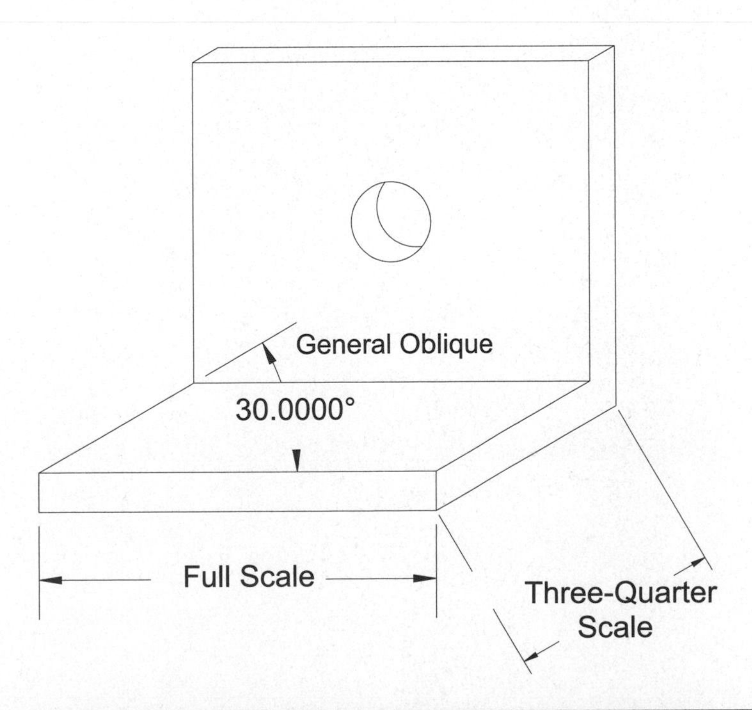


**Figure 1–24** *This figure illustrates a general oblique projection with the line-of-sight set at a 30° angle to the X-axis. The protruding lines are drawn at three-quarter scale.*

**Using AutoCAD to produce an oblique projection**

The creation of a two-dimensional oblique projection using AutoCAD is not as simple as creating a two-dimensional isometric projection. In AutoCAD, two-dimensional isometrics are achieved by changing the orientation of the cross hairs from the Standard mode to the Isometric mode, using the SNAP command. However, AutoCAD does not currently have an oblique command or even an oblique option. For this reason, the USER COORDINATE SYSTEM (UCS) command will be used to produce an oblique. The UCS command allows the technician to define a coordinate system that is different from the standard AutoCAD World Coordinate System by repositioning the origin of the UCS. The coordinate system can also be rotated about one of the three principle axes, which alters the orientation of the XY-axis. These two capabilities alone make the production of an oblique and three-dimensional objects a much easier task.

To create an oblique projection, start by drawing the front view of the object using standard AutoCAD commands at a scale of 1:1. The type of oblique projection to be produced must be determined before the perception of depth can be added to the object. Once the type of oblique projection is ascertained, the coordinate system can be changed to accommodate the decision. This is accomplished by using the UCS command. There are a couple of things to keep in mind before employing the UCS command. First, when AutoCAD is launched, it displays the ***W***orld ***C***oordinate ***S***ystem icon, alerting the technician to the orientation of the X- and Y-axis that defines the XY-plane. It is on this plane that AutoCAD creates two-dimensional objects. Therefore, it is imperative to rotate about the Z-axis only. Rotating about the X- or Y-axis will cause the XY-plane to be perpendicular to the screen. Second, once the XY-plane has been rotated about the Z-axis to match the angle of the selected oblique projection, standard polar and/or Cartesian coordinate systems (polar: Distance <Direction, Cartesian: X, Y, Z) can be used to generate the depth. The following example constructs the cavalier oblique shown in Figure 1–22.

## Step #1

Start by drawing the front face of the object. This is the edge that reveals the thickness of the part, as shown in Figure 1–25.

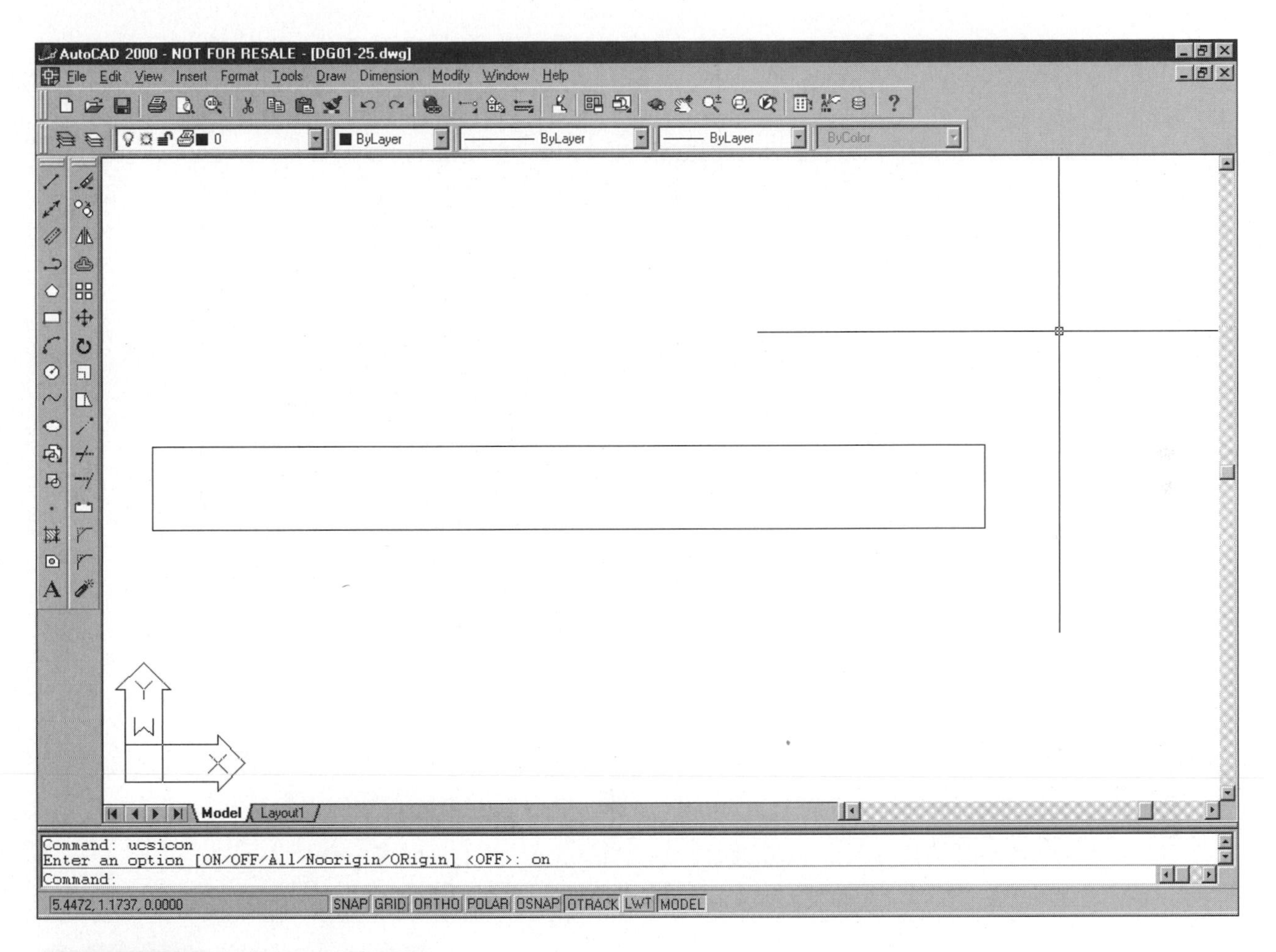


**Figure 1–25** *Front face of part.*

Command: RECTANG (ENTER)

First corner: 0,0 *(Press* ENTER *to start lower left-hand corner of rectangle at coordinates 0,0.)*

Other corner: @5,0.5 *(Press* ENTER *to create a rectangle, starting from the first corner selected, 5 units in length along the X-axis and 0.5 unit in length along the Y-axis.)*

## Step #2

Next, continue working on the front view, but now concentrate on drawing the face labeled "Back Face" (see Figure 1–26). This includes placing the hole in that face.

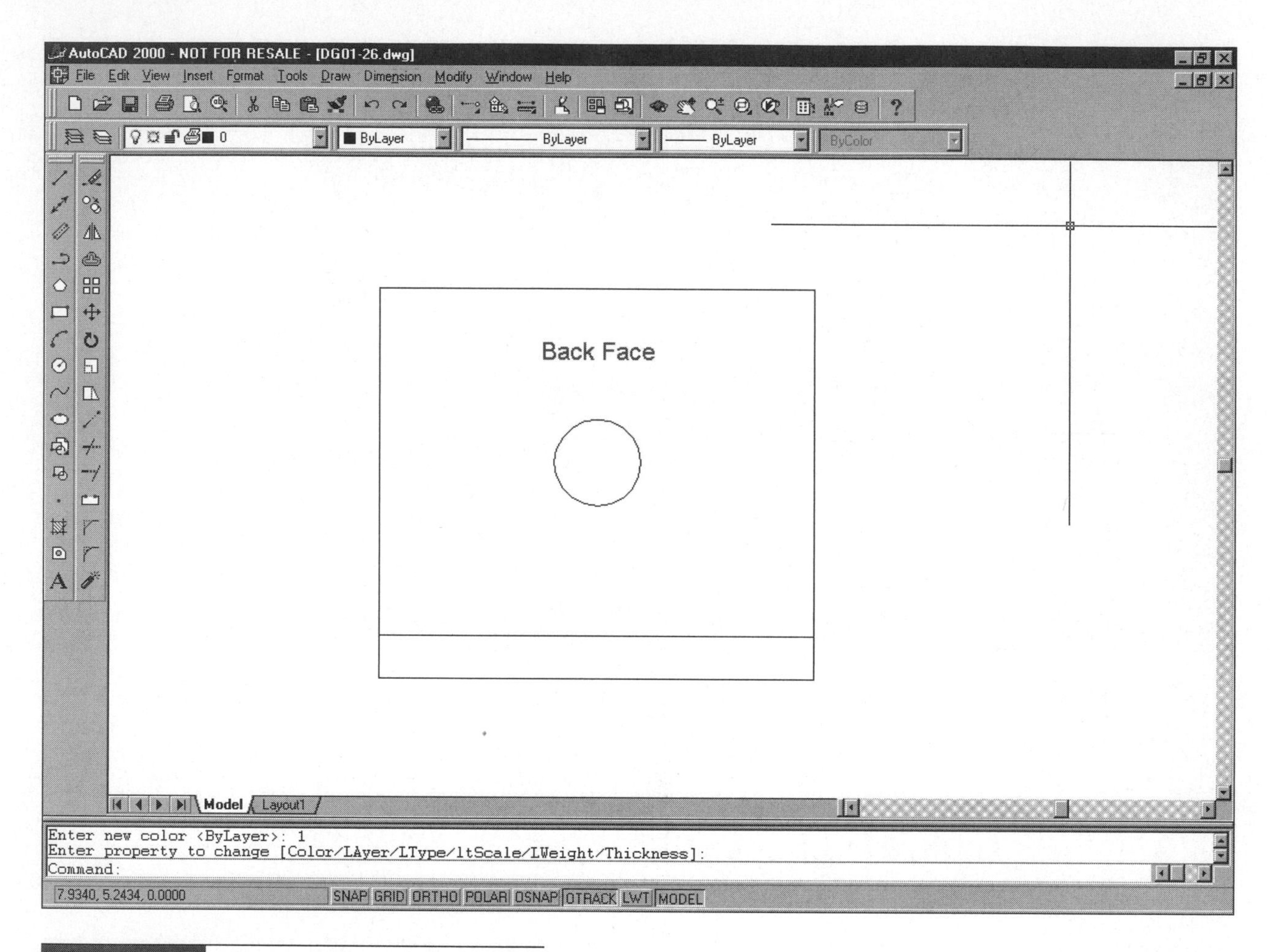


**Figure 1–26** *The back face is now added.*

Command: RECTANG (ENTER)

First corner: end of *(Select the upper left-hand corner of the rectangle drawn in step #1.)*

Other corner: @5,4 *(Press* ENTER *to create a rectangle starting from the first corner selected, 5 units in length along the X-axis and 4 units in length along the Y-axis.)*

Command: CIRCLE 3P/2P/TTR/<Center point>: .y of mid of (need XZ): .x of mid of (need Z): Diameter/<Radius> <0.5000>: D (ENTER)

Diameter <1.0000>: 1 *(Press* ENTER *to construct a circle in the center of the rectangle.)*

## Step #3

Execute the UCS command to rotate the cross hairs to a 45° angle, then generate the protruding lines to give the object the illusion of depth. Finally, the face containing the hole is moved to its proper location at the back of the part (see Figure 1–27). Resetting the UCS in this step is not necessary when using AutoCAD's Object SNAP command.

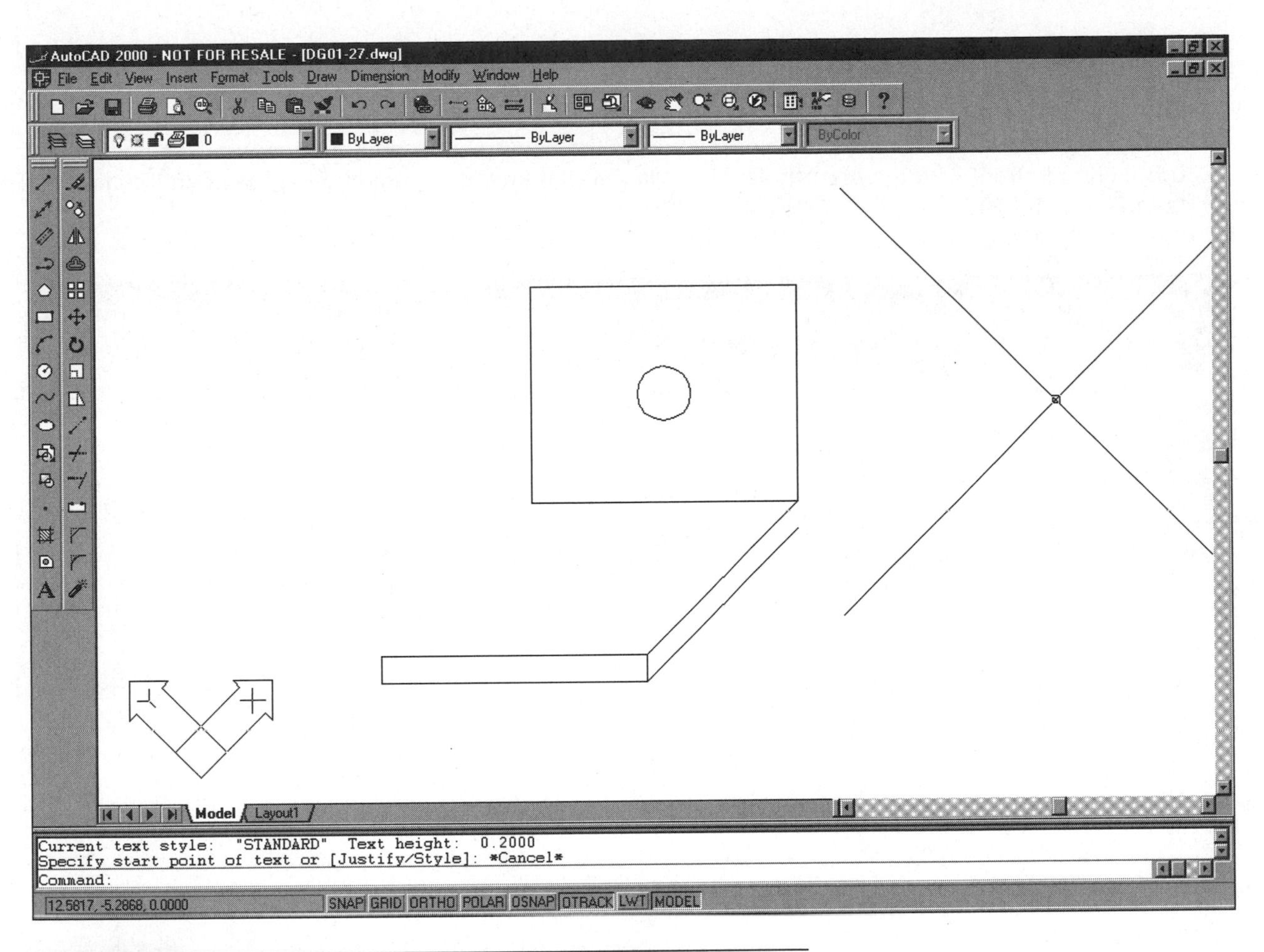


**Figure 1–27** *The back face is now moved into its proper position.*

Command: UCS (ENTER)

Origin/ZAxis/3point/OBject/View/X/Y/Z/Prev/Restore/Save/Del/?/<World>: Z (ENTER)

Rotation angle about Z axis <0>: 45 (ENTER)

Command: LINE (ENTER)

From point: end of *(Select the upper right-hand corner of the 5,0.5 rectangle.)*

To point: @4<<0 *(Press* ENTER *to use the polar coordinate system to draw a line 4 inches long.)*

To point: *(Press* ENTER *to terminate the* LINE *command.)*

Command: LINE (ENTER)

From point: end of *(Select the lower right-hand corner of 5,0 rectangle.)*

To point: @4<<0 *(Press* ENTER *to use the polar coordinate system to draw a line 4 inches long.)*

To point: end of *(Select the endpoint of the first protruding line.)*

To point: *(Press* ENTER *to terminate the* LINE *command.)*

Command: MOVE (ENTER)

Select objects: Other corner: 2 found *(Select the face containing the hole and the hole.)*

Select objects: *(Press* ENTER *to end selection process.)*

Base point or displacement: end of *(Using the* END OSNAP *command, select the lower right corner of the face containing the hole, and the end point of the line protruding from the upper right corner of the face showing the thickness.)*

Second point of displacement: end of

## Step #4

Draw a line from the upper left-hand corner of the face showing thickness, to the lower left-hand corner of the face containing the hole. Next, copy the hole from its present position to a position 0.5 inches in front of the current hole. This will give the hole the illusion of depth when it is finished (see Figure 1–28).

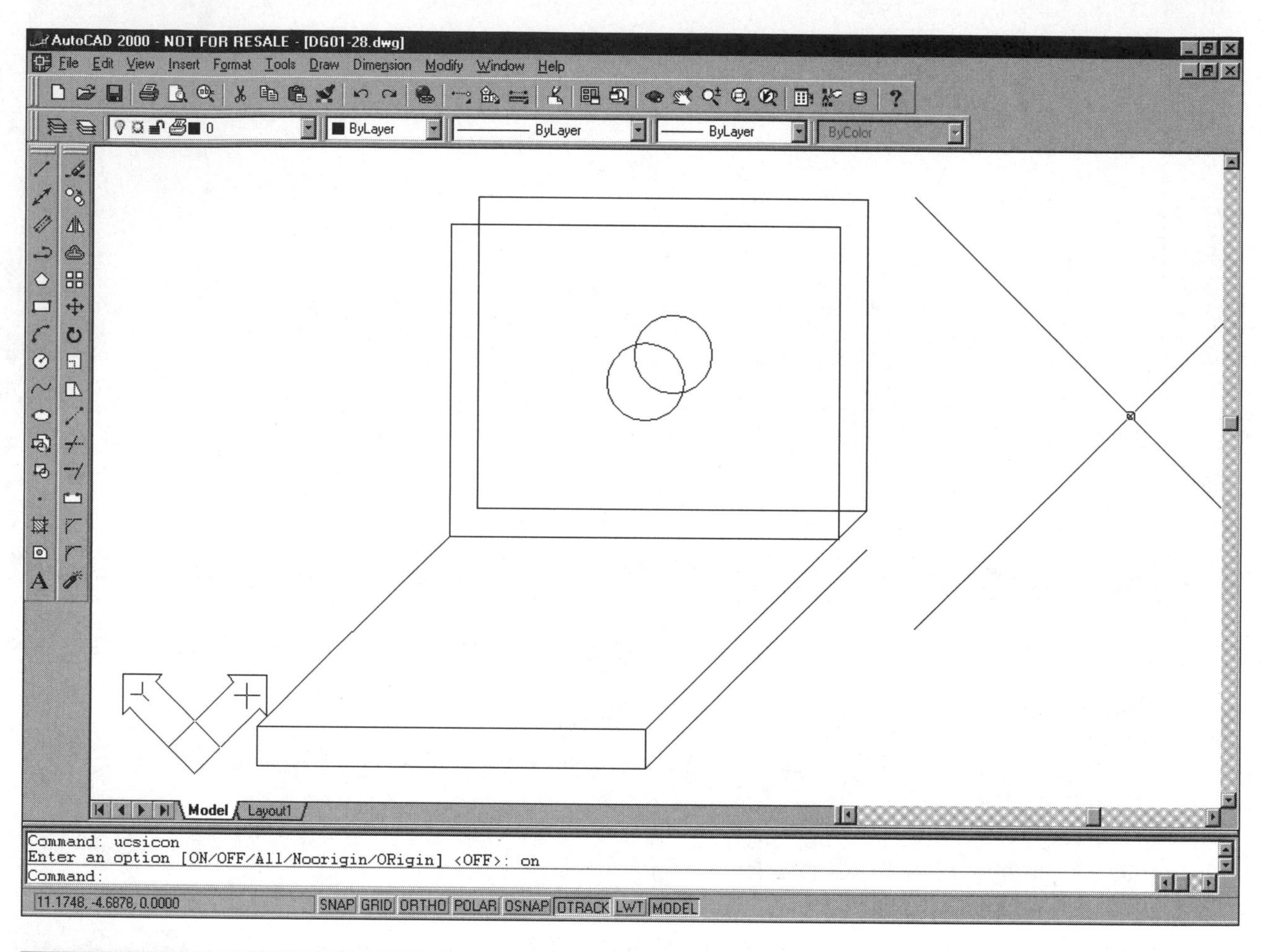


**Figure 1–28** *The protruding lines are added.*

Command: LINE (ENTER)

From point: end of *(Select the upper left-hand corner of the 0,0.5 rectangle.)*

To point: end of *(Select the lower left-hand corner of the face containing the hole.)*

To point: *(Press ENTER to terminate the LINE command.)*

Command: COPY (ENTER)

Select objects: Other corner: 2 found *(Select the face containing the hole and the hole.)*

Select objects: *(Press ENTER to end selection mode.)*

<Base point or displacement>/Multiple: *(Select any point as a base point)*

Second point of displacement: @0.5<<180 *(Press ENTER, use the polar coordinates to copy the selection 0.5 inches along the negative X-axis.)*

## Step #5

Complete the projection by adding any missing lines, and removing excess lines using the TRIM and ERASE commands (see Figure 1-29).

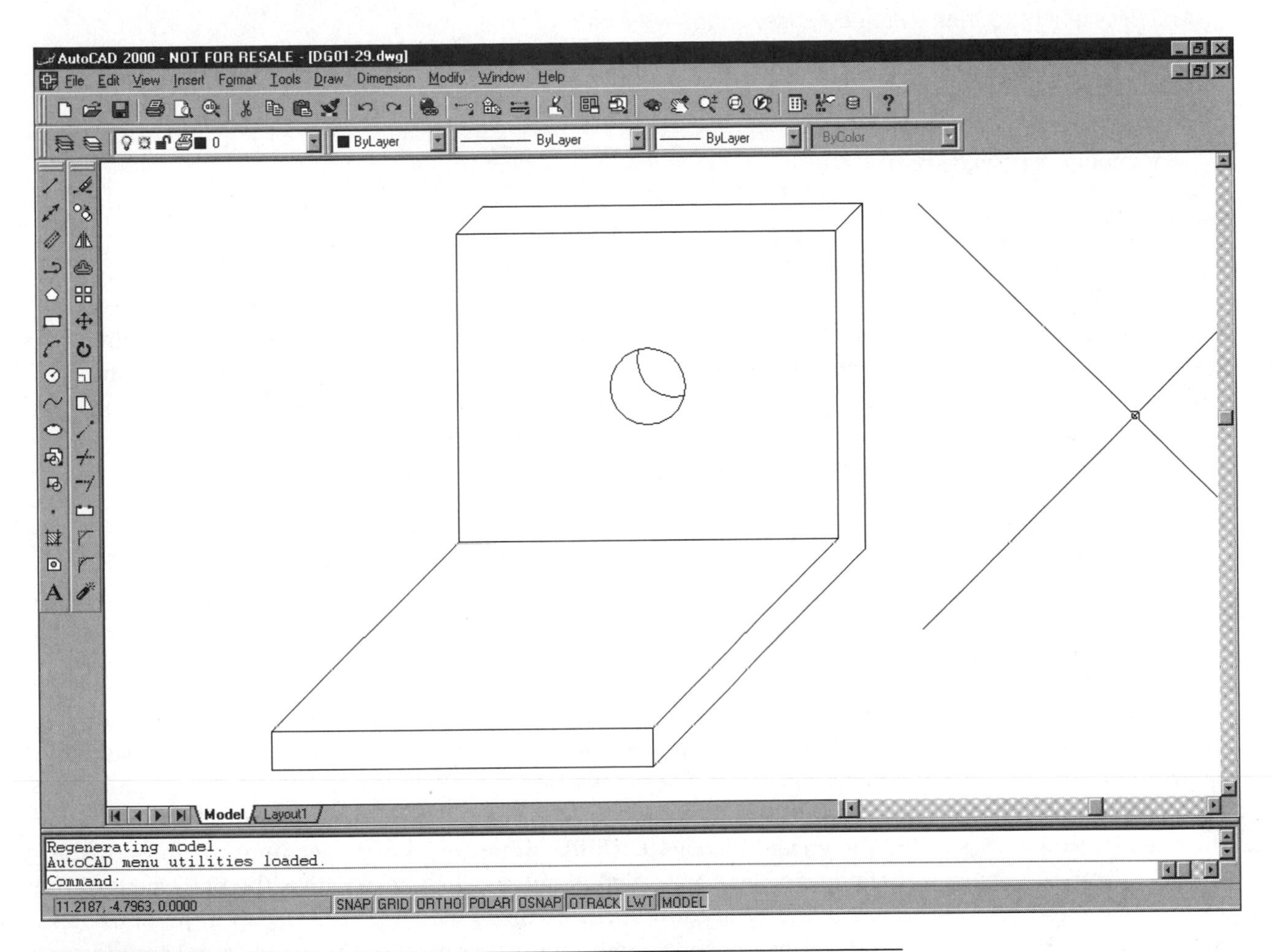


**Figure 1-29** *The object is cleaned up and the oblique projection finished.*

Command: LINE (ENTER)
From point: end of
To point: end of
To point: *(Press* ENTER *to terminate the* LINE *command.)*
Command: LINE (ENTER)
From point: end of
To point: end of
To point: *(Press* ENTER *to terminate the* LINE *command.)*
Command: TRIM (ENTER)
Select cutting edges: (Projmode = UCS, Edgemode = No extend)
Select objects: Other corner: 10 found
Select objects:
<Select object to trim>/Project/Edge/Undo:

```
<Select object to trim>/Project/Edge/Undo:
<Select object to trim>/Project/Edge/Undo:
<Select object to trim>/Project/Edge/Undo:
<Select object to trim>/Project/Edge/Undo:
<Select object to trim>/Project/Edge/Undo:
Command: ERASE (ENTER)
Select objects: 1 found
Select objects: 1 found (ENTER)
Command: REDRAW (ENTER)
```

## Perspective Projections

In the previous two forms of projections, the projection lines were drawn parallel to each other and/or parallel to the projection plane. This gives axonometric and oblique projections a somewhat distorted look, because the observer must be at an infinite distance from the object that they are viewing. This does not occur naturally; in nature the projection lines from the object converge into a single point—the eye of the observer. This form of projection is known as ***perspective projection*** (often referred to as perspectives). It should be noted that although axonometric and oblique projections distort the view of the object, they are still widely used in the engineering and architectural fields because of their ease of construction.

Perspective projections give the object a more natural appearance by allowing the projection lines to converge to a single point. This type of drawing is more readily used in the art world, because this illusion of lines appears more realistic. However, perspectives have found their way into the engineering and architectural world as presentation drawings. These drawings are especially helpful if a presentation is to be given to a group of non-technical persons by providing a better visualization of how the object will appear in reality.

Before generating a perspective in AutoCAD, a few basic terms and concepts should be considered. The location of the observer's eye is called the ***station point***, or ***SP***. The point in which all lines collapse into a single point is called the ***vanishing point***, or ***VP***. The ***ground plane*** or ***GP*** is the edge view of the ground. The ***picture plane*** is an imaginary plane on which the object is projected. The intersection of the ground plane with the picture plane is called the ***ground line***, or ***GL***. Unlike other forms of projections, the lines that are extended from the object to the picture plane are called ***visual rays***. In a perspective, the placement of the picture plane determines the size of the projection. If the picture plane is placed between the object and the observer, then the perspective will be foreshortened. This is how most of the perspectives are created. If the projection plane is placed behind the object the effect is reversed, and the object appears enlarged (see Figure 1–30). The percentage of enlargement or reduction depends entirely upon the distance of the picture plane from the object.

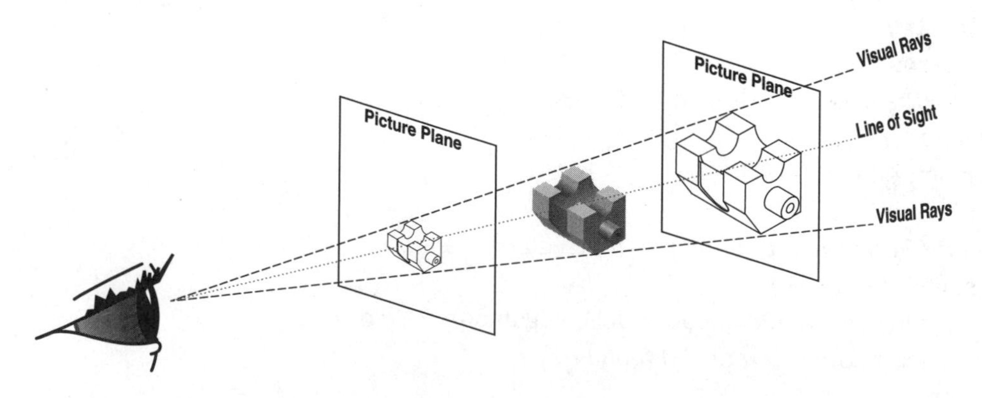


**Figure 1–30** *Placement of the projection plane.*

### One-Point Perspective

A ***one-point perspective*** is a projection in which all the visual rays converge into a single vanishing point. In this perspective, the front of the object is placed parallel to the picture plane, and the lines in the object that are parallel to the picture plane remain parallel. This form of perspective can easily be drawn in AutoCAD by locating a station point and extending the visual rays from the points and contours of the object to the station point to create the depth of the object (see Figure 1–31). One-point perspectives are popular in architectural design for showing long halls and corridors. By adding shading to these drawings, a simple yet effective rendering can be achieved.

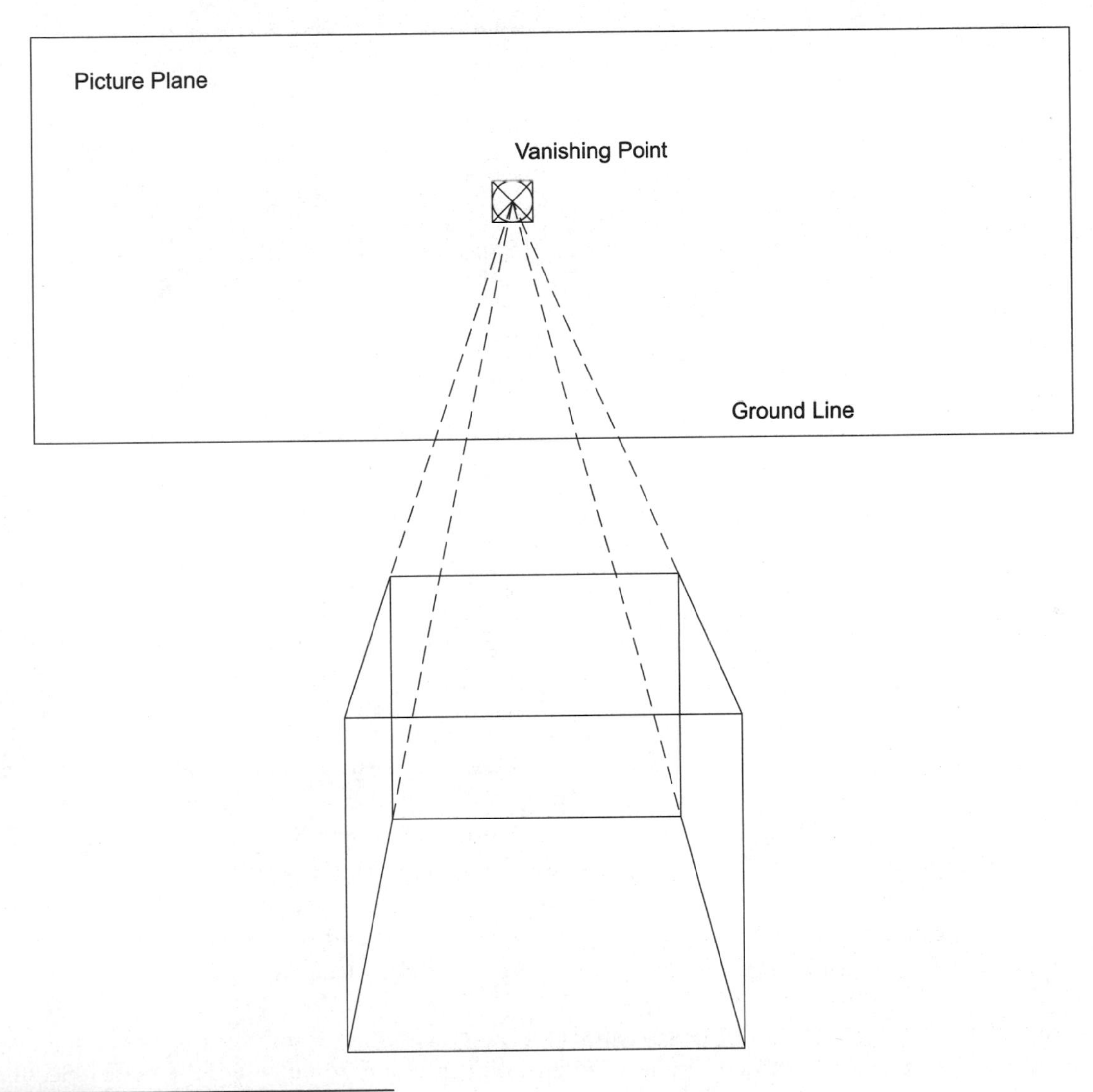


**Figure 1–31** *One-Point Perspective.*

### Two-Point Perspective

In this form of perspective, the object is constructed from two vanishing points. In construction, it is similar to the isometric projection because the object is rotated in relation to the picture plane. However, in an isometric projection the projection lines are kept parallel to one another. In a two-point perspective the visual rays all converge into two vanishing points. These drawings are more complex to generate but the result is a more suitable rendering for large architectural projects (see Figure 1–32).

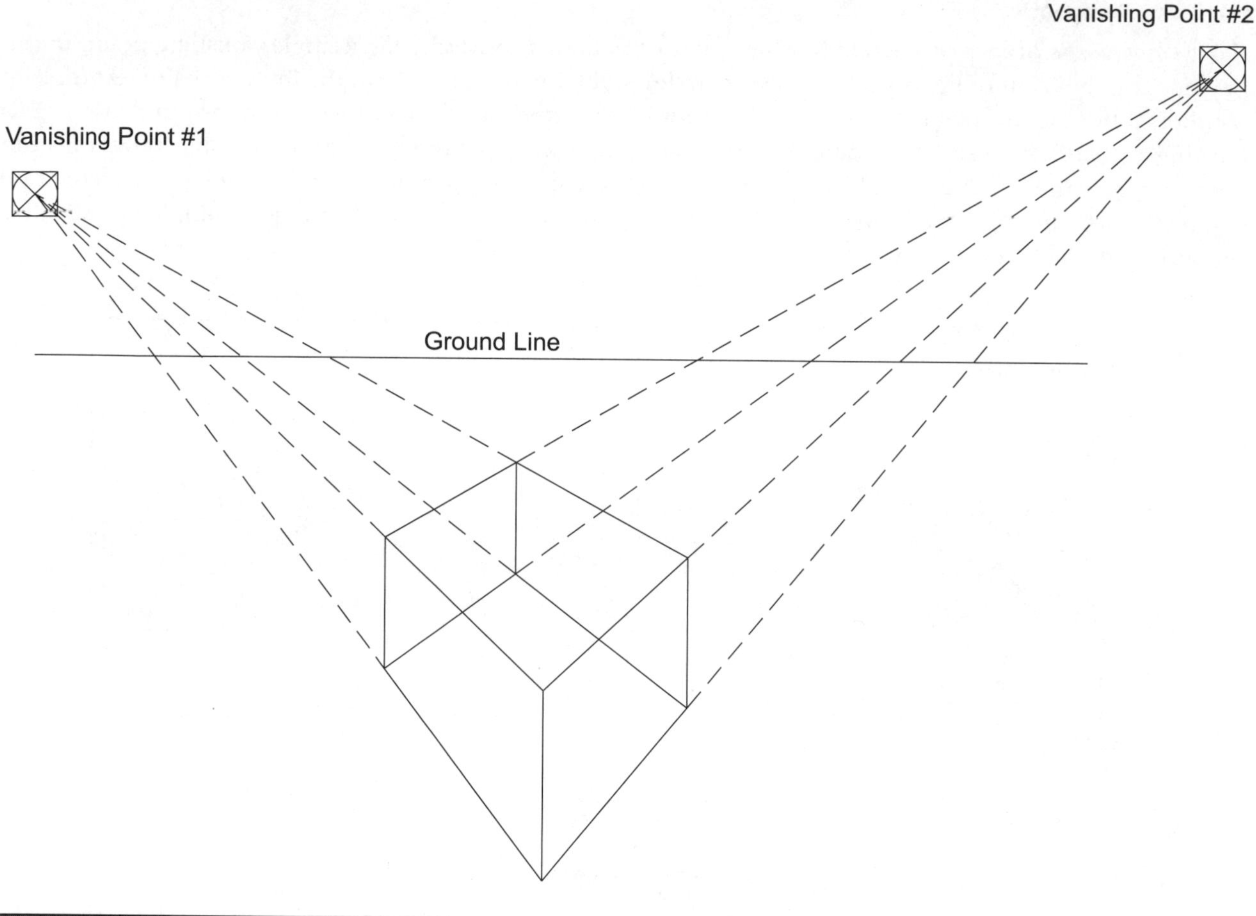


**Figure 1–32** *Two-Point Perspective.*

**Constructing a Perspective from a Multiview**

Perspectives can also be generated from multiview drawings. This form of perspective is called a ***multiview perspective.*** The front view is replaced with a perspective of the front of the object, derived from the points of the top and side views. In this form of perspective, the picture plane is parallel to the perspective, making it appear in edge view form in the top and side views. In the top view, the distance from the object to the picture plane and station point is projected to the side view using standard orthographic projection techniques. At first glance this type of projection appears to have two station points and two picture planes, but in reality there is only one, they are just being observed from different positions (see Figure 1–33).

When locating the station point for this type of projection, two guidelines should be followed. First, working in the top view, the station point should be placed slightly off center and to the left of the object. Second, the station point should be positioned at a distance that allows the entire object to fall within a 30° range (see Figure 1–34). This can be accomplished by drawing two rays that are rotated 30° from each other, originating from the station point in the top view. The object should be encompassed within these two rays; if it is not, the point must be moved back until the object falls within this range. It is much easier to correctly position the station point at the beginning rather than change it in the middle of a drawing, which would require all the angles of the object to be redrawn.

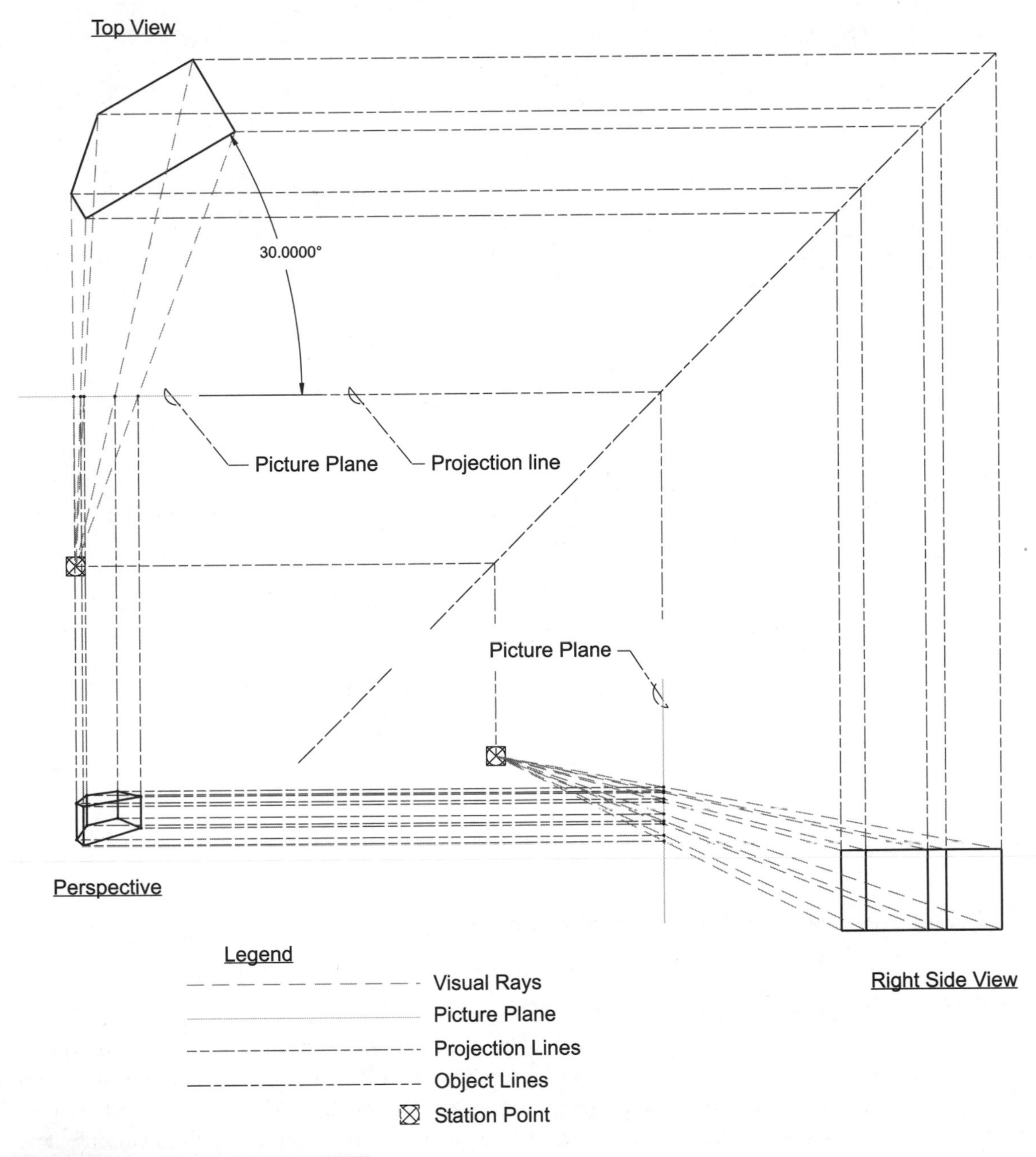


**Figure 1–33** *Multiview perspective.*

In a multiview perspective, the object is rotated with respect to the edge of the picture plane in the top view (see Figure 1–33). Any angle may be selected for this rotation; however, 30° is most commonly used. The visual rays in the top view should intersect the top edge of the picture plane and converge at the station point. The points at which the visual rays intersect the picture plane are transferred down to the location of the perspective. The visual rays that are extended from the side view intersect the edge of the picture plane to converge at the station point located off the side view. The points at which these visual rays pierce the picture plane in the side view are transferred across to the perspective. Here, they intersect with the lines extended from the top view. These intersections are the location of the points in the perspective (see Figure 1–33). Connecting the points in the perspective completes the view.

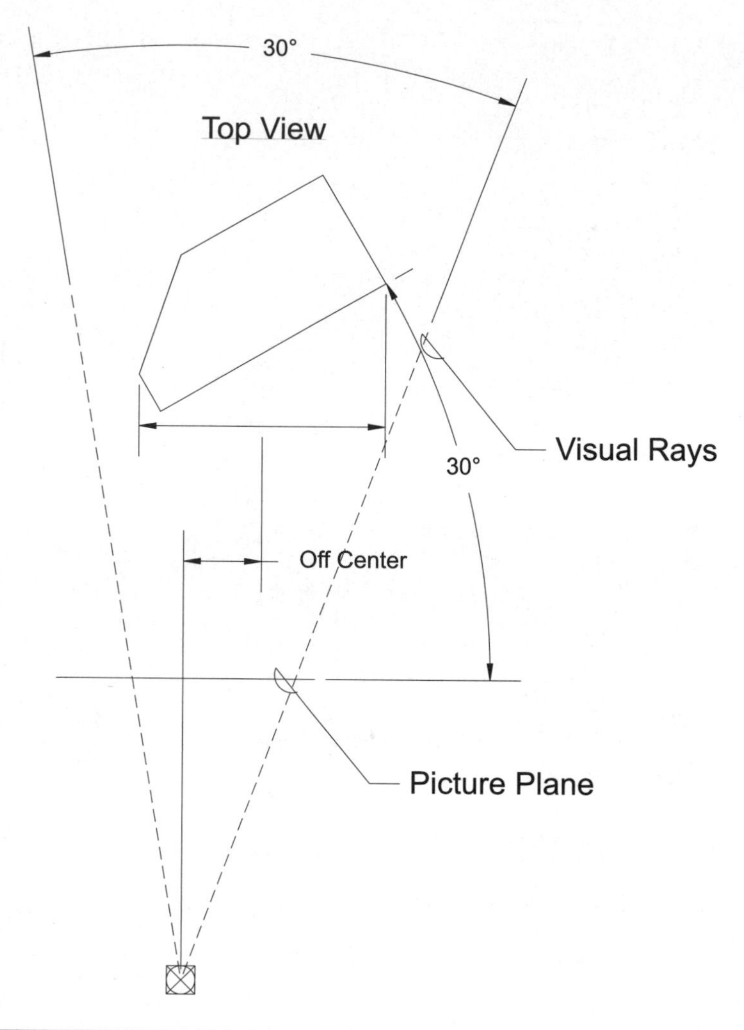


**Figure 1–34** *Rotating the top view in a multiview perspective.*

## Types of Three-Dimensional Views

The previous section focused on the creation of two-dimensional objects and views using techniques developed for manual drafting. Because of AutoCAD's increasing ability to construct three-dimensional models, the computer generation of most of these projections has become much easier and more cost-effective. For example, when creating a multiview drawing from a three-dimensional object, the technician only draws the object once. Rotating the position of the observer creates the additional views automatically. This reduces the time required to generate an orthographic and decreases the possibility of making costly errors. The following section will look at several different techniques used to generate orthographic and isometric projections of three-dimensional objects and solid models.

### Types of Three-Dimensional Objects

AutoCAD can create three different classes of three-dimensional models. They are wire frame models, surface models, and solid models. The ***wire frame models*** are the crudest of the three-dimensional models to construct and view. They are nothing more than a series of three-dimensional points connected by entities such as lines, arcs, and circles (see Figure 1–35). When completed, they give the appearance of being created from clothes hangers. The only information that can be retrieved from a wire frame model is the coordinates of all the endpoints, making this model ineffective for detailed object analysis.

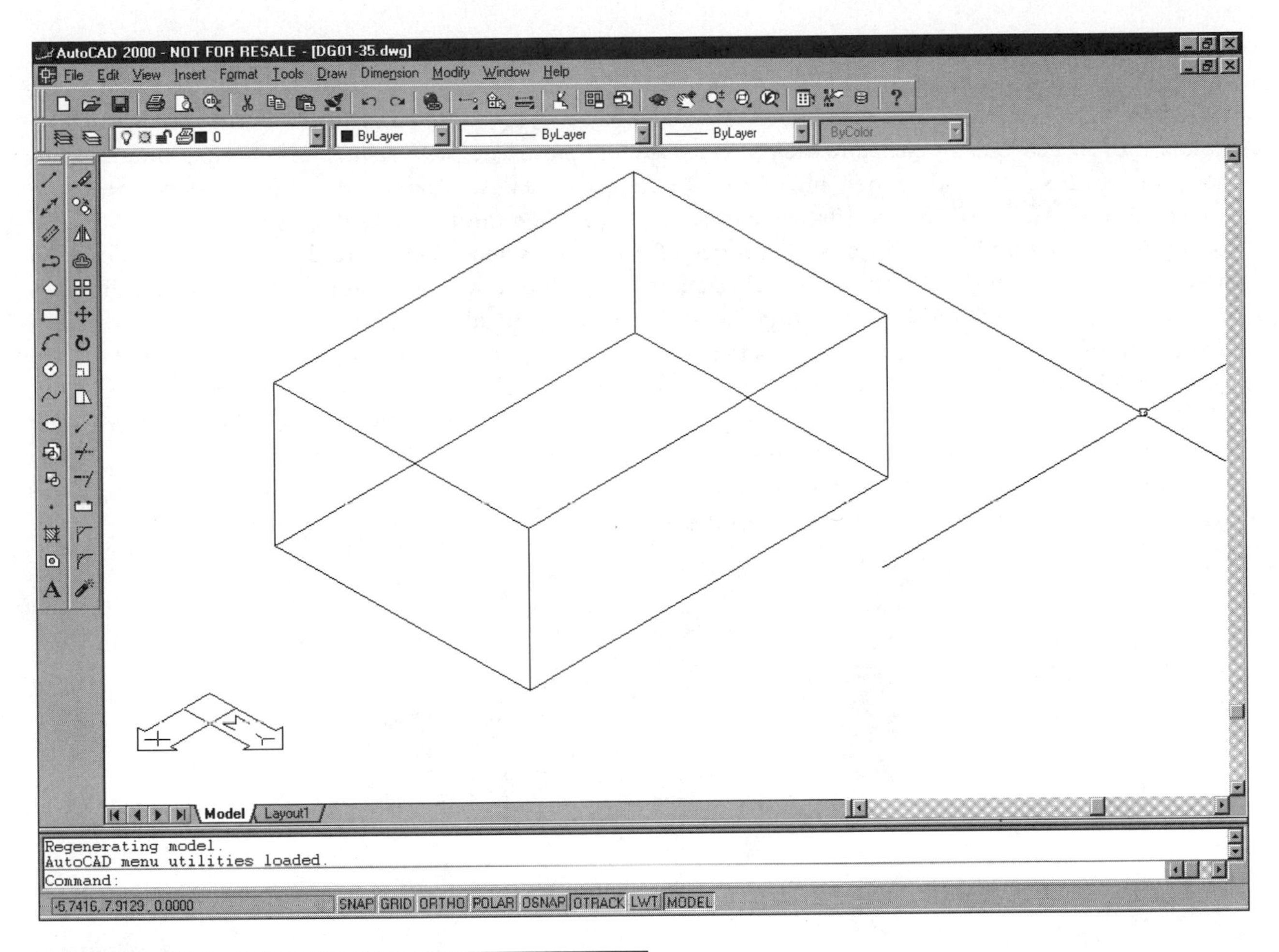


**Figure 1–35** *Three-Dimensional wire frame model.*

The next class of three-dimensional models is the ***surface model***. Surface models are similar to wire frame models because they are constructed from three-dimensional entities such as lines, circles, arcs, and so on, but differ because these models have a three-dimensional face applied to the object's surface (see Figure 1–36). These models not only enable the technician to obtain coordinates of endpoints, but also the location of coordinates along the surface of the object. Additionally, the faces hide lines that would normally be concealed in nature, giving the part a much more natural appearance. These models can be shaded or even rendered and used as presentational drawings.

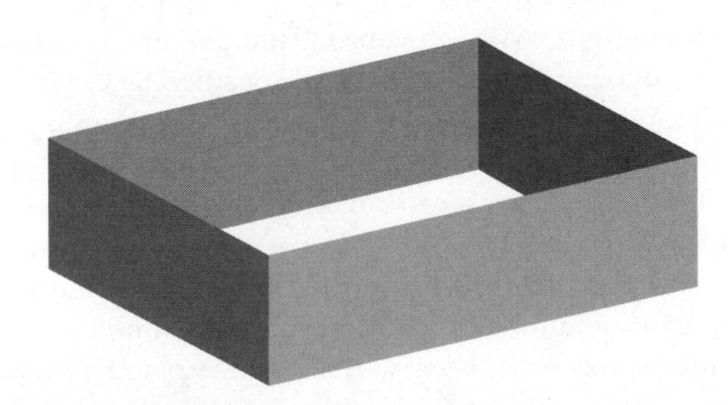

**Figure 1–36** *Three-Dimensional surface model.*

The third class of three-dimensional models is the ***solid model.*** These models are the most advanced of the three classes of the three-dimensional models; not only can lines be hidden and point information obtained, but also details concerning the mass of the object can be generated (see Figure 1–37). Solid models can be broken into two categories, ***nonparametric solid models*** and ***parametric solid models.*** The nonparametric solid models were the first solid models introduced by Autodesk. These models came into existence with the introduction of AutoCAD release 12 as an add-on package called *A*dvanced *M*odeling *E*xtension (AME). To construct a nonparametric solid model, a series of two-dimensional objects are drawn and extruded into three-dimensional solids. Adding and/or subtracting these solids from one another creates a composite solid of the part. If the size of the object needs to be increased, then an additional solid must be created with the dimensions of the amount to be added and the two models are joined. With the release of AutoCAD release 13 the AME package is now incorporated into the main software and a new add-on package has been released called ***mechanical desktop***. Mechanical desktop introduced a new type of solid model, the ***parametric solid model.*** This type of solid model is a dimension-driven solid, meaning that the object is defined using a procedure similar to the process in which a part is normally designed. Solids created in mechanical desktop can be changed by simply selecting the dimension of the feature on the object to be changed.

**Figure 1–37** *Three dimensional solid model.*

**Paper Space/Model Space**

Before an orthographic and/or isometric projection can be created from a three-dimensional model, a basic concept that is essential to the development of three-dimensional models and their projections must first be addressed. AutoCAD has two realities (or spaces, as they are most often called) in which entities are created. They are ***model space*** and ***paper space.*** Model space is a three-dimensional coordinate space used for the creation of objects. The majority of the technician's time is spent in this realm. While paper space, on the other hand, is a three-dimensional space used for the creation of views and drawing layouts. One way of visualizing this concept is to imagine placing a sheet of paper between the object and the line-of-sight, thereby completely concealing the object from the observer. The object is revealed only by cutting holes into the paper; each time a hole is cut a different view of the object emerges. Each view created in paper space can be customized to show the model from any angle and/or scale specified by the user. It is only by combining the two realities (paper space and model space) that the advantages of using three-dimensional models can be fully realized. For example, when a drawing of a three-dimensional object is created in AutoCAD it is first constructed in model space. This allows the user to construct the model at full scale (1:1), making the construction process must easier while reducing the possibility of errors (from working with scale factors). Once the geometry of the part has been completed, the user then toggles from model space to paper space where the actual view construction and annotation occurs. Projections are then created by establishing windows in which objects constructed in model space may be viewed. Finally, by placing a title block around the views and adding annotation and dimension, the drawing is completed and ready for plotting.

In releases previous to AutoCAD 2000, paper space and model space could be distinguished by the UCS displayed icon. For example, when model space is active, AutoCAD displays the standard XY-axis icon (see Figure 1–38). When paper space is active it takes on the appearance of a 30-60-90 triangle (see Figure 1–39). In AutoCAD 2000, the icons can still be used to distinguish the current mode. However, tabs have been added at the bottom of the graphic screen (labeled model and layout), much like Microsoft's Excel, where the user can distinguish which mode they are currently in by the tab that is enabled. The tabs also serve another purpose; they enable the user to switch from one mode to another by simply selecting one of the tabs. In previous releases of AutoCAD, only one paper space was allowed per drawing. This has changed in AutoCAD 2000, where multi-paper spaces can be configured that allow the user to create multi-layouts of a particular object all contained in one drawing file. In previous AutoCAD releases, toggling between paper space and model space was accomplished by setting the AutoCAD system variable TILEMODE. When TILEMODE is set to one, model space is made active, and when TILEMODE is set to zero, paper space is active. This same procedure can still be used in AutoCAD 2000 to toggle between the two modes.

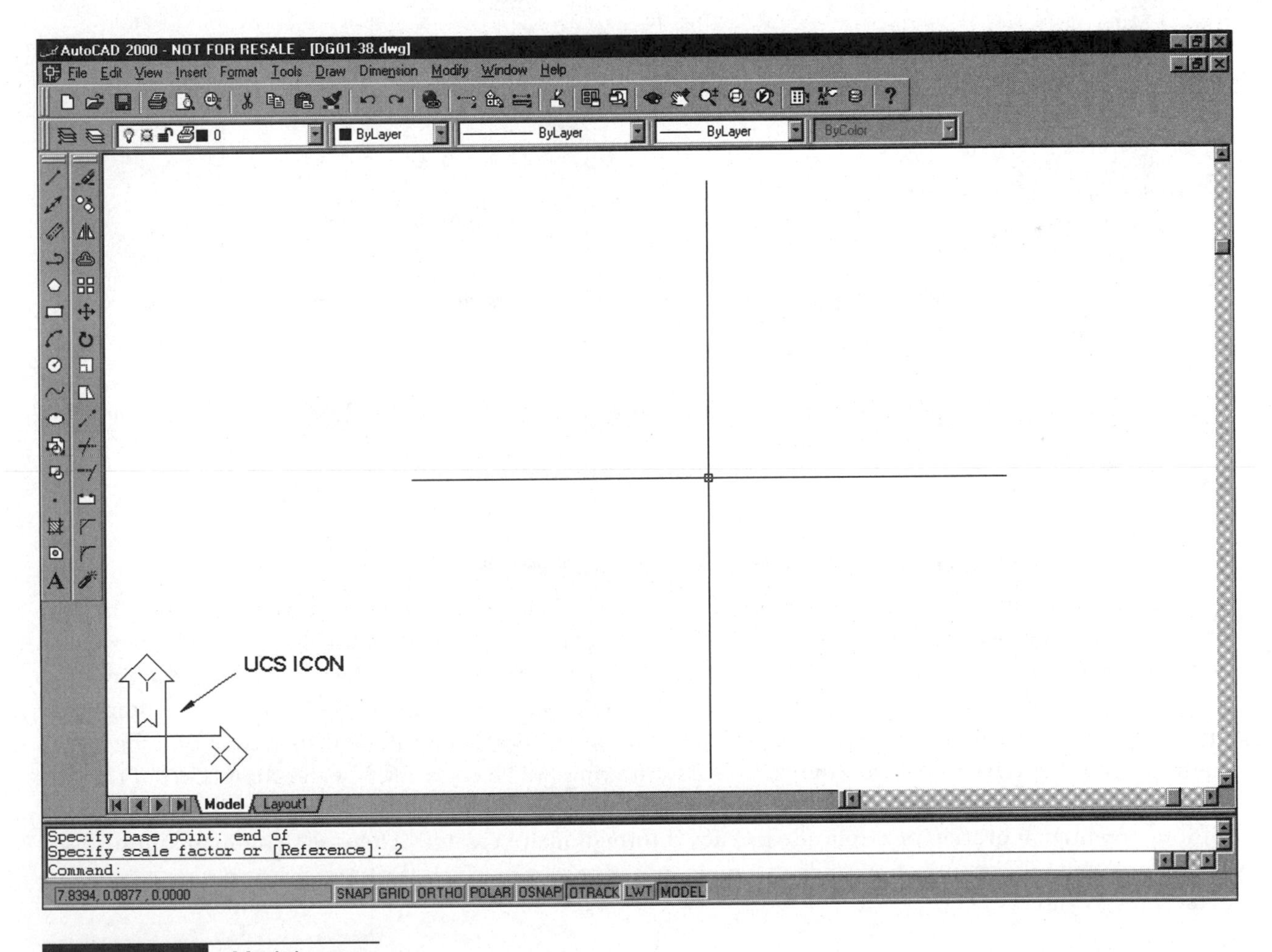


**Figure 1–38** *Model space.*

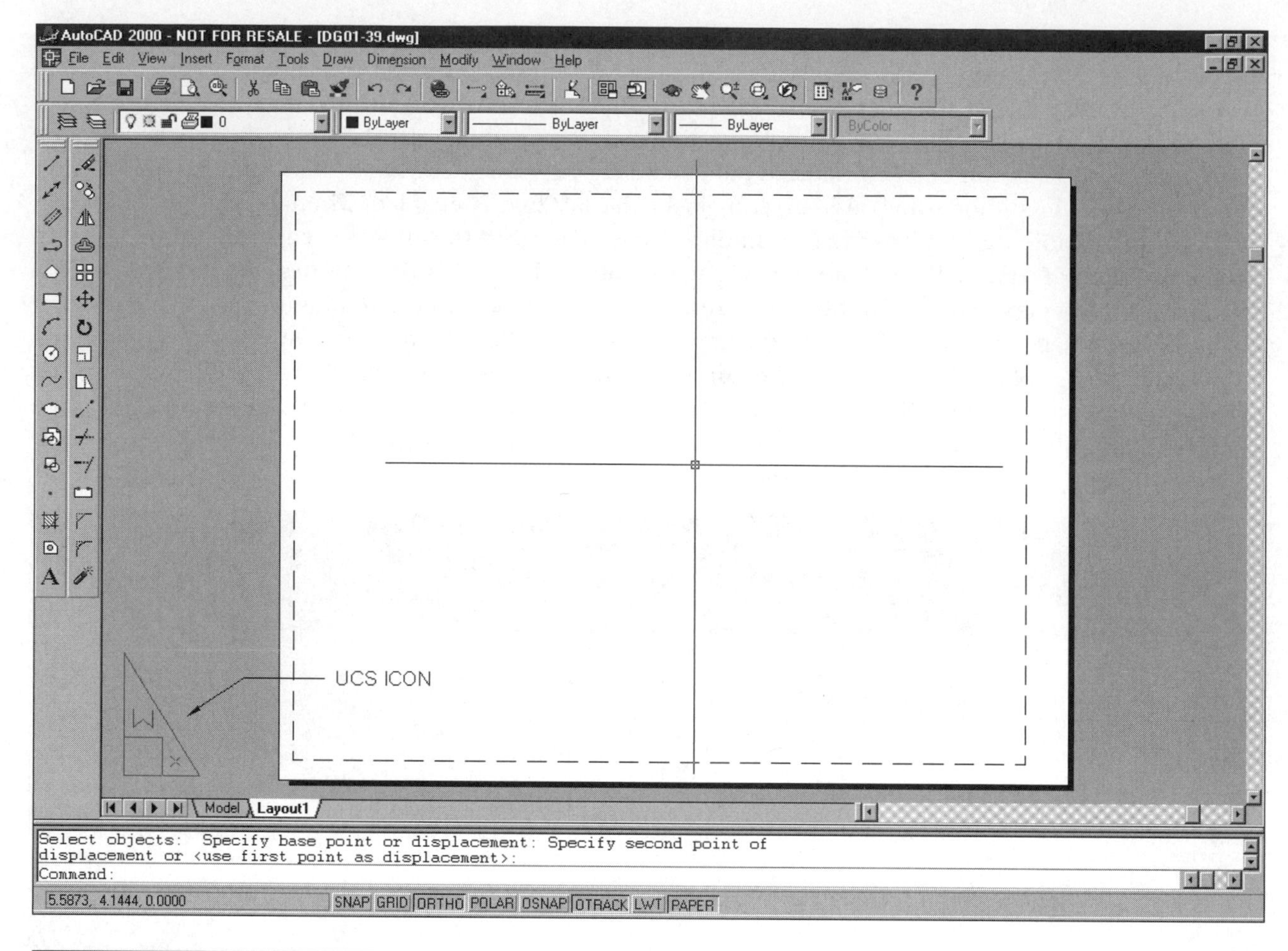


**Figure 1-39** *Paper space.*

### Orthographic Projections/Isometric Projections Using MVIEW-VPOINT

To create an orthographic or isometric projection of a three-dimensional object, a view port must be constructed in paper space using the MVIEW command. This command creates floating viewports in paper space. A ***floating viewport*** is one that can be easily modified and manipulated into any number of configurations (copied, erased, or moved). These viewports can be thought of as windows into model space. The number of windows or viewports that can be created is not limited, but only sixteen will be displayed at one time. Normally, the number of views employed in a multiview drawing will be controlled by the complexity of the part itself. For clarity and conservation of time, only the views required to fully convey the necessary information without creating an overlap or duplication of any information are created. The MVIEW command is only used for the creation of viewports; to modify the angle or position from which the part is viewed, the VPOINT command must be used.

The VPOINT command changes the observer's line-of-sight by repositioning it either along one of the axes or in relationship to and from the XY-plane. This command can only be executed in model space. To use this command, model space must either be active or activated in the viewport where the line-of-sight is to be repositioned. To activate model space in a paper space viewport, the MSPACE or MS command is used. This command switches from paper space to model space within a selected viewport. If more than one viewport is present, then the last viewport created or the last viewport made active becomes operative when the command is executed. The changes made to the viewing angle are only displayed in the viewport where the command is executed. Once the desired changes have been made to the view the PSPACE or PS command is executed to return to paper space.

The VPOINT command has two options that can be used to create the necessary views. These options are Rotate and Viewpoint. The Rotate option changes the observer's line-of-sight by specifying an angle in the XY-plane and from the XY-plane. For example, to create a view in which the line-of-sight is positioned 20° in the XY-plane and 20° from the XY-plane (see Figure 1–40), the following syntax would be used.

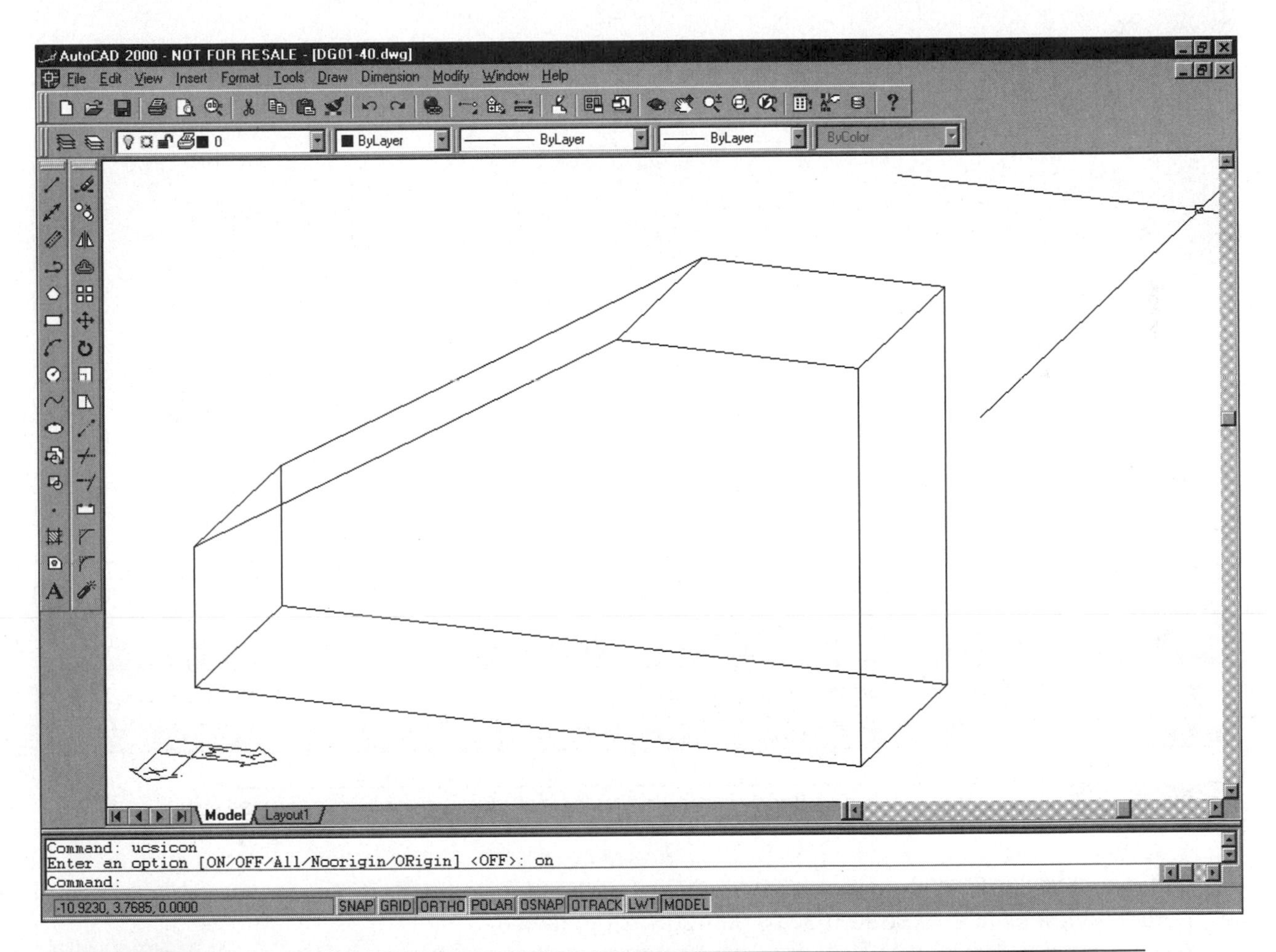


**Figure 1–40** VPOINT *command with a rotation of 20° in the XY-plane and 20° from the XY-plane.*

Command: VPOINT (ENTER)
Rotate/<View point> <0.0,0.0,0.0>: R (ENTER)
Enter angle in XY-plane from X axis <0>: 20 (ENTER)
Enter angle from XY-plane <0>: 20 (ENTER)
Regenerating drawing.

The Viewpoint option changes the position of the line-of-sight by placing the origin of an X-, Y-, and Z-axis at the center of the object and then repositioning the line-of-sight either along one or any combination of the three axes. For example, to create a view in which the line-of-sight is positioned along the positive X-axis, the following syntax would be used (see Figure 1–41).

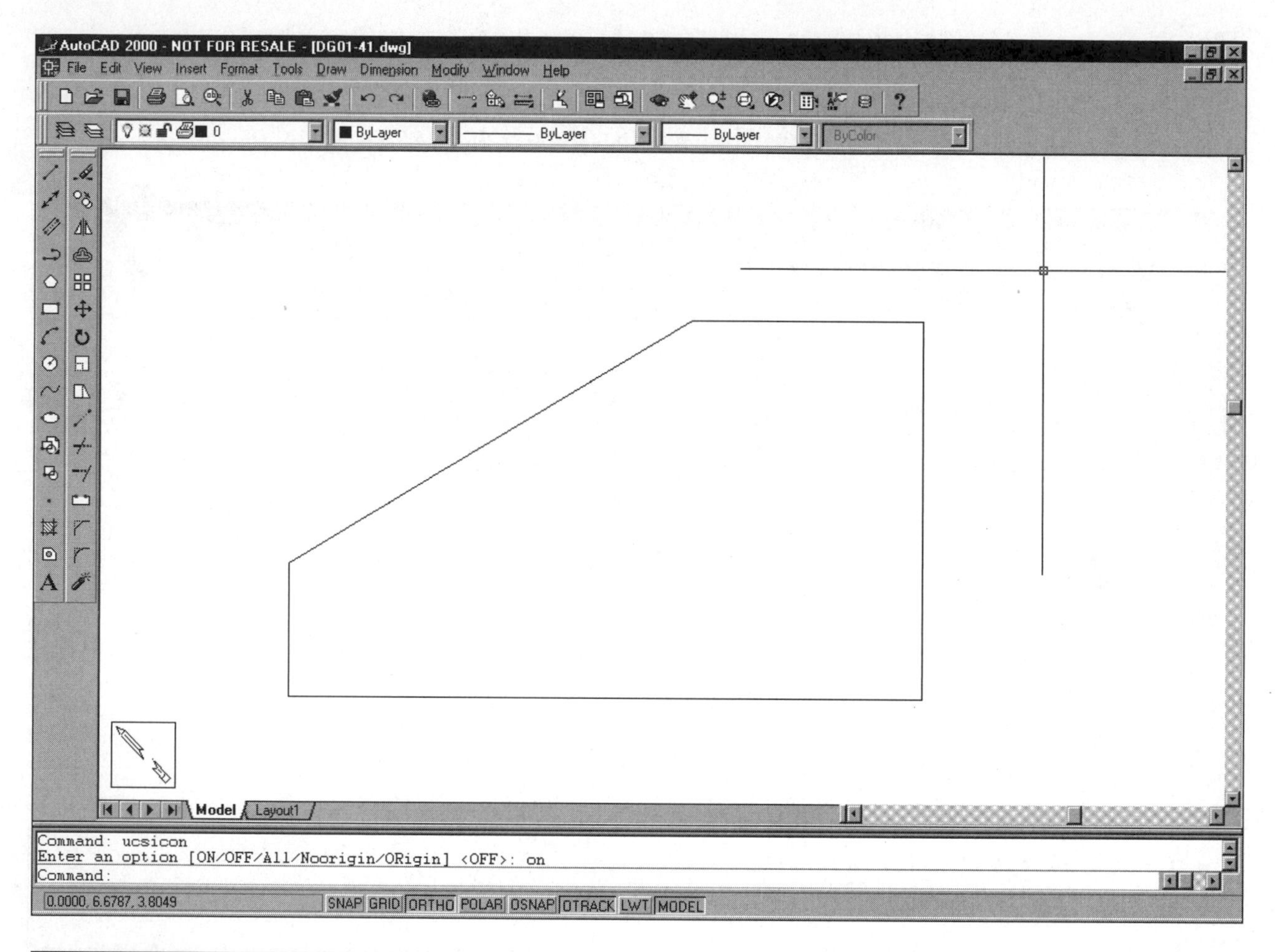


**Figure 1–41** *The* VPOINT *command used to position an object along the positive X-axis.*

Command: VPOINT (ENTER)

Rotate/<View point> <0.8830,0.3214,0.3420>: 1,0,0 (ENTER)

Regenerating drawing.

In this example, the 1,0,0 instructs AutoCAD to position the line-of-sight along the positive X-axis and not the Y- or Z-axis. The first number references the X-axis, while the second number references the Y-axis and the third number references the Z-axis. The use of a positive and negative number tells AutoCAD either to position the line-of-sight along the positive or negative axis. To position the line-of-sight along the negative X-axis, then a negative one (-1) would be used in place of the positive one (1). It should be noted that any non-zero integer value might be used in place of the one (1), and the results will be the same.

To generate an orthographic and/or isometric projection using the MVIEW/VPOINT combination, start with an existing or create a three-dimensional model of the part. Next, after paper space has been made active, create the required number of viewports using the MVIEW command. Using the MSPACE command, activate model space and change the line-of-sight to the position necessary to create the desired view using the VPOINT command. Repeat the process until all required views have been generated. The following table is provided to help set up orthographic views using the Rotate and Viewpoint options. When generating the different views, the right-hand rule may be used to help determine which axis to revolve to achieve the desired view (see Figures 1–42, 1–43, and Table 1–3).

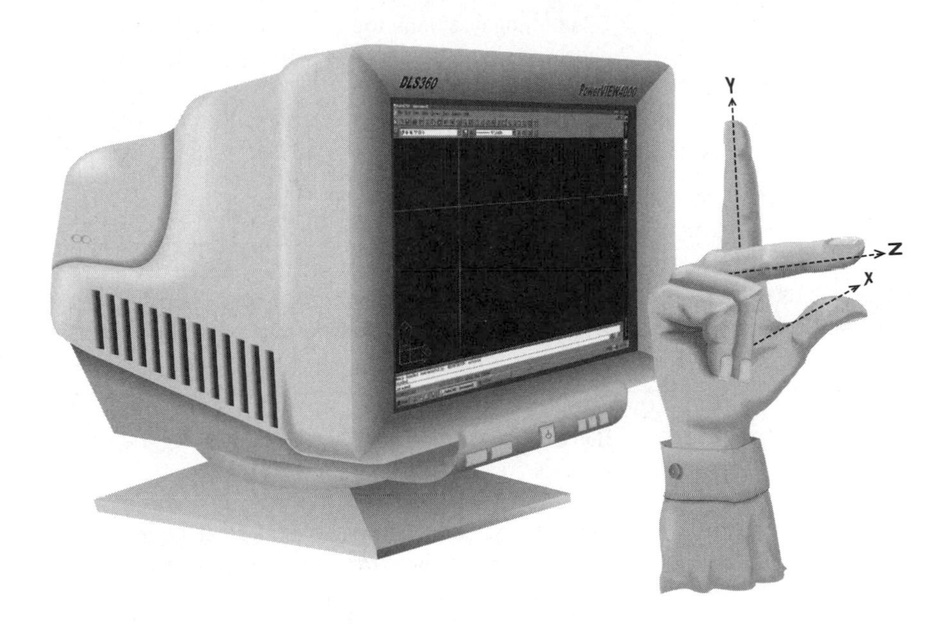


**Figure 1–42** *Standard AutoCAD view (model space) applying the right-hand rule.*

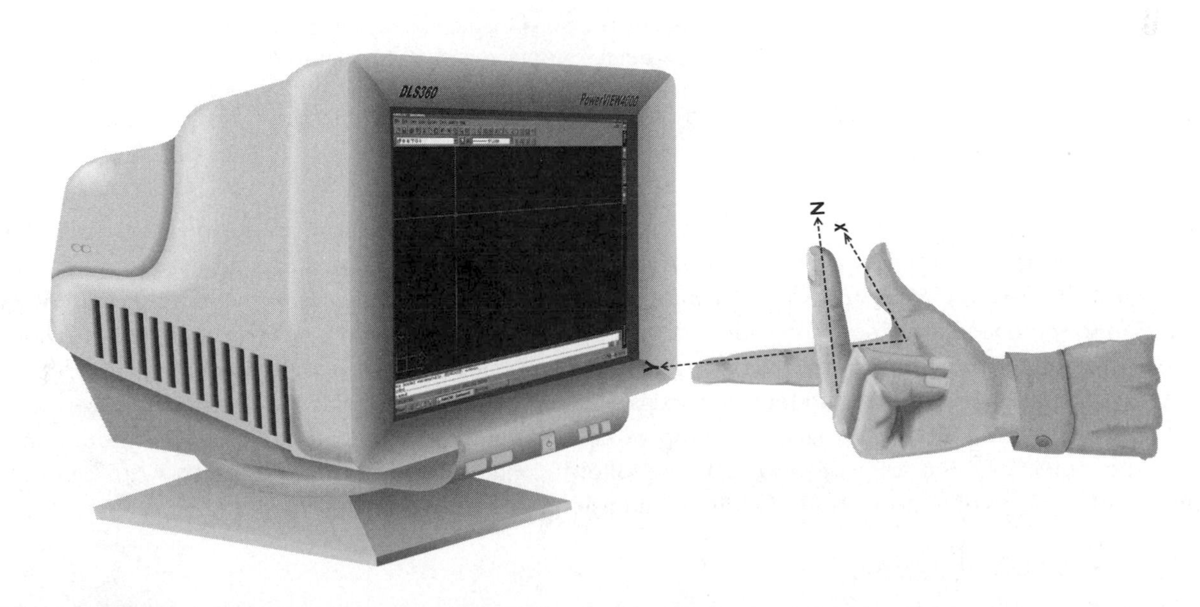


**Figure 1–43** *Rotated AutoCAD view (model space) applying the right-hand rule.*

| Table 1–3 VPOINT Command Using the Viewpoint Option | | |
|---|---|---|
| **In the XY Plane** | **From XY Plane** | **Description** |
| 270° | 90° | This setting positions the line of sight at an angle of 270° in the XY plane and 90° from the XY plane to produce a top view. |
| 270° | 0° | This setting positions the line of sight at an angle of 270° in the XY plane and 0° from the XY plane to produce a front view. |
| 0° | 0° | This setting positions the line of sight at an angle of 0° in the XY plane and 0° from the XY plane to produce a right side view. |
| 45° | 35°16' | This setting positions the line of sight at an angle of 45° in the XY plane and 35°16' from the XY plane to produce an isometric. |

| VPOINT Command Using the Rotate Option | |
|---|---|
| **Setting** | **Description** |
| 0,0,1 | This setting positions the line of sight along the positive Z-axis. This setting is used to generate a top view. |
| 0,1,0 | This setting positions the line of sight along the positive Z-axis. This setting is used to generate a rear view. If -1 is used then a front view is generated instead. |
| 1,0,0 | This setting positions the line of sight along the positive X-axis. This setting is used to generate a right side view. If -1 is used then a left side view is generated instead. |
| 1,1,1 | This setting positions the line of sight into a position that is 45 from the X-axis in the XY plane and 45 from the XY plane. This is similar to an isometric. |

#### Using MVSETUP

Using the AutoLISP routine, MVSETUP can automate the creation of orthographic projections. Although this is an AutoLISP program, simply entering the program name at the command prompt will execute the routine just like any other AutoCAD command. If this program is run before entities are created, then it allows the technician to setup vital drawing information: drawing units, scale factors, and it prompts for the creation of a drawing border. If the program is started after the model has been created, then the technician can: change AutoCAD's drawing mode from model space to paper space, create and scale viewports, and insert a title block. When using the MVSETUP program to setup a drawing, the program offers these options: none, single, standard engineering, and array of viewports. The following example illustrates how MVSETUP can be used to generate an orthographic projection of a three-dimensional solid model.

Command: MVSETUP (ENTER)
Initializing...
Enable paper space? (No/<Yes>): Y *(Press ENTER. If this prompt is answered No, then the program will setup the drawing scale and units.)*

Entering Paper space. Use MVIEW to insert Model space viewports.
Regenerating drawing.
Creating the default file mvsetup.dfs in the directory C:\AV2\AVWIN\.
Align/Create/Scale viewports/Options/
Title block/Undo: C *(Press ENTER. The Align option aligns a view of an object in one viewport with a base point in another viewport; the Create option creates viewports. The Scale option scales a viewport. The Options option sets layers, limits, units, and xref. The Title block option inserts a title block; the Undo option undoes the last change.)*
Delete objects/Undo/<Create viewports>: C (ENTER)
Available MVIEW viewport layout options:
0: None
1: Single
2: Std. Engineering
3: Array of Viewports
Redisplay/<Number of entry to load>: 2 (ENTER)
Bounding area for viewports. First point: *(Specifies one of the corners of the viewports to be created.)*
Other point: *(Specifies the diagonal corner of the viewport to be created.)*
Distance between viewports in X. <0.0>: *(Press ENTER. Sets the distance between viewports in the X direction.)*
Distance between viewports in Y. <0.0>: *(Press ENTER. Sets the distance between viewports in the Y direction.)*
Align/Create/Scale viewports/Options/Title block/Undo:
Command:

**Note: When using the standard engineering option, the program will automatically create a top, front, right side, and isometric projection.**

### Using AMDWGVIEW

AMDWGVIEW is similar to the AutoLISP program MVSETUP except that it is used in Mechanical Desktop. When activated, it displays a dialog box for selecting the type of projection to be created. The options are: Base, Ortho, Aux, Iso, Detail, and Broken views, shown in Figure 1–44.

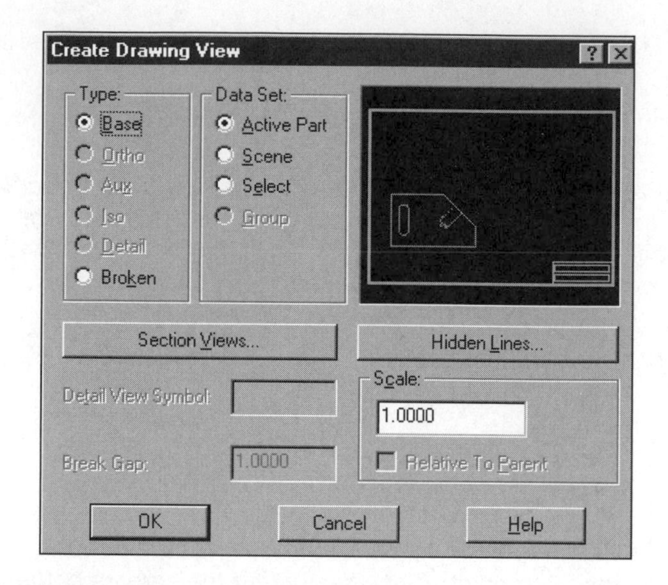


**Figure 1-44** AMDWGVIEW *dialog box.*

**Using DVIEW to Produce a Perspective Projection of a Three-Dimensional Model**

The Previous sections discussed constructing multiview drawings and isometric projections from a three-dimensional solid object. However, as stated earlier, sometimes a more realistic view, or perspective, of a part is required. Up to this point the perspectives created have been two-dimensional. Perspectives of three-dimensional objects are created with the DVIEW command. The DVIEW command generates a perspective projection by using a camera-target system. This system can view the object from any angle, and defines the line-of-sight as a line connecting the target and the camera.

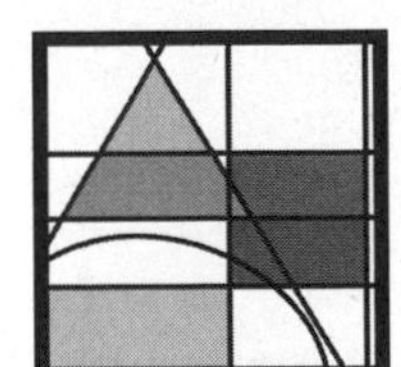

# advanced applications

## Using AutoLISP and DCL to Generate Orthographic and Isometric Views

One advantage of using AutoLISP programming is the ability to write and store programs on the hard drive that can be loaded into an AutoCAD drawing anytime. This ability can be used to create programs that will automate specific drawing routines. For example, it is essential when creating a three-dimensional drawing to rotate the viewpoint throughout the drawing process to coincide with the current area of work. To accomplish this, the VPOINT command must be executed along with one of the settings discussed earlier each time the view is changed. This process can be automated using AutoLISP programming and thereby save valuable time. The following AutoLISP program is provided as an example. When typing this program into the computer, any word processing package can be used. The only requirements are that the file must be saved as an ASCII text file, and it must have the extension LSP. To load the program once it has been saved on the computer, enter the following syntax at the AutoCAD command prompt (load "/directory/file.name"). Once loaded enter CVIEW at the AutoCAD command prompt to start the program. The program allows the technician to change the viewpoint of the object by selecting one of the choices provided (Top, Front, Bottom, Rear, Left, Right, Isometric Right Top, Isometric Right Bottom, Isometric Left Top, Isometric Left Bottom, and Hide Lines), as shown in Figure 1–45.

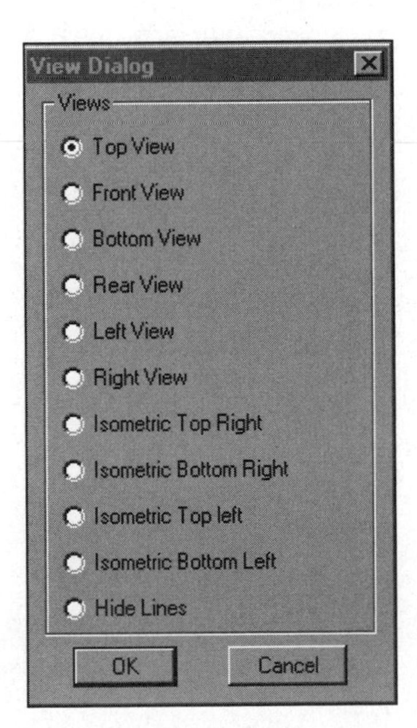


**Figure 1–45** *Dialog box for Chapter one AutoLISP program.*

## AutoLISP Program

```
;;;*************************************************************
;;;
;;;     Program Name: DG01.lsp
;;;
;;;     Program Purpose: This program allows the user to
;;;                      change their point of view in model
;;;                      space. It can also be used in the
;;;                      creation of multiview drawings, by
;;;                      allowing the user to change the
;;;                      the user to change the viewpoint in a
;;;                      viewport, to one of the six principle
;;;                      views.
;;;
;;;
;;;     Program Date: 10/25/98
;;;
;;;     Written By: James Kevin Standiford
;;;
;;;*************************************************************
;;;*************************************************************
;;;
;;;                    Main Program
;;;
;;;*************************************************************
(defun c:cview ()
 (setvar "cmdecho" 0)
 (setq
  dcl_id4
   (load_dialog
    "c:/windows/desktop/geometry/student/AutoLISP programs/DG01.dcl"
   )
 )
 (if (not (new_dialog "Dv" dcl_id4))
  (exit)
 )
 (set_tile "top" "1")
 (action_tile "accept" "(vie) (done_dialog)")
 (start_dialog)
 (if (= vi 1)
  (progn
   (if (= top "1")
       (progn
        (command "vpoint" "0,0,1" "zoom" "e" "Zoom" ".7x")
        (command "modemacro" "Top View")
       )
   )
   (if (= bot "1")
       (progn
        (command "vpoint" "0,0,-1" "zoom" "e" "Zoom" ".7x")
        (command "modemacro" "Bottom View")
       )
```

```
)
(if (= fro "1")
    (progn
     (command "vpoint" "0,-1,0" "zoom" "e" "Zoom" ".7x")
     (command "modemacro" "Front View")
    )
)
(if (= rea "1")
    (progn
     (command "vpoint" "0,1,0" "zoom" "e" "Zoom" ".7x")
     (command "modemacro" "Rear View")
    )
)
(if (= rig "1")
    (progn
     (command "vpoint" "1,0,0" "zoom" "e" "Zoom" ".7x")
     (command "modemacro" "Right Side View")
    )
)
(if (= lef "1")
    (progn
     (command "vpoint" "-1,0,0" "zoom" "e" "Zoom" ".7x")
     (command "modemacro" "Left Side View")
    )
)
(if (= isort "1")
    (progn
     (command "vpoint" "1,1,1" "zoom" "e" "Zoom" ".7x" "hide")
     (command "modemacro" "Isometric Right Top View")
    )
)
(if (= isorb "1")
    (progn
     (command "vpoint" "1,1,-1" "zoom" "e" "Zoom" ".7x" "hide")
     (command "modemacro" "Isometric Right Bottom View")
    )
)
(if (= isolt "1")
    (progn
     (command "vpoint" "1,-1,1" "zoom" "e" "Zoom" ".7x" "hide")
     (command "modemacro" "Isometric Left Top View")
    )
)
(if (= isolb "1")
    (progn
     (command "vpoint" "1,-1,-1" "zoom" "e" "Zoom" ".7x" "hide")
     (command "modemacro" "Isometric Left Bottom View")
    )
)
(if (= hid "1")
    (progn
     (command "hide")
     (command "modemacro" "Hide Line")
```

```
          )
        )
      )
    )
    (setq vi 0)
    (princ)
  )
  (defun vie ()
    (setq top  (get_tile "top")
          bot  (get_tile "bot")
          lef  (get_tile "lef")
          rig  (get_tile "rig")
          rea  (get_tile "rea")
          fro  (get_tile "fro")
          isort (get_tile "isort")
          isorb (get_tile "isorb")
          isolt (get_tile "isolt")
          isolb (get_tile "isolb")
          hid  (get_tile "hid")
          vi  1
    )
  )
(princ "\nTo excute enter cview at the command prompt ")
(princ)
```

## Dialog Control Language Program

```
//%%%%%%%%%%%%%%%%%%%%%%%%%%%%%%%%%%%%%%%%%%%%%%%%%%%%%%%%%%%%
//
//      Activates dialog box
//
//      Descriptive Geometry Chapter 1 DCL File View Dialog
//
//
//
//%%%%%%%%%%%%%%%%%%%%%%%%%%%%%%%%%%%%%%%%%%%%%%%%%%%%%%%%%%%%
Dv : dialog {
label = "View Dialog";
  : boxed_column {
   label = "Views";
   children_fixed_width = true;
   children_alignment = left;
  : radio_column {
  : radio_button {
   key = "top";
   label = "Top View";
  }
  : radio_button {
   key = "fro";
   label = "Front View";
  }
  : radio_button {
   key = "bot";
```

```
   label = "Bottom View";
  }
  : radio_button {
   key = "rea";
   label = "Rear View";
  }
  : radio_button {
   key = "lef";
   label = "Left View";
  }
  : radio_button {
   key = "rig";
   label = "Right View";
  }
  : radio_button {
   key = "isort";
   label = "Isometric Top Right";
  }
  : radio_button {
   key = "isorb";
   label = "Isometric Bottom Right";
  }
  : radio_button {
   key = "isolt";
   label = "Isometric Top left";
  }
  : radio_button {
   key = "isolb";
   label = "Isometric Bottom Left";
  }
  : radio_button {
   key = "hid";
   label = "Hide Lines";
  }
 }
 }
  is_default = true;
  ok_cancel;
}
```

## Review Questions

Answer the following questions on a separate sheet of paper. Your answers should be as complete as possible.

1. What is the purpose of an isometric projection?
2. List the three categories of axonometric projections.
3. What is the purpose of folding lines?
4. Describe how an isometric scale is constructed.
5. List the six views produced in an orthographic projection.
6. Define the following terms: station point, ground plane, vanishing point, and picture plane.
7. What is the difference between nonparametric and parametric solid models?
8. List the three types of three-dimensional objects that AutoCAD can produce.
9. What is layout space/model space and what is it used for?
10. Describe how to construct an isometric projection using AutoCAD.
11. How do isometric projections differ from dimetrics and trimetrics?
12. Define the following terms: adjacent view, frontal plane, ground line, horizontal plane, miter line, plane, profile plane, and visual ray.
13. At what angle is the front of the object viewed in relation to the projection plane in an oblique projection?
14. The protruding lines in a cavalier oblique are drawn at what angle?
15. How does the cabinet oblique differ from the general oblique?

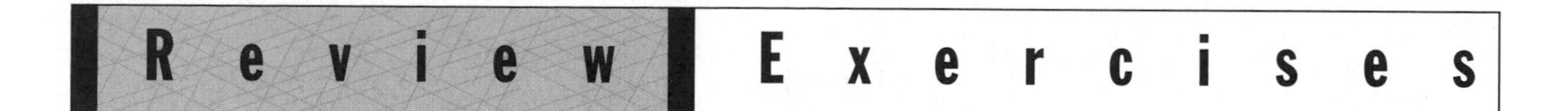


## Two-Dimensional Orthographic Projections

Using AutoCAD, create the missing views as indicated in problems 1 through 4. All dimensions are supplied in inches and all problems marked with an * can be found on the student CD-ROM.

***#1**

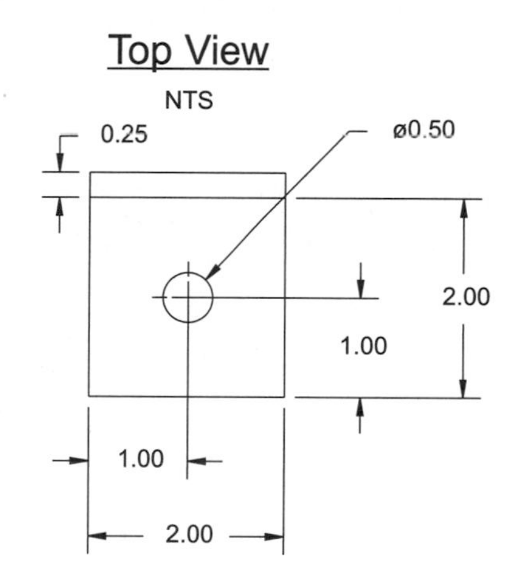


Rear View
NTS

Left Side View
NTS

Front View
NTS

1.75

Right Side View
NTS

Bottom View
NTS

*#2

## Top View
NTS

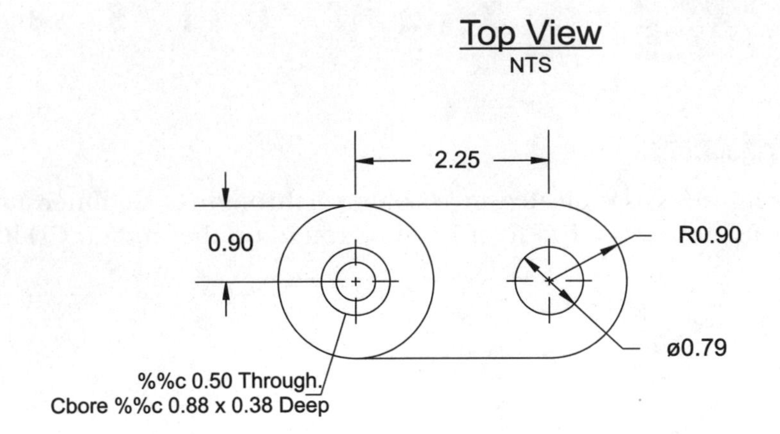


## Right Side View
NTS

## Front View
NTS

## Right Side View
NTS

0.68

0.22

***#3**

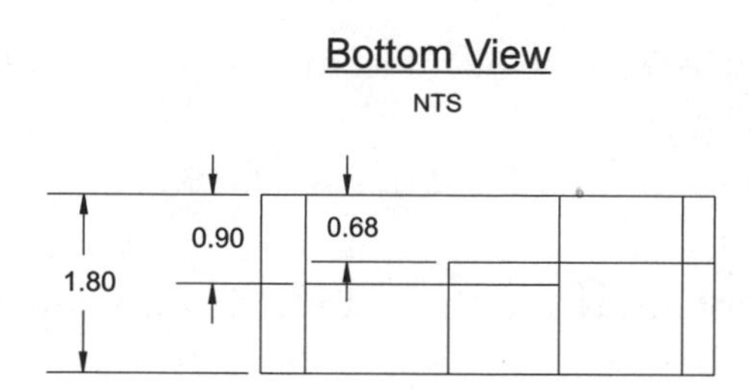


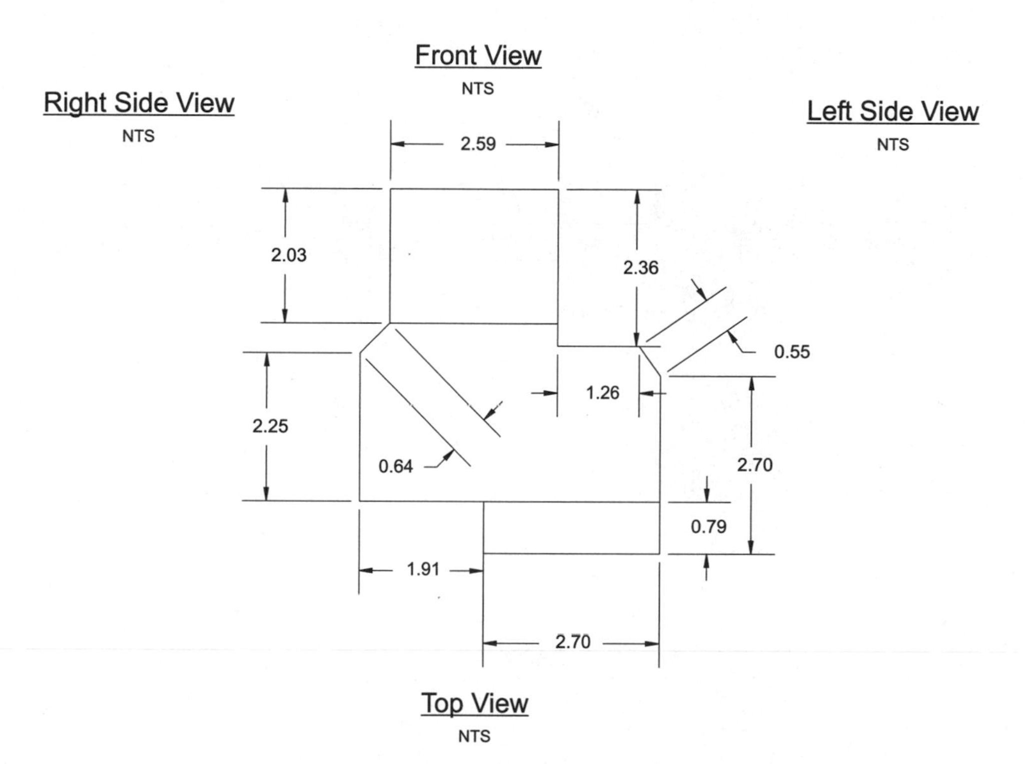


***#4**

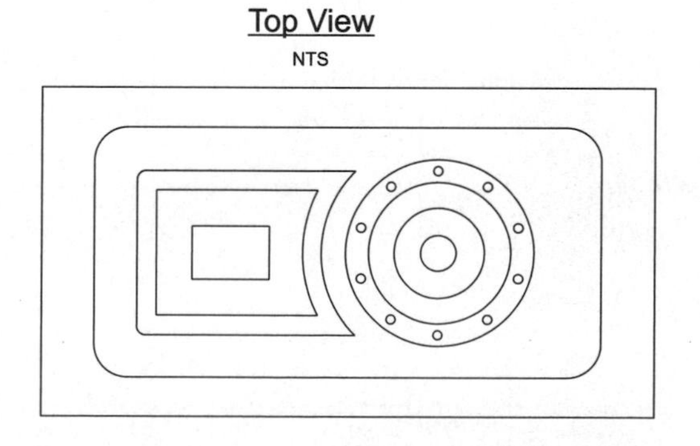


Left Side View
NTS

Front View
NTS

Right Side View
NTS

## 2D Isometric Projections

For problems 5 through 8, use AutoCAD to create an isometric projection for the objects featured in problems 1 through 4.

## 2D Oblique Projections

For problems 9 through 11, use AutoCAD to create the following oblique projections.

***#9**

Cavalier Oblique

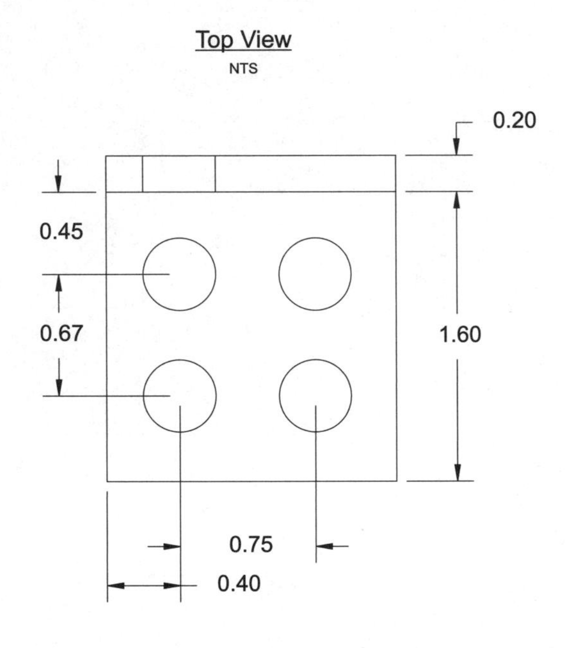


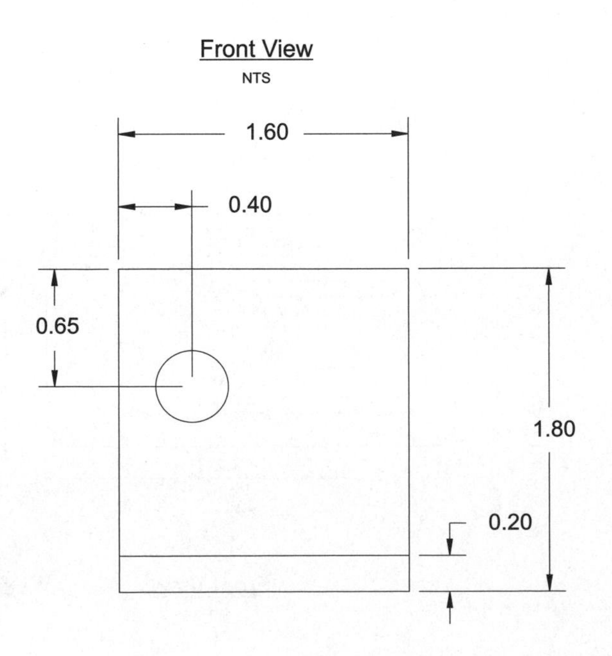

***#10**

General Oblique

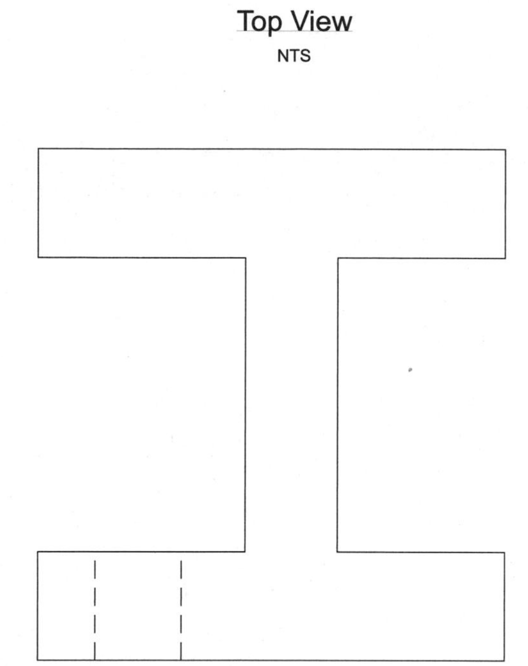


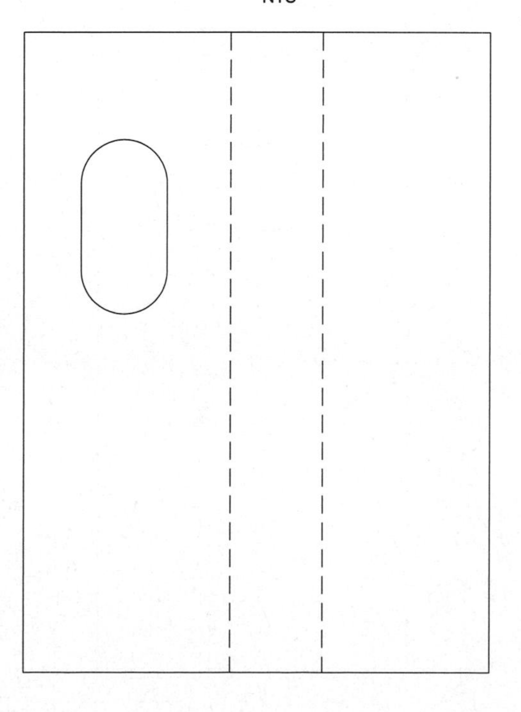

***#11**

Cabinet Oblique

Top View

NTS

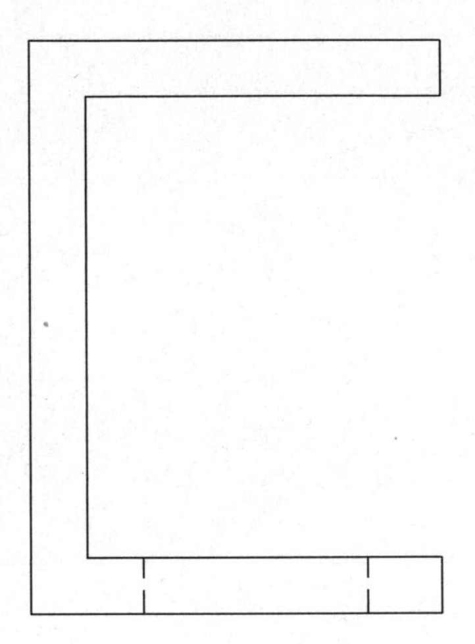

Front View

NTS

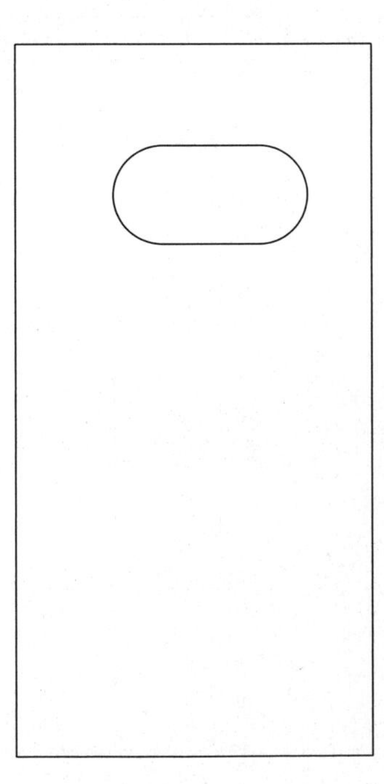

## 2D Perspectives

**#12**

Using AutoCAD, create a one-point perspective of the object in problem 9.

**#13**

Using AutoCAD, create a multiview perspective of the object in problem 1.

**#14**

Using AutoCAD, create a two-point perspective of the object in problem 11.

## 3D Views

***#15**

Using the MVIEW-VPOINT method, create the six principal views of the objects shown in problem 3. A 3D version of this object is furnished on the student CD-ROM.

# Points and Lines

## Objectives

***Upon completion of this chapter the student will be able to do the following:***

- *Define the following terms: point, line, bearing, foreshortened line, oblique line, inclined line, bearing, azimuth, locus, longitude and latitude.*
- *Determine the equivalent distance in statute miles and feet for a given degree of latitude.*
- *Determine the equivalent in hours of a given degree of longitude.*
- *Use the AutoCAD commands* LIST, ID, *and* PROPERTIES *to determine the location of a point.*
- *Describe the difference between a bearing and an azimuth.*
- *Determine the bearing and azimuth of a given line.*
- *Convert from bearings to azimuths and from azimuths to bearings.*
- *Use AutoCAD to determine the bearing and azimuth of a line.*

## Key Words and Terms

| | | |
|---|---|---|
| **Azimuth** | **Horizontal line** | **Point** |
| **Statute mile** | **Inclined line** | **Points** |
| **Bearing** | **Level line** | **Profile line** |
| **Foreshortened line** | **Locus** | **Slope** |
| **Frontal line** | **Oblique line** | **True length** |
| **Grade** | | |

## Introduction to Points and Lines

ll objects, whether they are man-made or the result of natural conditions and/or forces, contain points and lines. They are the basic building blocks for all two- and three-dimensional objects. Both points and lines have been widely studied in almost every technical field. In mathematics, for example, formulas and equations have been derived to pinpoint their exact position in space. In computer numeric control, points are used to define the center of holes, the end of a contour,

or the beginning of a slot, plus much more. In physics, points and lines can be used to illustrate the exact location of a force or even its magnitude. Whatever the application of points and/or lines, the fact remains the same; any person entering a technical field must have a good understanding of them.

## Points

By definition, a point is a theoretical exact position in space, which will be defined by an X-, Y-, and Z-coordinate. Although points are defined by coordinates, the actual point itself does not contain a width, height, or depth. On paper, a point is normally depicted by a small cross, each side measuring 1/8" in length (see Figure 2–1). In AutoCAD, points (or nodes, as they are referred to in computer applications) can be represented by a variety of different objects or shapes (see Figure 2–2). Often, when working with machinery, a point is used to define the position of a moving part with respect to time. In this particular case, the moving part may have several distinct positions (points) that define its motion; these points maybe connected to show the parts range of motion. The line used to connect these points,is known as a ***locus.*** The locus is used to represent all possible locations of the moving part (each possible point), as shown in Figure 2–3.

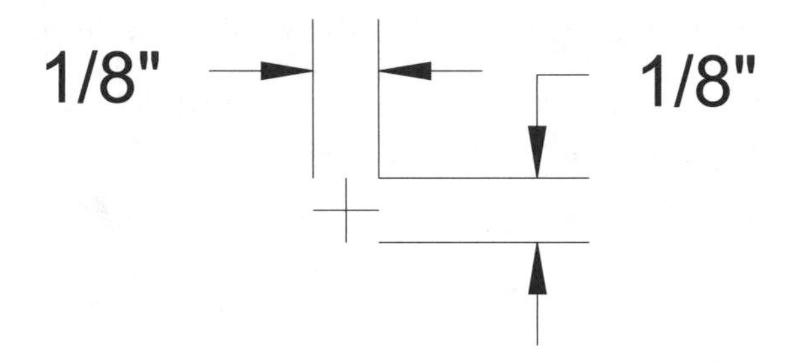


**Figure 2–1** *Dimensions of a point (manually drawn).*

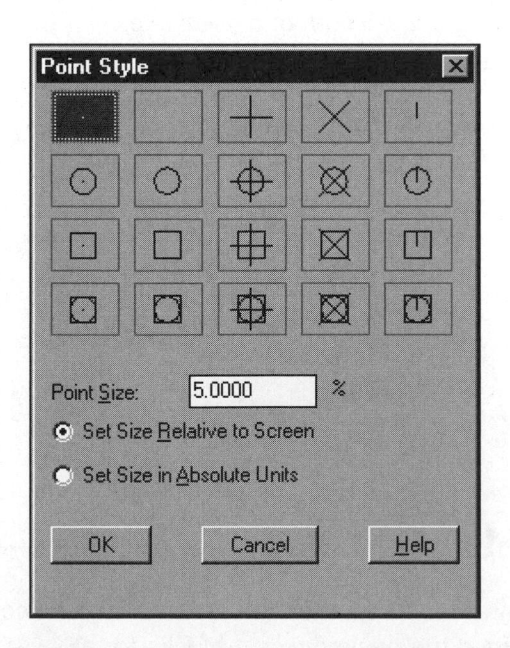


**Figure 2–2** *All possible methods of representing a point using the* PDMODE *command.*

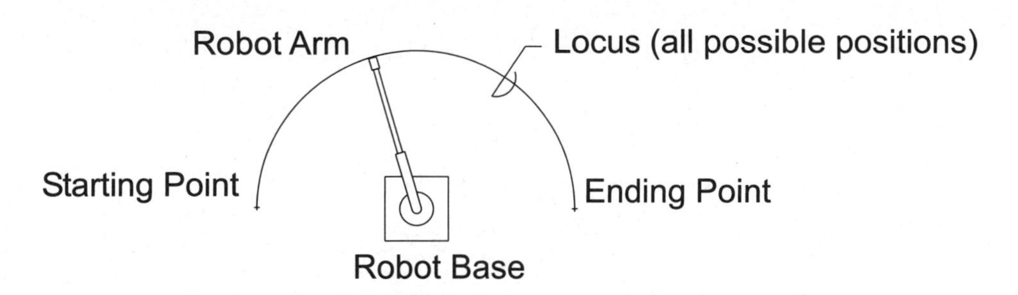


**Figure 2–3** *A locus used to represent all possible locations of a robotic arm.*

In descriptive geometry, a point is used to describe the ends of a line segment when that line is set parallel to the projection plane, as shown in Figure 2–4. They are used to represent a line when that line is positioned perpendicular to the projection plane (see Figure 2–5). They illustrate the intersection of two lines, as shown in Figure 2–6, or the intersection of a line and a plane, as shown in Figure 2–7.

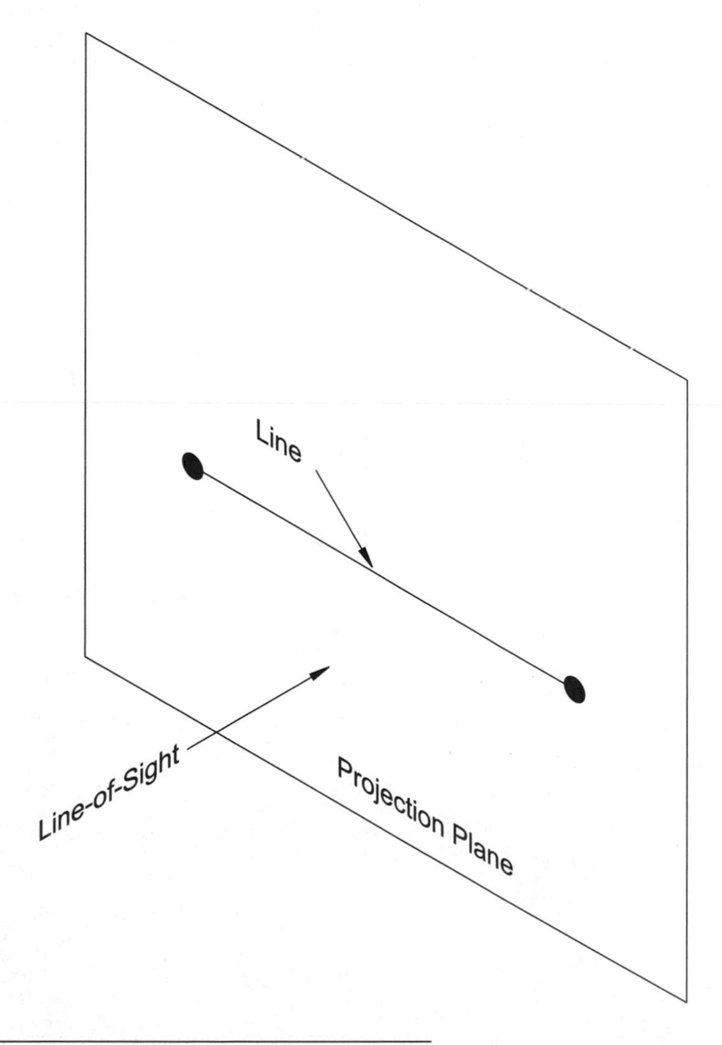


**Figure 2–4** *A line drawn parallel to the projection plane.*

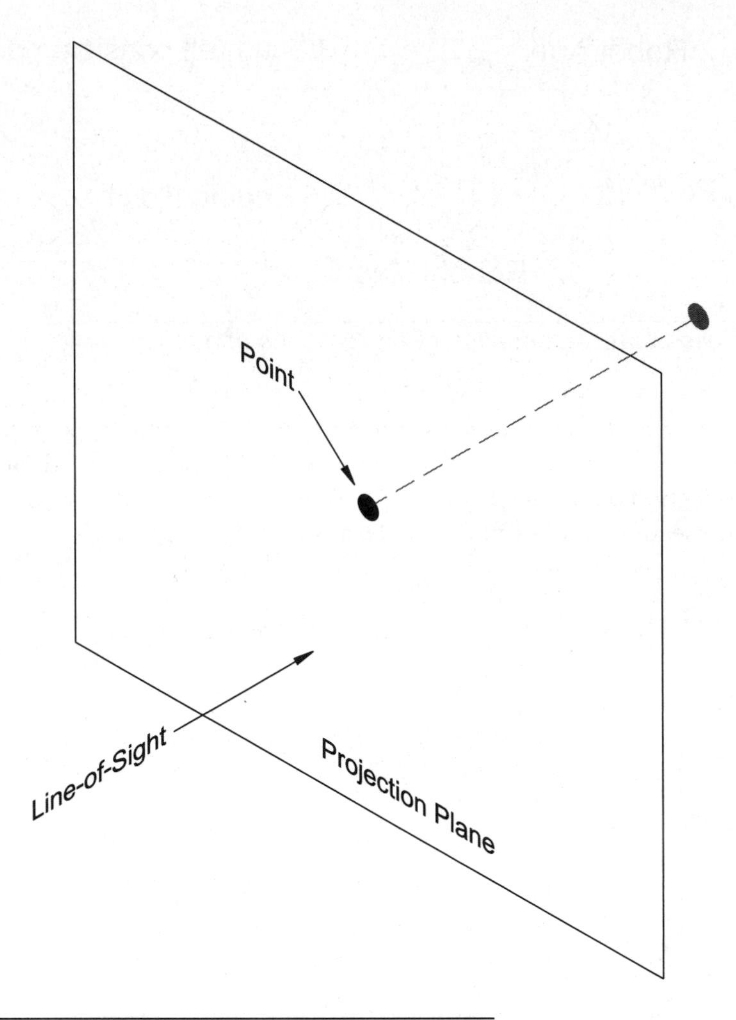


**Figure 2-5** *A line drawn perpendicular to a projection plane.*

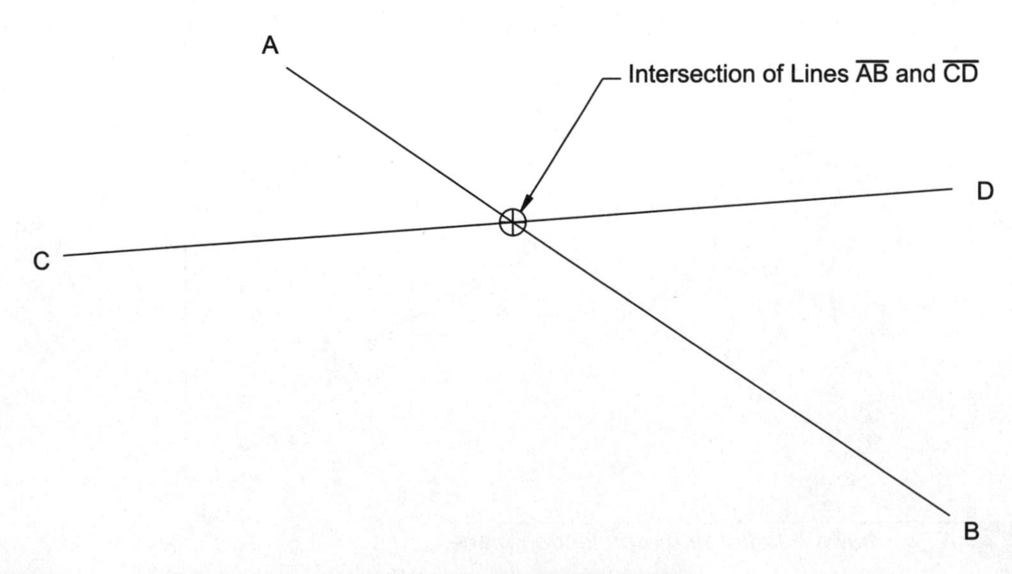


**Figure 2-6** *The intersection of two lines is illustrated with a point.*

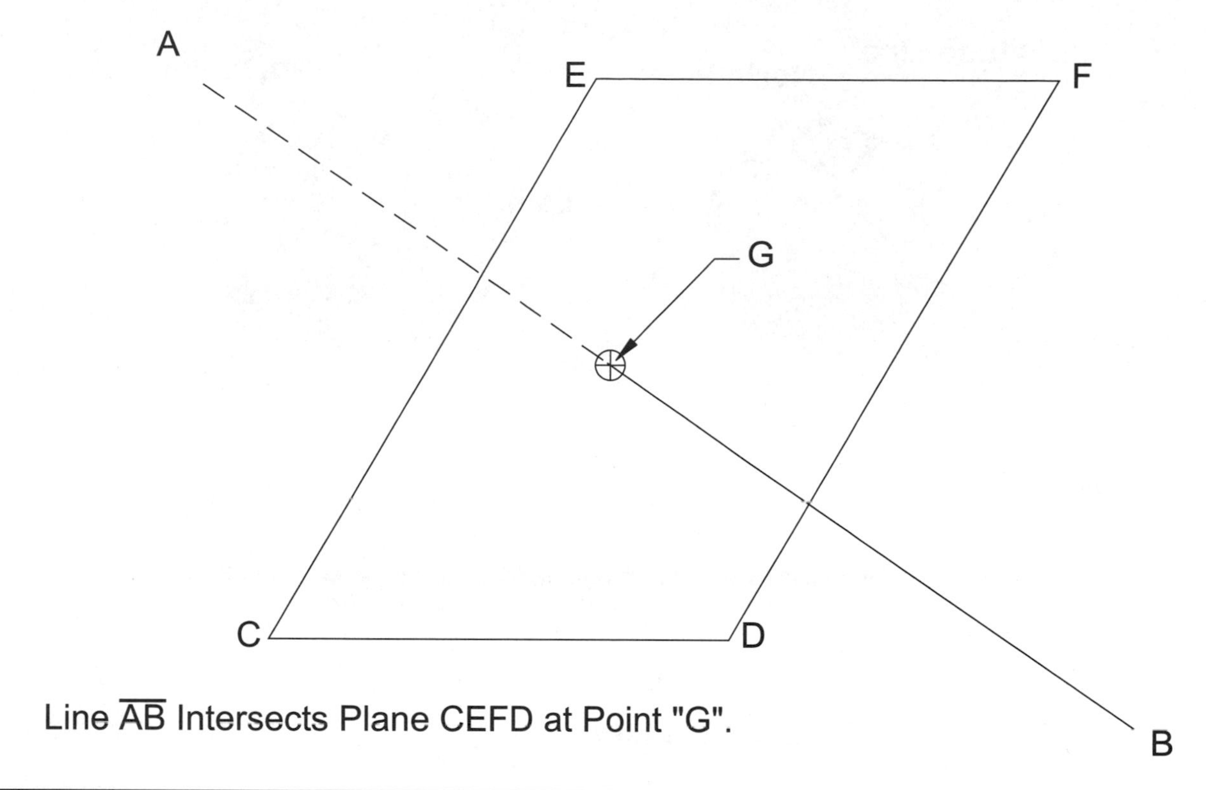


**Figure 2–7** *The intersection of a point and a plane.*

## Finding the Location of a Point

Before the location of a point can be determined, a reference system must be established. Regardless of the application, this fact remains unchanged. For example, when locating a point on the earth's surface, lines circling the earth running west to east (latitude) and lines starting at the poles emanating out (longitude) are used as a reference (see Figures 2–8, 2–9, and 2–10). In this system, both latitude and longitude are measured in degrees. The zero reference for latitude is the equator (dividing the earth into two halves, a northern and southern hemisphere). The zero reference mark for longitude is the prime meridian (a line connecting the north and south poles, running through Greenwich, England). There are 90° of latitude to the north and south of the equator, putting the 90° mark at both poles. For longitude there are 180° to the east and west of the prime meridian. To determine the location of a point using this system, the latitude is found by multiplying 69 (this is equal to the number of statute miles in one degree of latitude) times the cosine of the angle, which is formed between the point in question, the center of the earth and the equator. Next, the longitude is determined by measuring the difference in time at which noon occurs (at the point where the longitude is needed) against the standard reference longitude of the prime meridian. The difference in time must then be corrected using the time of "Ephemeris Transit" provided in the Astronomical Almanac for the particular day of the year in which the measurement is made. This correction is needed to compensate for the variation in the earth's speed due to its elliptical orbit around the sun. Once this correction has been made, the corrected time difference is then multiplied by 360°/24 hours (there are 360° in a 24-hour period) or 15°/hours, thus yielding the longitude for the particular point in degrees.

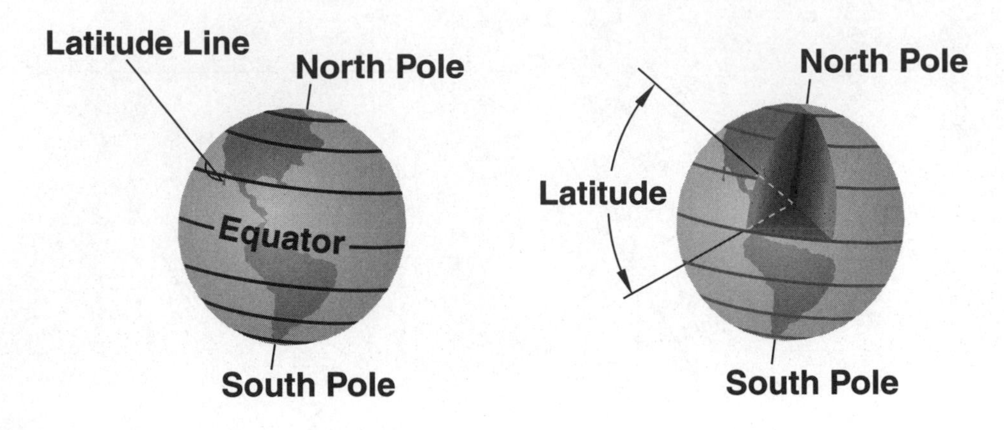


**Figure 2-8** *Lines of latitude circling the earth's surface.*

**Note: A degree in latitude is equal to approximately 69 statute miles or 364,566 feet.**

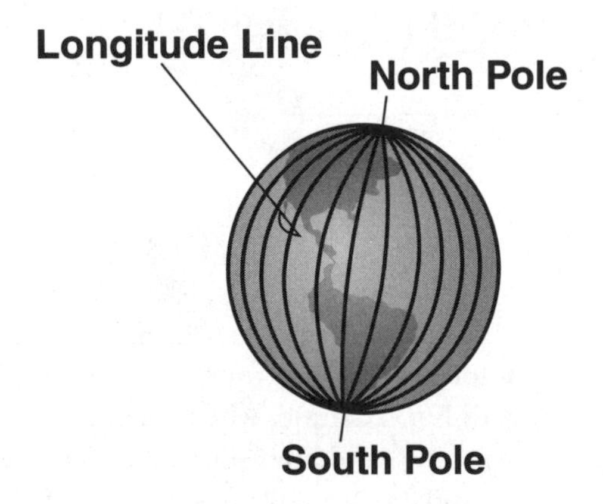


**Figure 2-9** *Lines of longitude circling the earth's surface.*

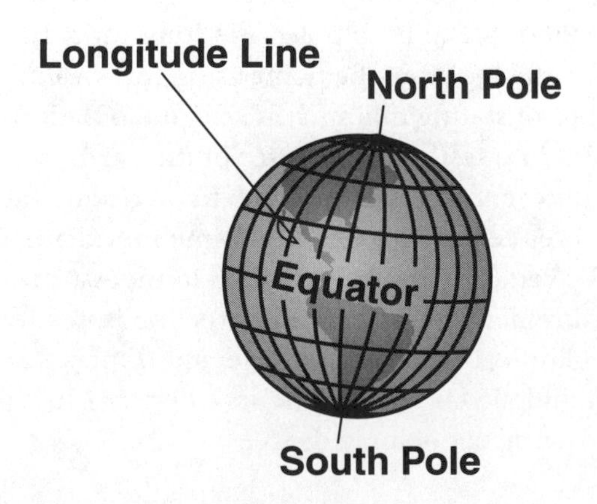


**Figure 2-10** *Both lines of latitude and longitude.*

In descriptive geometry, the system used to establish the location of a point is the orthographic and/or auxiliary view method. The folding lines produced from these methods are used as a reference to locate the position of points. These measurements are then transferred to the corresponding views using the folding lines. In short, the measurement is taken from a parallel folding line and transferred to a view where that measurement would be parallel again (see Figure 2–11).

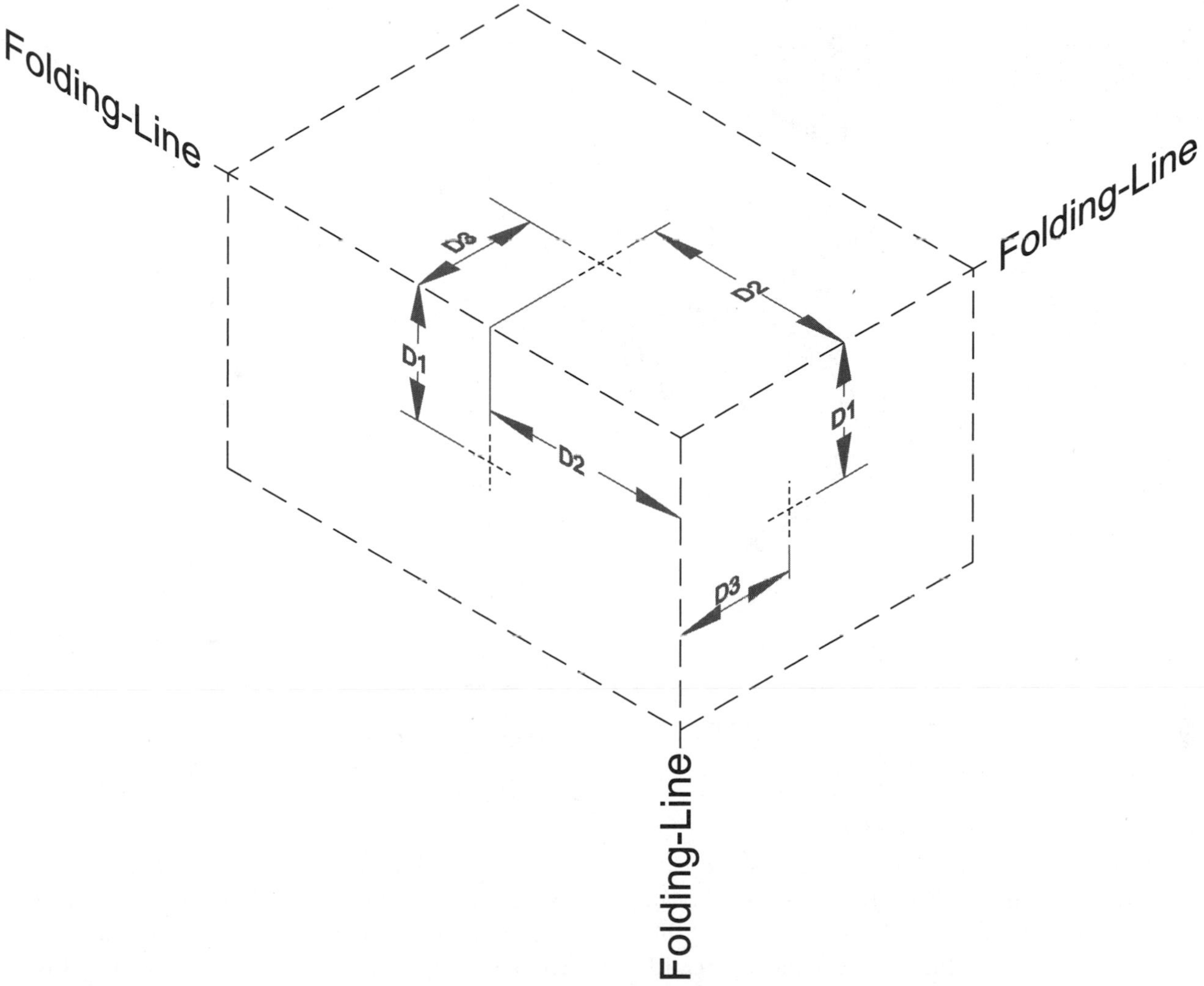

**Figure 2–11** *Locating a point on a glass cube using the folding line method.*

Another method of determining the position of a point is with the use of Cartesian coordinates. In this method, the X-, Y-, and Z-axes are imposed onto a corner of the entire object. The point is then located by drawing a line perpendicular from the axis to the point. The position where the line intersects the axis determines the location of the point for that axis. For example, using the object featured in Figure 2–11, the Cartesian coordinates for the point specified would be X5, Y4, and Z4, (see Figure 2–12).

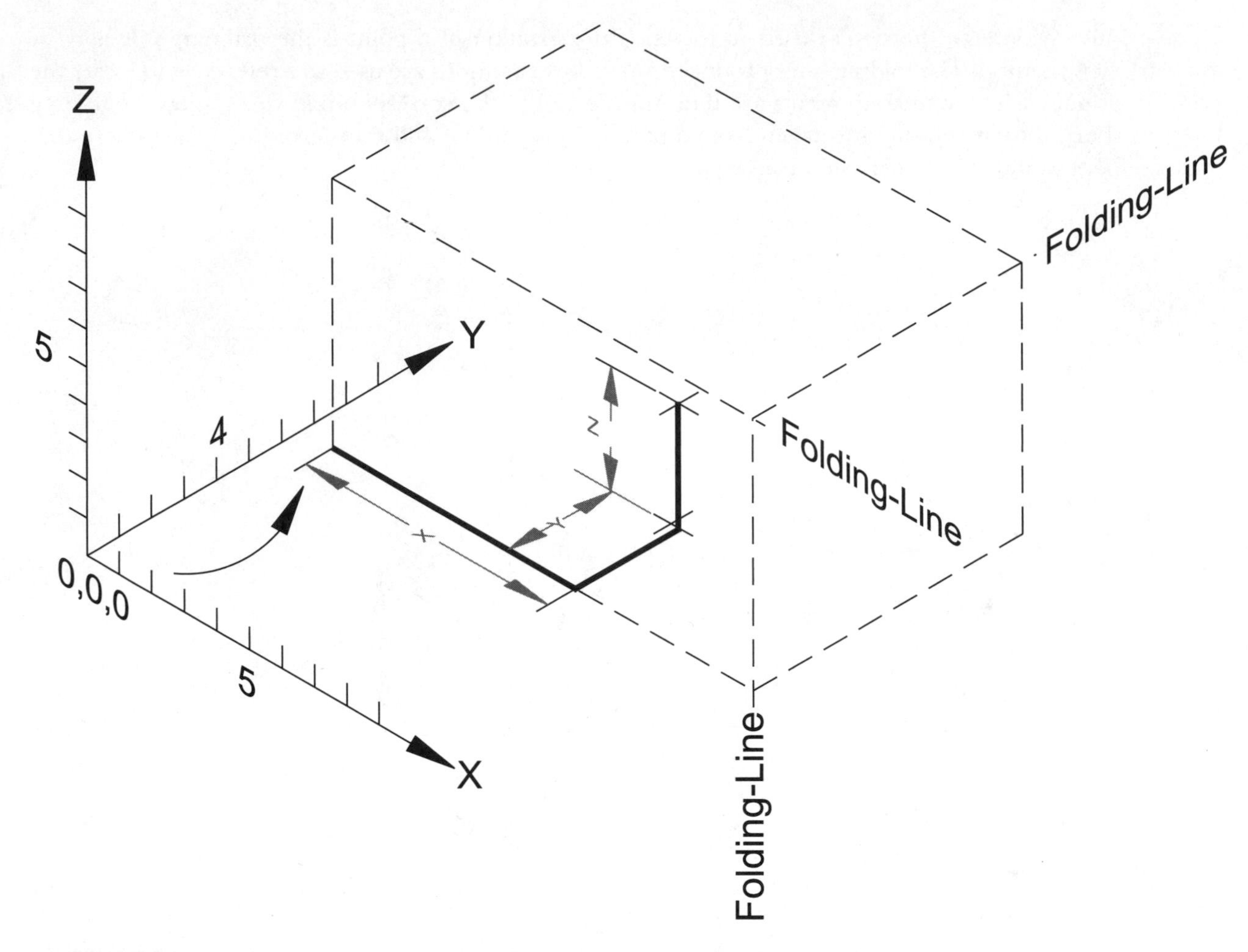


**Figure 2–12** *Locating a point in a glass cube using Cartesian coordinates.*

#### Using AutoCAD to Locate a Point

In AutoCAD, points are located using the Cartesian coordinates system. To transfer a point from one location to another, projection lines are drawn from that point into the view where the point is to be located (as described in Chapter one). It is the intersection of two projection lines that determines the location of a point. However, in some applications the actual Cartesian coordinate may be needed. In this case, AutoCAD has three different commands that may be used to determine these coordinates. These commands are ID, LIST, and DDMODIFY.

The ID command was developed strictly to identify the coordinates of points. The only information that this command returns is the actual X-, Y-, and Z-coordinates. For example, once the point in Figure 2–13 has been selected, the ID command can be executed to retrieve the exact coordinates of that point.

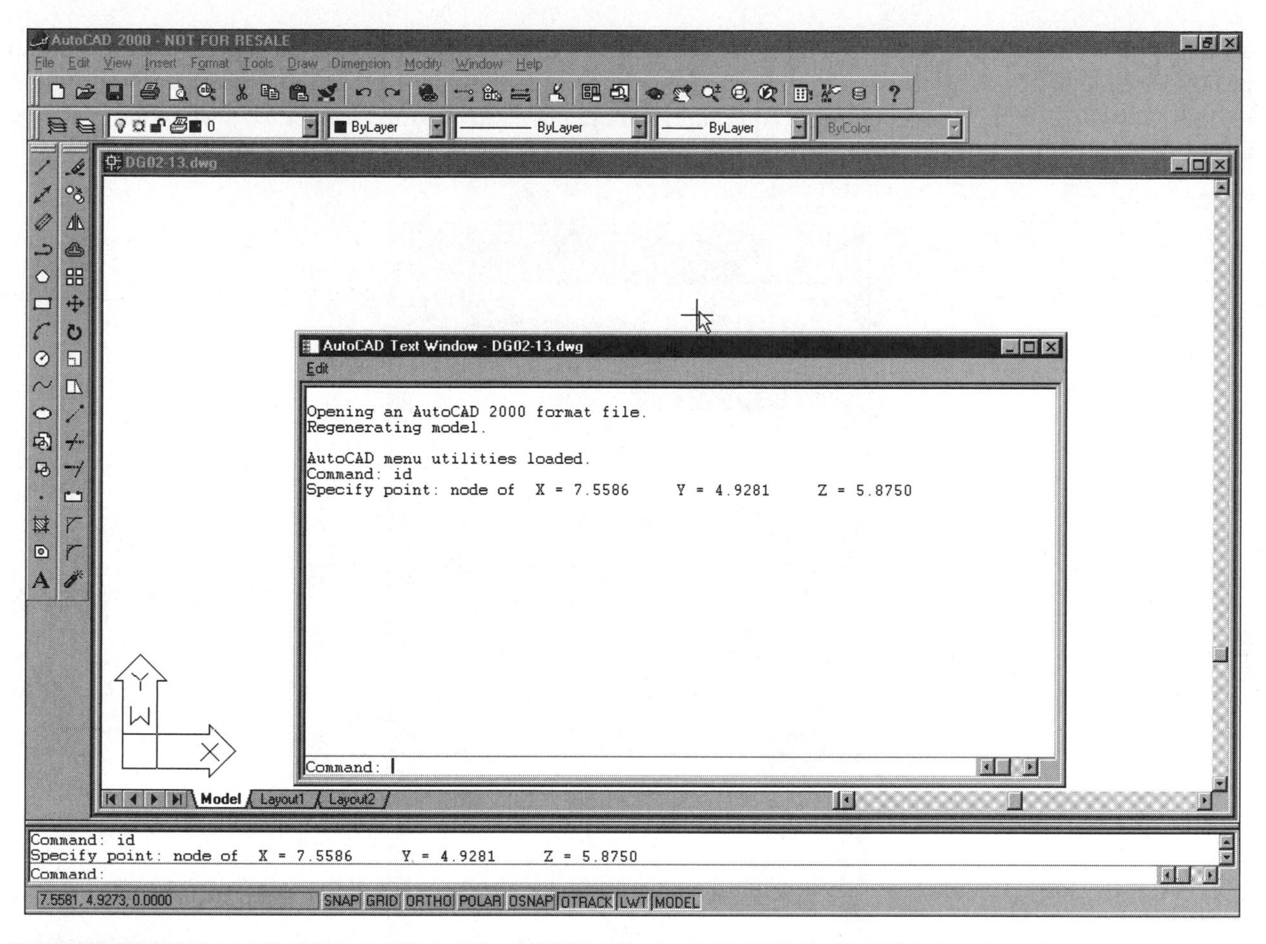


**Figure 2–13** *Using the* ID *command to determine the Cartesian coordinates of a point.*

Command: ID (ENTER)
Specify point: node
of X = 7.5586 Y = 4.9281 Z = 5.8750

**Note: The X-axis is used to denote width, while the Y-axis is used for the height, and the Z-axis is used for the depth. The three axes intersect at the lower left-hand corner of the screen. This makes the positive X-axis run from left to right (facing the computer screen), the positive Y-axis run from bottom to top, and the positive Z-axis comes out of the monitor toward the user.**

The LIST command determines the coordinates regarding a point, as well as other important information such as layer, entity type, layer color, and so on. Using the LIST command to determine the coordinates of the point from Figure 2–13 yields the following.

Command: LIST (ENTER)
Select objects: Specify opposite corner: 1 found
Select objects: (ENTER)
POINT Layer: "0" *(This line identifies the entity type, as well as the layer on which entity exist.)*
Space: Model space *(This line informs the user that the entity was constructed in model space.)*
Handle = 2A *(This line identifies the entity handle.)*
at point, X= 7.5586 Y= 4.9281 Z= 0.0000
Command:

The last command is the DDMODIFY command. Not only does the DDMODIFY command determine more than just the coordinates, it also offers the user an opportunity to modify that information. Selecting the point in Figure 2–13 and executing the DDMODIFY command will open a dialog box like the one shown in Figure 2–14, where the user can modify the returned information.

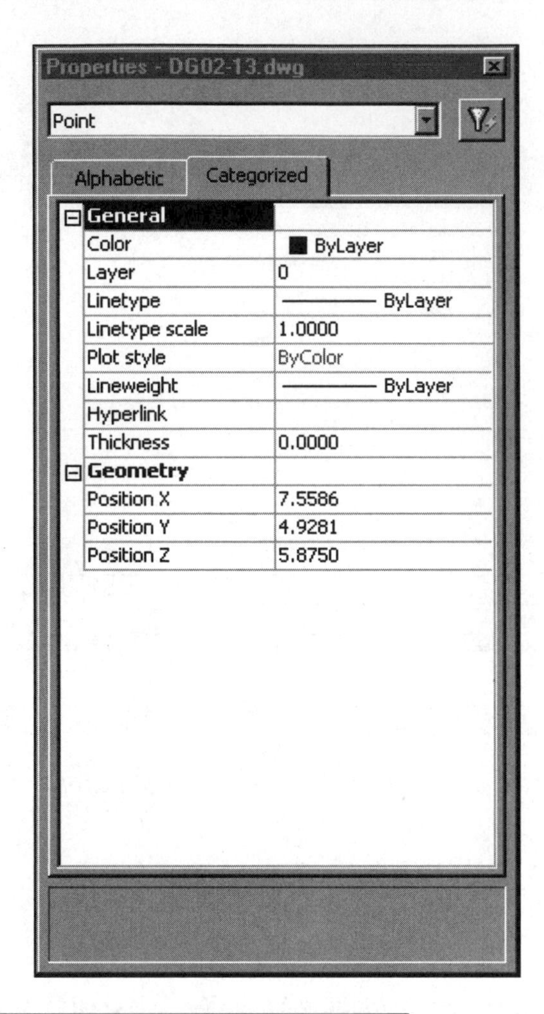


**Figure 2–14** *Identifying a point using the* DDMODIFY *command.*

## Lines

A ***line*** is a non-curved entity that is void of any width and stretches out to infinity in both directions; therefore completely absent of a starting or ending point. A ***line segment*** is a portion of a line that is also void of any width, but is constrained by two end points. Throughout time, various formulas and equations have been derived in advanced mathematics that describe almost every attribute of both lines and line segments, and a good deal of time is required to master these concepts and methodologies. The main function of descriptive geometry is to categorize and locate the position in space of these entities. These attributes are paramount to understanding descriptive geometry as a whole. That is why the following sections are crucial subject matter for this discipline.

## Types of Lines

A line can be placed into one of five categories in descriptive geometry, ***frontal***, ***horizontal***, ***profile***, ***vertical***, and ***oblique***, each having its own unique characteristics. However, a preliminary understanding of some basic terms is necessary to define these larger concepts. When a line segment is said to be shown in ***true length***, then the exact distance between the endpoints is revealed. A line segment that is not shown as true length is ***foreshortened***. The ***true angle*** of a line is the angle formed by a line shown in true length and the edge view of one of the principal planes. Finally, when a line appears inclined in one of the principal planes and parallel to the other two projection planes, then that line is known as an ***inclined line***.

### Frontal Lines

A line that appears as true length when the observer's line-of-sight is set perpendicular to the frontal plane is known as a ***frontal line***. These lines will either reside in or be parallel to the frontal plane. Their true angle will be observed in the front or rear view, and is determined by the angle formed between the true length of the frontal line and the adjacent principal plane that is shown in edge view. For example, the true angle of the frontal line AB shown in Figure 2–15 is 30°; the angle formed by the frontal line and the horizontal plane. It is in this view that the true length of the frontal line is revealed, along with the edge view of the adjacent principal plane (horizontal plane).

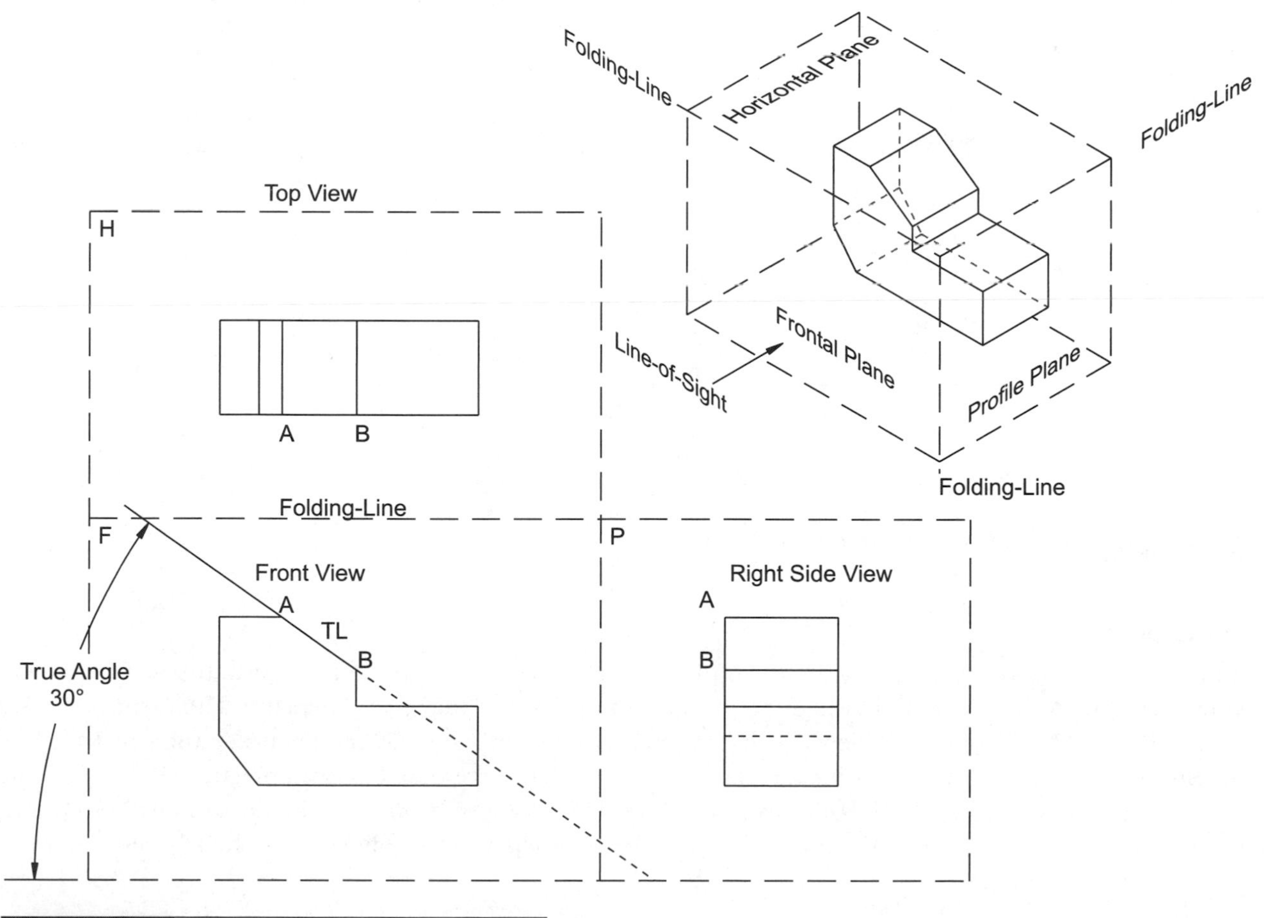


**Figure 2–15** *Frontal line and its true angle.*

#### Horizontal Lines

A line that appears as true length when the observer's line-of-sight is set perpendicular to the horizontal plane is known as a ***horizontal line*** (also referred to as a ***level line***). These lines will either reside in or be parallel to the horizontal plane. Their true angle will be observed in the plan or bottom view and is determined by the angle formed between the true length of the horizontal line and the adjacent principal plane that is shown in edge view. For example, the true angle of the frontal line shown in Figure 2–16 is 67°; the angle formed by the horizontal line and the frontal plane. It is in this view that the true length of the horizontal line is revealed along with the edge view of the adjacent principal plane (frontal plane).

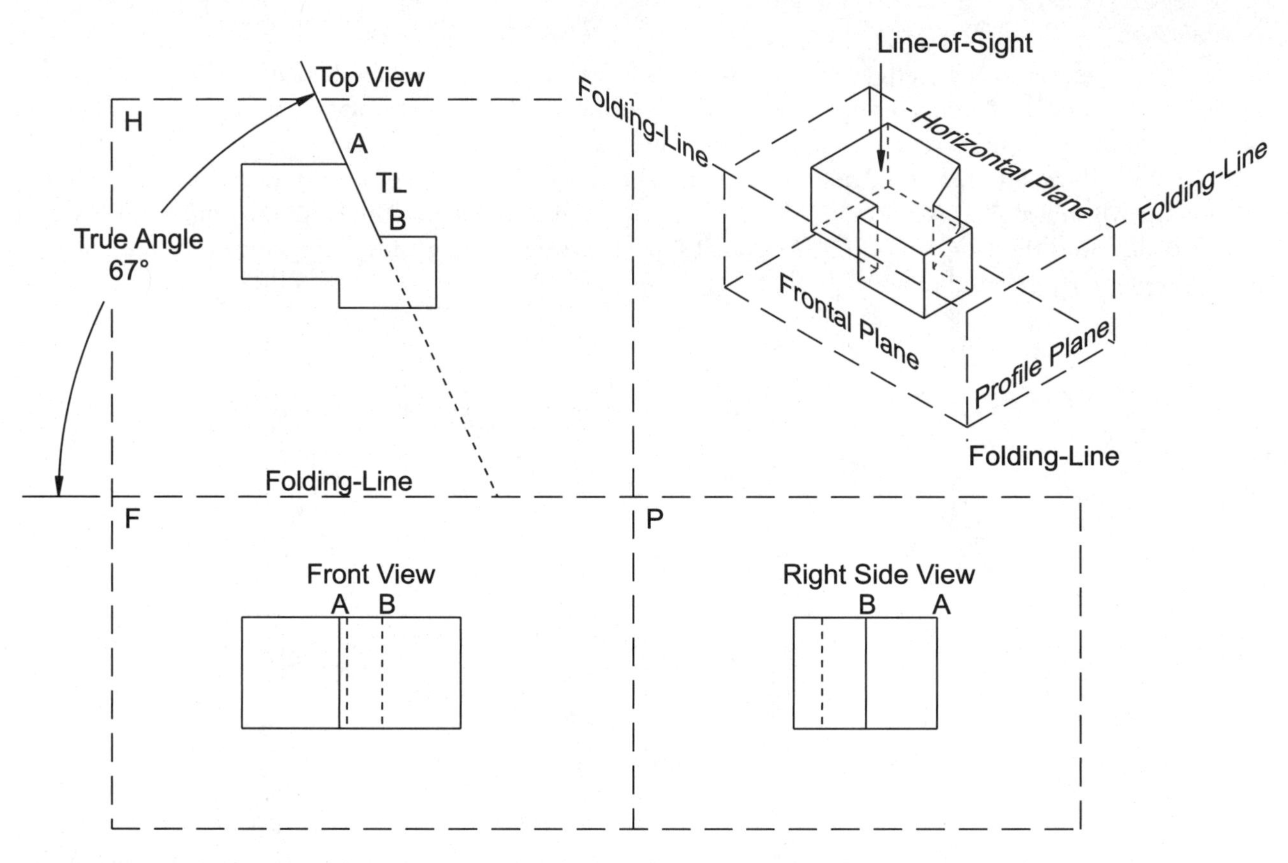


**Figure 2–16** *Horizontal line and its true angle.*

#### Profile Lines

A line that appears as true length when the observer's line-of-sight is set perpendicular to the profile plane is known as a ***profile line***. These lines will either reside in or be parallel to the profile plane. Their true angle will be observed in the right or left side views and is determined by the angle formed between the true length of the profile line and the adjacent principal plane that is shown in edge view. For example, the true angle of the profile line shown in Figure 2–17 is 45°; the angle formed by the profile line and the horizontal plane. It is in this view that the true length of the profile line is revealed along with the edge view of the adjacent principal plane (horizontal plane).

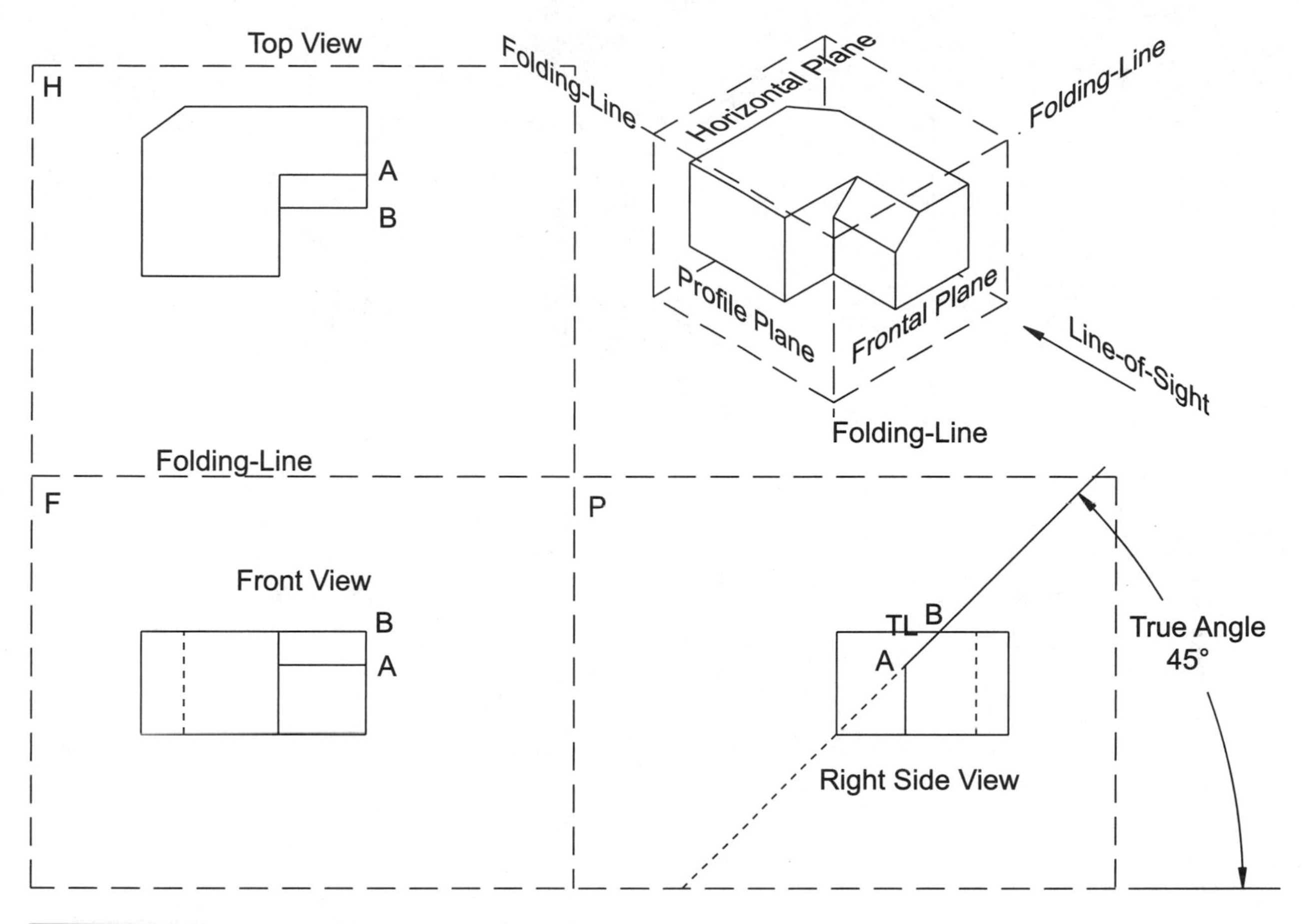


**Figure 2–17** *Profile line and its true angle.*

**Vertical Lines and Oblique Lines**

When an orthographic projection is constructed and a line appears in true length in any of the frontal or profile views (in civil engineering these are known as elevations), that line is known as a ***vertical line*** (see Figure 2–18). Their true angle is found just like the other types of lines previously mentioned; it is determined in any view in which the line appears in true length and the adjacent principal plane appears in edge view. On the other hand, when a line does not appear in true length in any of the principal views produced by the principal planes, then this line is called an ***oblique line***, as shown in Figure 2–19.

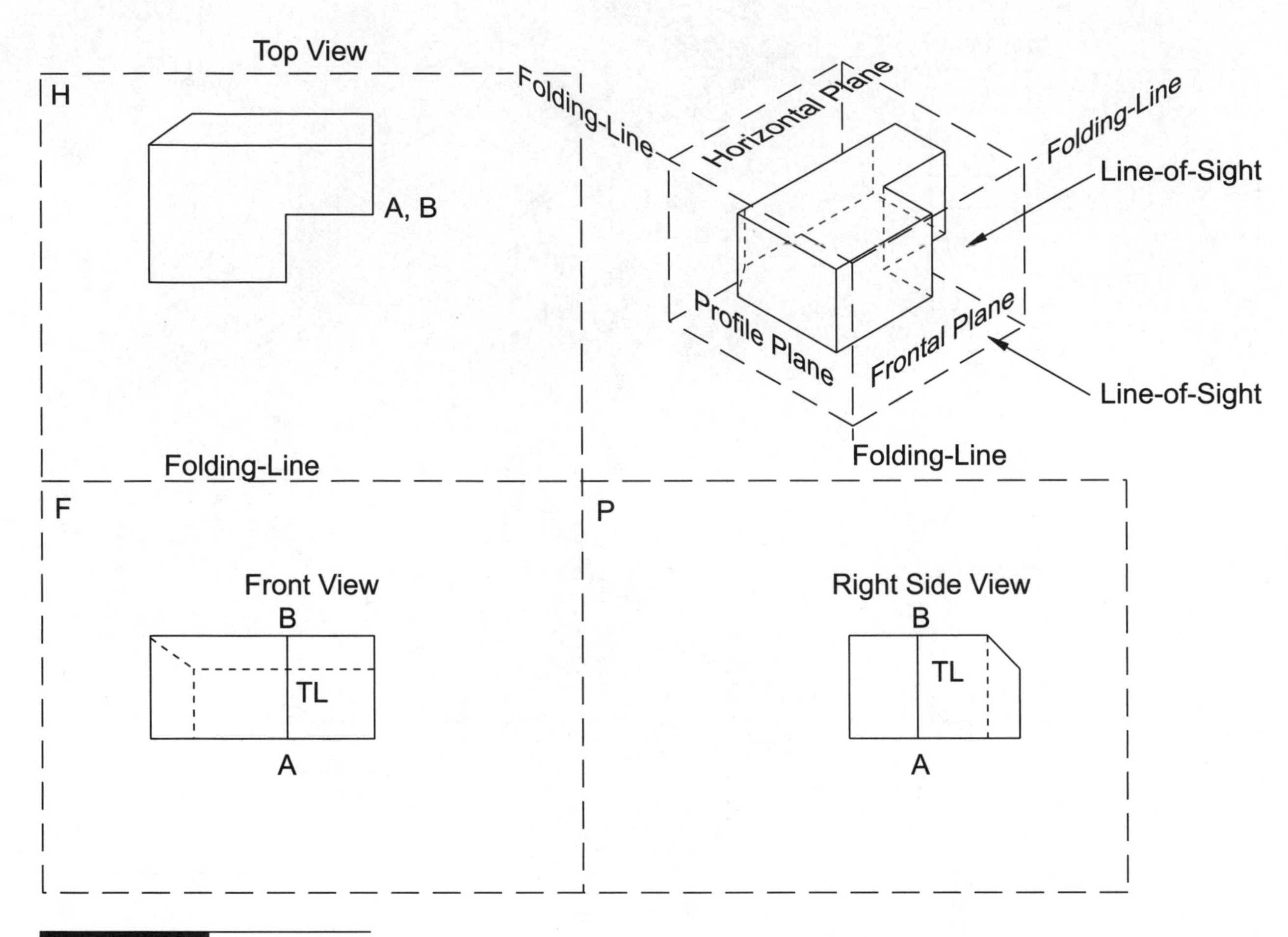


**Figure 2–18** *Vertical line.*

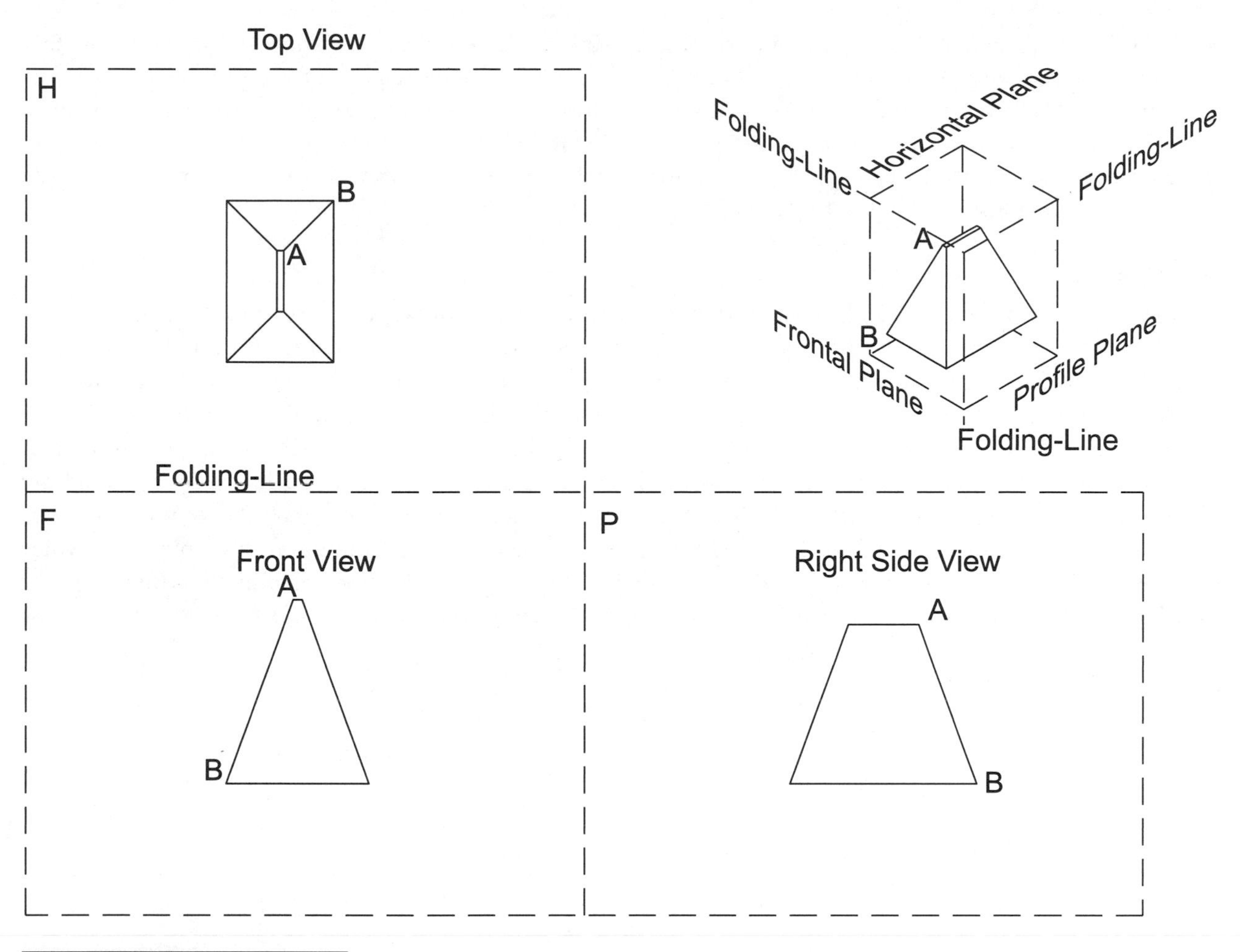


**Figure 2-19** *Oblique line.*

## Determining the Location of a Line

Much like determining the location of a point, the location of a line can be found by using one of three different methods. They can be found by describing their position in and from the horizontal plane using a system of bearings and slopes. They can be described in terms of the location of their endpoints in relation to an X-, Y-, and Z-axis, using Cartesian coordinates, or their location can be determined by using the folding line method described in Chapter one.

### Bearings, Azimuths, Grade, and Slope

As stated earlier, the location (position in space) of a line segment can be defined using the Cartesian (also known as rectangular X, Y, Z) or polar (distance, direction) coordinate systems. However, these methods of defining the location of a line are not readily used in the civil engineering, surveying, and navigational industries. These industries are concerned primarily, although not exclusively, with the course of lines along the surface of the earth. In these fields, the attributes of a line are described with ***bearings***, ***azimuths***, ***grades***, and ***slope***. Recall from the previous sections that a line segment is a non-curved entity connecting two or more points. Therefore, in order to define the position of a line along the surface of the earth, we must either redefine the definition of a line or flatten out the earth's surface. If the definition of a line is redefined to include curved paths, then our line quickly becomes an arc. It is possible to flatten a small portion of the earth's surface without compromising the line's location. On the average, the earth's surface curves approximately 8 inches per mile (5,280 feet). If the line segment is viewed from the frontal position (in civil engineering this is called a profile), the curvature will affect the position of the line. If this same line segment is viewed from the top or plan position, this curving has no effect on the horizontal position whatsoever. For this reason, bearings and azimuths are ideal for describing this type of line.

#### Bearings

A bearing is a term used to describe the location of a line in the horizontal plane. It is measured in respect to its position from due north or due south, and is always denoted in angular degrees. Due north can be assumed to be the top of the drawing unless otherwise specified. The annotation for a bearing typically lists the reference point first (north or south), followed by the angle the line deviates from the reference. Finally, the annotation lists the direction (east or west) that the line inclines toward. For example, the line in Figure 2–20 has a bearing of N60°0'0"E from point A to point B. This tells us that the line starts in the south (quadrant three) and is heading in a northeast direction. It also tells us that the line is running 60°0'0" to the east of due north (this puts the line in quadrant one). This line can also be described from point B to point A; in which case the bearing becomes S60°0'0"W (see Figure 2–21). The angle of the line remains the same, only the description changes. It should be noted that the angle specified in a bearing will never exceed 90°.

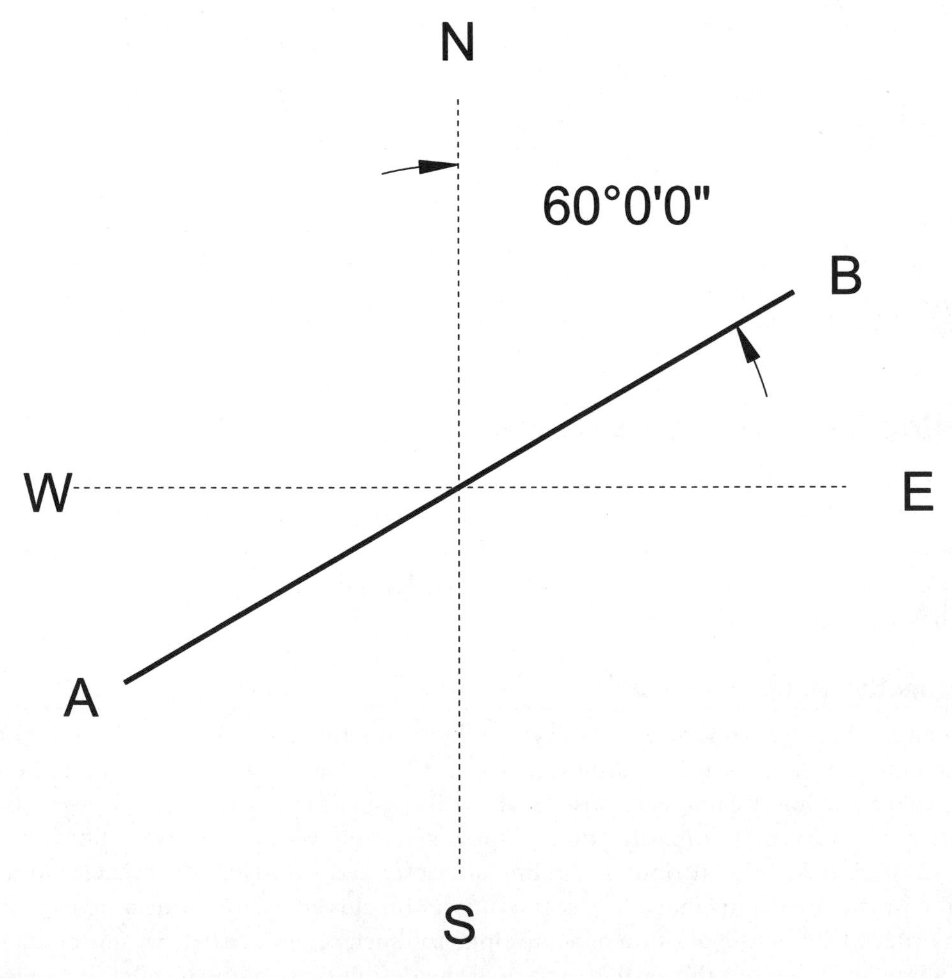


**Figure 2–20** *Line AB has a bearing of N60°0'0"E.*

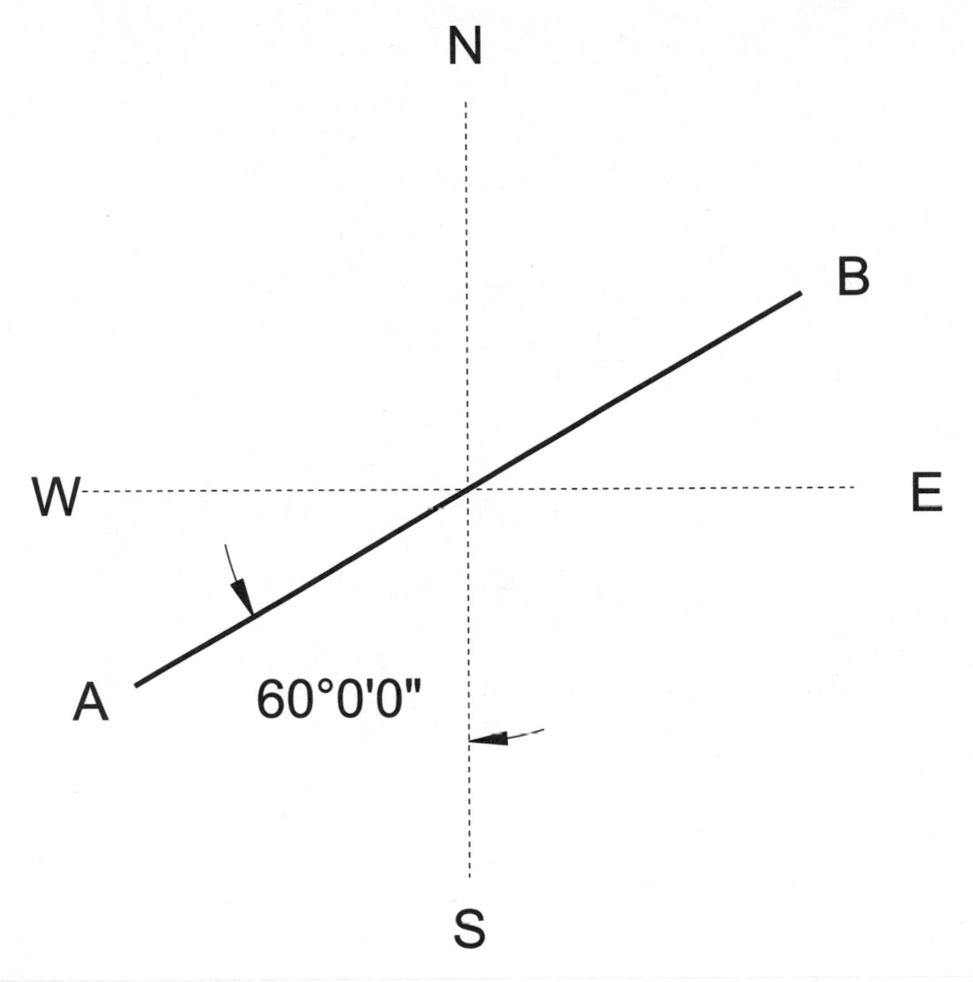


**Figure 2–21** *Line BA has a bearing of S60°0'0"W.*

**Azimuth**

Another method of specifying the location or position of a line in the horizontal plane is with azimuths. An azimuth is an angle measured clockwise, typically from due north. Astronomers, the military, and the National Geological Survey normally use this. However, they can also be measured from due south. Azimuths differ from bearings because an azimuth does not require any letters designating where the line starts. Only a numeric value, always given in degrees, is required. Unless otherwise specified, all azimuths are assumed to be measured from the north, clockwise, instead of from the south, clockwise. Four examples of azimuths are shown in Figure 2–22.

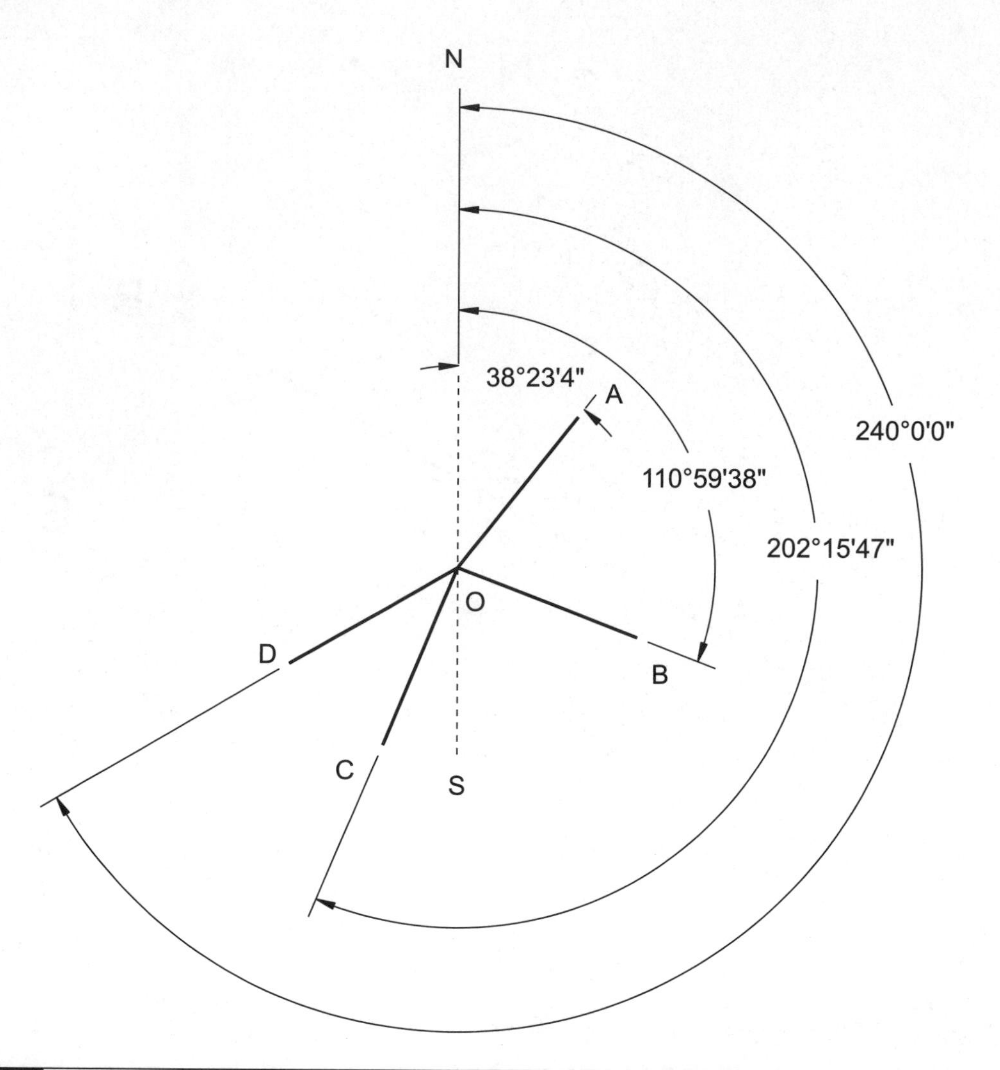


**Figure 2–22** *The azimuth for line AO = 38°23'04". The azimuth for line BO = 110°59'38". The azimuth for line CO = 202°15'45". The azimuth for line DO = 220°00'00".*

The difference between bearings and azimuths can be summarized as follows:

1. The angle of a bearing will range from 0° to 90°. The angle of an azimuth will range from 0° to 360°.
2. Bearings require that two directions be given in conjunction with the angular value. Azimuths require only the angular value (unless a base other than north is used; then it should be noted on the drawing).
3. Bearings are measured from either a clockwise or counterclockwise direction. Azimuths are measured from a clockwise direction only.
4. Bearings are measured from either north or south. Azimuths are measured typically from the north (unless otherwise specified).

**Note: See Figures 2–23, 2–24, and 2–25 for a comparison of azimuths and bearings.**

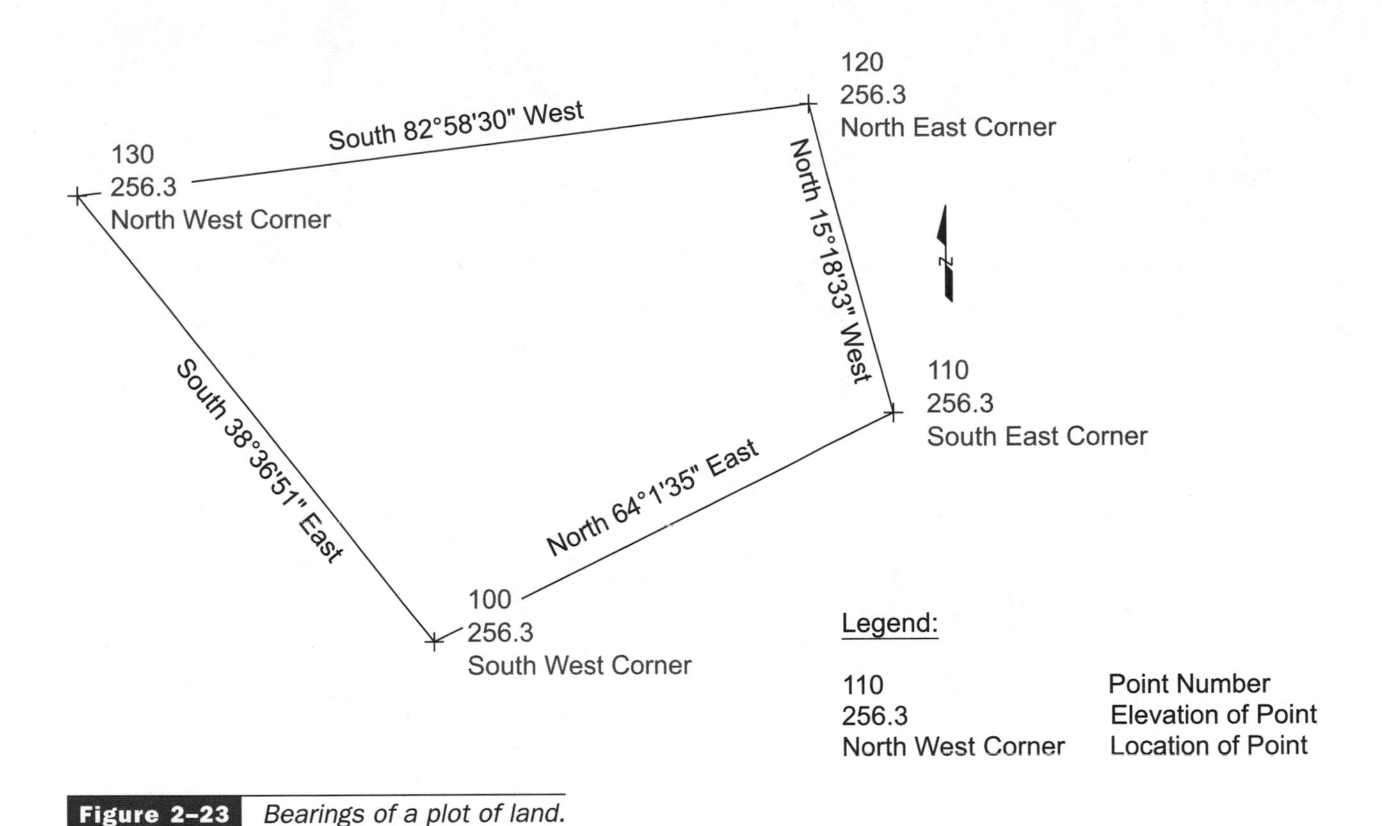


**Figure 2–23** *Bearings of a plot of land.*

120
256.3
North East Corner
262°58'30"
130
256.3
North West Corner
344°41'26"
N
110
256.3
South East Corner
141°23'8"
64°1'35"
100
256.3
South West Corner
Legend:
110 Point Number
256.3 Elevation of Point
North West Corner Location of Point

**Figure 2–24** *Azimuth of the same plot of land using north as a reference.*

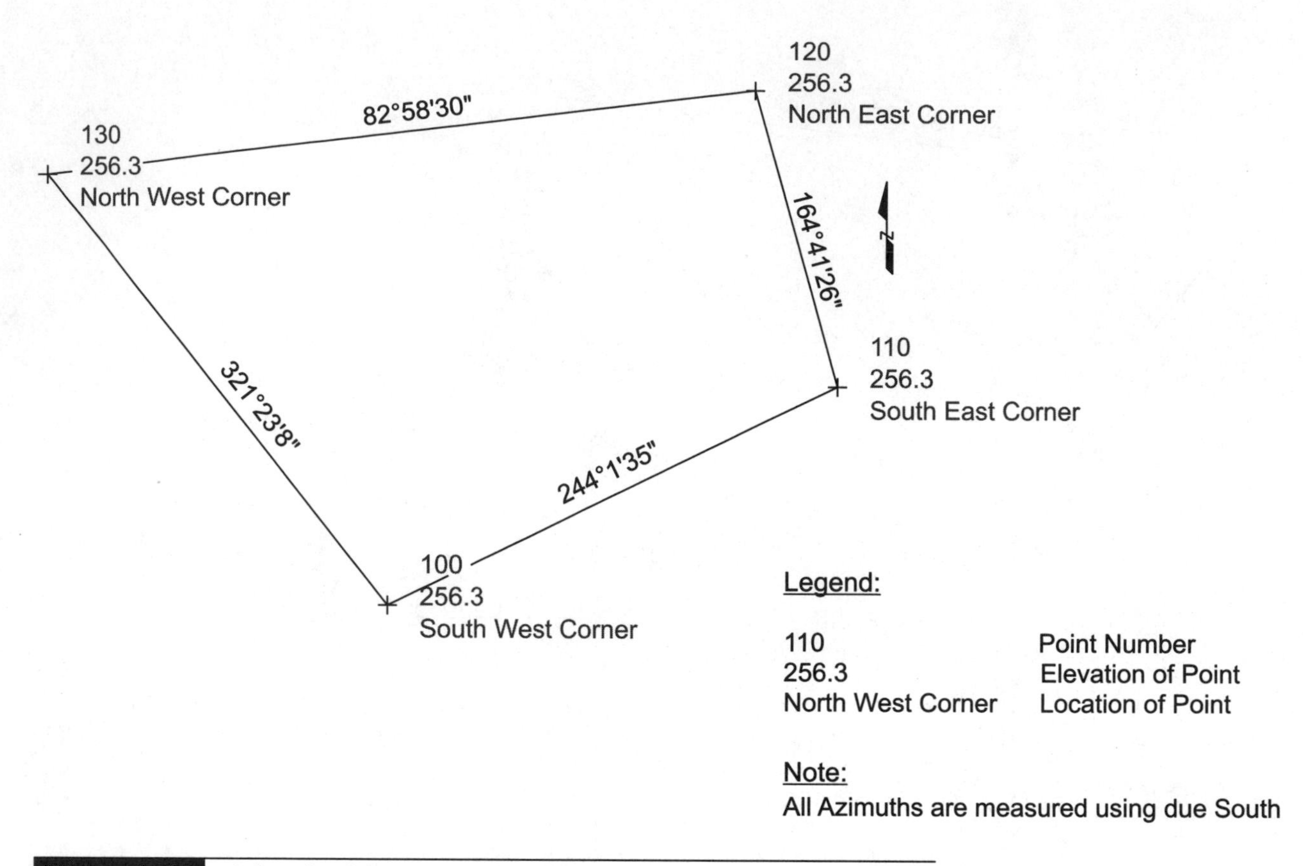


**Figure 2–25** *Azimuth of the same plot of land using south as a reference.*

***Converting from Bearings to Azimuths*** Often it becomes necessary to convert from one method to the other (bearings to azimuths, azimuths to bearings). This conversion can be easily accomplished by following these four simple guidelines.

1. For all lines in the first quadrant, the angle associated with the bearing will be the same for the azimuth.
2. For all lines in the second quadrant, the azimuth is calculated by subtracting the bearing from 360°.
3. For all lines in the third quadrant, the azimuth is calculated by adding the bearing to 180°.
4. For all lines in the fourth quadrant, the azimuth is calculated by subtracting the bearing from 180°.

For Azimuths using North as their Reference Please refer to Figure 2–26.

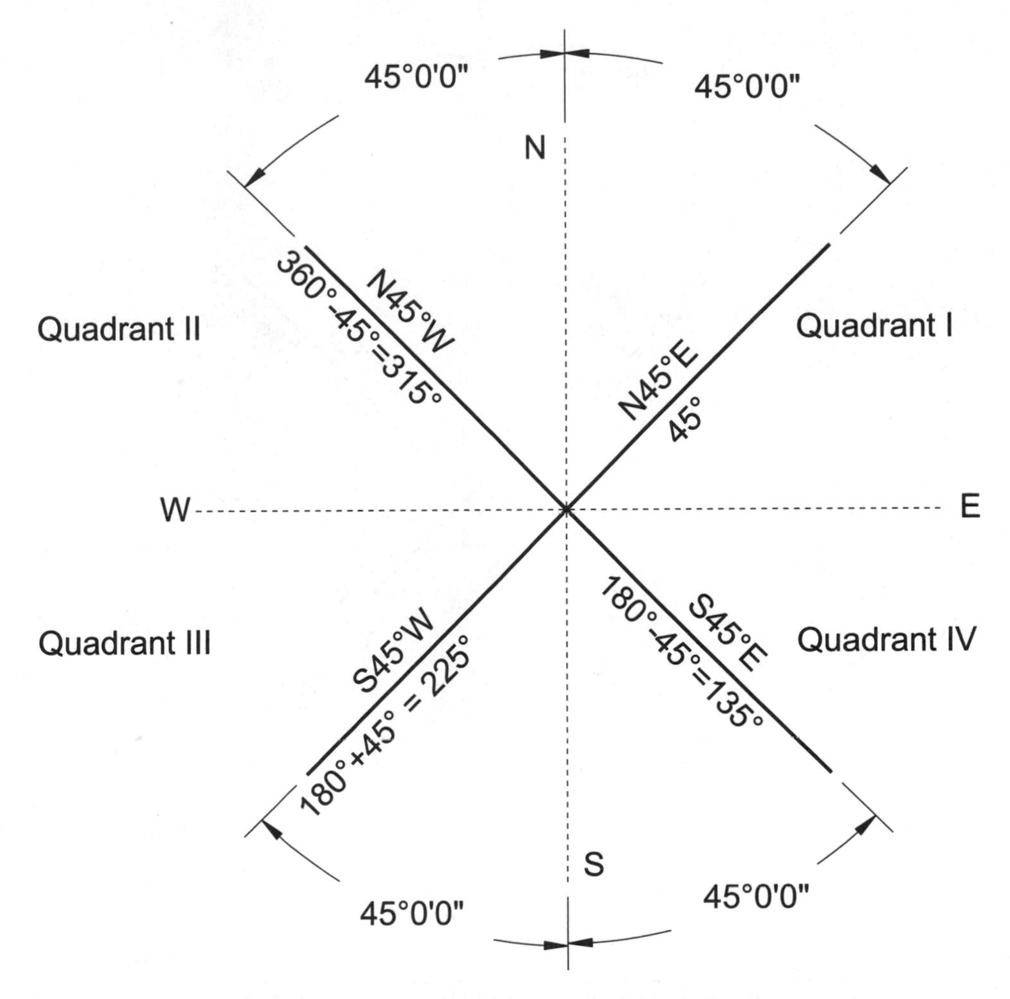


**Figure 2–26** *Converting bearings to azimuths using north as a reference.*

1. For all lines in the first quadrant, the azimuth is calculated by adding the bearing to 180°.
2. For all lines in the second quadrant, the azimuth is calculated by subtracting the bearing from 180°.
3. For all lines in the third quadrant, the angle associated with the bearing will be the same for the azimuth.
4. For all lines in the fourth quadrant, the azimuth is calculated by subtracting the bearing from 360°.

For Azimuths using South as Their Reference please refer to Figure 2–27.

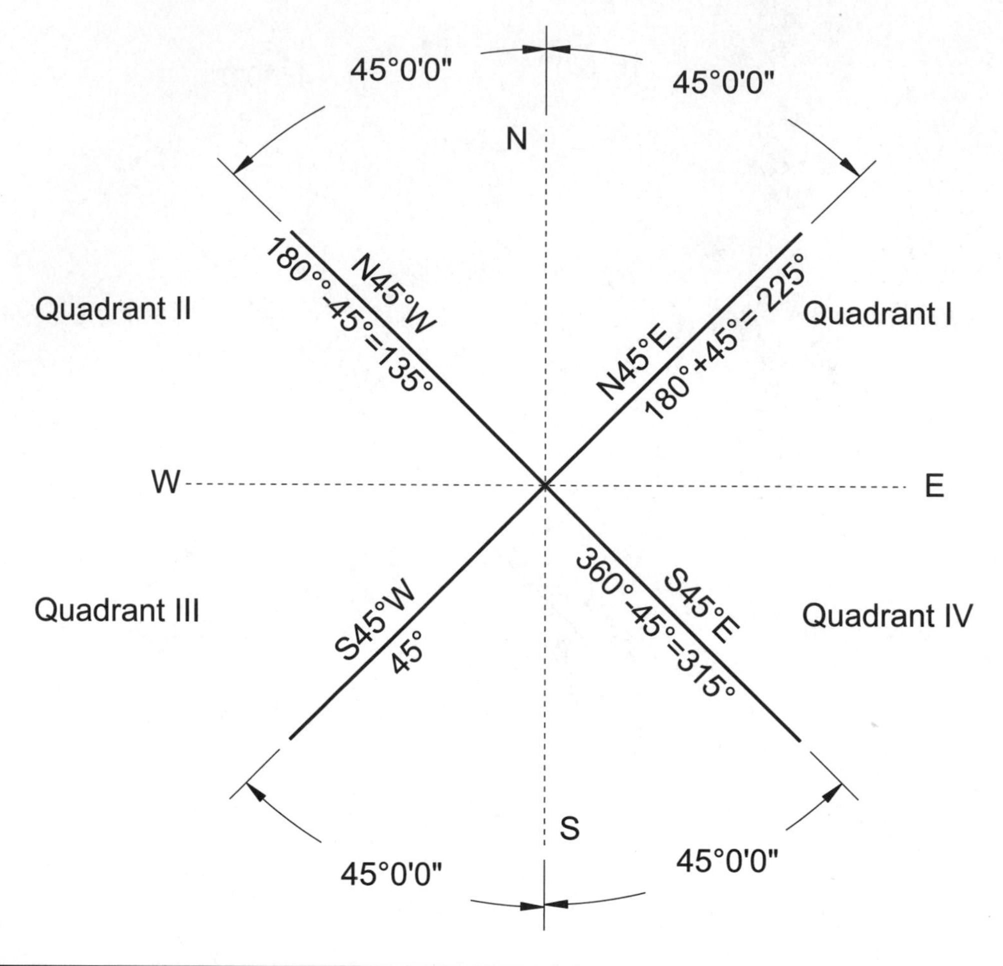


**Figure 2–27** *Converting bearings to azimuths using south as a reference.*

***Using AutoCAD to Determine Bearings and Azimuths of a Line*** The bearings of a line are easily found using AutoCAD. AutoCAD has the ability to display the angle created by the line and the XY-axis. The angle can be displayed in decimal degrees, degrees, minutes, and seconds, gradients, radians, or surveyor's units. To determine the bearings of a line, change the angle display format using the UNITS command. Using this command offers the option of not only controlling the units used when displaying the length of a line, but also the type, direction, and starting position from which angles are measured with respect to the current UCS setting. To illustrate this point, the following steps would be used to determine the bearing of the line shown in Figure 2–28.

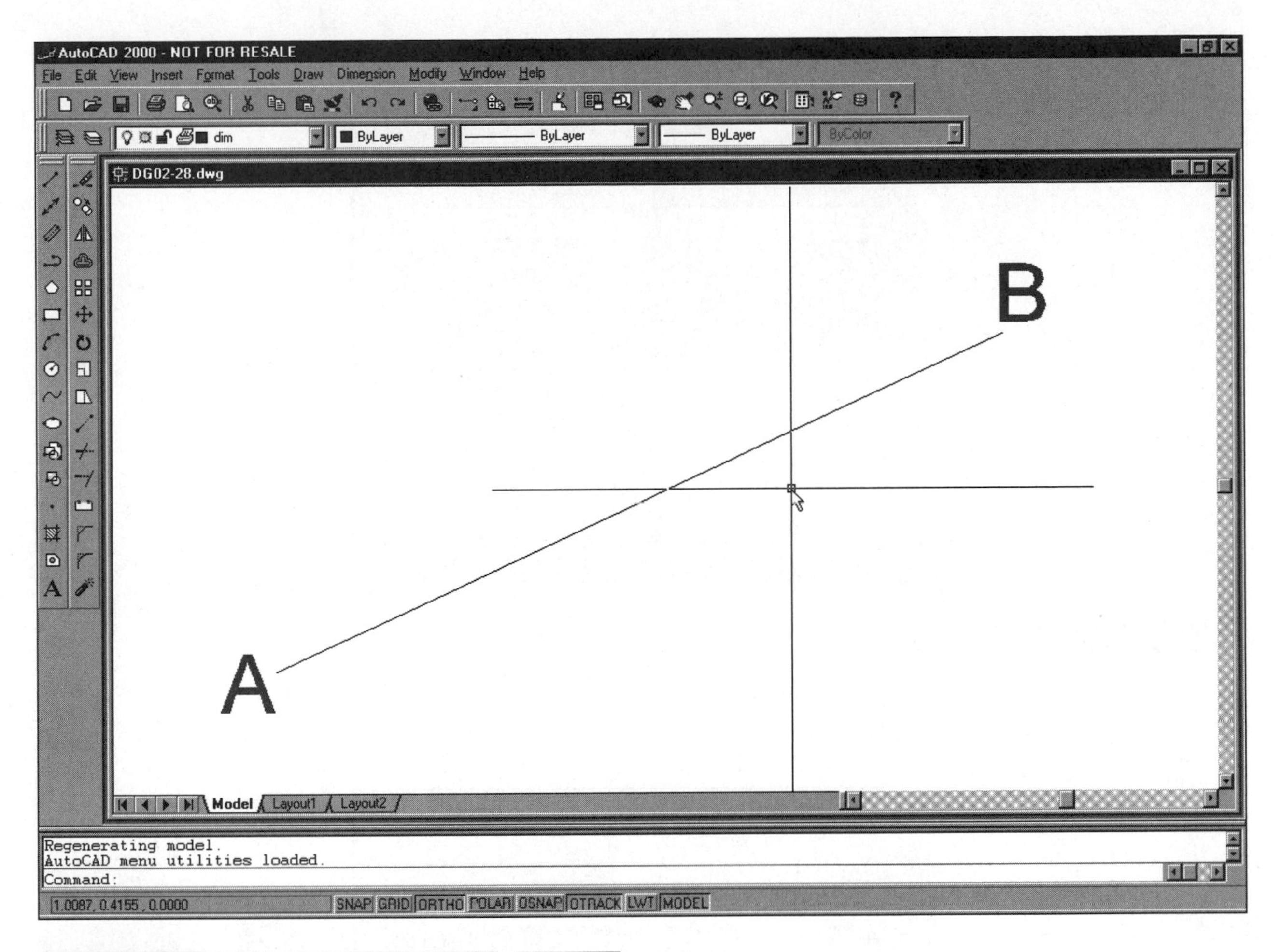


**Figure 2–28** *Determining the bearing of line AB.*

## Step #1

Using the UNITS command, the angular display is changed from decimal degrees to surveyor's units, along with the precision of the units set for degrees, minutes, and seconds. See Figures 2–29, 2–30, and 2–31.

Command: UNITS (ENTER)

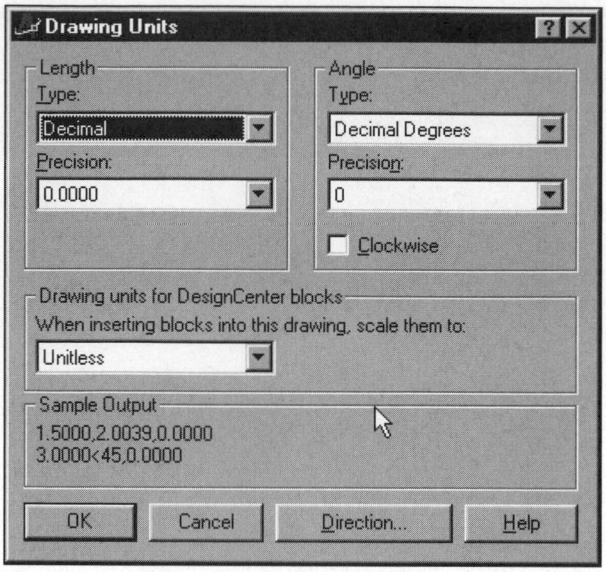

**Figure 2–29** UNITS *dialog box.*

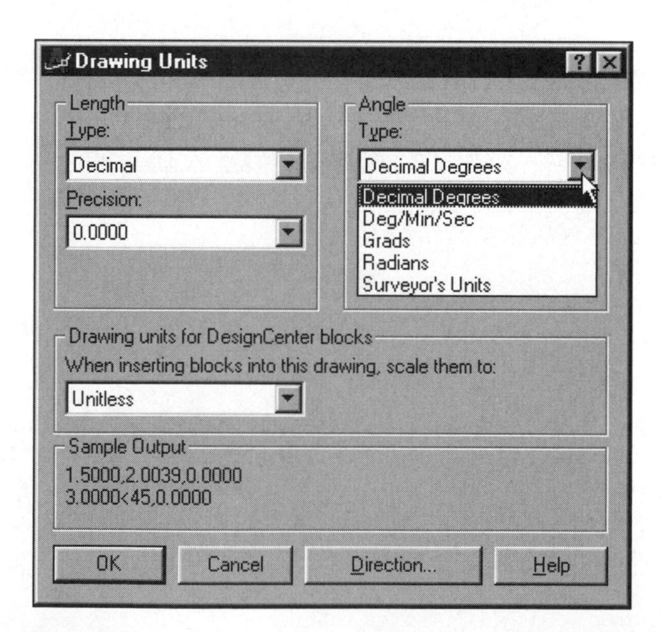


**Figure 2–30** *Setting angle type.*

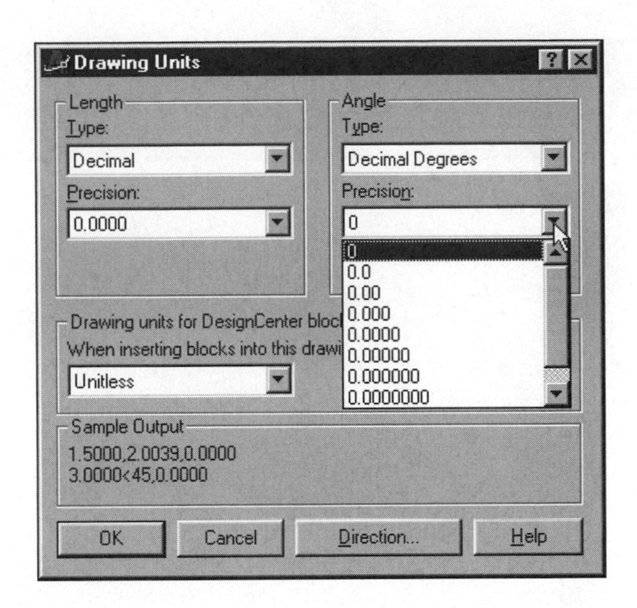


**Figure 2-31** *Setting angle precision.*

## Step #2

The bearing of the line can now be acquired by using any of the query commands (LIST, PROPERTIES, etc.). To use the LIST command to determine the bearings of the line, the command is entered at the command prompt and the object selected. If the PROPERTIES command is used, then the line can either be selected before or after the command is executed or after.

### The List Command

Command: LIST (ENTER)
Select objects: 1 found
Select objects: (ENTER)
LINE Layer: "0"
Space: Model space
Handle = 2B
from point, X= 3.6996 Y= 2.9262 Z= 0.0000
to point, X= 8.8347 Y= 5.2705 Z= 0.0000
Length = 5.6449, Angle in ***XY Plane = N 65d28' E***
Delta X = 5.1351, Delta Y = 2.3443, Delta Z = 0.0000

### The Properties Command

Command: PROPERTIES (ENTER)
Command:

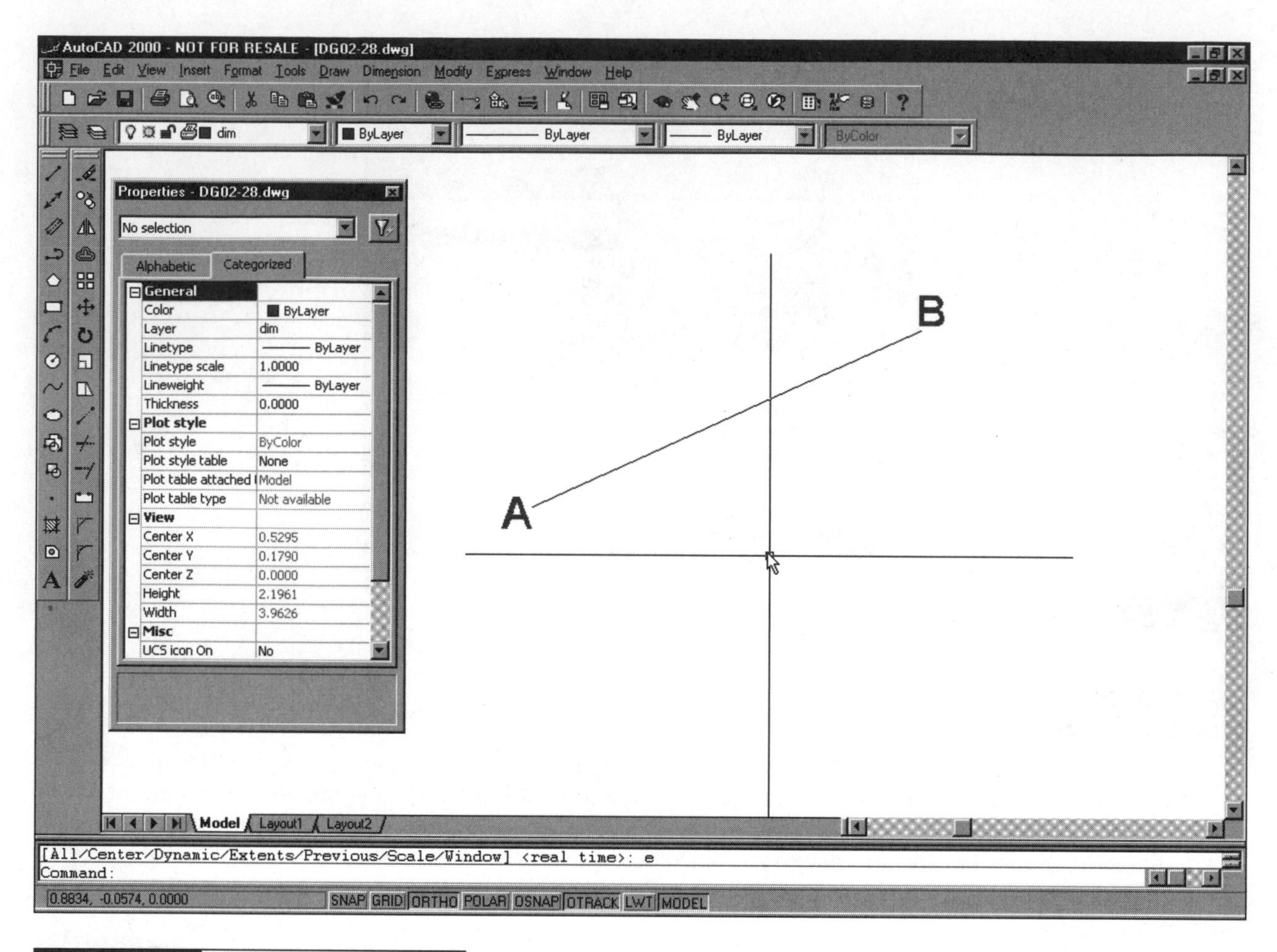


**Figure 2–32** PROPERTIES *dialog box.*

To determine the azimuth of a line, the same basic procedure is repeated with only a few minor adjustments. The units must be set to degree/minutes/seconds, the direction of the measurement must be set to clockwise, and the starting position of the measurements must either be set to north or south. The direction of measurement is changed by selecting the clockwise toggle located in the UNITS dialog box, as shown in Figure 2–33. The starting point of the measurement is changed by first selecting the direction button located at the bottom of the UNITS dialog box. A new dialog box will appear in which the user is able to specify the reference point for the measurements (see Figure 2–34). If north is selected, then all angular measurements will be made starting from north and proceeding in a clockwise direction.

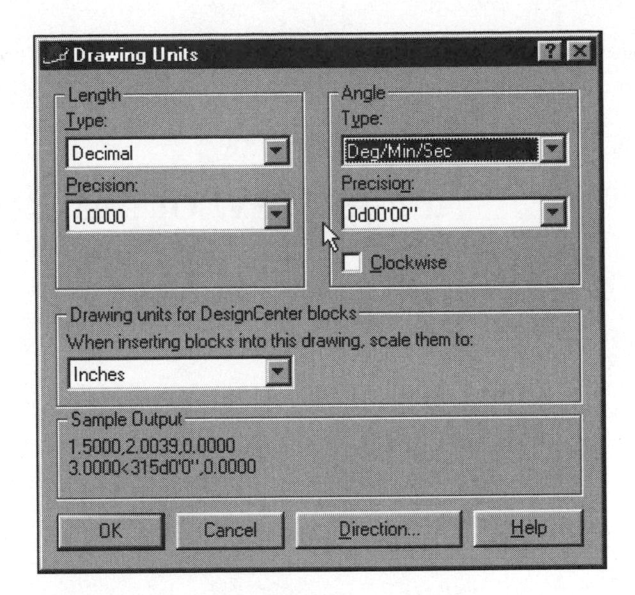


**Figure 2–33** *Clockwise toggle and direction button.*

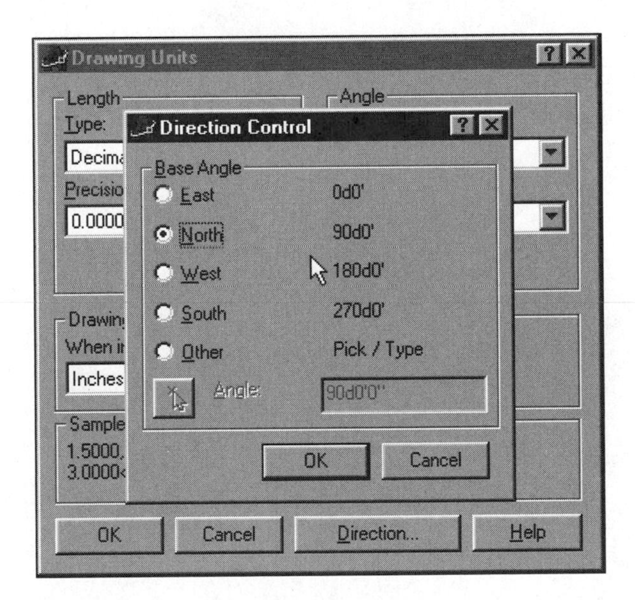


**Figure 2–34** *Direction dialog box.*

### Slope

Recall from the previous discussion that bearings and azimuths are the angles that are formed between a line and the north/south line located in the horizontal plane. In a real world application this is only half the solution to the process of defining the exact position of a line in space. The remainder is to define the position of the line in the front view with respect to an adjacent principal plane. In most cases this will be the angle formed by the line and the edge view of the horizontal plane. This angle is known as the ***slope*** and is always defined in angular degrees with either the word degrees or the ° symbol following the numeric value (see Figure 2–35). For an orthographic projection, the slope is typically measured and labeled in any view where the line appears as true length and is adjacent to the horizontal plane (see Figure 2–36). Because the line in this drawing does not appear in true length in any of the principal views, an auxiliary view (see Chapter three) was created in which the line would appear as true length. An auxiliary view must be created for any line that does not appear in true length in any of the principal views (other than the top or bottom views).

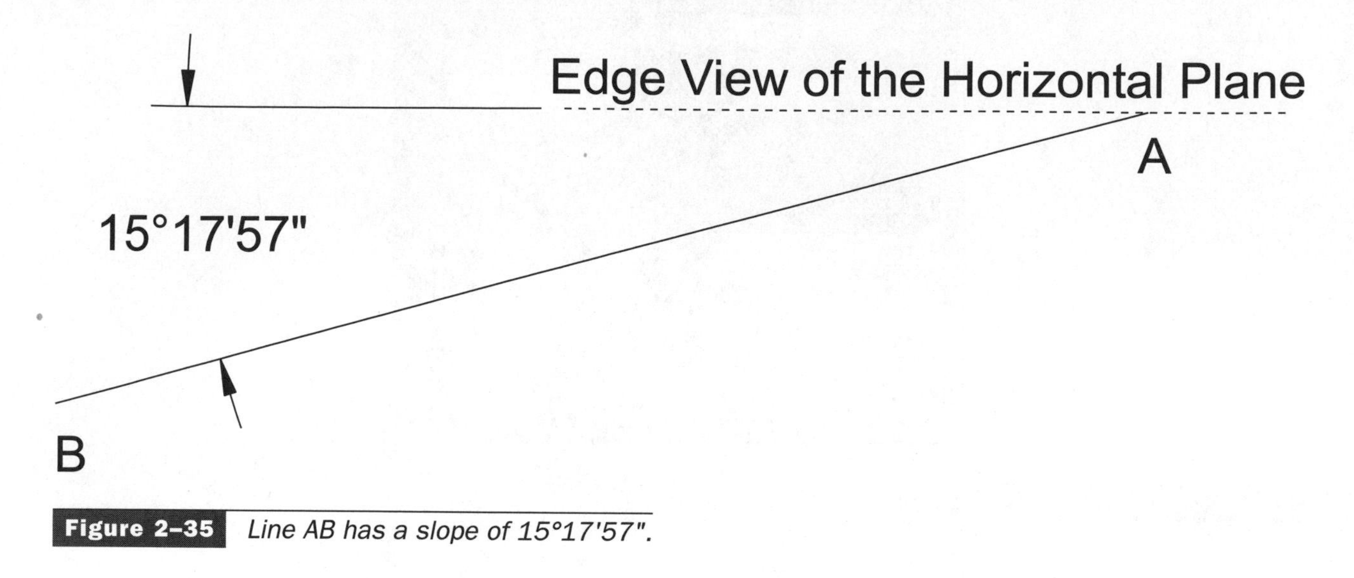


**Figure 2-35** *Line AB has a slope of 15°17'57".*

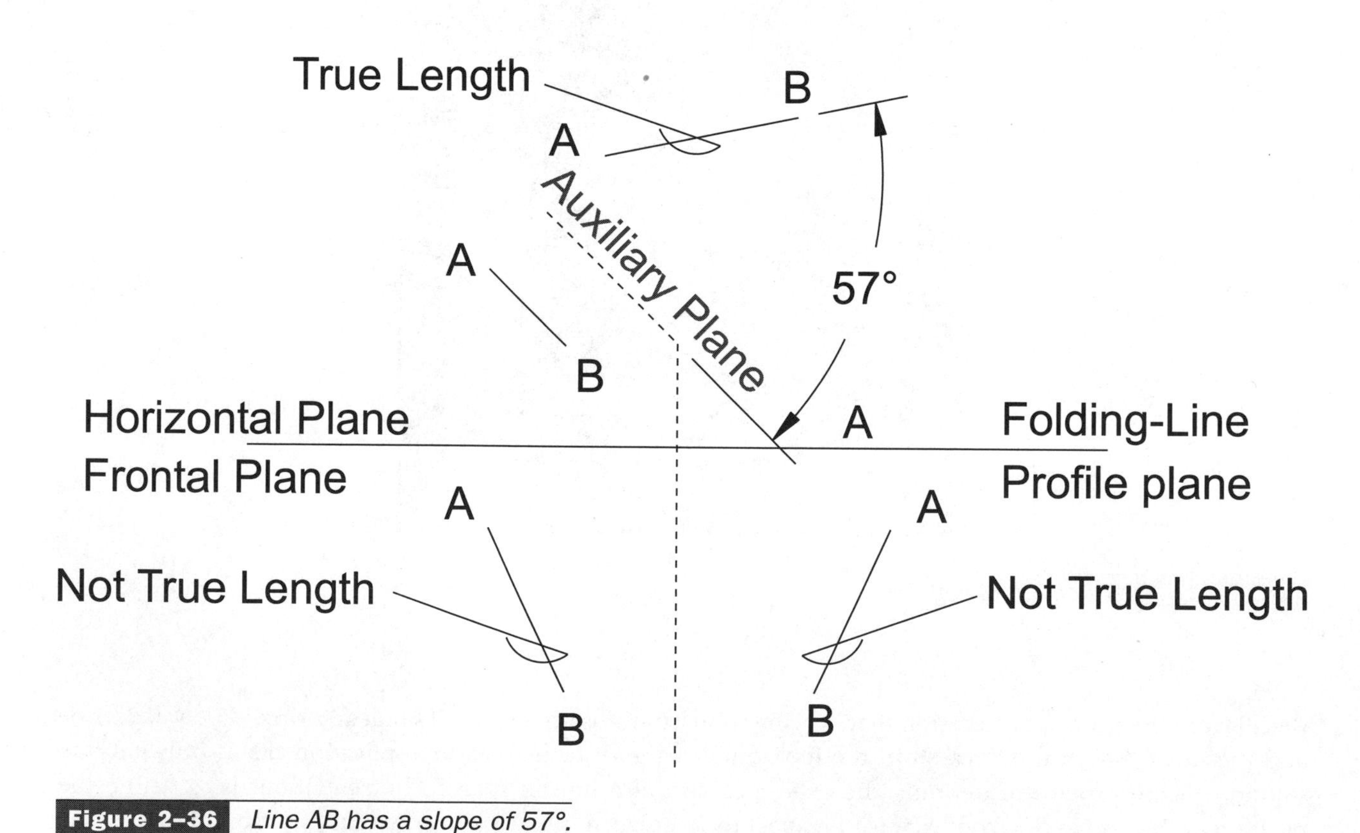


**Figure 2-36** *Line AB has a slope of 57°.*

**Grade**

Another method used to describe the angle of a line with respect to the horizontal plane is ***grade***. The grade is the percentage of incline between the line and the edge view of the horizontal plane. It is defined as the vertical rise of the line divided by its horizontal run, with the quotient multiplied by 100. The grade in Figure 2–37 was calculated by dividing 0.8305 by 1.0668. Its quotient was then multiplied by 100, producing a grade of 77.8496%. Grade is always defined as the percentage of incline and will be followed by either the word percent or the % symbol. Another example of grade is shown in Figure 2–38. The grade was calculated by dividing 0.8239 by 3.0119. Its quotient was then multiplied by 100, producing a grade of 27.3548%. Like slope, grade can only be calculated using the true length of a line that is contained within a view adjacent to the horizontal plane. Therefore, if the grade cannot be determined in one of the principal views (other than the top or bottom view), then an auxiliary view must be created.

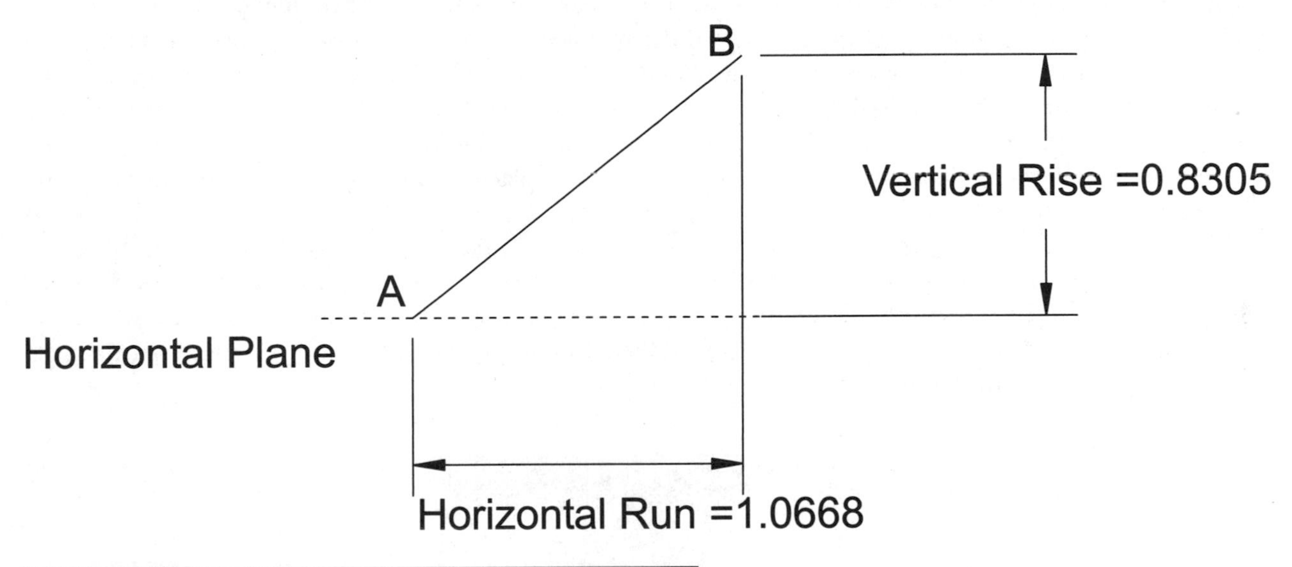


**Figure 2–37** *Line AB has a grade of 77.8496 percent.*

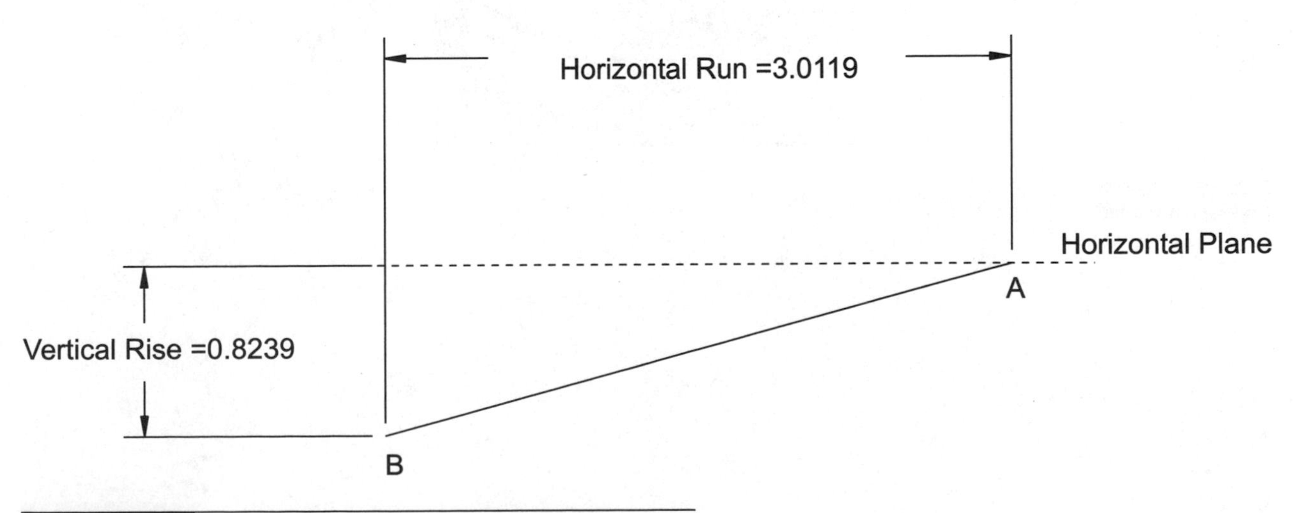


**Figure 2–38** *Line AB has a grade of 27.3548 percent.*

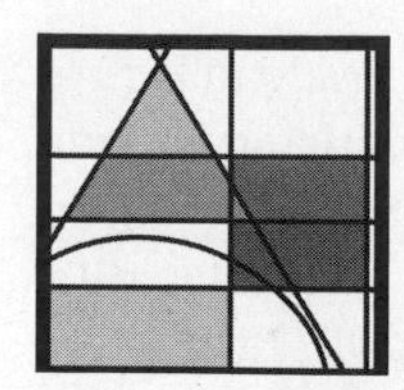

# advanced applications

## *Using AutoLISP to Label the Bearings and Azimuths of a Line*

As mentioned earlier, AutoCAD has the ability to determine the bearings and azimuths of a line; unfortunately, AutoCAD lacks the ability to list these findings on the drawing itself. This problem can be overcome by developing an application in AutoLISP that uses the principles and theories previously covered. The following AutoLISP program demonstrates this ability. When the program is loaded, the user is prompted to select the line that requires a bearing/azimuth notation. Once a line has been selected, the program displays a dialog box where the user is provided with the option of labeling either the bearings or azimuths of the line selected (see Figure 2–39). This program also allows the user to place the notation either above or below the selected line (see Figure 2–40).

The program calculates the bearings for the selected line by determining the angle formed when an XY-axis is superimposed over the starting and ending points of the line. Depending upon the quadrant where the calculated angle resides, the program follows the guidelines established earlier in this chapter regarding the conversion of bearings to azimuths and vice versa.

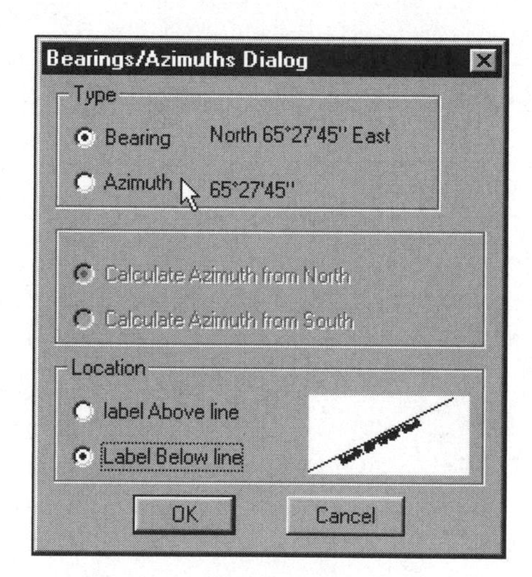


**Figure 2–39** *Label dialog box.*

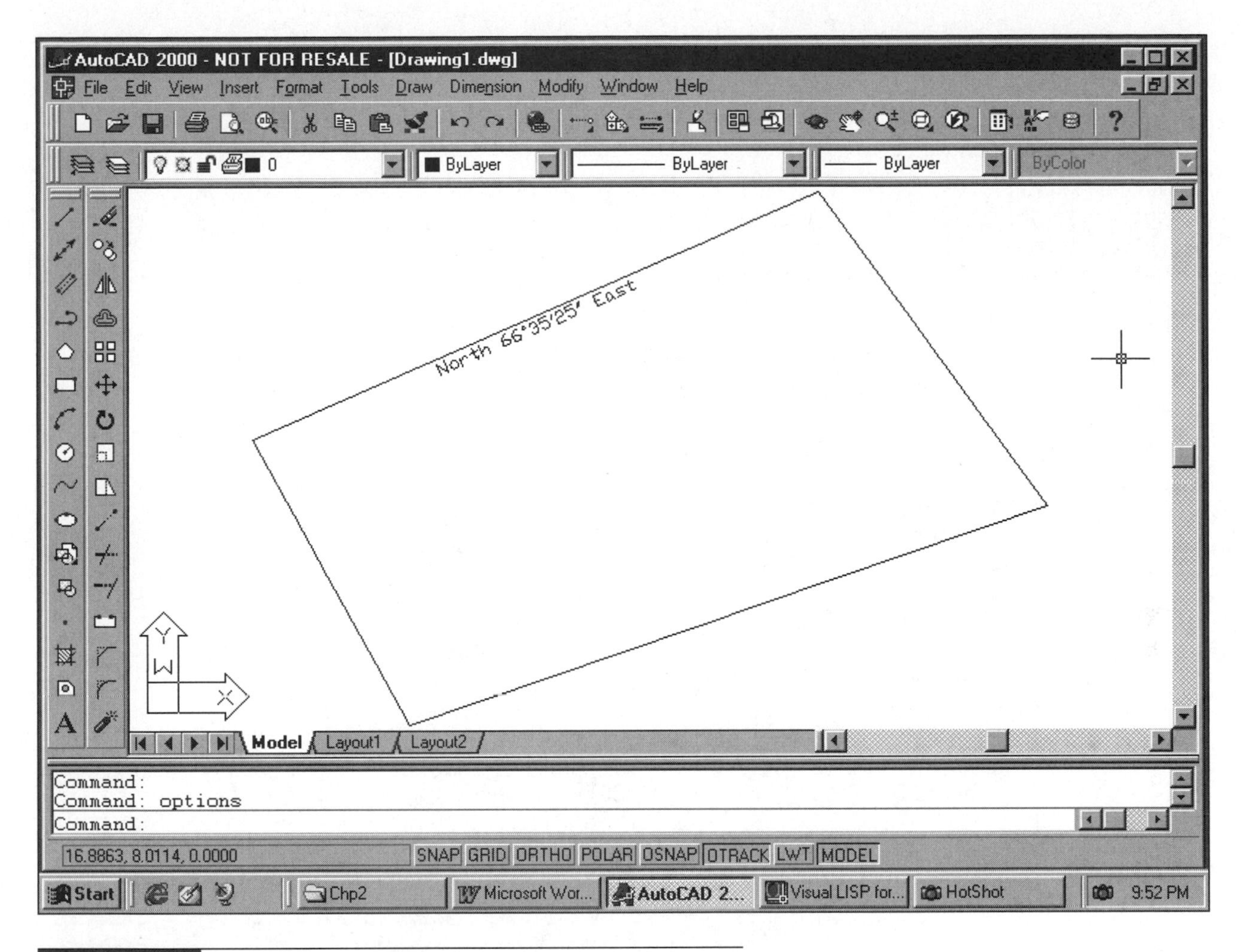


**Figure 2-40** *AutoCAD drawing containing lines to be labeled.*

```
;;;***********************************************************
;;;
;;;    Program Name: DG02.lsp
;;;
;;;    Program Purpose: This program allows the user to label
;;;                     the bearing or azimuth of a line
;;;
;;;    Program Date: 04/14/99
;;;
;;;    Written By: James Kevin Standiford
;;;
;;;***********************************************************
;;;***********************************************************
;;;
;;;                 Main Program
;;;
;;;***********************************************************
(defun c:label ()
  (setqangbase (getvar "angbase")
       angdir  (getvar "angdir")
       aunits  (getvar "aunits")
```

```
      cmdecho (getvar "cmdecho")
      old_snap (getvar "osmode")
)
(setvar "angbase" 0)
(setvar "angdir" 0)
(setvar "aunits" 0)
(setvar "cmdecho" 0)
(setvar "osmode" 0)
(setq ent1 (entsel "\nSelect line : "))
(if (/= ent1 nil)
  (progn
    (if (/= (cdr (assoc 0 (entget (car ent1)))) "LINE")
     (alert "Only Lines maybe selected : ")
    )
    (if (= (cdr (assoc 0 (entget (car ent1)))) "LINE")
     (progn
       (setqent  (entget (car ent1))
            fir  (cdr (assoc 10 ent))
            sec  (cdr (assoc 11 ent))
            ang  (angle fir sec)
            angl (* ang (/ 180 pi))
       )
       (if (and (> angl 0) (> 90 angl))
         (progn
           (setq angl1 (abs (- angl 90))
                 angl2 (abs (- angl 90))
                 angl3 (+ (abs (- angl 90)) 180)
           )
           (info)
           (setq str            (strcat "North "
                                (itoa first_num)
                                "°"
                                (itoa second_num)
                                "'"
                                (itoa third_num)
                                "\""
                                " East"
                         )
                 rotation angl
           )
         )
       )
       (if (and (> angl 90) (> 180 angl))
         (progn
           (setq angl1 (abs (- 90 angl))
                 angl2 (abs (- (- angl 90) 360))
                 angl3 (abs (- angl 270))
           )
           (info)
           (setq str            (strcat "North "
                                (itoa first_num)
                                "°"
                                (itoa second_num)
                                "'"
                                (itoa third_num)
```

```
                        "\""
                        " West"
                 )
         rotation (+ angl 180)
    )
  )
)
(if (and (> angl 180) (> 270 angl))
  (progn
    (setq angl1 (abs (- 270 angl))
         angl2 (abs (+ (- 270 angl) 180))
         angl3 (abs (- 270 angl))
    )
    (info)
    (setq str            (strcat "South "
                        (itoa first_num)
                        "°"
                        (itoa second_num)
                        "'"
                        (itoa third_num)
                        "\""
                        " West"
                 )
         rotation (+ angl 180)
    )
  )
)
(if (and (> angl 270) (> 360 angl))
  (progn
    (setq angl1 (abs (- 270 angl))
         angl2 (abs (+ (- 360 angl) 90))
         angl3 (abs (+ (- 360 angl) 270))
    )
    (info)
    (setq str            (strcat "South "
                        (itoa first_num)
                        "°"
                        (itoa second_num)
                        "'"
                        (itoa third_num)
                        "\""
                        " East"
                 )
         rotation angl
    )
  )
)
(setqstr1   (strcat(itoa first_num_a)
                   "°"
                   (itoa second_num_a)
                   "'"
                   (itoa third_num_a)
                   "\""
            )
     str2   (strcat(itoa first_num_ab)
```

```
                                "o"
                                (itoa second_num_ab)
                                "'"
                                (itoa third_num_ab)
                                "\""
                        )
                dcl_id4
                        (load_dialog
                          "c:/windows/desktop/geometry/student/AutoLISP Programs/DG02.dcl"
                        )
          )
          (if (not (new_dialog "lab" dcl_id4))
            (exit)
          )
          (action_tile "accept" "(vi) (done_dialog)")
          (action_tile
            "cancel"
            "(setq bea nil azi nil) (done_dialog)"
          )
          (set_tile "field1" str)
          (set_tile "field2" str1)
          (set_tile "north" "1")
          (action_tile "north" "(set_tile \"field2\" str1)")
          (action_tile "south" "(set_tile \"field2\" str2)")
          (start_image "photo")
          (setqx (dimx_tile "photo")
               y (dimy_tile "photo")
          )
          (slide_image
            0 0x y "above.sld")
          (end_image)
          (action_tile
            "above"
            "(progn (start_image \"photo\")
             (fill_image 0 0 x y 0)
             (slide_image 0 0 x y \"c:/windows/desktop/geometry/student/autolisp programs/
above.sld\")
             (end_image))"
          )
          (action_tile
            "below"
            "(progn (start_image \"photo\")
            (fill_image 0 0 x y 0)
             (slide_image 0 0 x y  \"c:/windows/desktop/geometry/student/autolisp programs/
below.sld\")
             (end_image))"
          )
          (set_tile "bea" "1")
          (set_tile "above" "1")
          (mode_tile "north" 1)
          (mode_tile "south" 1)
          (action_tile
            "bea"
            "(progn (mode_tile \"south\"  (atoi $value))(mode_tile \"north\"  (atoi
$value)))"
```

```
        )
        (action_tile
          "azi"
          "(progn (mode_tile \"south\" (- 1 (atoi $value)))(mode_tile \"north\" (- 1 (atoi
$value))))"
        )
        (start_dialog)
        (if (= bea "1")
          (bearing)
        )
        (if (= azi "1")
          (azimuth)
        )
      )
    )
  )
 )
 (setvar "angbase" angbase)
 (setvar "angdir" angdir)
 (setvar "aunits" aunits)
 (setvar "cmdecho" cmdecho)
 (setvar "osmode" old_snap)
)
(defun bearing ()
 (command "text" "j" "m" midpt "" rotation str)
)
(defun azimuth ()
 (if (= north "1")
   (progn
     (if (and (> angl2 0) (< angl2 180))
      (command "text" "j" "m" midpt "" angl str1)
     )
     (if (and (> angl2 180) (< angl2 360))
      (command "text" "j" "m" midpt "" (+ angl 180) str1)
     )
   )
 )
 (if (= south "1")
   (progn
     (if (and (> angl2 0) (< angl2 180))
      (command "text" "j" "m" midpt "" angl str2)
     )
     (if (and (> angl2 180) (< angl2 360))
      (command "text" "j" "m" midpt "" (+ angl 180) str2)
     )
   )
 )
)
(defun info ()
 (setqfirst_num     (fix angl1)
      second_num    (fix (setq second (* (- angl1 first_num) 60)))
      third_num     (fix (setq third (* (- second second_num) 60)))
      first_num_a   (fix angl2)
      second_num_a  (fix (setq second_a (* (- angl2 first_num_a) 60)))
      third_num_a   (fix (setq third_a (* (- second_a second_num_a) 60)))
```

```
            first_num_ab  (fix angl3)
            second_num_ab (fix (setq second_ab (* (- angl3 first_num_ab) 60)))
            third_num_ab  (fix (setq third_ab (* (- second_ab second_num_ab) 60))
                          )
            x1            (car fir)
            x2            (car sec)
            y1            (car (cdr fir))
            y2            (car (cdr sec))
            midx          (/ (+ x1 x2) 2)
            midy          (/ (+ y1 y2) 2)
  )
)
(defun vi ()
  (setqbea   (get_tile "bea")
        azi   (get_tile "azi")
        above (get_tile "above")
        below (get_tile "below")
        north (get_tile "north")
        south (get_tile "south")
  )
  (if (= above "1")
    (setq midpt (list midx (+ (getvar "textsize") midy)))
  )
  (if (= below "1")
    (setq midpt (list midx (- midy (getvar "textsize"))))
  )
)
Dialog Box Control
//%%%%%%%%%%%%%%%%%%%%%%%%%%%%%%%%%%%%%%%%%%%%%%%%%%%%%%%%%%%%%%%%%
//
//      Activates dialog box
//
//      Descriptive Geometry Chapter 2 DCL File bearing or
//      azimuth of a line
//
//
//
//%%%%%%%%%%%%%%%%%%%%%%%%%%%%%%%%%%%%%%%%%%%%%%%%%%%%%%%%%%%%%%%%%
lab : dialog {
label = "Bearings/Azimuths Dialog";
       : boxed_row {
       fixed_width = true;
       children_fixed_width = true;
        label = "Type";
         : column {
              : radio_button {
              key = "bea";
       label = "Bearing";
       width = 7;
       }
       : radio_button {
       key = "azi";
       label = "Azimuth";
       width = 7;
       }
```

```
        }
        : column {
              : text {
              key = "field1";
              alignment = left;
              width = 20;
              }
              : text {
              key = "field2";
              alignment = left;
              width = 20;
              }
        }
        }
       : boxed_column {
       : radio_button {
       key = "north";
       label = "Calculate Azimuth from North";
       }
       : radio_button {
       key = "south";
       label = "Calculate Azimuth from South";
       }
       }
       : boxed_row {
       label = "Location";
       children_fixed_width = true;
       children_alignment = left;
        : column {
              : radio_button {
              key = "above";
              label = "label Above line";
              }
              : radio_button {
              key = "below";
              label = "Label Below line";
              }
        }
        : image {
        key = "photo";
        color = 0;
        alignment = right;
        width = 15;
        height = 3;
        }
       }
    is_default = true;
    ok_cancel;
}
```

# Review Questions

Answer the following questions on a separate sheet of paper. Your answers should be as complete as possible.

1. Define the following terms: point, line, bearing, foreshortened line, oblique line, inclined line, bearing, azimuth, locus, longitude, latitude, and prime meridian.
2. List the five categories of lines used in descriptive geometry, describe their characteristics, and draw an example of each.
3. What is a statute mile?
4. How is the location of a point established in descriptive geometry?
5. What three methods can be used to determine the location of a line in space?
6. Describe the difference between a line and a line segment.
7. Convert the following bearings to azimuths using north as a reference: N23°W, N90°, N13.2°E, S11.98°W, N17°S, S11°W, and S79.89°E.
8. Convert the bearings in question 7 to azimuths using south as a reference.
9. Convert the following azimuths to bearings using north as a reference: 15.6°, 25.9°, 98.8°, 325.47°, 5°, 68.7°, and 55.9°.
10. How can the bearing of a line be determined using AutoCAD?
11. Which AutoCAD commands may be used to determine the location of a point?
12. How is the location of a line determined using AutoCAD?
13. Given the following longitudes, find the statute miles: 15.6°, 25.9°, 48.8°, 25.47°, 5°, 68.7°, and 55.9°
14. How is the slope of a line determined?
15. How is the grade of a line determined?
16. Convert the azimuths in question 9 to bearings using south as a reference.

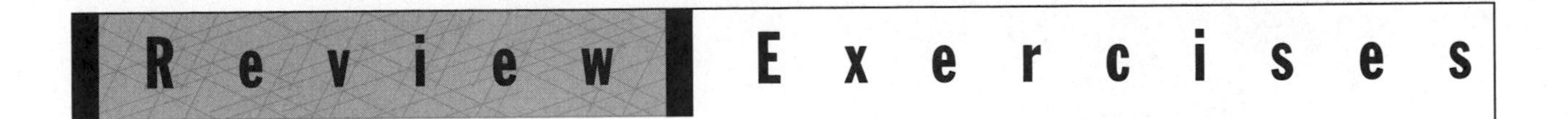

# Review Exercises

## Bearing

Determine the bearings of the following lines. All problems marked with an * can be found on the student CD-ROM.

**#1**

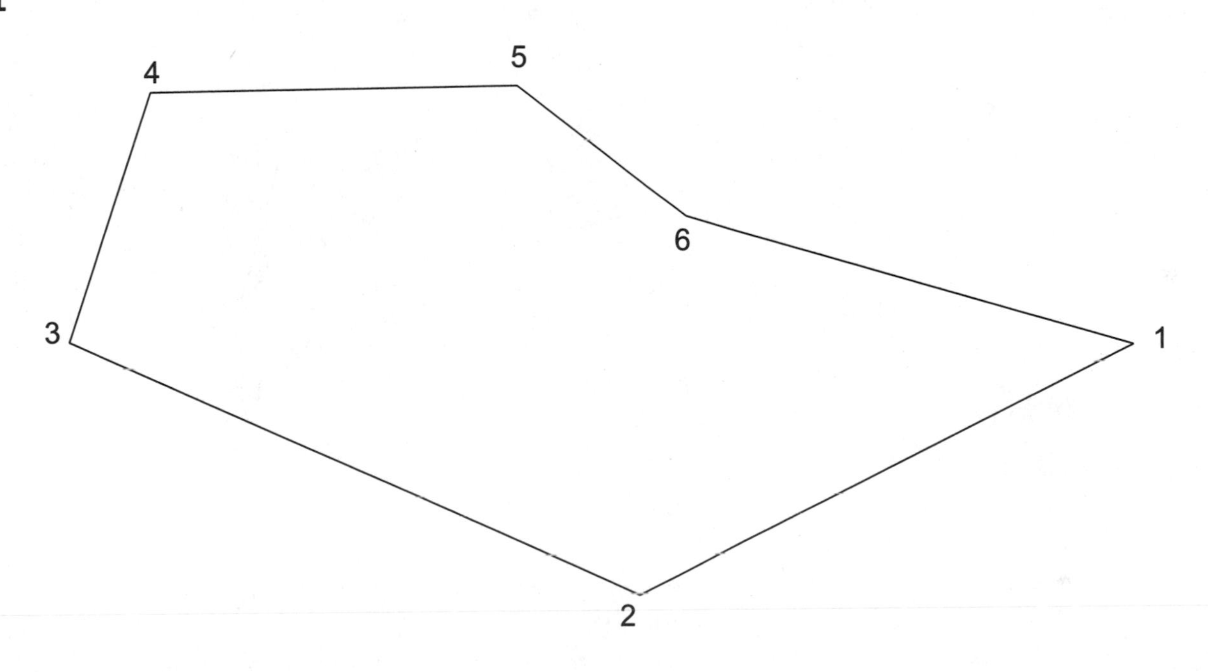


**#2**

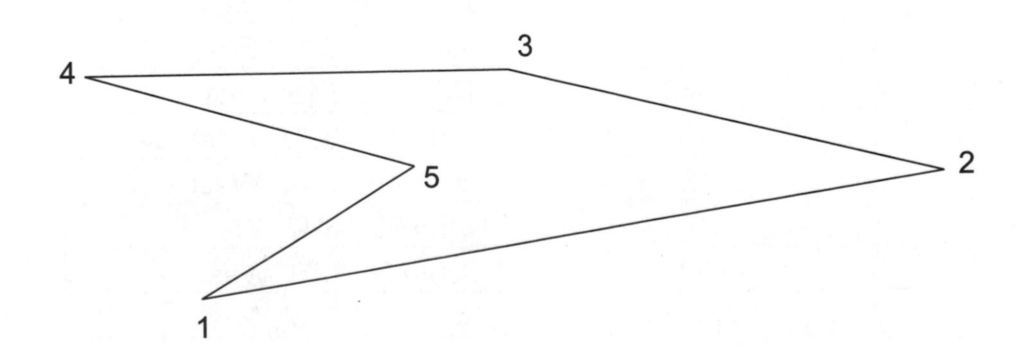

#3

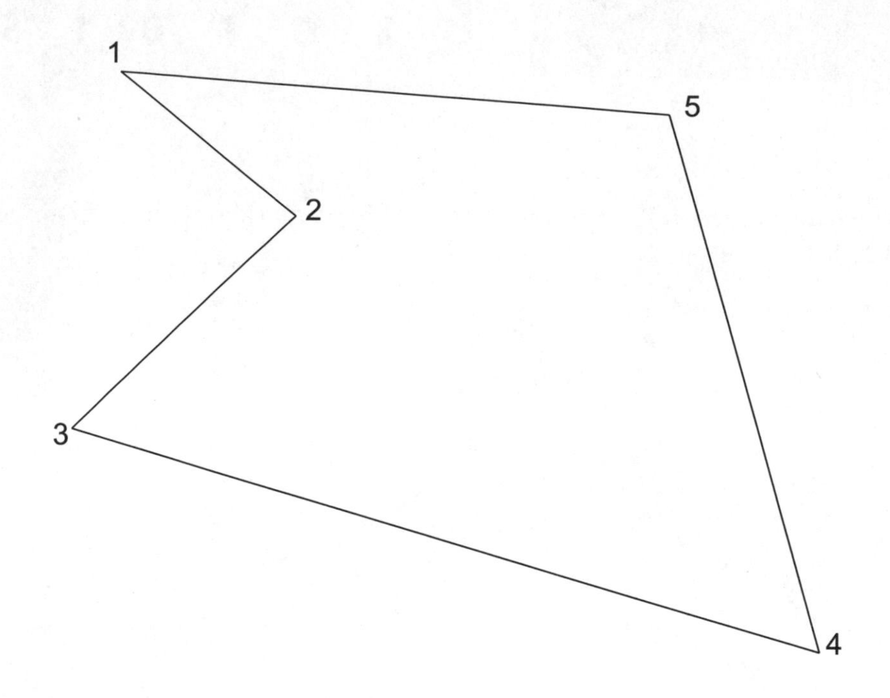


## Using AutoCAD to Construct Bearings

Using AutoCAD, construct the bearings listed in the charts below.

#4

| Plotting Bearings | | |
|---|---|---|
| Line # | Length | Bearing |
| 1 | 2.7643 | South 50°15'33" East |
| 2 | 4.5826 | South 79°54'16" East |
| 3 | 2.8124 | North 54°45'18" East |
| 4 | 4.0064 | North 12°3'16" East |
| 5 | 3.6162 | North 70°40'29" West |
| 6 | 3.1181 | South 40°19'44" West |
| 7 | 2.6189 | North 63°36'47" West |
| 8 | 3.9328 | South 81°22'9" West |
| 9 | 2.4633 | South 30°51'8" West |
| 10 | 3.1420 | South 85°48'39" East |

Note: Start line #1 at X = 3.1336 Y = 2.5902 Z = 0.0000

#5

| Plotting Bearings | | |
|---|---|---|
| Line # | Length | Bearing |
| 1 | 2.5324 | South 53°26'53" West |
| 2 | 2.8781 | North 65°47'19" West |
| 3 | 2.3383 | North 18°49'46" West |
| 4 | 3.1471 | North 63°43'10" East |
| 5 | 1.5840 | South 25°47'12" East |
| 6 | 2.2344 | North 82°24'40" East |
| 7 | 3.6380 | South 52°31'53" East |
| 8 | 2.0441 | South 51°51'51" West |
| 9 | 2.0726 | North 50°9'29" West |

Note: Start line #1 at X = 8.5886 Y = 4.2541 Z = 0.0000

#6

| Plotting Bearings | | |
|---|---|---|
| Line # | Length | Bearing |
| 1 | 5.4483 | North 80°59'54" East |
| 2 | 2.0442 | North 41°46'12" West |
| 3 | 3.5729 | North 64°9'29" West |
| 4 | 2.4351 | South 75°58'27" West |
| 5 | 2.3434 | South 39°3'25" West |
| 6 | 3.3965 | South 63°19'44" East |

Note: Start line #1 at X = 4.6347 Y = 1.9262 Z = 0.0000

## Azimuths

Determine the azimuths of the following lines.

**#7**

Using north as a reference

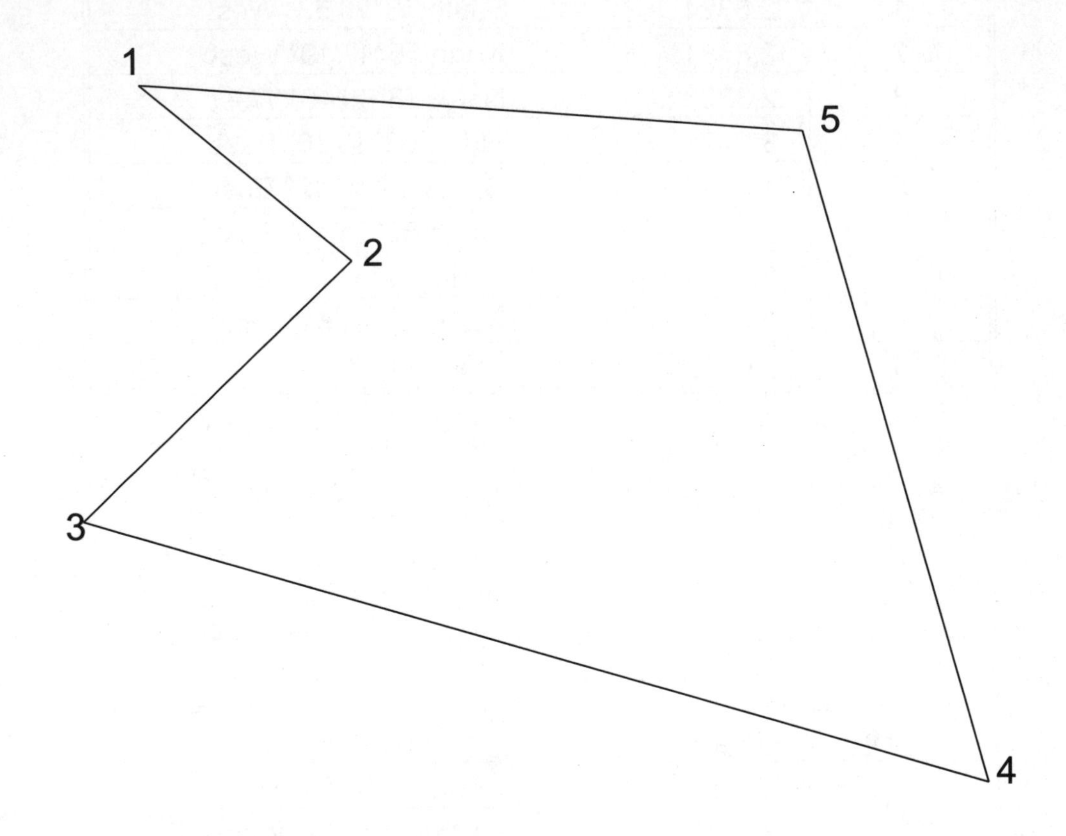


**#8**

Using north as a reference

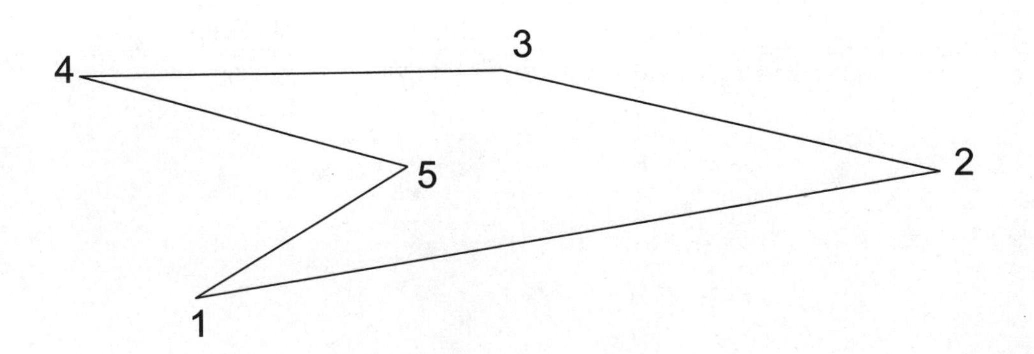

**#9**

Using north as a reference

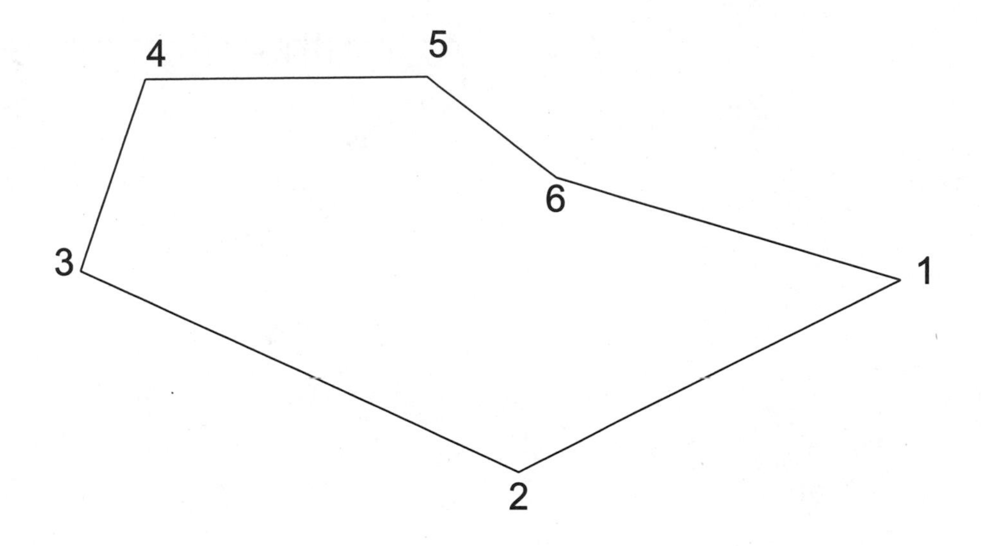


**#10**

Using south as a reference

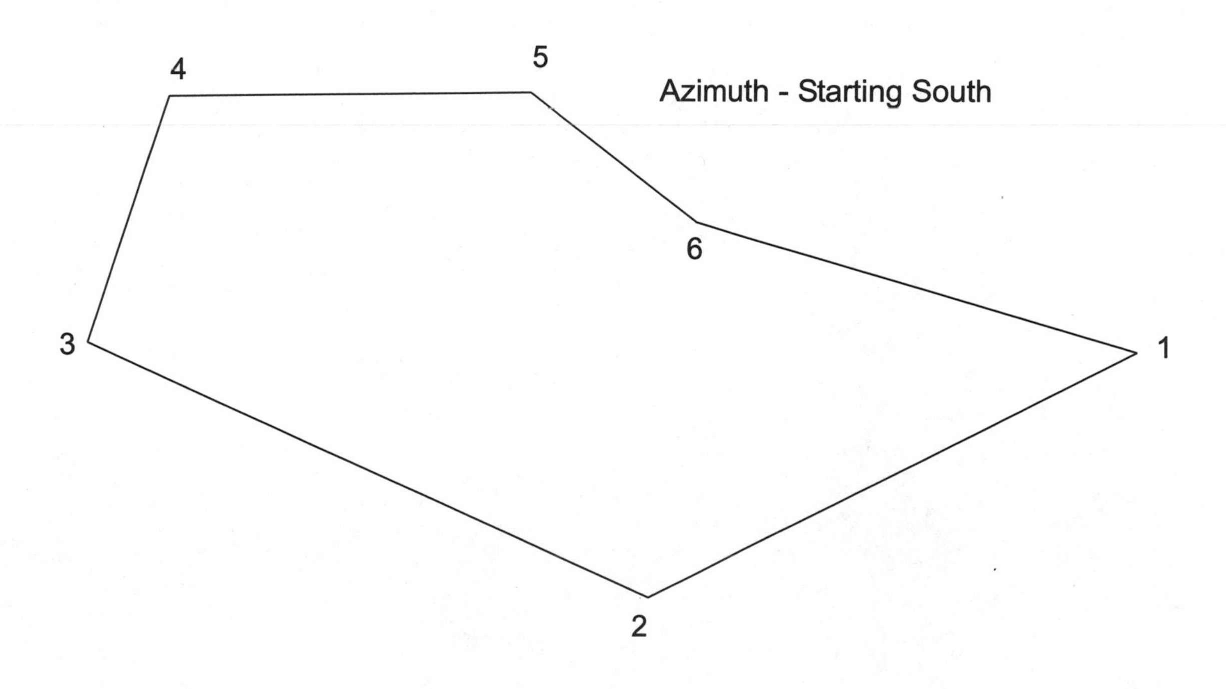

**#11**

Using south as a reference

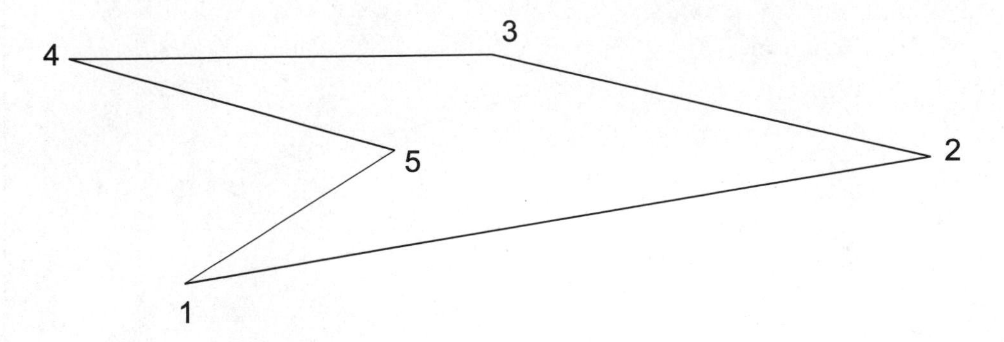


**#12**

Using south as a reference

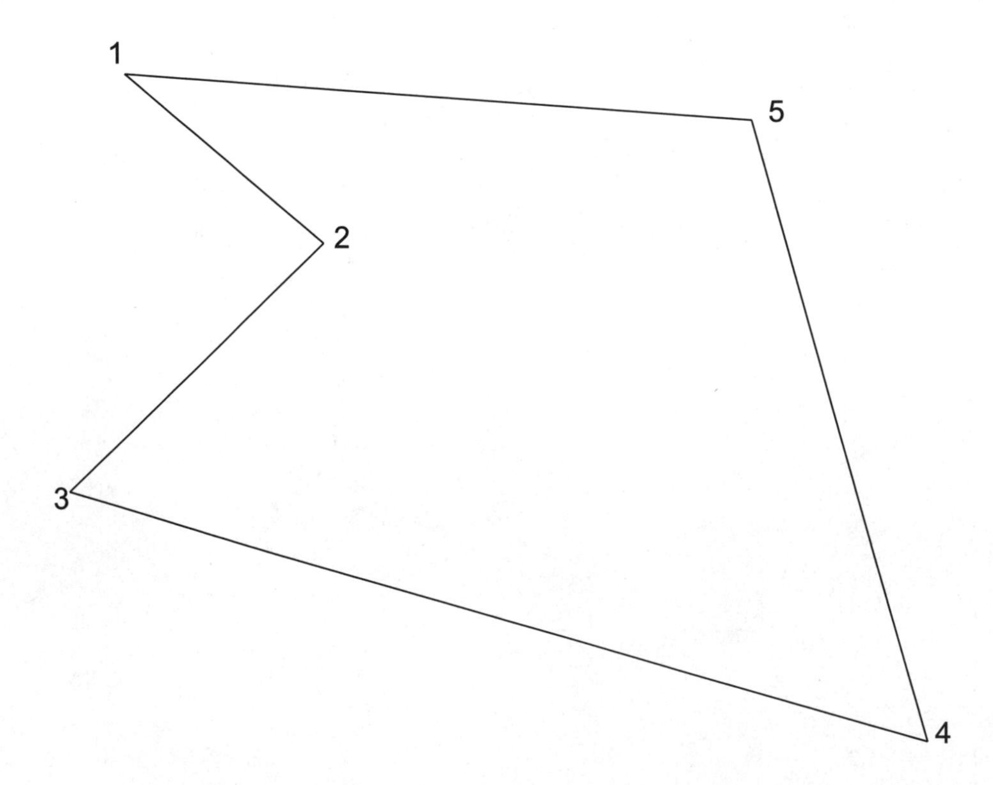

## Using AutoCAD to Construct Azimuths

Using AutoCAD, construct the azimuths listed in the charts below.

**#13**

Using north as a reference

| Plotting Azimuths (North) | | |
|---|---|---|
| Line # | Length | Bearing |
| 1 | 5.4483 | 68°9'39" |
| 2 | 1.5540 | 40°55'31" |
| 3 | 3.5729 | 295°50'30" |
| 4 | 2.4351 | 255°58'27" |
| 5 | 2.3434 | 219°3'25" |
| 6 | 3.3965 | 116°40'15" |

Note: Start line #1 at X = 4.6347 Y = 1.9262 Z = 0.0000

**#14**

Using north as a reference

| Plotting Azimuths (North) | | |
|---|---|---|
| Line # | Length | Bearing |
| 1 | 2.2244 | 148°4'50" |
| 2 | 5.1765 | 93°26'6" |
| 3 | 2.4354 | 55°50'49" |
| 4 | 4.3630 | 15°26'54" |
| 5 | 3.1514 | 275°6'20" |
| 6 | 2.8734 | 234°7'38" |
| 7 | 3.1854 | 257°55'27" |
| 8 | 2.9952 | 290°18'26" |
| 9 | 3.2366 | 211°42'54" |
| 10 | 3.5951 | 83°28'0" |

Note: Start line #1 at X = 3.1336 Y = 2.5902 Z = 0.0000

**#15**

Using north as a reference

| Plotting Azimuths (North) | | |
|---|---|---|
| Line # | Length | Bearing |
| 1 | 3.1085 | 240°48'33" |
| 2 | 1.1911 | 3°58'37" |
| 3 | 3.5556 | 308°29'36" |
| 4 | 3.1471 | 63°43'10" |
| 5 | 0.7056 | 77°34'36" |
| 6 | 2.5596 | 120°4'53" |
| 7 | 2.6917 | 135°8'0" |
| 8 | 1.6855 | 201°33'8" |
| 9 | 2.0726 | 309°50'30" |

Note: Start line #1 at X = 8.5886 Y = 4.2541 Z = 0.0000

**#16**

Using south as a reference

| Plotting Azimuths (South) | | |
|---|---|---|
| Line # | Length | Bearing |
| 1 | 2.5855 | 103°45'36" |
| 2 | 1.7980 | 153°3'36" |
| 3 | 2.2126 | 218°7'26" |
| 4 | 2.9879 | 288°34'24" |
| 5 | 1.9824 | 319°50'53" |
| 6 | 2.6173 | 55°15'27" |

Note: Start line #1 at X = 4.6347 Y = 1.9262 Z = 0.0000

**#17**

Using south as a reference

| Plotting Azimuths (South) | | |
|---|---|---|
| Line # | Length | Bearing |
| 1 | 3.4008 | 281°39'44" |
| 2 | 2.6269 | 238°5'56" |
| 3 | 2.1107 | 118°59'15" |
| 4 | 1.6086 | 118°59'15" |
| 5 | 3.4992 | 115°20'35" |
| 6 | 2.0372 | 145°34'36" |
| 7 | 2.3379 | 64°3'12" |
| 8 | 2.2824 | 10°8'54" |
| 9 | 2.6506 | 279°31'12" |
| 10 | 2.4144 | 299°38'12" |

Note: Start line #1 at X = 3.1336 Y = 2.5902 Z = 0.0000

**#18**

Using south as a reference

| Plotting Azimuths (South) | | |
|---|---|---|
| Line # | Length | Bearing |
| 1 | 2.6873 | 79°21'2" |
| 2 | 2.6344 | 114°17'16" |
| 3 | 1.5858 | 144°6'58" |
| 4 | 1.4222 | 231°39'14" |
| 5 | 3.2639 | 265°22'31" |
| 6 | 2.3463 | 263°33'40" |
| 7 | 2.4292 | 318°12'26" |
| 8 | 2.7133 | 83°26'53" |
| 9 | 1.2109 | 343°17'19" |

Note: Start line #1 at X = 8.5886 Y = 4.2541 Z = 0.0000

## Slope

Determine the slopes of the following lines.

**#15**

| Plotting Azimuths (North) | | |
|---|---|---|
| Line # | Length | Bearing |
| 1 | 3.1085 | 240°48'33" |
| 2 | 1.1911 | 3°58'37" |
| 3 | 3.5556 | 308°29'36" |
| 4 | 3.1471 | 63°43'10" |
| 5 | 0.7056 | 77°34'36" |
| 6 | 2.5596 | 120°4'53" |
| 7 | 2.6917 | 135°8'0" |
| 8 | 1.6855 | 201°33'8" |
| 9 | 2.0726 | 309°50'30" |

Note: Start line #1 at X = 8.5886 Y = 4.2541 Z = 0.0000

**#16**

| Plotting Azimuths (South) | | |
|---|---|---|
| Line # | Length | Bearing |
| 1 | 2.5855 | 103°45'36" |
| 2 | 1.7980 | 153°3'36" |
| 3 | 2.2126 | 218°7'26" |
| 4 | 2.9879 | 288°34'24" |
| 5 | 1.9824 | 319°50'53" |
| 6 | 2.6173 | 55°15'27" |

Note: Start line #1 at X = 4.6347 Y = 1.9262 Z = 0.0000

#17

| Plotting Azimuths (South) | | |
|---|---|---|
| Line # | Length | Bearing |
| 1 | 3.4008 | 281°39'44" |
| 2 | 2.6269 | 238°5'56" |
| 3 | 2.1107 | 118°59'15" |
| 4 | 1.6086 | 118°59'15" |
| 5 | 3.4992 | 115°20'35" |
| 6 | 2.0372 | 145°34'36" |
| 7 | 2.3379 | 64°3'12" |
| 8 | 2.2824 | 10°8'54" |
| 9 | 2.6506 | 279°31'12" |
| 10 | 2.4144 | 299°38'12" |

Note: Start line #1 at X = 3.1336 Y = 2.5902 Z = 0.0000

#18

| Plotting Azimuths (South) | | |
|---|---|---|
| Line # | Length | Bearing |
| 1 | 2.6873 | 79°21'2" |
| 2 | 2.6344 | 114°17'16" |
| 3 | 1.5858 | 144°6'58" |
| 4 | 1.4222 | 231°39'14" |
| 5 | 3.2639 | 265°22'31" |
| 6 | 2.3463 | 263°33'40" |
| 7 | 2.4292 | 318°12'26" |
| 8 | 2.7133 | 83°26'53" |
| 9 | 1.2109 | 343°17'19" |

Note: Start line #1 at X = 8.5886 Y = 4.2541 Z = 0.0000

## Using AutoCAD to Determine the Slopes of the Following Lines

Using AutoCAD, determine the bearings of the following problems.

***#19**

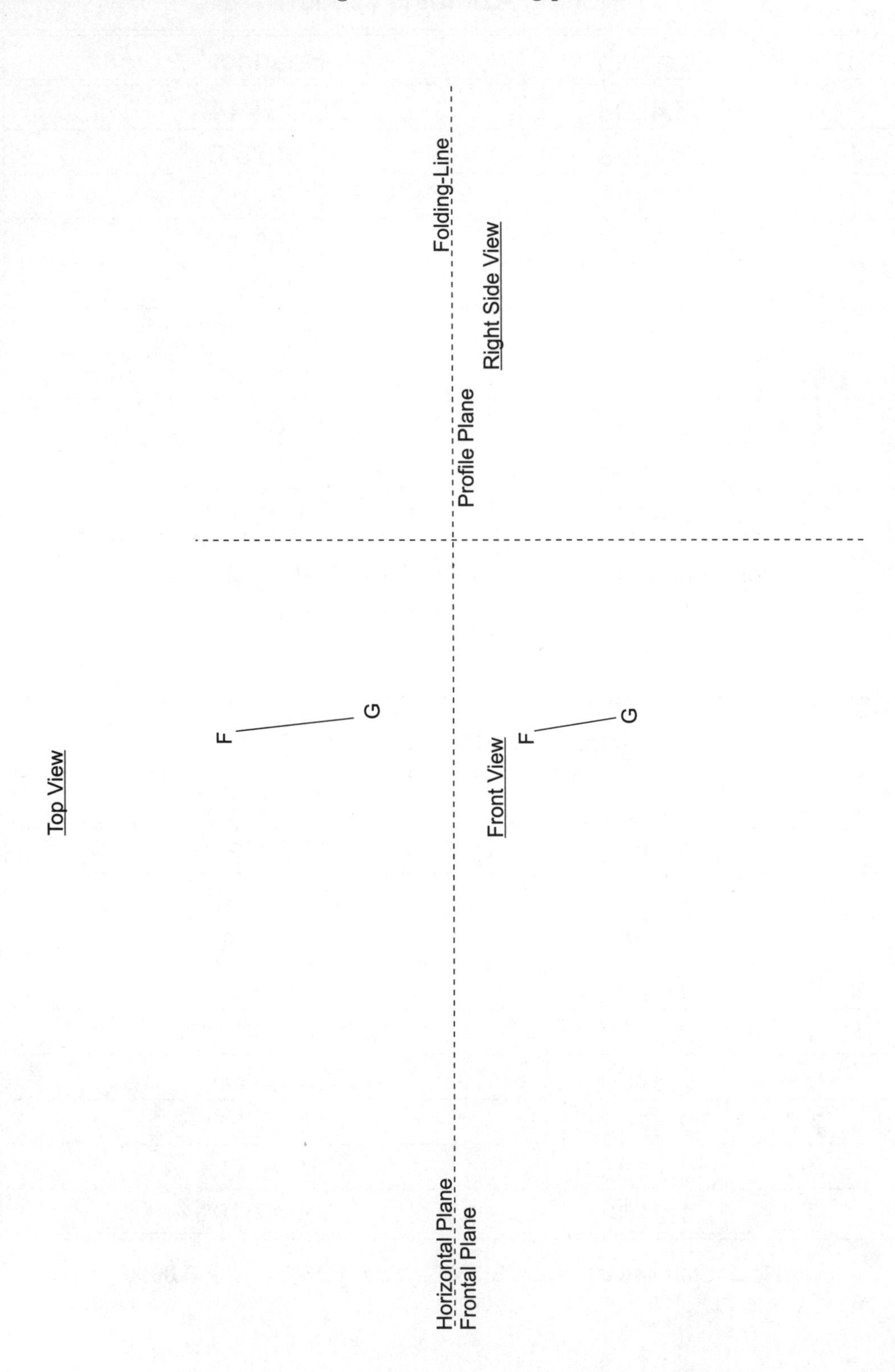

***#20**

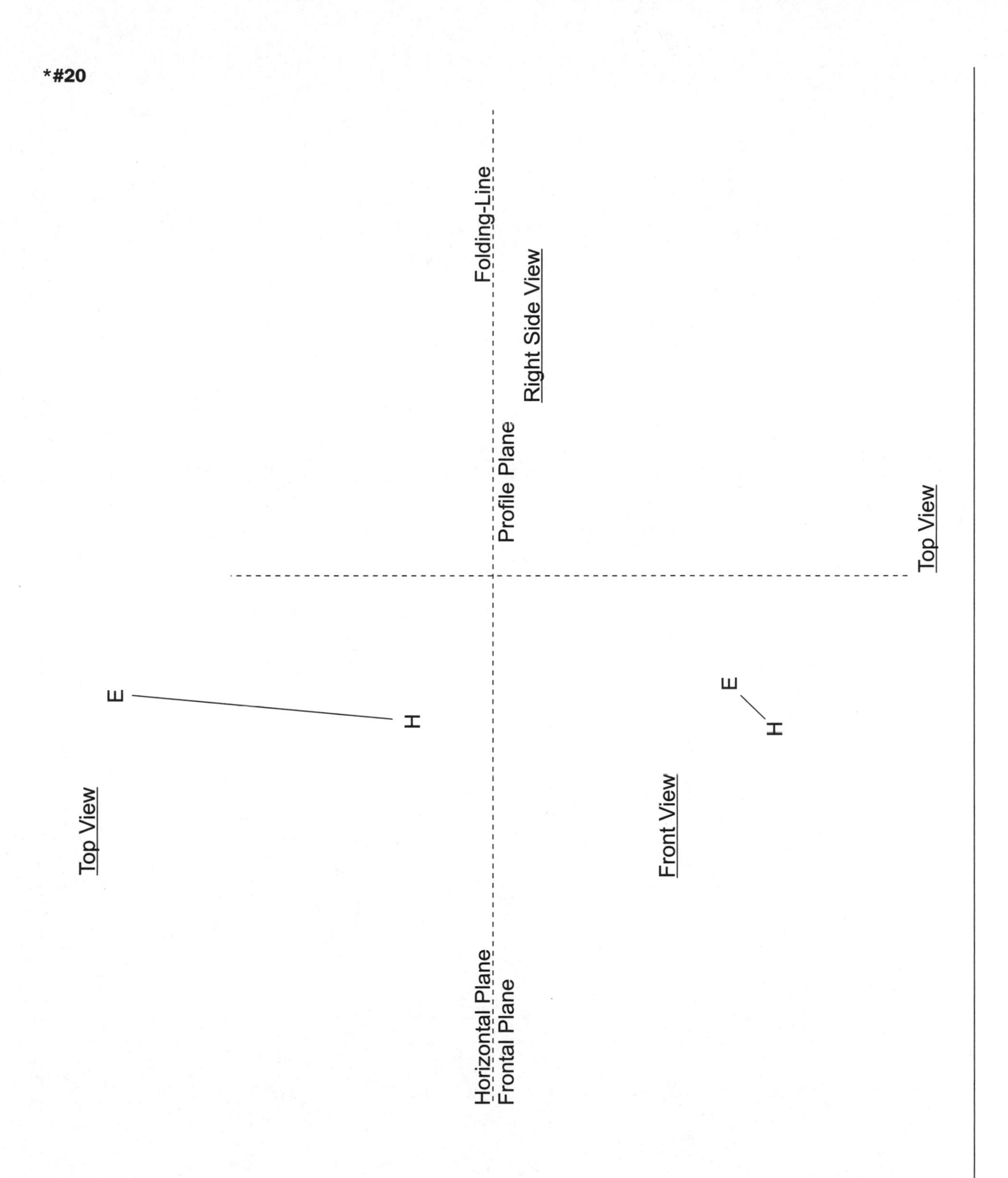
Folding-Line
Right Side View
Profile Plane
Top View
E
H
Top View
E
H
Front View
Horizontal Plane
Frontal Plane

***#21**

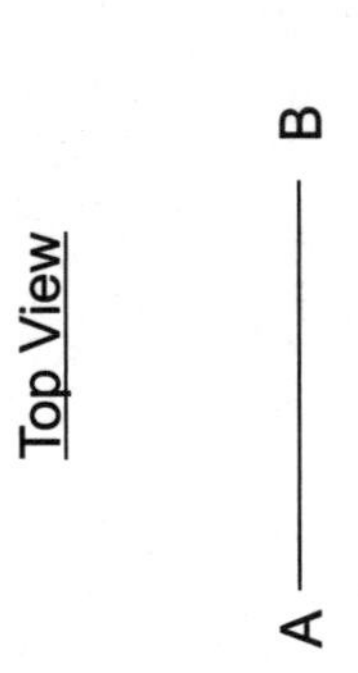


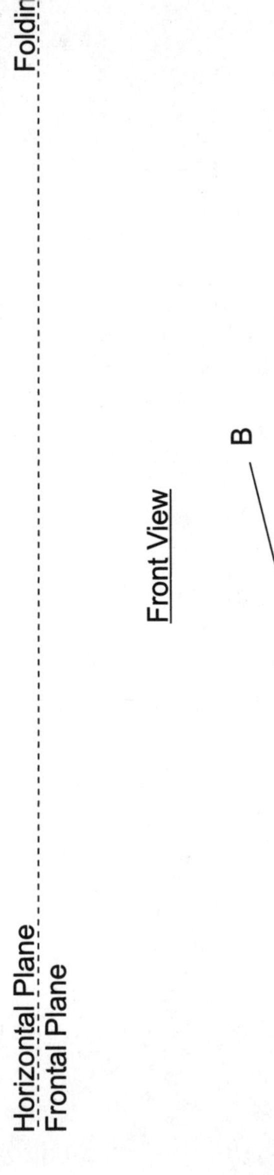


## Grade

Determine the grades of the lines shown in problems 19 through 21.

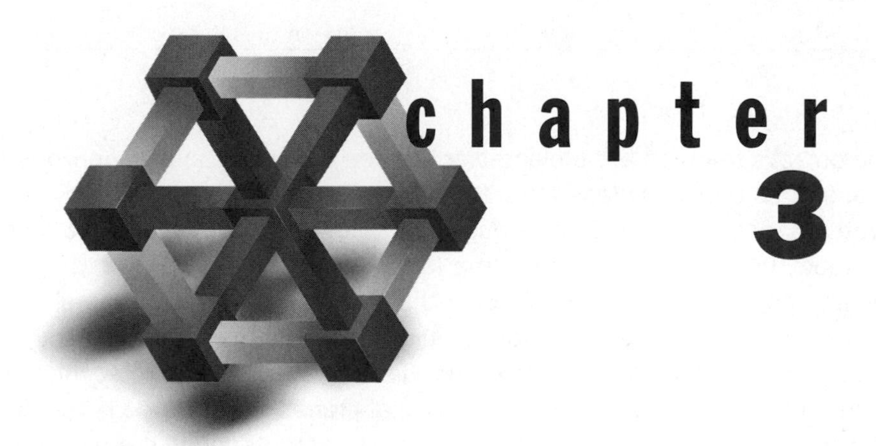


# Auxiliary Views

## Objectives

***Upon completion of this chapter the student will be able to do the following:***

- *Define auxiliary view, primary auxiliary view, secondary auxiliary view, and successive auxiliary view.*
- *Describe the function of auxiliary views.*
- *Identify and describe the difference between the two methods used for the construction of auxiliary views.*
- *Construct primary, secondary, and successive auxiliary views using the folding-line method.*
- *Apply the folding line method to AutoCAD for the construction of auxiliary views.*
- *Explain what the* UCSFOLLOW *system variable is use for.*
- *Generate an auxiliary view using the mechanical desktop* AMDWGVIEW *command.*
- *Convert a nonparametric solid model into a parametric solid model using the* AMNEWPART *command.*

## Key Words and Terms

| | | |
|---|---|---|
| **Auxiliary view** | **Secondary auxiliary view** | **MVIEW - UCSFOLLOW - UCS** |
| **Auxiliary plane** | **Successive auxiliary view** | **MVIEW - VPOINT** |
| **Normal plane** | **Folding-line method** | **AMDWGVIEW** |
| **Primary auxiliary view** | **Reference plane method** | |

## Introduction to Auxiliary Views

Chapter one covered the fundamentals of viewing two- and three-dimensional objects using orthographic, axonometric, oblique, and perspective projection techniques. The purpose of these projections, except the orthographic, is to illustrate to a nontechnical person exactly how the object will appear. The orthographic projection shows the exact location of lines, surfaces, and contours, and projects them from one view to another. Recall that in an orthographic projection,

the object's features are projected onto three principal planes—horizontal, frontal, and profile. It is from these three principal planes that the six principal views are produced. However, when an object contains features that are inclined or oblique, then a view other than one of the six principal views must be derived to show those features in their true form. In other words, in an orthographic projection these features would appear foreshortened (not as the actual size or shape), as shown in Figure 3–1. The new view shown in Figure 3–2 is necessary to fully describe the sloping face of this part; this type of projection is called an ***auxiliary view***. An ***auxiliary view*** uses orthographic techniques to produce a projection on a plane other than one of the three principal planes. Its purpose is to show the true shape, size and/or length of features that would otherwise be foreshortened in the principal views. The plane on which an auxiliary view is projected is called the ***auxiliary plane***. The ***auxiliary plane*** will be adjacent to two of the principal planes and perpendicular to the third. For example, in Figure 3–3 the plane marked 'auxiliary' is perpendicular to the frontal plane and adjacent to both the profile and horizontal planes. If the auxiliary plane is perpendicular to two of the principal planes (the frontal and horizontal, frontal and profile, or horizontal and profile), it is a principal plane (also known as a ***normal plane***) and not an auxiliary plane.

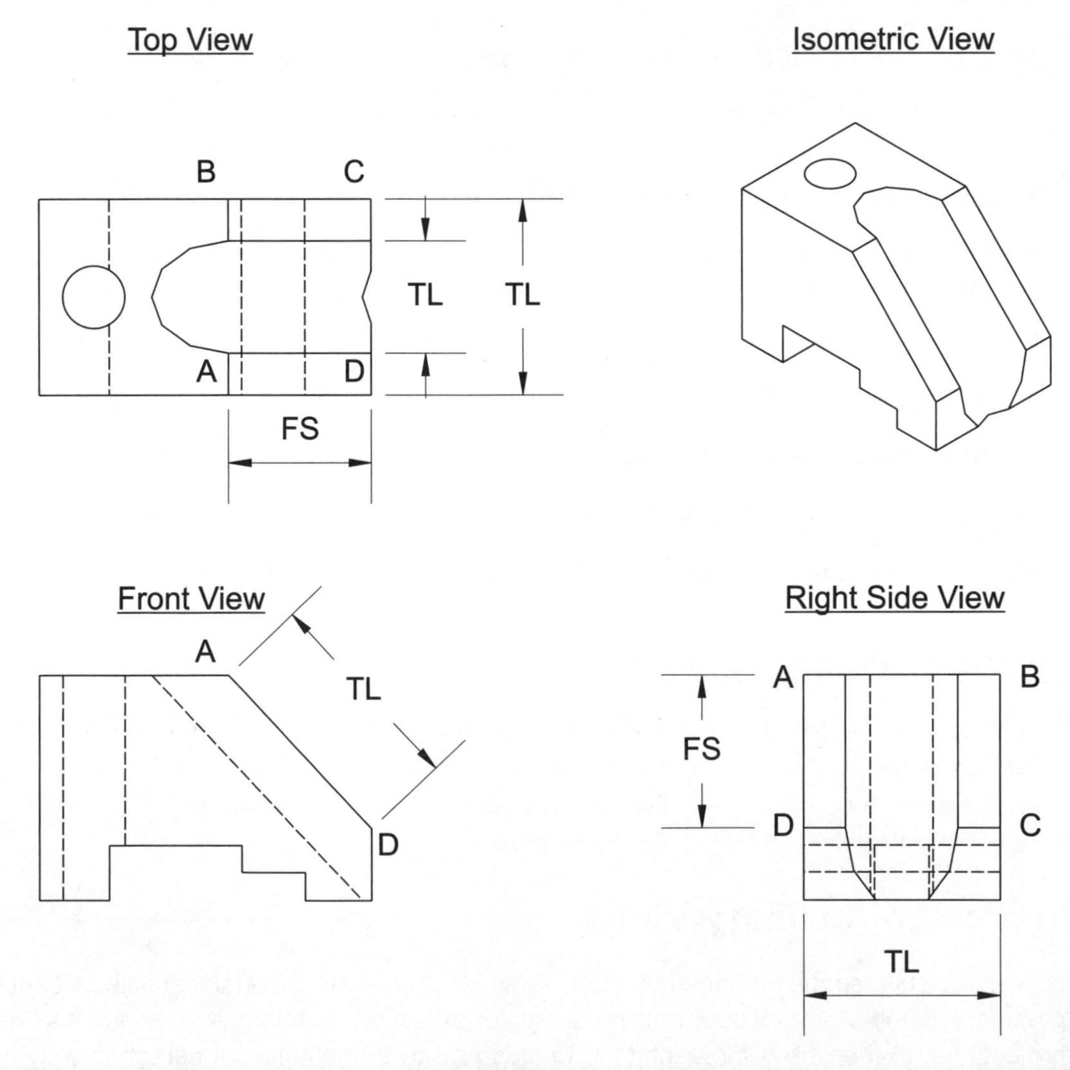


**Figure 3–1** *Orthographic projection of a mechanical part. In this drawing line AD is shown in true length in the front view, while lines AB and DC are shown in true length in the top and right side views. The true shape of the inclined surface is never revealed.*

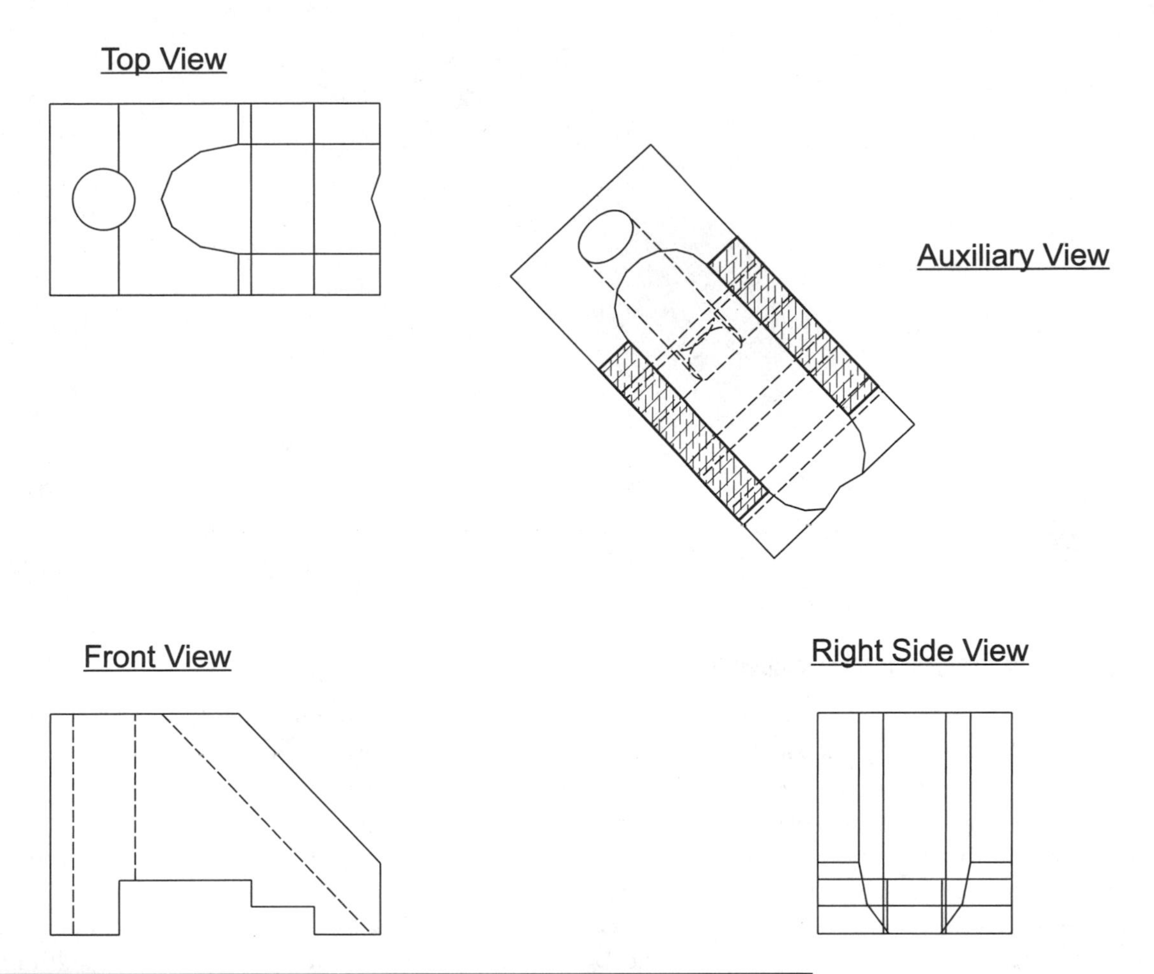


**Figure 3–2** *Auxiliary view showing plane ABDC in true shape and size.*

There are three classifications of auxiliary view: ***primary auxiliary view***, ***secondary auxiliary view***, and ***successive auxiliary view***, but their method of construction is essentially the same. The ***primary auxiliary view*** is produced by positioning the auxiliary plane parallel to the inclined surface of the object and perpendicular to one of the principal planes. Its construction is dependent upon two principal views. The first view is used to establish the auxiliary folding lines (these lines indicate the edge of the auxiliary plane), while the second view is used for the transfer of measurements to the auxiliary view. A third principal view does not add any additional information to the auxiliary view and its presence is not necessary for the construction. The ***secondary auxiliary view*** is constructed by positioning the projection plane parallel to the inclined surface of the object featured in the primary auxiliary view, because the primary auxiliary view is used as a basis for the secondary auxiliary view's folding-lines. It also requires the use of a principal view for the transfer of measurements. When an auxiliary view is created from a secondary auxiliary view, then it is known as a ***successive auxiliary view***. Successive auxiliary views are also constrained by two other views, either a primary auxiliary view and a secondary view or two secondary views.

When creating an orthographic view, the line-of-sight is stationed perpendicular to the principal plane on which the surface of the object is projected. In an auxiliary view, the line-of-sight is positioned perpendicular to the auxiliary plane on which the surface of the object is projected (see Figure 3–3). The result is a view where the surfaces that would normally appear foreshortened (in a principal view) are now shown in their true shape, size, and dimension, as shown in Figure 3–5.

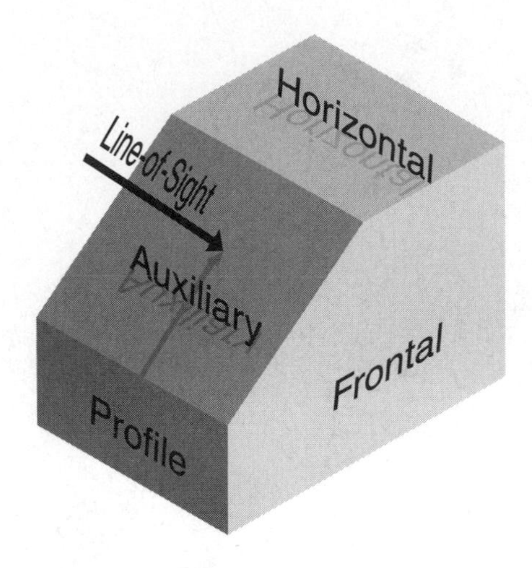


**Figure 3–3** *Line-of-sight set perpendicular to the auxiliary plane.*

## Creation of Auxiliary Views

There are two methods for generating auxiliary views, the ***folding-line*** and the ***reference-plane methods.*** As indicates by its name, the ***folding-line method*** is based on the theory used for the production of orthographic drawings. This method positions the object inside a glass cube, but this time the cube vaguely reflects the shape of the object (see Figure 3–4). Once the features have been projected onto the surface, the cube is unfolded to reveal the principal and auxiliary views (see Figure 3–5). This method allows for the transfer of points and distances from the principal views to the auxiliary view via the folding-lines (see Figure 3–6). For example, the object in Figure 3–7 contains the inclined surface ABDC. When an orthographic drawing is created of this part (see Figure 3–8), lines AB and DC of plane ABDC are shown in their true length in the front view. Also, the true length of lines AC, EG, FH, and BD are revealed in the top, left, and right views. Yet the true shape of the plane ABDC is still not revealed in any of these views. Therefore, an auxiliary view must be created in which its true shape is shown.

This is accomplished by constructing an auxiliary folding line that is set parallel to the inclined surface ABDC (see Figure 3–9A). Once this has been accomplished, the proximity of the endpoints of the object lines are established by drawing projection lines from the object in the front view perpendicular to the auxiliary folding line and onto the auxiliary plane (points A, B, C, D, E, F, G, H). The width of the object is then found by transferring the lengths of lines AC, EG, BD, and FH from either the profile or horizontal planes (top or side views) to the auxiliary plane. The distances from the folding line to points A, B, E, and F in the horizontal and profile planes (top and side views) are the same as the distances from the auxiliary folding line to points A, B, E, and F in the auxiliary plane (see Figure 3–9A). Showing line EG as being a hidden line (using a hidden line type) completes the object. Now surface ABDC's true shape and size is depicted (notice that this plane appears as a square in right side view, but its true shape is actually a rectangle).

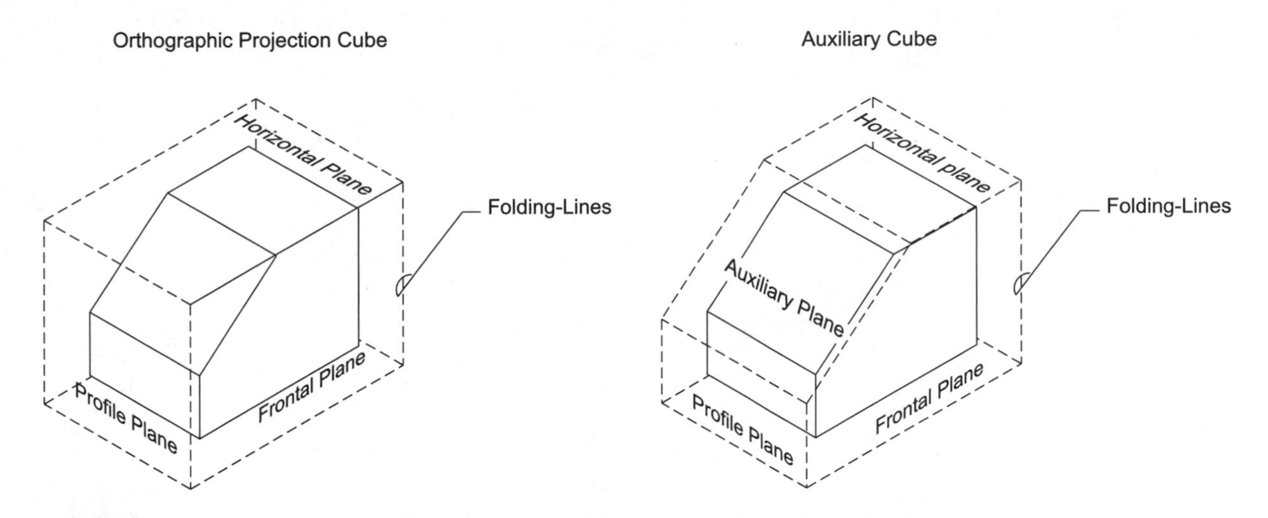


**Figure 3-4** *The object is positioned in a glass cube similar to the cube used in an orthographic projection. This glass cube is then deformed vaguely to imitate the shape of the object inside.*

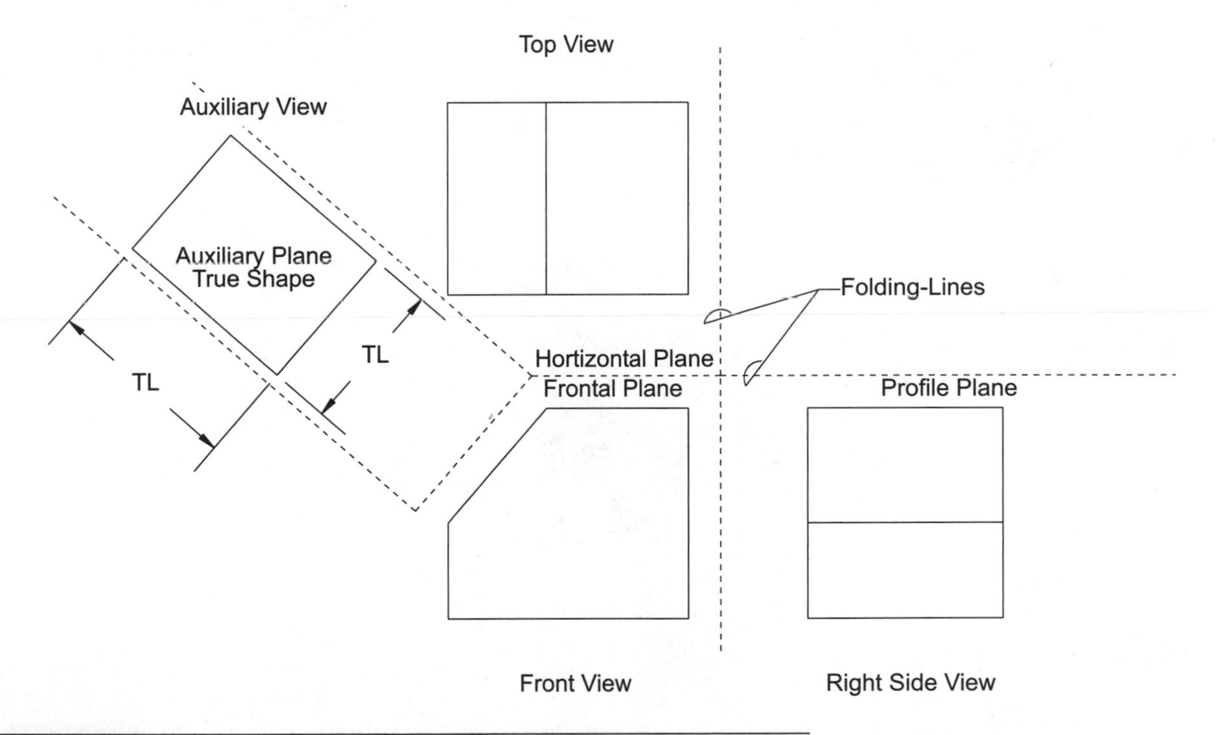


**Figure 3-5** *Next, the cube is unfolded to reveal the different views.*

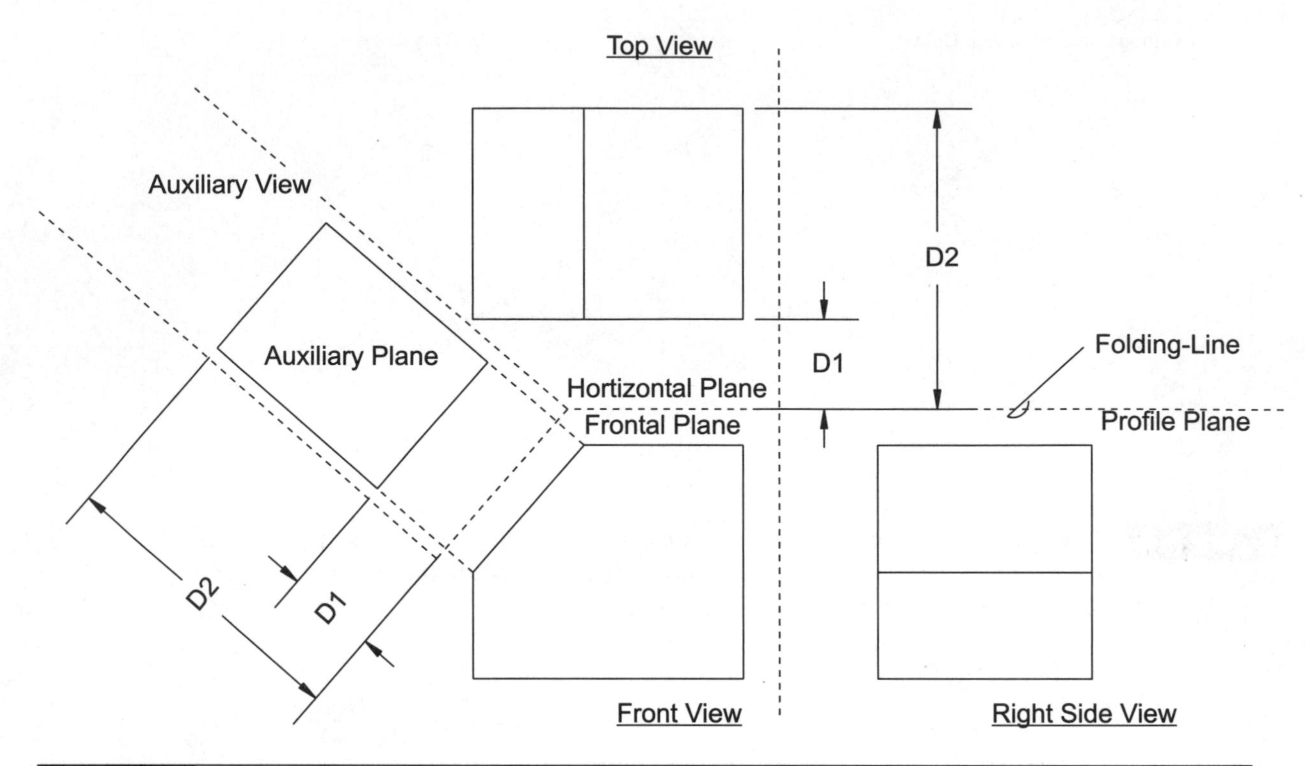


**Figure 3–6** *Points can now be transferred from one view to another using the same techniques used for an orthographic projection.*

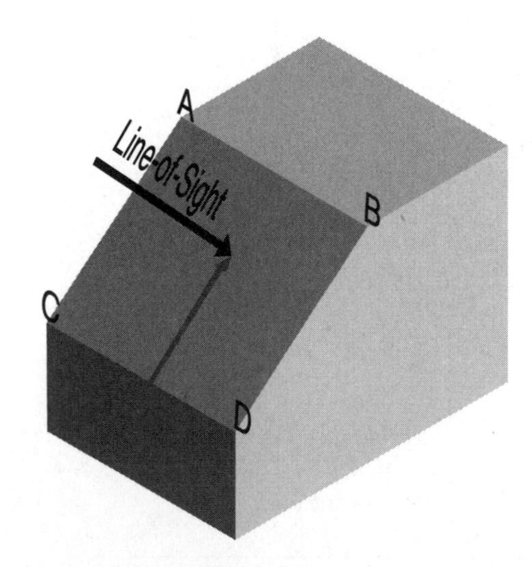


**Figure 3–7** *Block containing an inclined surface.*

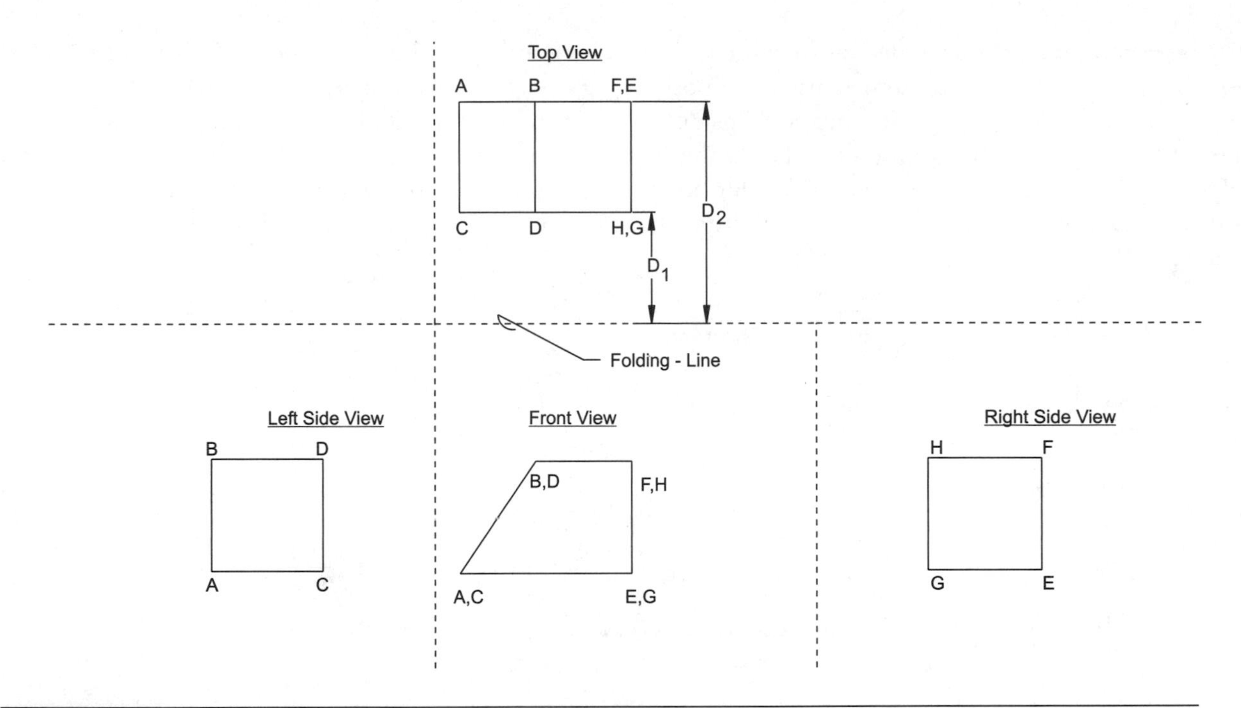


**Figure 3-8** *Orthographic projection of block containing the top, front, right, and left side views. The true shape of plane ABDC is not revealed in any of these views.*

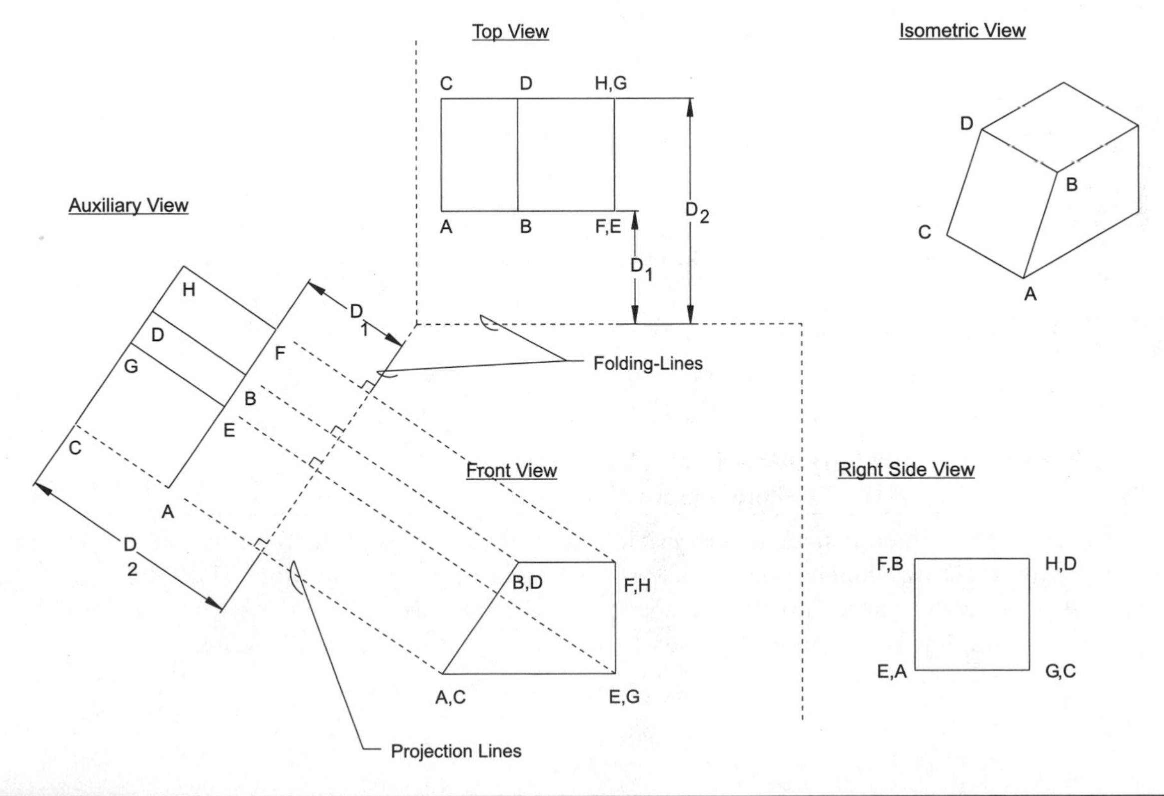


**Figure 3-9A** *Surface ABDC is in edge view in the front view where it is shown in true length, but not true shape.*

The ***reference plane method*** is a derivative of the folding-line method. It differs from the folding-line method by establishing a plane as its reference instead of positioning the object inside a cube. The reference plane can be established at any convenient location (inside, onto, or about) in relation to the object. For example, if the reference plane is positioned parallel to the frontal plane (top or bottom view) then it will appear in edge view (as a line) in the horizontal, profile, and auxiliary planes (front, back, side, and auxiliary views). Any of these views may be used to transfer distances to the reference plane using orthographic techniques (see Figure 3–9B).

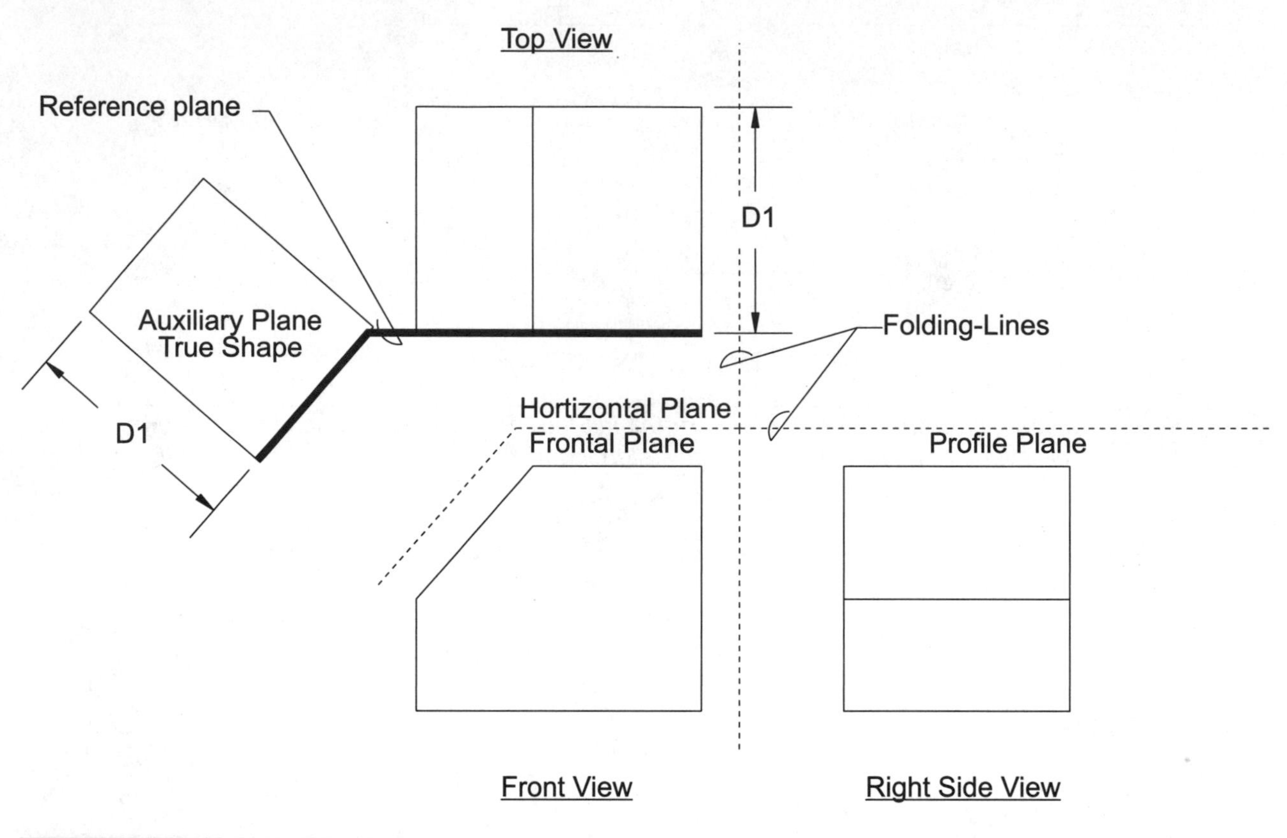


**Figure 3–9B** *Reference plane method.*

## Creation of a Primary Auxiliary View from a Two-Dimensional AutoCAD Drawing Using the Folding-Line Method

Once it has been determined that an auxiliary view is needed, the five steps used for its construction are the same for an AutoCAD two-dimensional object as those used in manual drafting. The following example illustrates the steps involved by constructing an auxiliary view of the object shown in Figure 3–10. The following example can be completed using drawing DG03-10_student from the student CD-ROM.

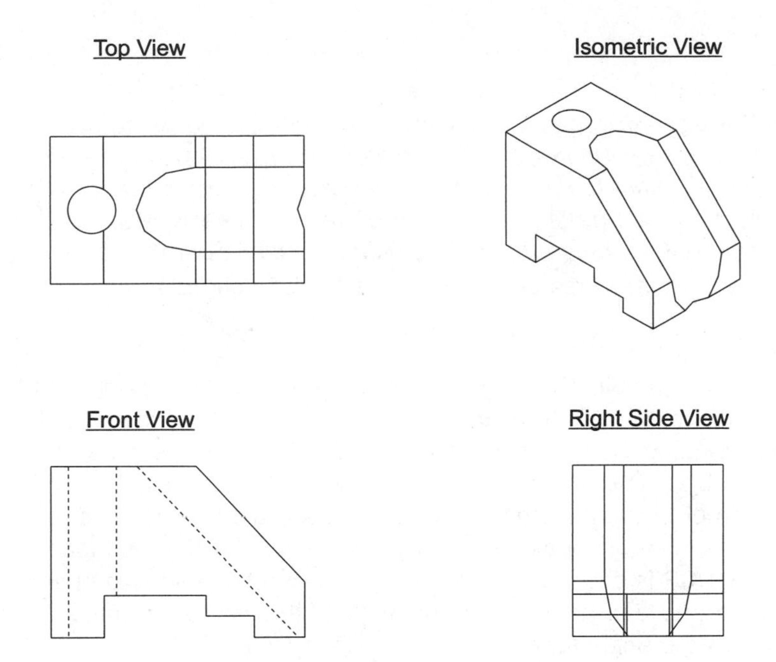


**Figure 3–10** *Orthographic projection of a block spacer.*

## Step #1

Draw the auxiliary view folding line. This line can be positioned at any convenient location from the object. Its only requirement is that it must be parallel to the inclined surface for which the auxiliary view is being produced. The auxiliary folding line can be established by using either the OFFSET, UCS-LINE, or COPY methods. The offset method utilizes the AutoCAD OFFSET command to generate a folding line parallel at a point or distance specified by the technician (see Figure 3–11).

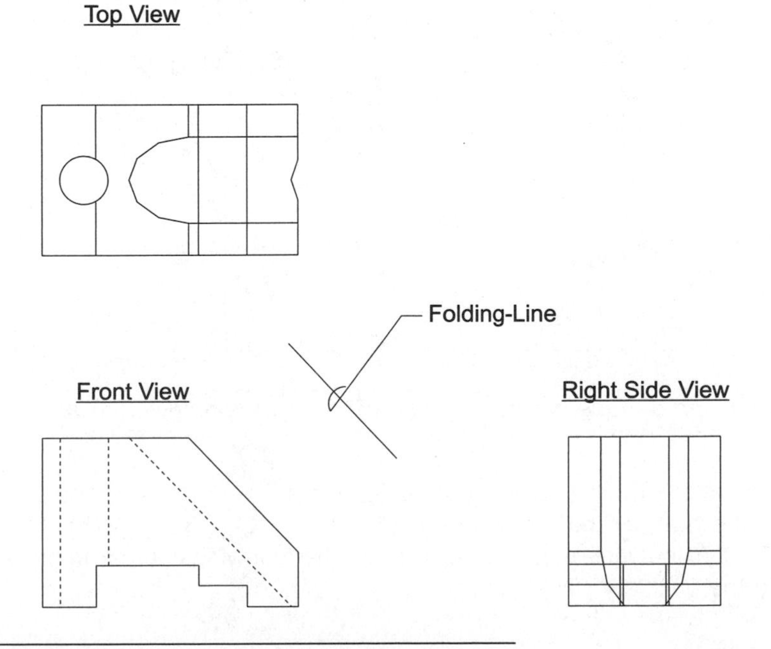


**Figure 3–11** *Auxiliary plane is copied into its proper position.*

### The Offset Method

Command: OFFSET (ENTER)

Offset distance or Through <Through>: *(Pressing* ENTER *enables the random selection of a distance from the object to the location where the auxiliary folding line will be positioned.)*

Select object to offset: *(Select the inclined line in the view from which the auxiliary is to be taken.)*

Through point: *(Select a random point to the right of the front view where, if the line being offset were extended, it would pass through the point picked but still be parallel to its original counterpart.)*

Select object to offset: *(Press* ENTER *to terminate the* OFFSET *command.)*

### The UCS-Line Method

The UCS-LINE method is a combination of AutoCAD's UCS and LINE commands. Start by using the UCS command to rotate the cross-hairs to match the angle of the inclined feature. This method has two main advantages over using the OFFSET command. First, it enables the use of ORTHO when projecting the endpoints of the object into the auxiliary view. Second, the Cartesian and polar coordinate systems can be used without having to add or subtract the angle of the inclined feature to the direction of the line created. For example, suppose that a part contains an inclined surface that is at a 35° angle from the X-axis. If the UCS is not rotated, then to construct a line that is perpendicular to the inclined surface, 90° would first have to be added to the angle of the inclined surface (35°). Their product would then be used as the angle for drawing the line (see Figure 3–12).

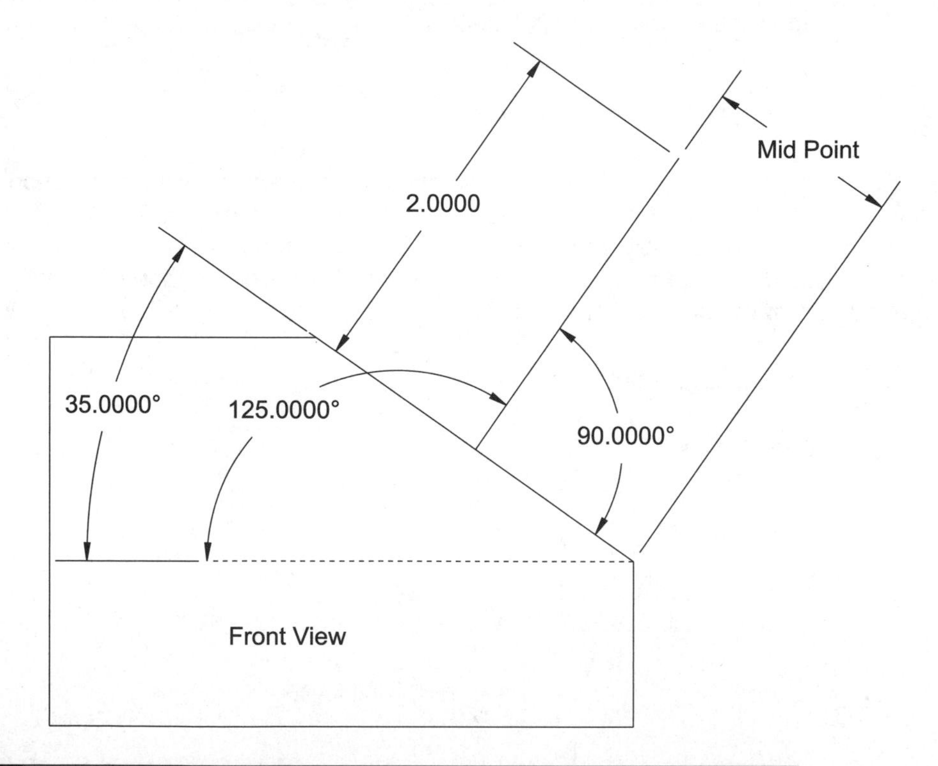


**Figure 3–12** *The inclined surface in this figure is 35° from the positive X-axis.*

If the UCS is not reset to match the inclined angle, then a line drawn perpendicular to the inclined surface would require that 90° be added to the 35° angle (90° + 35° = 125°) to calculate the position of the line from the X-axis. Once this has been done, the following command sequence would be followed to create the auxiliary view folding-line.

Command: LINE (ENTER)

From point: MID of

To point: @2<<125 (ENTER)
Command: UCS (ENTER)
Origin/ZAxis/3point/OBject/View/X/Y/Z/Prev/Restore/Save/Del/?/<World>: E (ENTER)
Select object to align UCS: *(Select the inclined line to transfer the object's angle of rotation to the UCS.)*

Once the cross-hairs have been rotated to match the angle of the inclined surface, the folding-line is now constructed using the standard AutoCAD line command (with ORTHO enabled). The length of the folding-line is not critical and can be drawn to any length using either the Cartesian or polar coordinate systems. It should be noted that when using the Entity (for releases under R13) or Object (for releases over R12) option of the ucs command, AutoCAD not only rotates the X-, Y-, or Z-axes (causing the cross-hairs to rotate), it also automatically resets the origin of the UCS to the closest endpoint of the entity selected. After the axis has been rotated and the origin has been reset, the different coordinate systems may be used as if they were set in the world position (default), see Figure 3–13.

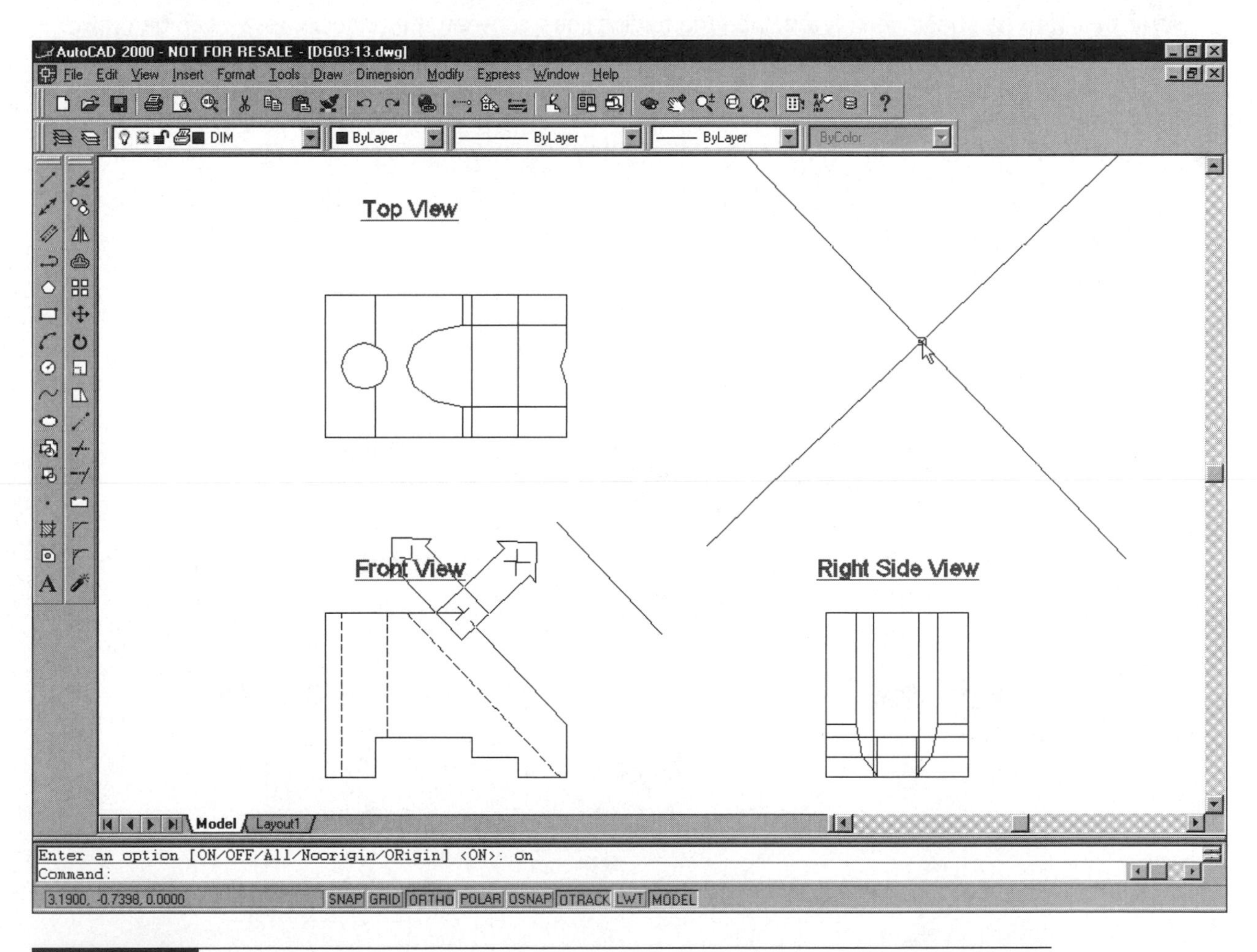


**Figure 3–13** *The auxiliary projection plane is created using the* UCS *and* LINE *commands.*

Command: LINE (ENTER)
From point: end of
To point: @2<<0 *(Press* ENTER *to construct a line from the last point selected 2 units long along the rotated UCS axis.)*
To point:

### The Copy Method

The third method of creating the auxiliary view folding-line is by using the COPY command. In this method, the inclined line is simply selected as the object to copy with the location of the folding-line (copied entity) specified by the technician (see Figure 3–11).

Command: COPY (ENTER)
Select objects: 1 found *(Select inclined surface in front view.)*
Select objects: *(Pressing the* ENTER *key terminates the copy selection mode.)*
<Base point or displacement>/Multiple:
Second point of displacement: *(Select arbitrary point to the right of the front view.)*
Command:

## Step #2

After this step has been completed, then the folding-lines between the principal views can be drawn. The easiest way to generate these lines is with the OFFSET command. Select one of the vertical or horizontal object lines in the principal view from which the folding line is to be positioned. For example, if an auxiliary view is taken from the front view, then the folding-line between the front and top views, along with the folding-line between the side and front views, should be created as shown in Figure 3–14.

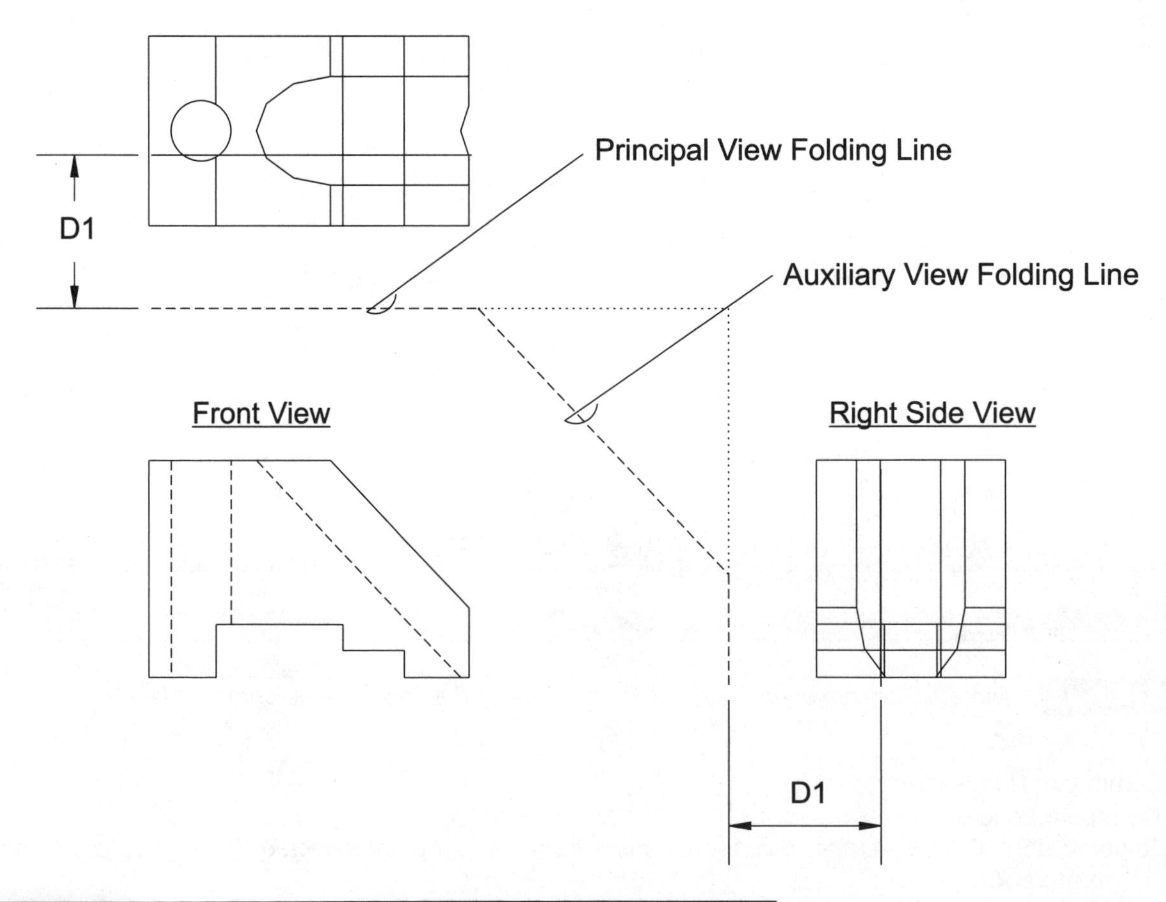


**Figure 3–14** *The remaining folding lines are added to the drawing.*

## Step #3

Using projection lines, the endpoints of the object are extended from the view where the auxiliary is to be taken onto the auxiliary plane. The projection lines will form 90° angles (perpendicular) to the auxiliary folding-line. These projection lines are produced using standard AutoCAD two-dimensional commands, see Figure 3–15.

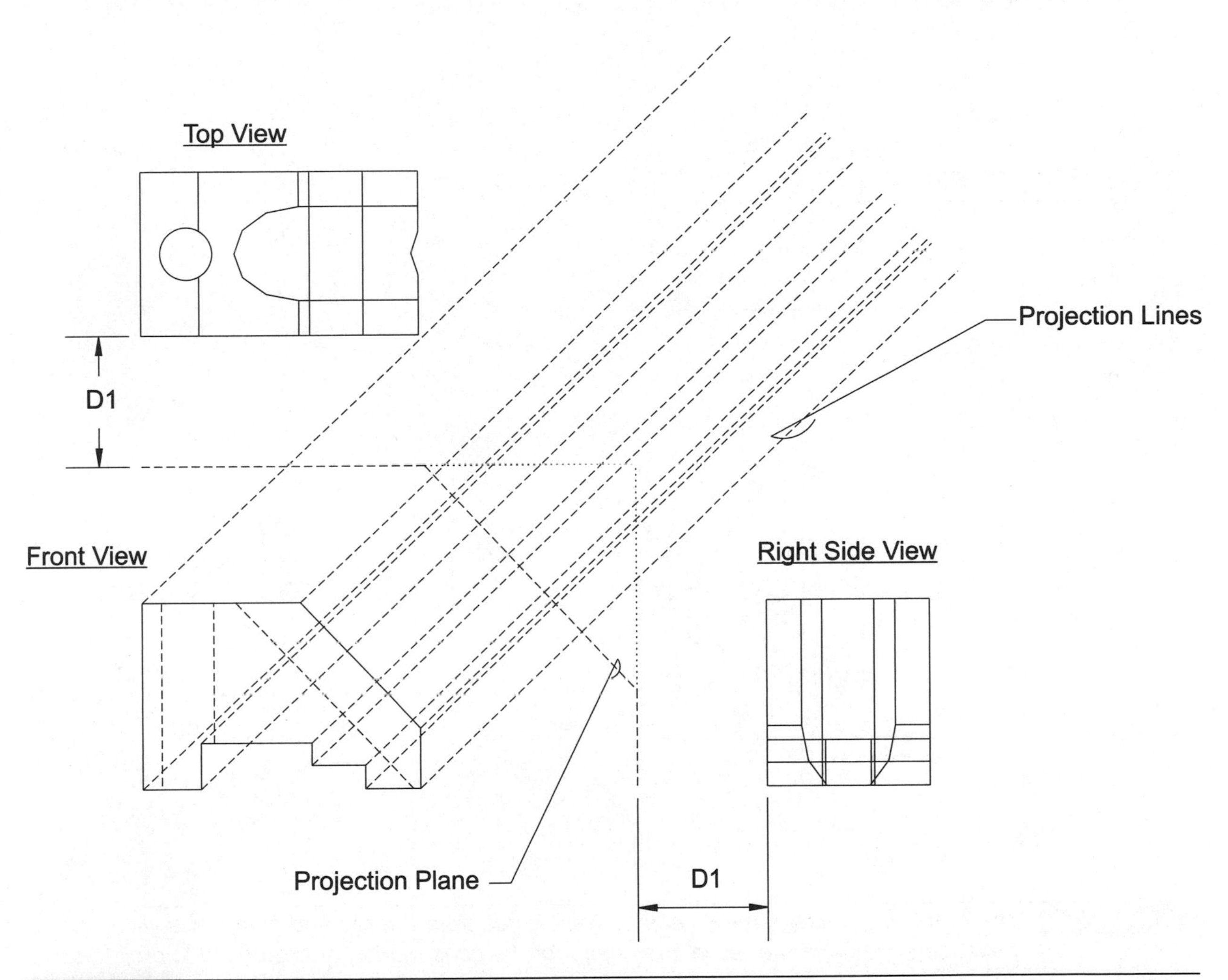


**Figure 3–15** *Projection lines are extended from each corner of the object into the auxiliary area. These projection lines will be drawn perpendicular to the auxiliary projection plane.*

## Step #4

The true length of each line and its distance from the folding-lines are then transferred from one of the principal views onto the auxiliary plane using the orthographic techniques that were employed with the production of the principal views (see Chapter one). This step locates the exact position of points in the auxiliary view by intersecting the transferred measurement lines with the projection lines produced in Step #3 (see Figure 3–16).

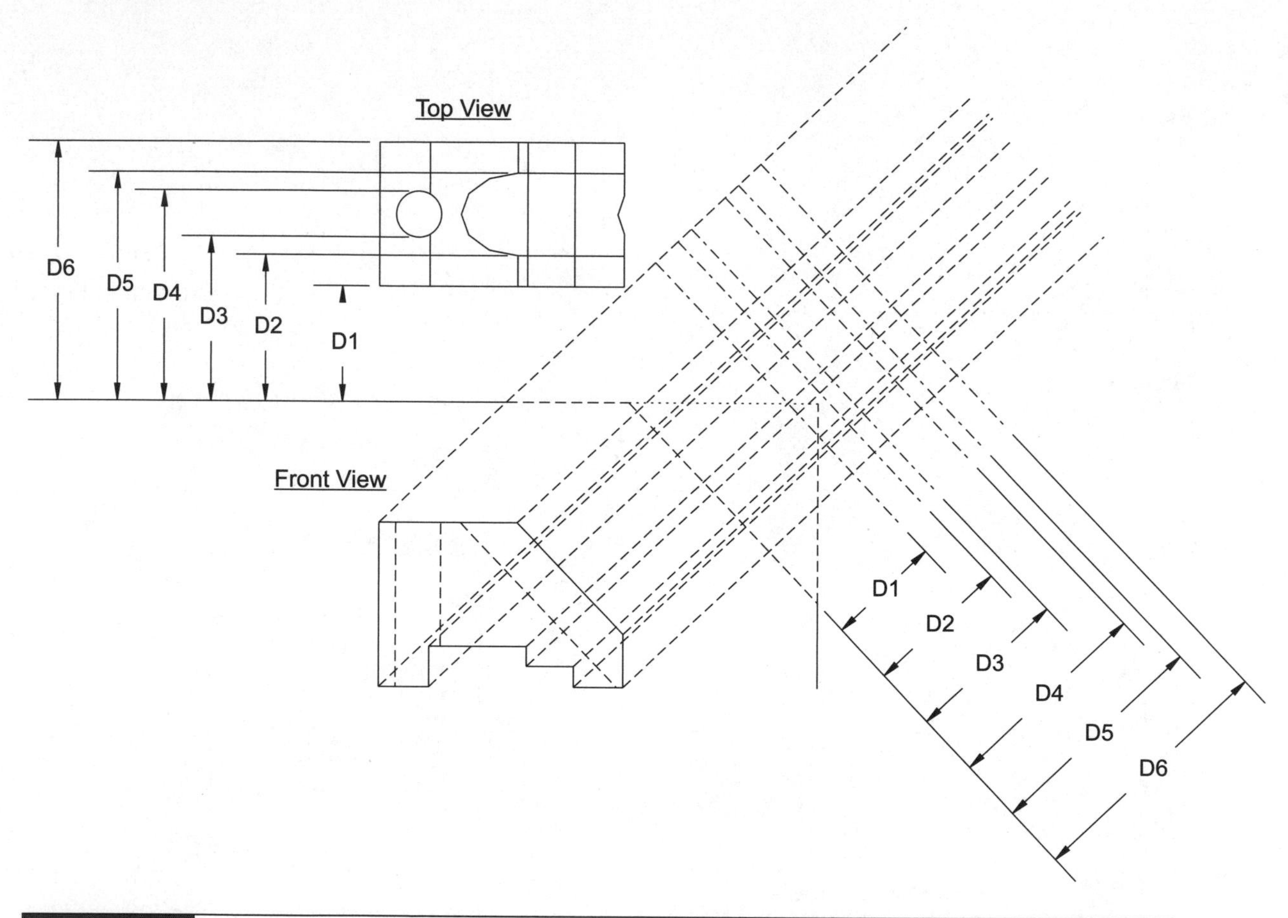


**Figure 3–16** *The remaining measurements can be transferred from the top and side views via projection lines and the techniques used for the construction of orthographic projections.*

## Step #5

The object lines are now drawn in the auxiliary view by connecting the points previously established in Step #4. Finally, adding the required annotation and dimensions completes the view (see Figure 3–17).

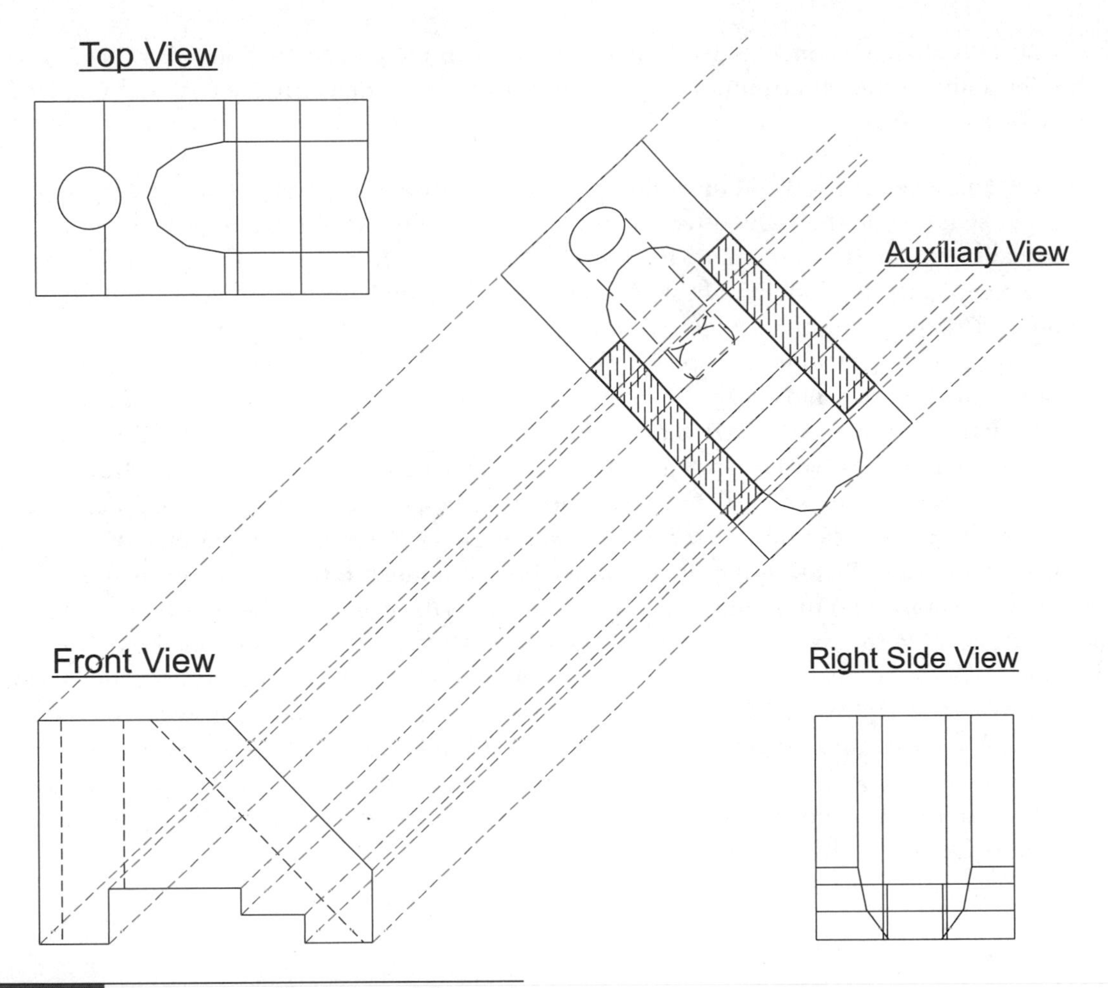


**Figure 3–17** *Finished auxiliary view of the part.*

## Creation of an Auxiliary View from an AutoCAD Three-Dimensional Model

The creation of an auxiliary view from a three-dimensional model is considerably easier than from a two-dimensional view. This is because the object has already been created; by repositioning the line-of-sight (leaving the object stationary) to a point perpendicular to the surface where the auxiliary is required, the view is automatically generated. This can be accomplished by using one of two different methods: the MVIEW - UCSFOLLOW - UCS, or the MVIEW - VPOINT. In both methods, the line-of-sight is repositioned and not the object. If the object is rotated in an active floating viewport instead of repositioning the line-of-sight, then the object will also rotate in the remaining floating viewports.

### Using MVIEW - UCSFOLLOW - UCS

This method is the least complex for creating an auxiliary view from a three-dimensional model because it employs the UCSFOLLOW variable in conjunction with the UCS command to reposition the observer's line-of-sight. The UCSFOLLOW is a system variable that, when activated, causes AutoCAD to automatically generate a top view whenever changes are made to the User Coordinate System. This variable is set and controlled separately in paper and model space as well as floating viewports. This allows it to be set in one mode (paper space, model space, or floating viewports) without affecting the settings in the others. After the variable has been set and changes are made to the UCS, AutoCAD repositions the line-of-sight to a point perpendicular to the new UCS. This change is only made in the mode in which the UCSFOLLOW system variable has been enabled; the viewpoint in the other modes remains unchanged (provided that the UCSFOLLOW variable is not active in the other modes).

Once the model has been created and the technician is ready to use this technique, AutoCAD is toggled from model to paper space using the TILEMODE command to enable the MVIEW command for creating the required number of floating viewports. Remember, the required number of floating viewports for an object is determined by the complexity of the design. In short, a good rule of thumb for both auxiliary and principal views is the number of floating viewports required will equal the minimum number of views necessary to fully describe the object. Once the object has been described, adding more views does not enhance the clarity of the description; instead, it could make the interpretation of the design more difficult. After the floating viewport has been generated, AutoCAD is again changed from paper to model space, this time using the MSPACE command so that the object can be edited from within the viewport. Once the mode has been changed, select the floating viewport in which the auxiliary view is to be generated. Then set the UCSFOLLOW system variable. The UCS command can now be used to reposition the UCS either on or parallel to the surface that requires an auxiliary view. By using the 3P option of the UCScommand, the user coordinate system can be repositioned by selecting a point on the surface of the object as the origin for the coordinate system and two other points from which the X- and Y-axes are defined. The UCSFOLLOW system variable should be set back to zero after the auxiliary view has been generated, and before any other floating viewport is made current. If the UCSFOLLOW system variable is not disabled before changing to a different floating viewport, then the changes made to the user coordinate system will cause AutoCAD to update the view in any viewport subsequently selected. This process is repeated until all auxiliary views have been generated. For example, the object shown in Figure 3–18 requires that four auxiliary views of the object be produced in which the true shapes of the inclined surfaces are revealed.

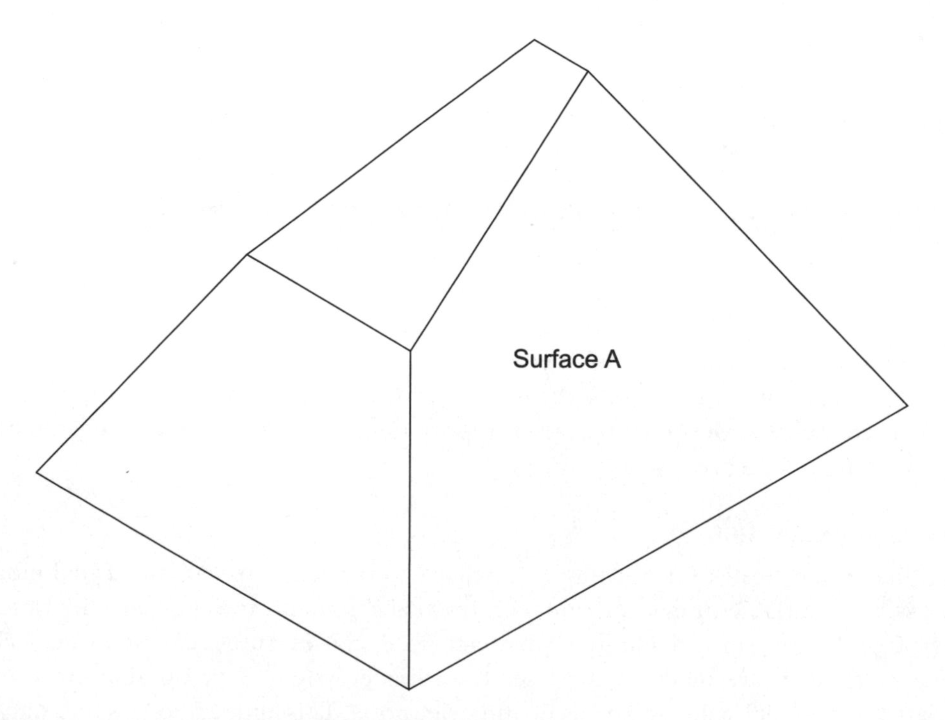


**Figure 3–18** *Object containing mult-inclined surfaces.*

## Step #1

Open file DG03-18 located on the student CD-ROM. Using the TILEMODE system variable, AutoCAD is toggled from model to paper space, for the creation of four floating viewports (these viewports are created using the MVIEW command).

Command: TILEMODE (ENTER)
New value for TILEMODE <1>: 0 *(Pressing* ENTER *sets AutoCAD's current drawing mode from model space to paper space.)*
Regenerating drawing.
Command: MVIEW
ON/OFF/Hideplot/Fit/2/3/4/Restore/<First Point>: 4 *(Pressing* ENTER *creates four floating viewports.)*
Fit/<First Point>: *(Select an arbitrary point; this point will define the lower left-hand corner of the lower left-hand floating view port.)*
Second point: @35,25 *(Pressing* ENTER *selects a point for the upper right-hand corner of the upper right-hand floating viewport at a location that is 35 units in the X-axis and 25 units in the Y-axis from the last point entered.)*
Regenerating drawing.

## Step #2

After the floating viewports have been created, start with the floating viewport in the upper left-hand corner, and enable model space by using the MSPACE command. Set the UCSFOLLOW system variable to one by using the UCSFOLLOW command. Next, reposition the UCS onto the plane where the auxiliary view is needed, by using the UCS command. Finally, scale the view using the XP option of the ZOOM command and reset the UCSFOLLOW system variable to zero. This step is repeated until all auxiliary views have been created.

Command: MS *(Pressing* ENTER *activates model space in the last floating viewport created.)*
Command: VPOINT (ENTER)

**Optional step:** Changing the viewpoint to an isometric view can make it easier for the technician to select the correct endpoints of the plane that will define the position of the UCS.

Rotate/<View point> <1.0000,1.0000,1.0000>: 1,1,1 (ENTER)
Regenerating drawing.
Command: UCSFOLLOW (ENTER)
New value for UCSFOLLOW <0>: 1 *(Pressing* ENTER *enables AutoCAD's* UCSFOLLOW *mode.)*
Command: UCS (ENTER)
Origin/ZAxis/3point/OBject/View/X/Y/Z/Prev/Restore/Save/Del/?/<World>: 3P (ENTER)
Origin point <0,0,0>: end of *(Select lower left-hand corner of surface A.)*
Point on positive portion of the X-axis <15.8502,3.1215,0.0000>: end of *(Select lower right-hand corner of surface A.)*
Point on positive-Y portion of the UCS XY plane <13.8502,3.1215,0.0000>: end of *(Select upper right-hand corner of surface A.)*
Regenerating drawing.
Command: ZOOM (ENTER)
All/Center/Dynamic/Extents/Left/Previous/Vmax/Window/<Scale(X/XP)>: E *(Pressing* ENTER *causes AutoCAD to perform a zoom extents.)*
Regenerating drawing.

**Note: Before using a zoom scale factor always perform a zoom extent first to ensure the object is scaled correctly.**

Command: ZOOM (ENTER)

All/Center/Dynamic/Extents/Left/Previous/Vmax/Window/<Scale(X/XP)>: 1XP *(Pressing* ENTER *performs a zoom scale of 1:1.)*

Command: UCSFOLLOW (ENTER)

New value for UCSFOLLOW <1>: 0 *(Pressing* ENTER *turns off* UCSFOLLOW.*)*

**Optional Step:** Once the view(s) have been generated, the VPOINT command can be employed to align the view(s) with the projection lines that extend from the object onto the auxiliary plane. To accomplish this the Rotate option is selected to rotate the object in tzhe XY plane (see Figure 3–19).

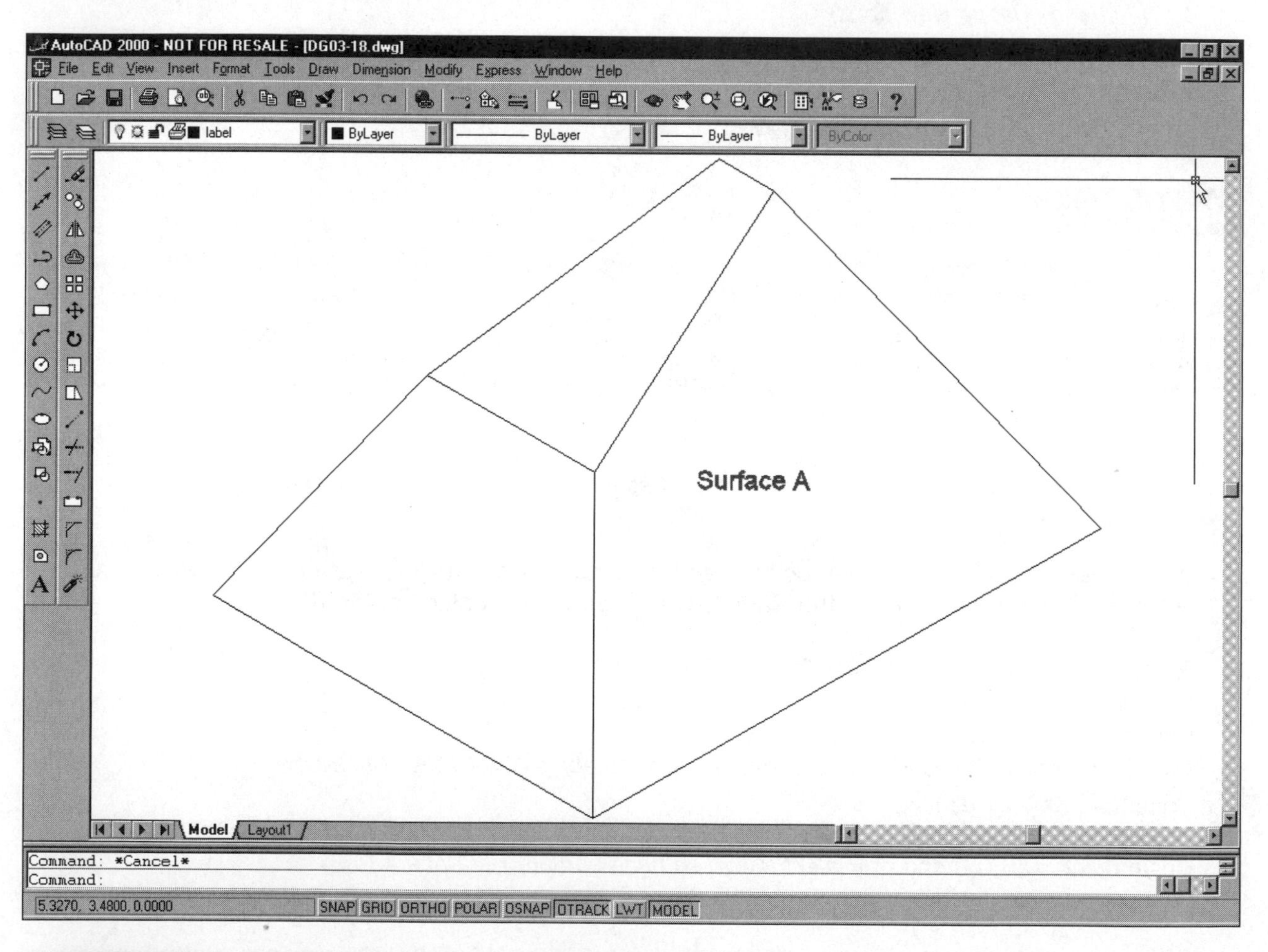


**Figure 3–19** *The object is rotated into an isometric view (1,1,1).*

**Using MVIEW - VPOINT**

The MVIEW - VPOINT method differs from the MVIEW - UCSFOLLOW - UCS method because it relies on the technician to reposition the observer's line-of-sight by rotating it in, and relative to, the XY-plane. Yet the basic procedure is still the same. Starting with the three-dimensional model DG03-21, located on the student CD-ROM, change from model space to paper space (by setting TILEMODE to zero). Using the MVIEW command, construct the required number of floating viewports. Again, using the MSPACE command, AutoCAD is changed from paper to model space in the current floating viewport. Next, using the Rotate option of the VPOINT command, the line-of-sight is rotated first, with respect to the X-axis in the XY-plane and then from the XY-plane (when AutoCAD is first started it defaults the vpoint rotation to 270° from the X-axis in the XY-plane and 90° from the XY-plane, as shown in Figure 3–20). For example, to produce an auxiliary view of the surface labeled A, as shown in Figure 3–21; a rotation of 0° from the X-axis in the XY-plane and 29° from the XY-plane would have to be used.

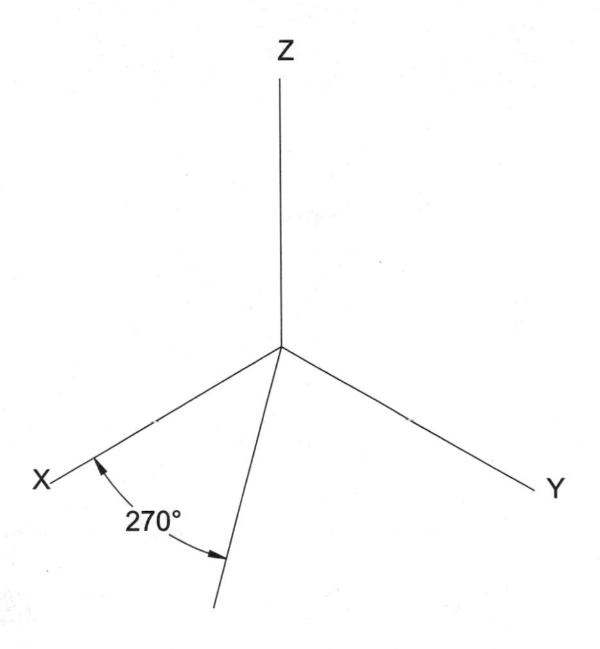


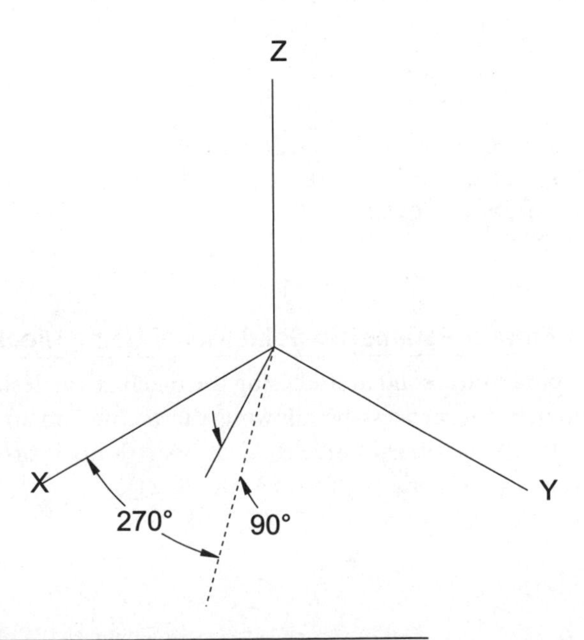


**Figure 3–20** *AutoCAD default setting for the VPOINT command.*

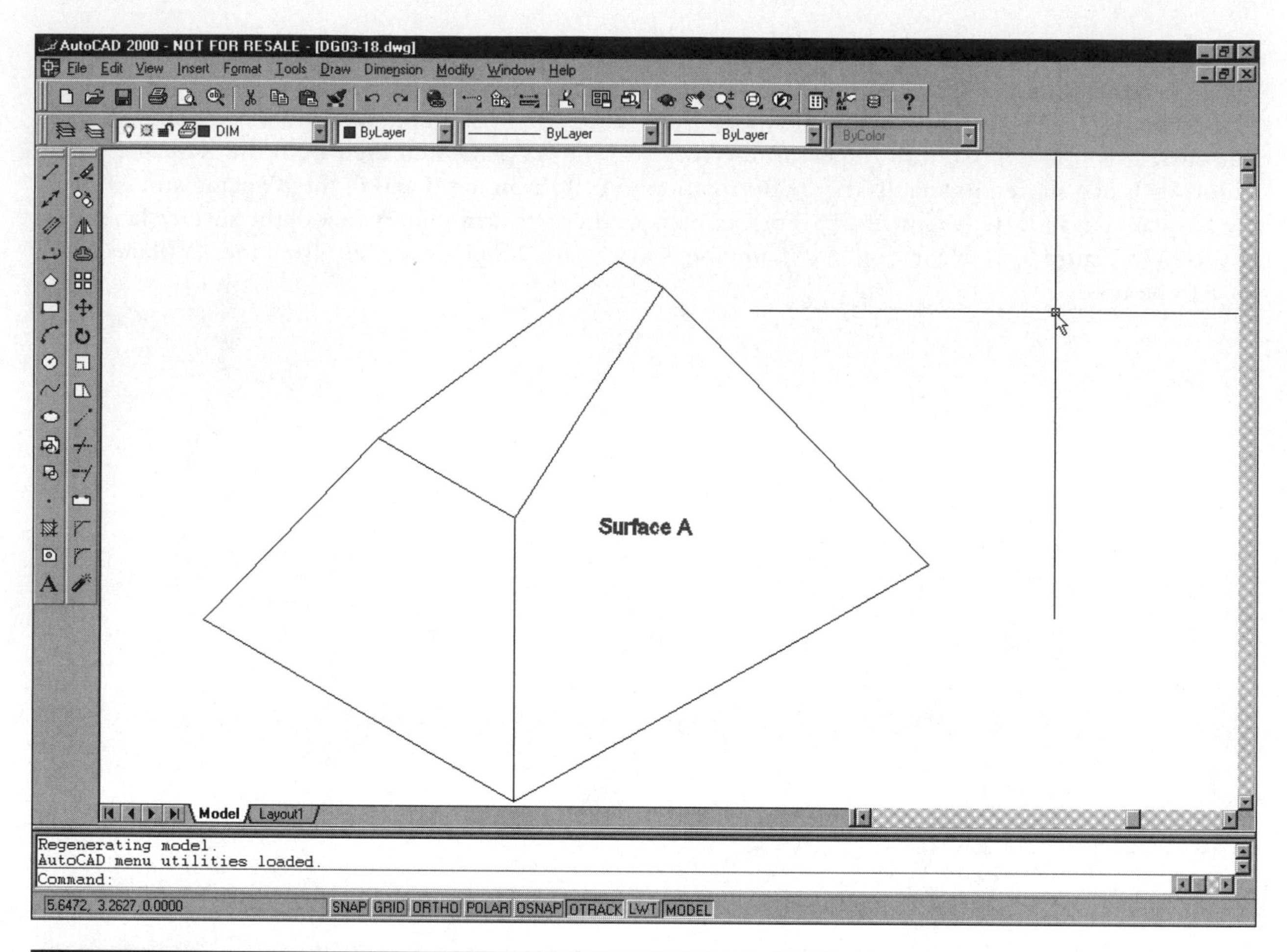


**Figure 3–21** *Using AutoCAD **R**otate option of the* VPOINT *command.*

Command: VPOINT (ENTER)
*** Switching to the WCS ***
Rotate/<View point> <0.0000,0.0000,1.0000>: R (ENTER)
Enter angle in XY plane from X axis <270>: 0 (ENTER)
Enter angle from XY plane <90>: 29 (ENTER)
*** Returning to the UCS ***
Regenerating drawing.

## Creating an Auxiliary View From a Parametric Solid Model Using Mechanical Desktop

To create an auxiliary view of a parametric solid model using the mechanical desktop, the command AMDWGVIEW is used. This command automates the process by allowing the technician to select a straight edge in the parent view from which to create the auxiliary. Once this has been done, it prompts the technician to select the location of the auxiliary view (see Figures 3–22, 3–23, and 3–24).

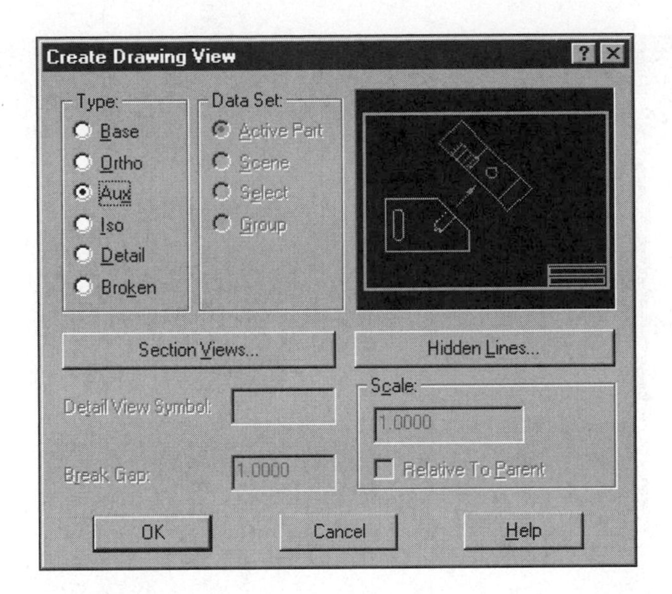


**Figure 3–22** AMDWGVIEW *dialog box.*

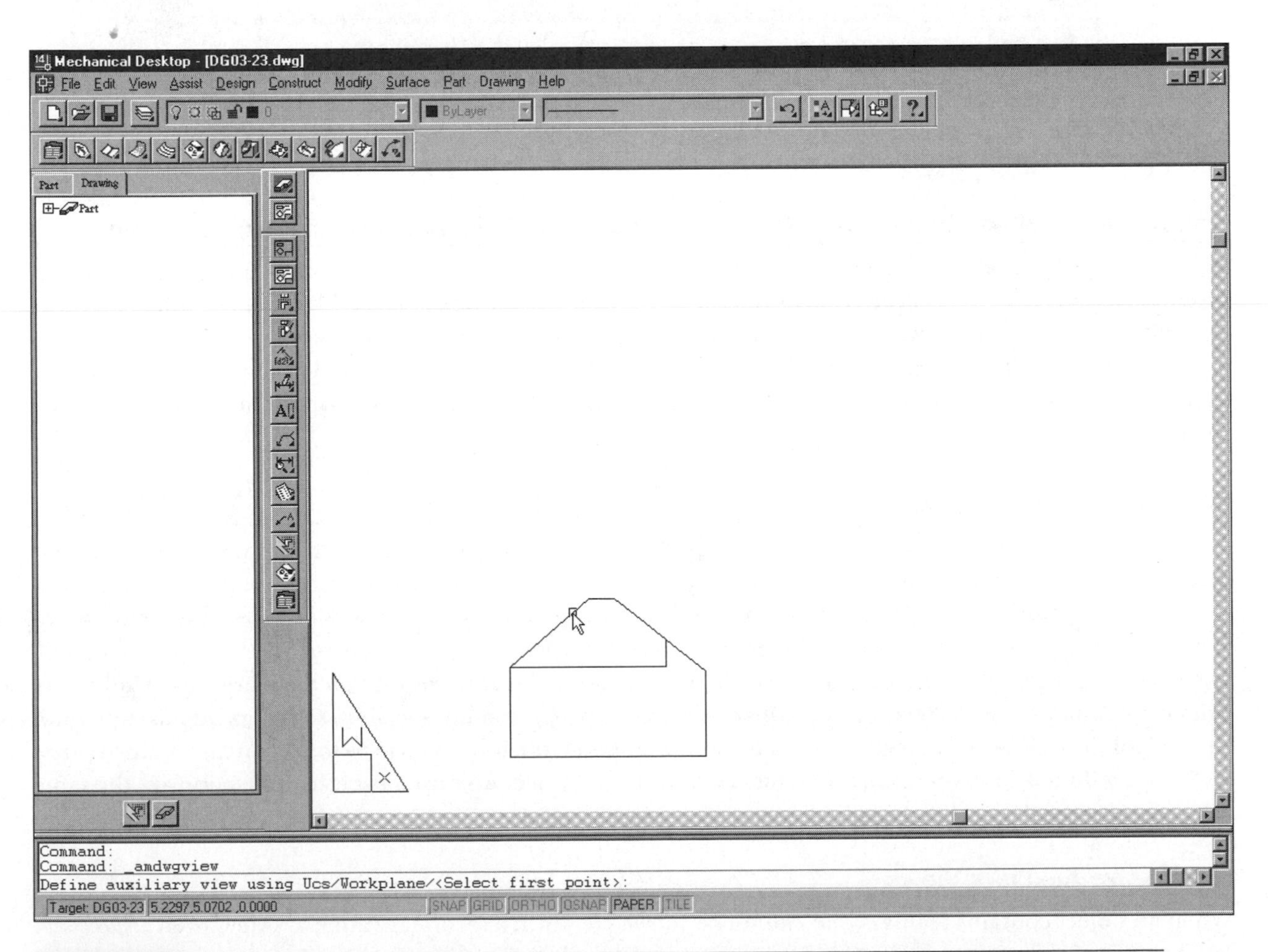


**Figure 3–23** *Selection of a straight edge in the parent view. The auxiliary plane will be positioned perpendicular to this edge.*

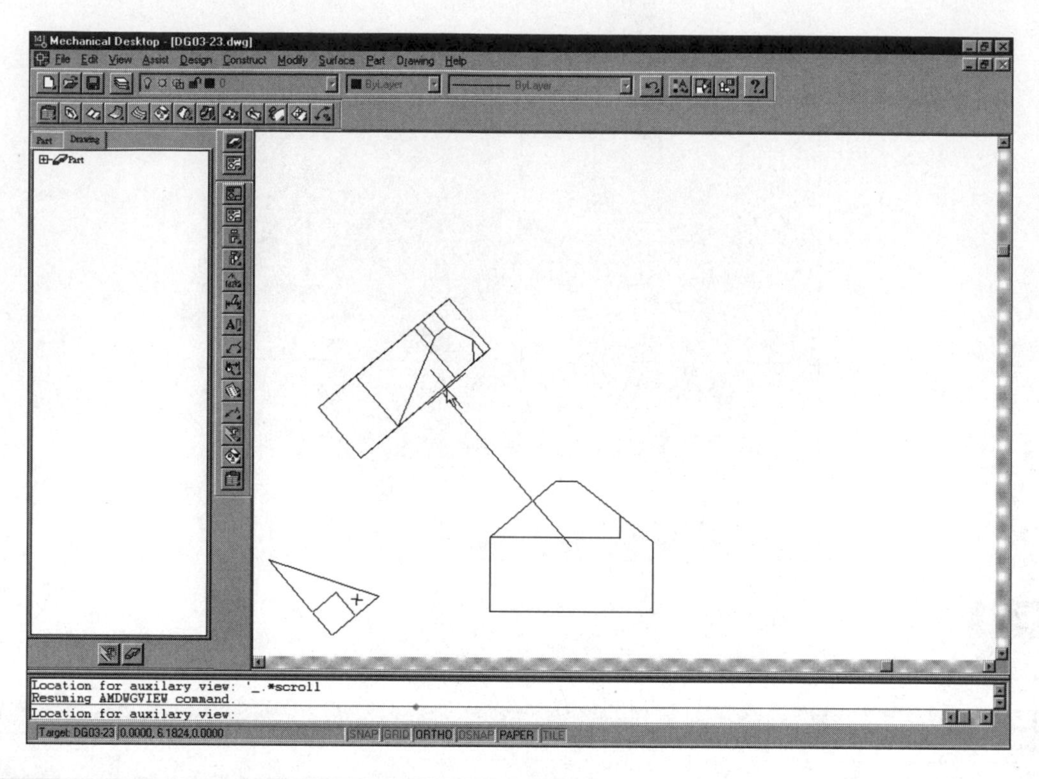


**Figure 3-24** *The location of the auxiliary view is selected.*

## Creating an Auxiliary View From a Nonparametric Solid Model Using Mechanical Desktop

Even when a solid model (nonparametric) is created outside of the mechanical desktop, the program can still be employed to create any of the types of views offered by the AMDWGVIEW command (orthographic, isometric, or auxiliary). To use the AMDWGVIEW command on a nonparametric model the solid must first be converted to a parametric solid. This is accomplished by using the AMNEWPART command. This command prompts the user to select the solid to convert to a parametric solid. After the conversion, the AMDWGVIEW command can be used to generate the required views.

## Secondary Auxiliary Views

When an object contains features that cannot be defined clearly in a primary auxiliary view, then a second auxiliary called a ***secondary auxiliary view*** must be drawn where the inclined features are presented in their true shape and size. The secondary auxiliary view, like the primary auxiliary view, is only dependent upon two views (a third principal view doesn't add any additional information) for its construction. The primary auxiliary view, plus a principal view, are the two views necessary to create a secondary auxiliary view. Unlike the primary auxiliary view, the secondary auxiliary receives its projection lines from the primary auxiliary and not a principal view. However, a principal view is used to supply all the information to the secondary auxiliary view concerning dimensions and measurements. Therefore, the procedures used for its construction are the same as those used for the primary auxiliary view.

## Successive Auxiliary Views

When an object contains features that cannot be shown clearly in a secondary auxiliary view, then a ***successive auxiliary view*** is constructed. These views, like the auxiliary views mentioned earlier, are also constructed from two views, either a primary auxiliary view and a secondary view or two secondary views. Their method of construction is the same as those used in both the primary auxiliary view and the secondary auxiliary view.

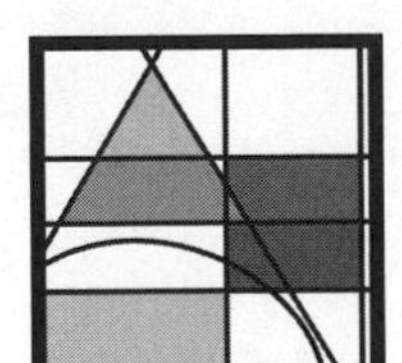

# advanced applications

## Using AutoLISP and DCL to Generate Auxiliary Views

One advantage to writing AutoLISP programs is the freedom to enhance and customize previously written programs by adding components as needed. For example, the programs presented in Chapter one can be further enhanced to incorporate the material covered in this chapter, producing a powerful tool for generating the different views of a drawing. By inserting just a few lines of code to both the AutoLISP program and the DCL program, the routine can be capable of producing auxiliary views of three-dimensional objects. To illustrate this point, copy the programs DG01.lsp and DG01.dcl from the folder marked DG01 to a folder marked DG03. Next, rename both programs from DG01.??? to DG03.???. After this has been completed, open the file DG03.lsp and insert the following syntax as indicated.

Proceeding the syntax

```
(if (= lef "1")
      (progn
        (command "vpoint" "-1,0,0" "zoom" "e" "Zoom" ".7x")
        (command "modemacro" "Left Side View")
      )
    )
```

in the main section of the program, add the lines:

```
(if (= auxil "1")
      (progn
        (setq pt1 (getpoint "\nSelect origin of auxiliary view : "))
        (setq pt2 (getpoint pt1 "\nSelect endpoint of X axis : "))
        (setq pt3 (getpoint pt1 "\nSelect endpoint of y axis : "))
        (command "ucsfollow" "1")
        (command "ucs" "3p" pt1 pt2 pt3)
        (command "modemacro" "Auxiliary View")
        (command "ucsfollow" "0")
        (command "hide")
      )
    )
```

Next, following the syntax, `isolb (get_tile "isolb")` in the subroutine of the program DG03.lsp, insert the following:

```
auxil (get_tile "auxil")
```

After the changes have been made to the AutoLISP program, save the file and open the DCL program DG03.dcl, and after the following syntax:

```
: radio_button {

key = "isolb";
```

```
        label = "Isometric Bottom Left";
    }
insert the lines:
    : radio_button {
        key = "auxil";
        label = "Auxiliary View";
    }
```

Again, after completing the changes to the program, save the file. The AutoLISP program, when loaded and executed, should now provide the on-screen prompt illustrated in Figure 3–25.

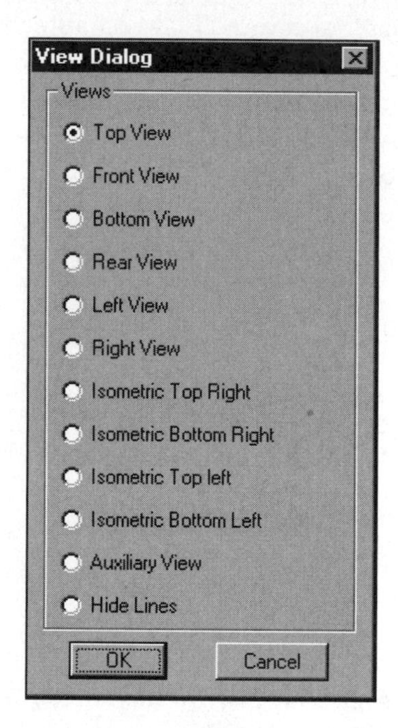


**Figure 3–25** *Chapter three dialog box.*

Once completed, the programs in their entirety should read as follows.

## AutoLISP Program

```
;;;*************************************************************************
;;;
;;;    Program Name: DG03.lsp
;;;
;;;    Program Purpose: This program allows the user to change their point
;;;                     of view in.  It can also be used in the creation of
;;;                     multiview drawing, by allowing the user to change
;;;                     the viewpoint in a view port, to one of the six
;;;                     principle views.  It is a revised version of the
;;;                     program featured in chapter one.  The revised version
;;;                     now includes auxiliary view creation capability.
;;;
;;;    Program Date: 10/25/98
;;;
```

```
;;;    Written By: James Kevin Standiford
;;;
;;;*************************************************************************
;;;*************************************************************************
;;;
;;;                  Main Program
;;;
;;;*************************************************************************
(defun c:cview ()
  (setvar "cmdecho" 0)
  (setqdcl_id4
        (load_dialog
          "c:/windows/desktop/geometry/student/autolisp programs/DG03.dcl"
        )
  )
  (if (not (new_dialog "Dv" dcl_id4))
    (exit)
  )
  (set_tile "top" "1")
  (action_tile "accept" "(vie) (done_dialog)")
  (start_dialog)
  (if (= vi 1)
    (progn
      (if (= top "1")
       (progn
         (command "vpoint" "0,0,1" "zoom" "e" "Zoom" ".7x")
         (command "modemacro" "Top View")
       )
      )
      (if (= bot "1")
       (progn
         (command "vpoint" "0,0,-1" "zoom" "e" "Zoom" ".7x")
         (command "modemacro" "Bottom View")
       )
      )
      (if (= fro "1")
       (progn
         (command "vpoint" "0,-1,0" "zoom" "e" "Zoom" ".7x")
         (command "modemacro" "Front View")
       )
      )
      (if (= rea "1")
       (progn
         (command "vpoint" "0,1,0" "zoom" "e" "Zoom" ".7x")
         (command "modemacro" "Rear View")
       )
      )
      (if (= rig "1")
       (progn
         (command "vpoint" "1,0,0" "zoom" "e" "Zoom" ".7x")
         (command "modemacro" "Right Side View")
       )
      )
```

```
      (if (= lef "1")
       (progn
         (command "vpoint" "-1,0,0" "zoom" "e" "Zoom" ".7x")
         (command "modemacro" "Left Side View")
       )
      )
      (if (= auxil "1")
       (progn
         (setq pt1 (getpoint "\nSelect origin of auxiliary view : "))
         (setq pt2 (getpoint pt1 "\nSelect endpoint of X axis : "))
         (setq pt3 (getpoint pt1 "\nSelect endpoint of y axis : "))
         (command "ucsfollow" "1")
         (command "ucs" "3p" pt1 pt2 pt3)
         (command "modemacro" "Auxiliary View")
         (command "ucsfollow" "0")
         (command "hide")
       )
      )
      (if (= isort "1")
       (progn
         (command "vpoint" "1,1,1" "zoom" "e" "Zoom" ".7x" "hide")
         (command "modemacro" "Isometric Right Top View")
       )
      )
      (if (= isorb "1")
       (progn
         (command "vpoint" "1,1,-1" "zoom" "e" "Zoom" ".7x" "hide")
         (command "modemacro" "Isometric Right Bottom View")
       )
      )
      (if (= isolt "1")
       (progn
         (command "vpoint" "1,-1,1" "zoom" "e" "Zoom" ".7x" "hide")
         (command "modemacro" "Isometric Left Top View")
       )
      )
      (if (= isolb "1")
       (progn
         (command "vpoint" "1,-1,-1" "zoom" "e" "Zoom" ".7x" "hide")
         (command "modemacro" "Isometric Left Bottom View")
       )
      )
      (if (= hid "1")
       (progn
         (command "hide")
         (command "modemacro" "Hide Line")
       )
      )
    )
  )
  (setq vi 0)
  (princ)
)
```

```
(defun vie ()
  (setqtop   (get_tile "top")
       bot   (get_tile "bot")
       lef   (get_tile "lef")
       rig   (get_tile "rig")
       rea   (get_tile "rea")
       fro   (get_tile "fro")
       isort (get_tile "isort")
       isorb (get_tile "isorb")
       isolt (get_tile "isolt")
       isolb (get_tile "isolb")
       auxil (get_tile "auxil")
       hid   (get_tile "hid")
       vi    1
  )
)
(princ "\nTo excute enter cview at the command prompt ")
(princ)
```

## Dialog Control Language Program

```
//%%%%%%%%%%%%%%%%%%%%%%%%%%%%%%%%%%%%%%%%%%%%%%%%%%%%%%%%%%%%%%%%%%%%%%%%%
//
//      Activates dialog box
//
//      Descriptive Geometry Chapter 3 DCL File Auxiliary Views
//
//
//
//%%%%%%%%%%%%%%%%%%%%%%%%%%%%%%%%%%%%%%%%%%%%%%%%%%%%%%%%%%%%%%%%%%%%%%%%%
Dv : dialog {
label = "View Dialog";
     : boxed_column {
       label = "Views";
       children_fixed_width = true;
       children_alignment = left;
     : radio_column {
     : radio_button {
       key = "top";
       label = "Top View";
     }
     : radio_button {
       key = "fro";
       label = "Front View";
     }
     : radio_button {
       key = "bot";
       label = "Bottom View";
     }
     : radio_button {
       key = "rea";
       label = "Rear View";
     }
```

```
      : radio_button {
        key = "lef";
        label = "Left View";
      }
      : radio_button {
        key = "rig";
        label = "Right View";
      }
      : radio_button {
        key = "isort";
        label = "Isometric Top Right";
      }
      : radio_button {
        key = "isorb";
        label = "Isometric Bottom Right";
      }
      : radio_button {
        key = "isolt";
        label = "Isometric Top left";
      }
      : radio_button {
        key = "isolb";
        label = "Isometric Bottom Left";
      }
      : radio_button {
        key = "auxil";
        label = "Auxiliary View";
      }
      : radio_button {
        key = "hid";
        label = "Hide Lines";
      }
    }
   }
      is_default = true;
      ok_cancel;
}
```

# Review Questions

Answer the following questions on a separate sheet of paper. Your answers should be as complete as possible.

1. How do auxiliary views differ from the six principal views?
2. In the mechanical desktop, how are auxiliary views created?
3. What are the two methods used to create an auxiliary view? How do these methods differ?
4. What are successive auxiliary views and when are they used?
5. What is the difference between a primary auxiliary view and a secondary auxiliary view? When are they used?
6. What is the UCSFOLLOW system variable used for?
7. When is an auxiliary view used?
8. What is an auxiliary plane? How does this differ from a normal plane?
9. Describe how a non-parametric solid is transformed into a parametric solid in mechanical desktop.
10. Describe how an auxiliary view of a non-parametric solid model is created using the AutoCAD MVIEW, UCS and UCSFOLLOW commands.
11. Describe how an auxiliary view of a non-parametric solid is created using the AutoCAD MVIEW and VPOINT commands.

# Review Exercises

## Two-Dimensional Primary Auxiliary Views

Complete the missing views in the following drawings. All problems marked with an * can be found on the student CD-ROM.

**#1**

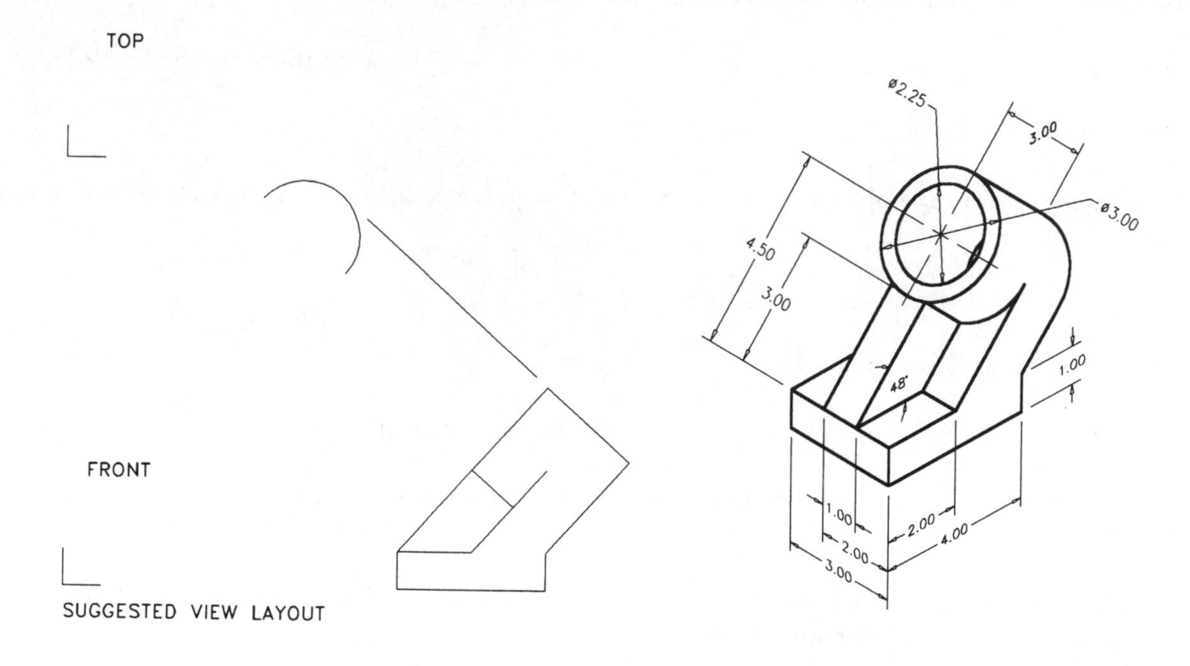


**#2**

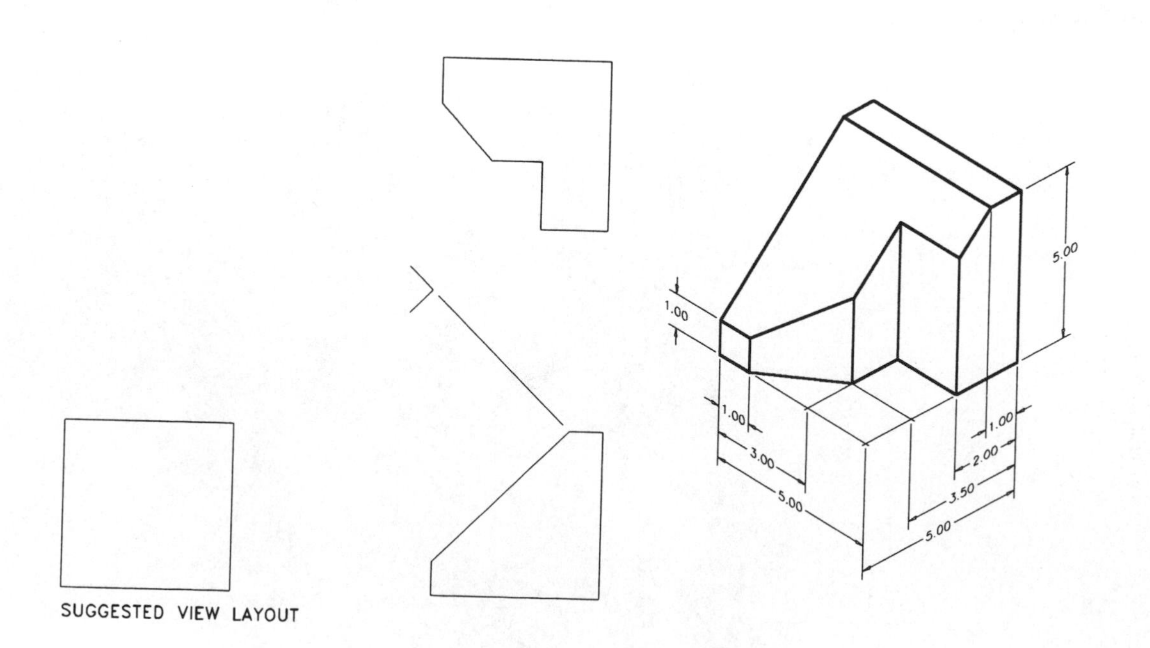

## Two-Dimensional Primary, Secondary and Successive Auxiliary Views

Given the following orthographic views, construct the indicated auxiliary views of the following parts.

***#3**

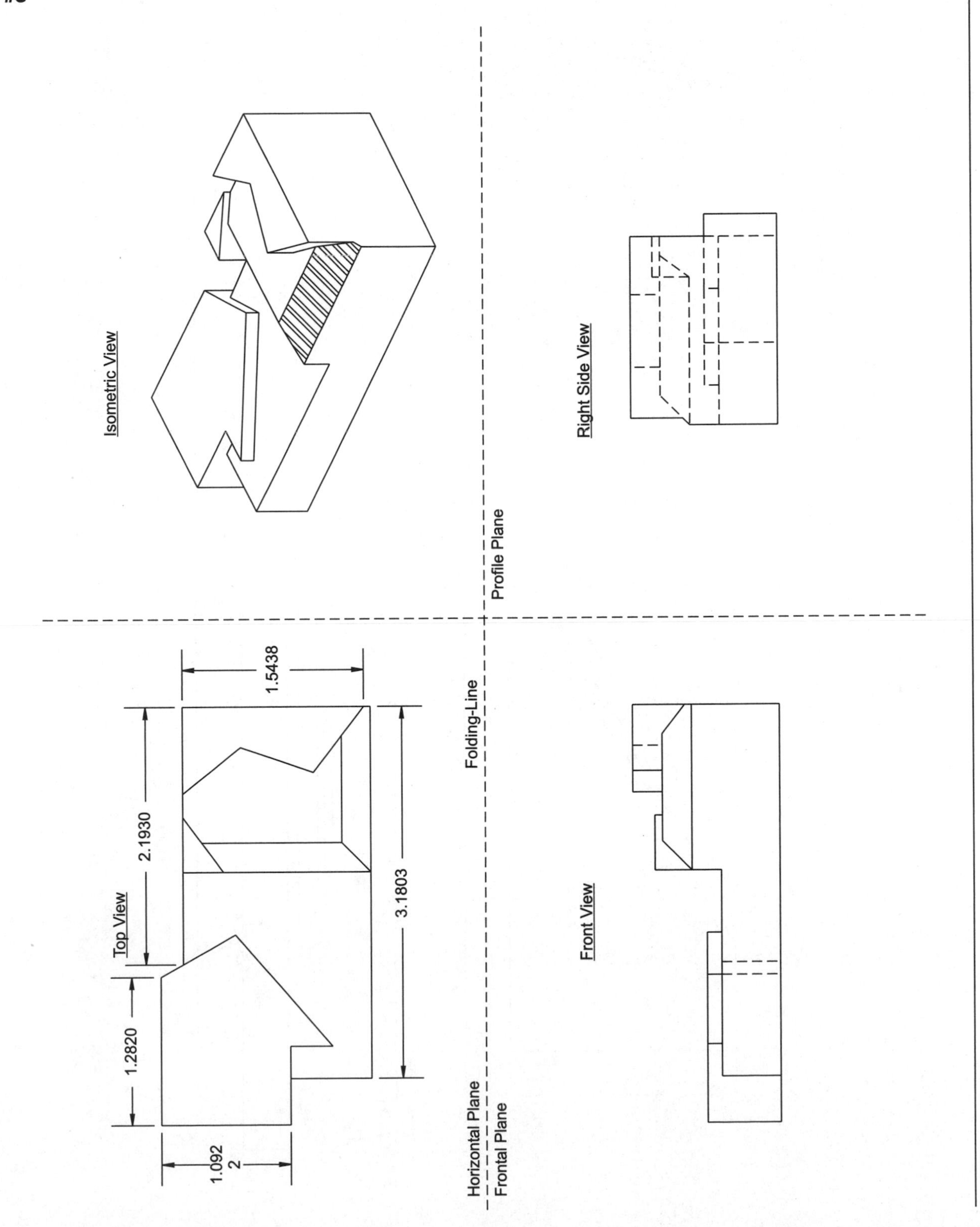

*#4

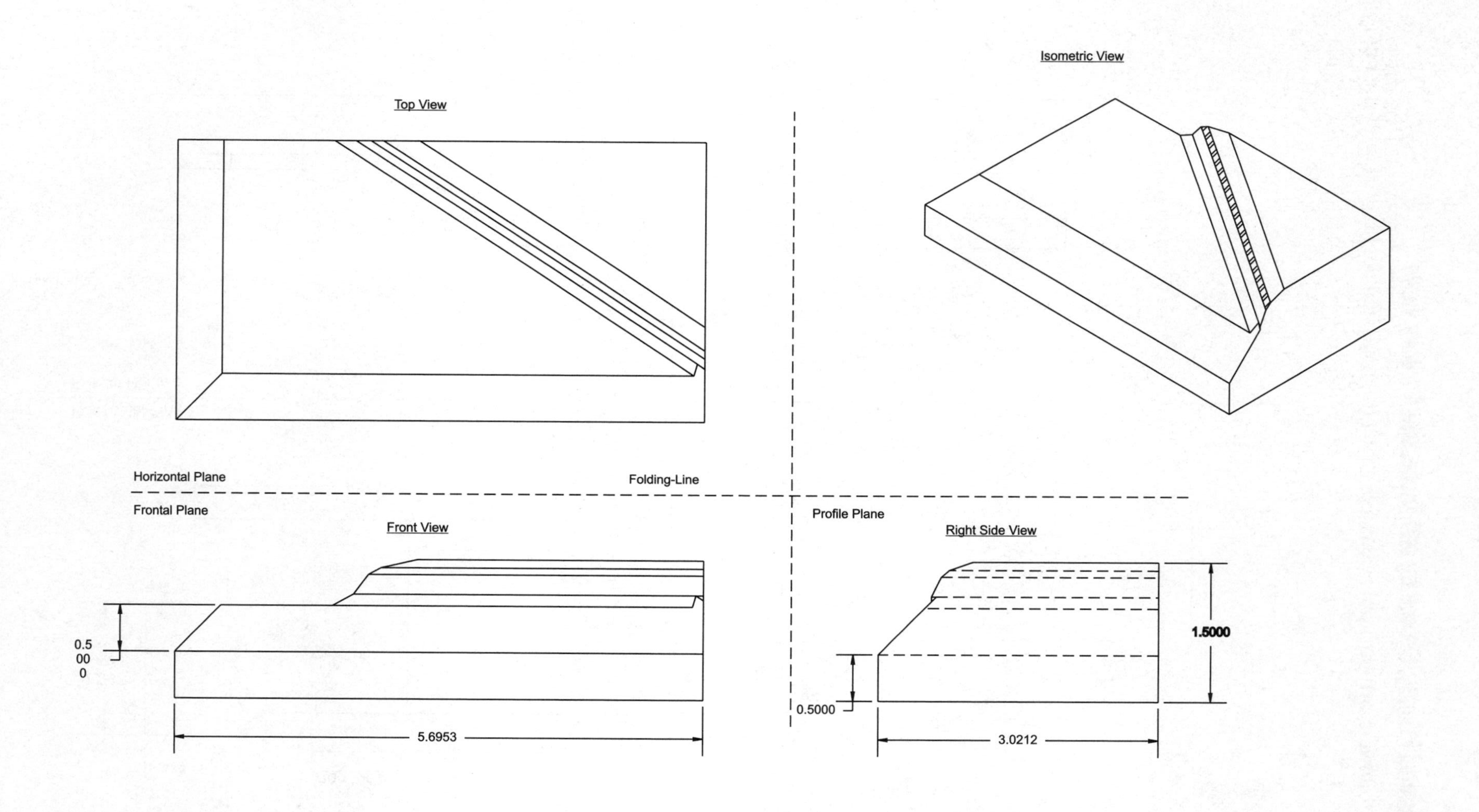
Isometric View
Top View
Horizontal Plane
Folding-Line
Frontal Plane
Front View
Profile Plane
Right Side View
0.5
00
0
5.6953
0.5000
3.0212
1.5000

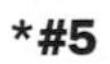

*#5

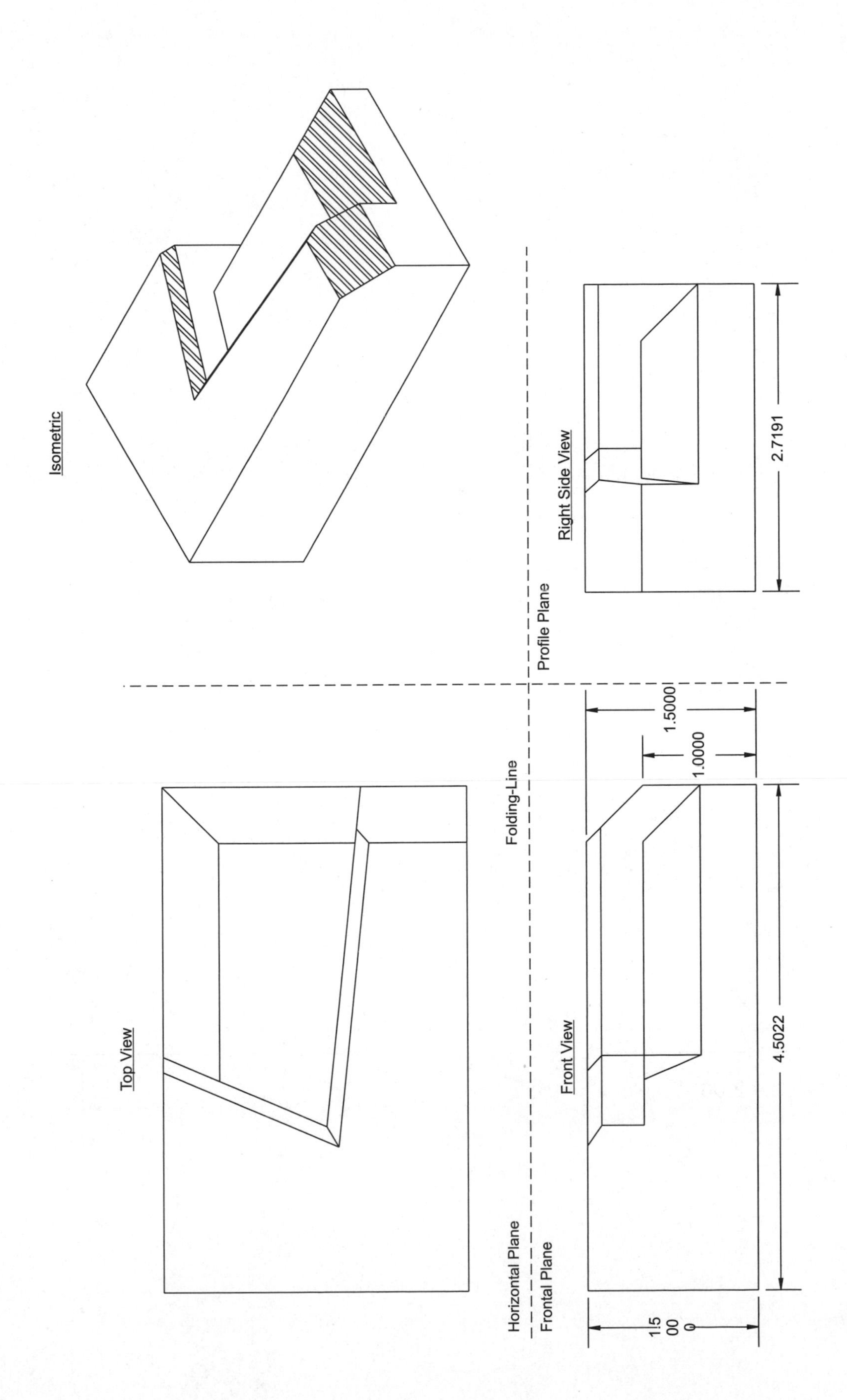
Isometric
Top View
Right Side View
Front View
Profile Plane
Folding-Line
Horizontal Plane
Frontal Plane
2.7191
1.5000
1.0000
4.5022
1.5000

*#6

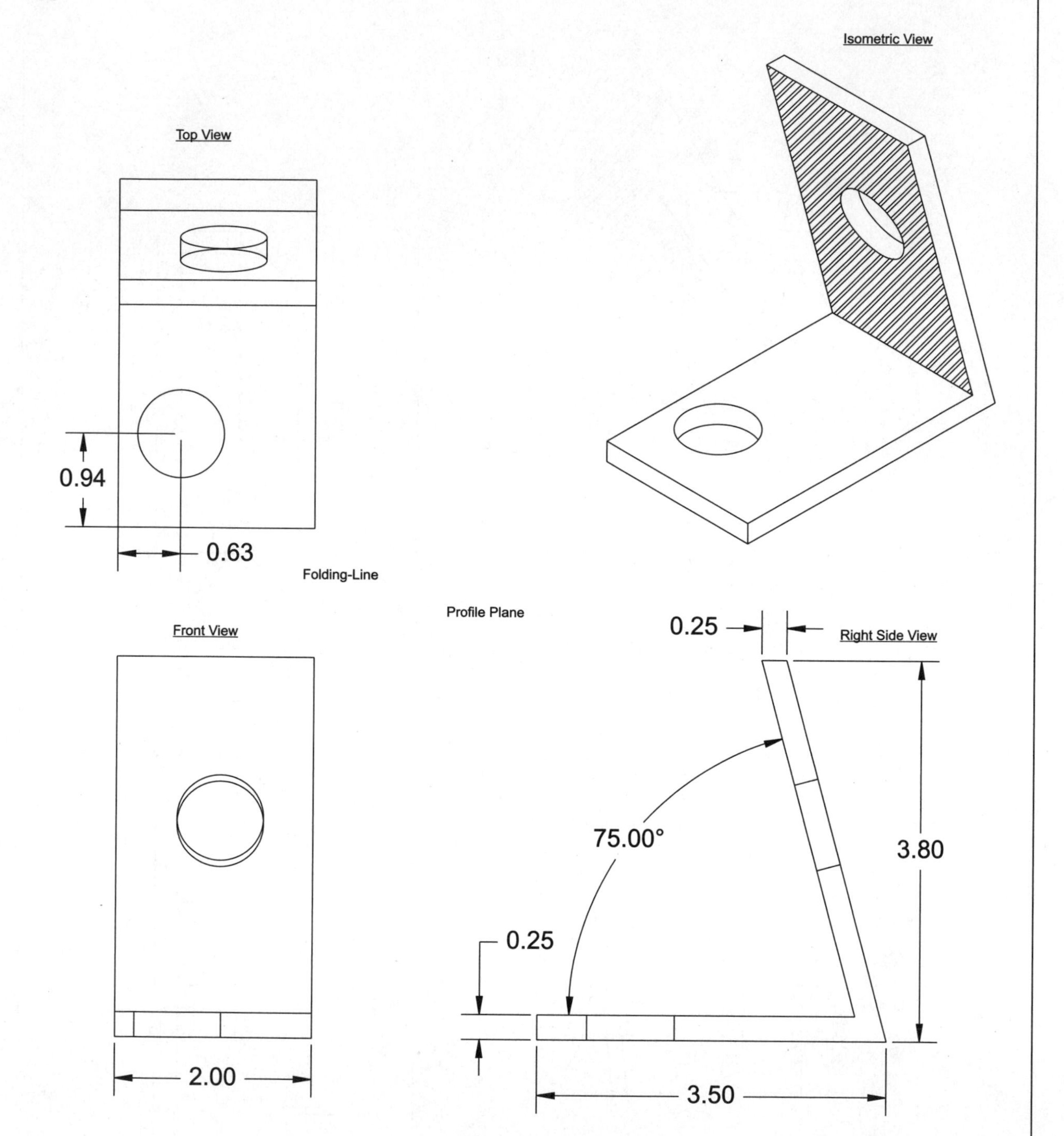

Isometric View
Top View
0.94
0.63
Folding-Line
Profile Plane
Front View
0.25
Right Side View
75.00°
3.80
0.25
2.00
3.50

***#7**

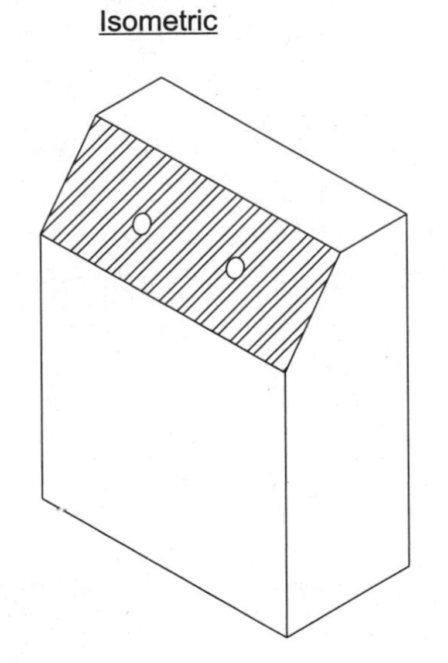


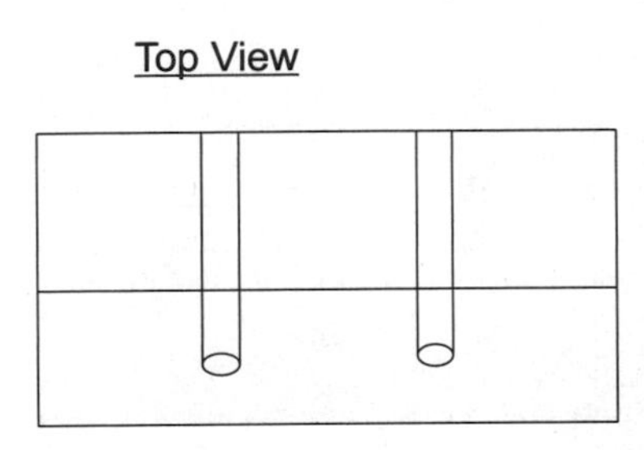


Horizontal Plan

Folding-Line

Frontal Plane

Profile Plane

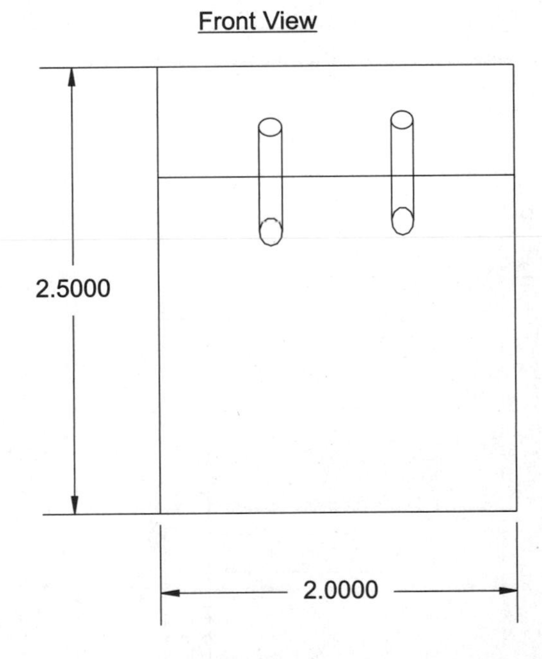


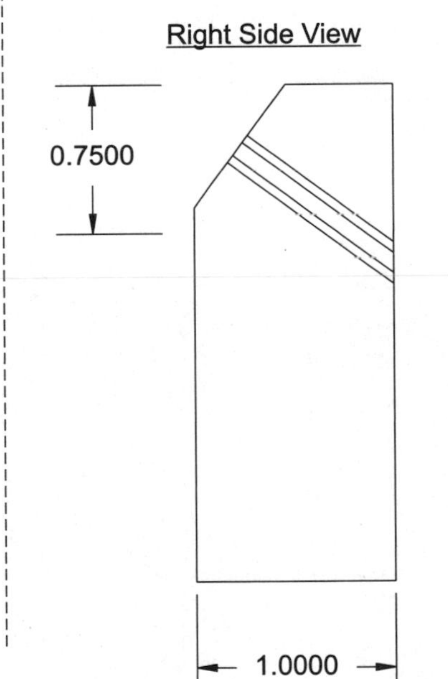

*#8

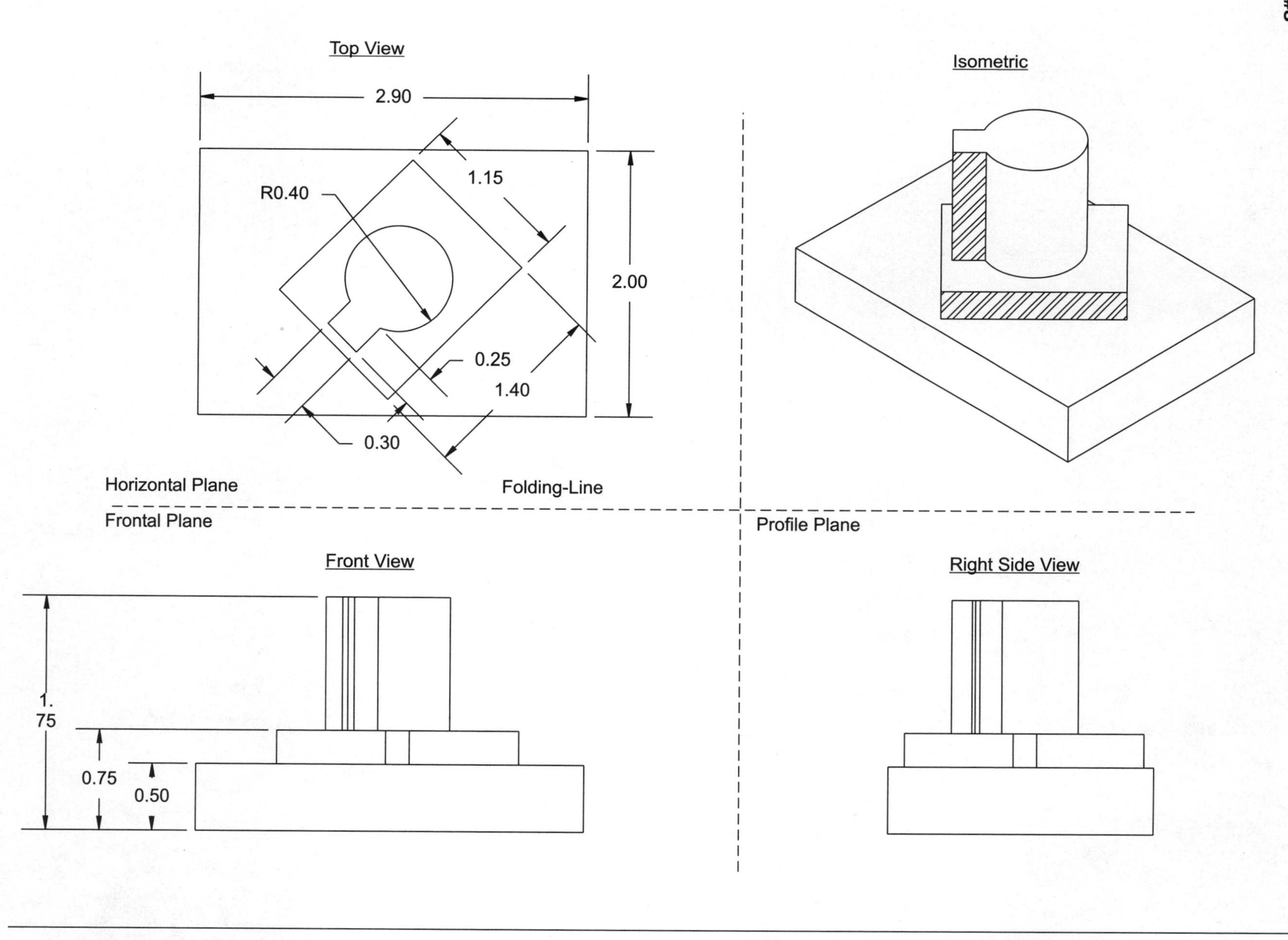
Top View
2.90
2.00
1.15
R0.40
0.25
1.40
0.30
Isometric
Horizontal Plane
Frontal Plane
Folding-Line
Profile Plane
Front View
Right Side View
1.
75
0.75
0.50

## Three-Dimensional Primary Auxiliary Views

First, using the following sketches, create a three-dimensional non-parametric solid model. Then, using paper space, create a multiview drawing of the part along with the necessary primary auxiliary view.

**#9**

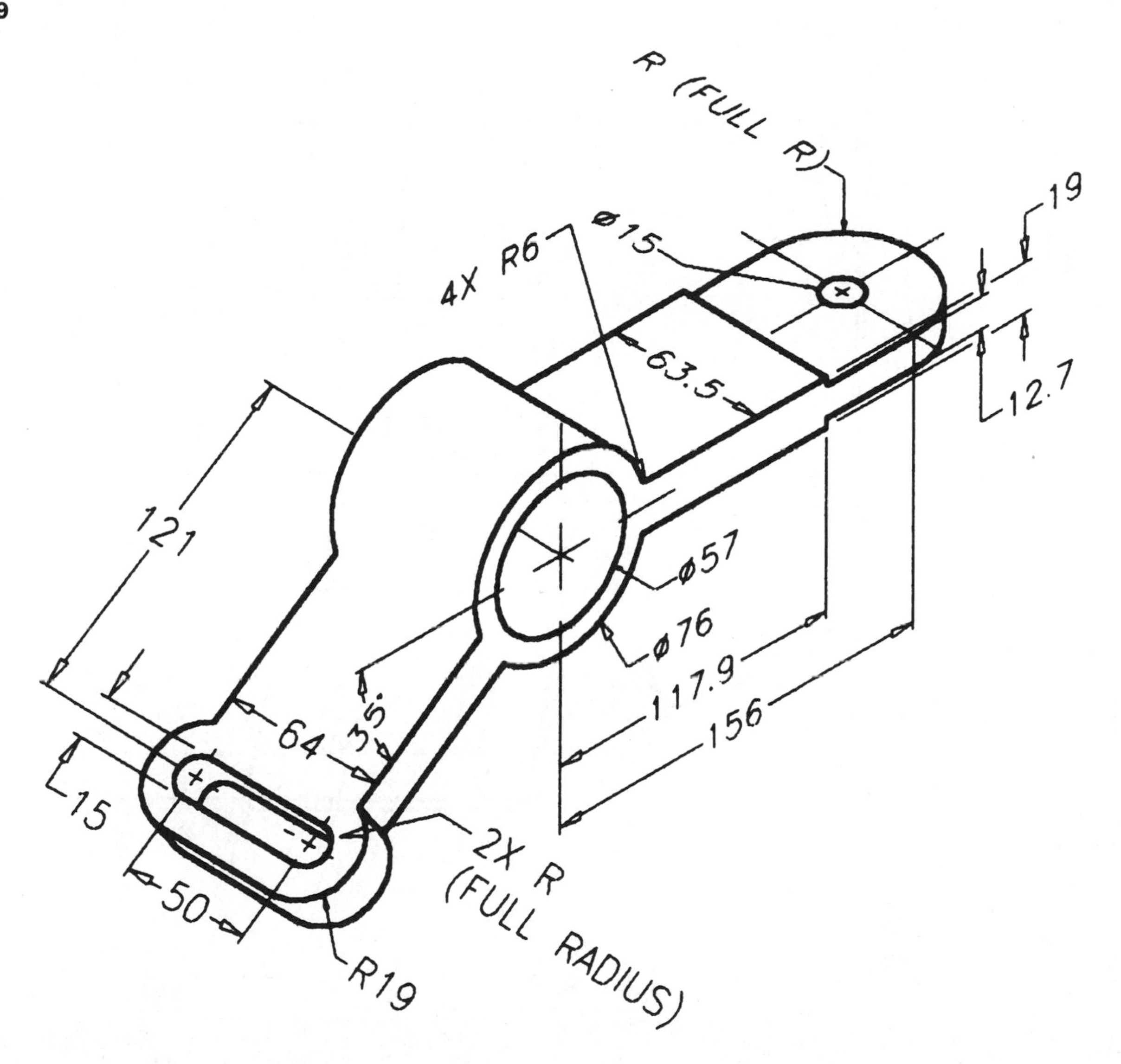

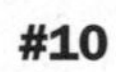

**#10**

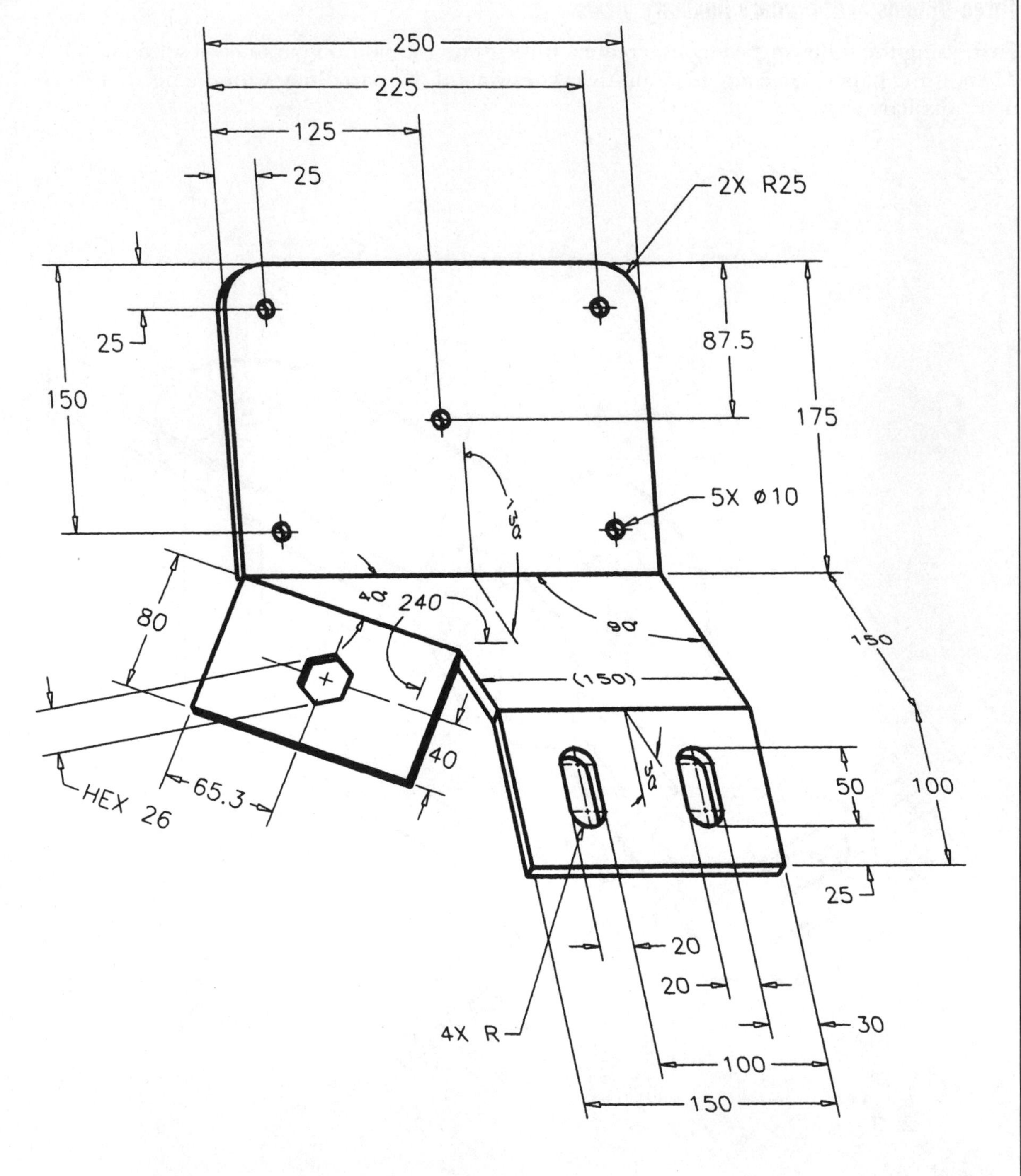
250
225
125
25
2X R25
25
150
87.5
175
5X ø10
130°
40°
240
90°
80
(150)
150
40
HEX 26
65.3
30°
50
100
25
20
20
30
4X R
100
150

# Planes

## Objectives

***Upon completion of this chapter the student will be able to do the following:***

- *Describe the difference between a surface and a plane.*
- *List the four ways that the boundaries of a plane may be defined.*
- *List the three types of planes encountered in descriptive geometry.*
- *Use AutoCAD to determine if two objects reside on the same plane.*
- *Determine the location of a point on a plane.*
- *Determine the location of a line on a plane.*
- *Locate the piercing point of a line and a plane using the auxiliary view and cutting plane methods.*
- *Use the cutting plane method in AutoCAD to determine the intersection of two planes.*
- *Locate the piercing point using the* POINT *command and* APP OSNAP *option.*
- *Locate the intersection of two planes.*
- *Determine the angle between two planes.*

## Key Words and Terms

**Normal plane**
**Oblique plane**
**Inclined plane**
**Frontal plane**
**Horizontal plane**
**Profile plane**
**Auxiliary view method**
**Cutting plane**
**Dihedral angle**
**Piercing point**
**Plane**
**Surface**
**Sketch planes**
**Work planes**

## Introduction to Planes

In Chapter one it was determined that descriptive geometry is the branch of mathematics that precisely describes three-dimensional objects by projecting their essential reality onto a two-dimensional surface such as paper or a computer screen. Until now these objects have been described using either points, lines, or a combination of the two, and illustrated using a variety of different projection techniques ranging from multiview drawings to isometric projections. In reality, all objects are composed of more than just lines and points, they also contain surfaces and/or planes. To fully describe an object, all attributes associated with it must be examined and interpreted. It is for this reason that the study of planes and their interactions are an integral part of descriptive geometry.

When dealing with planes, the term 'surface' is often used or implied. A plane and a surface may or may not be the same, and care must be taken when using these terms so that they are used in the correct context. By definition, a ***surface*** is a bound region in space that is constrained by two dimensions only. A surface has length and breadth, but does not contain thickness. A surface has the ability to show contours. For example, the outer portion of the earth's upper crust is referred to as a surface. In short, a surface can assume any shape and is most often used to describe the exterior of an object. A ***plane***, on the other hand, is defined as a non-curved region in space that may or may not extend indefinitely in two directions. A plane can have length and breadth but will never contain a thickness. Any two points on a plane may be connected with a straight line segment, and the entire segment will be contained on that plane. Also, any two lines on a plane will either intersect or be parallel. When a line is perpendicular to a plane it will be perpendicular to every line contained on that plane, as well as any line parallel to the plane. When a line is parallel to a plane it will be parallel to every line contained on the plane.

## Defining a Plane

Because a plane is a concept and not an entity, the very presence of a plane is established merely by defining the boundaries of that plane. There are four ways in which a plane's boundaries can be secured. They are: a line and a point, three points, two intersecting lines, and two parallel lines.

### Line and a Point

The boundaries of a plane can be determined using a line and a point when the following criteria is met. First, the line and the point must both reside on the plane that they are defining. Second, the point cannot reside on the line. If the point is on the line then only one possible boundary can be established. Once it is determined that these conditions have been satisfied, then the boundaries are obtained by drawing line segments that connect the endpoints of the line to the point (see Figure 4–1).

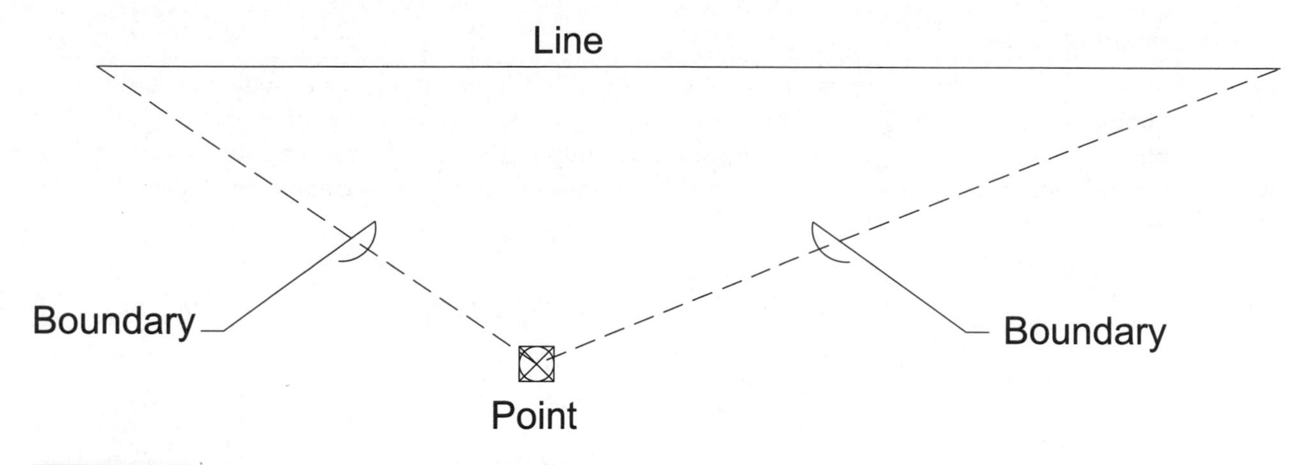


Figure 4–1 *Plane formed by a line and a point.*

### Three Points

The boundaries of a plane can be established by using the three-point method when the following conditions exist. First, all three points must reside on the plane that they are defining. Second, it must not be possible to connect all three points with a single, straight, line segment. If all three points can be connected with a straight line, then again, only one possible boundary of the plane has been established. When both of these factors are present, the boundaries are secured by drawing three line segments connecting the three points. This is illustrated in Figure 4–2.

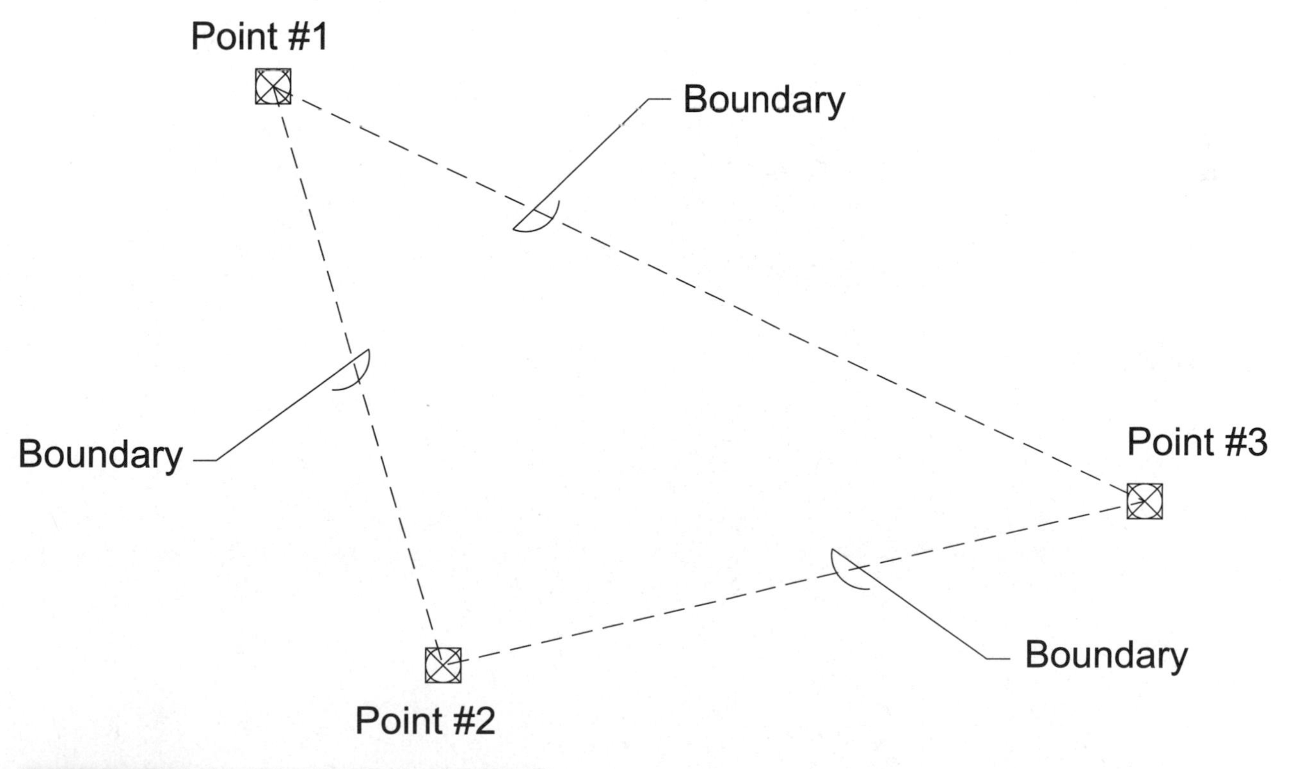


Figure 4–2 *Plane formed by three points.*

## Two Intersecting Lines

Two intersecting lines or line segments drawn at any length and intersecting at any angle may be used to define the boundaries of a plane if both lines reside on the plane that they are defining. Once this fact has been established, drawing line segments connecting the endpoints of the intersecting line segments produces the boundaries. Three examples of planes defined by intersecting line segments are depicted in Figures 4–3, 4–4, and 4–5.

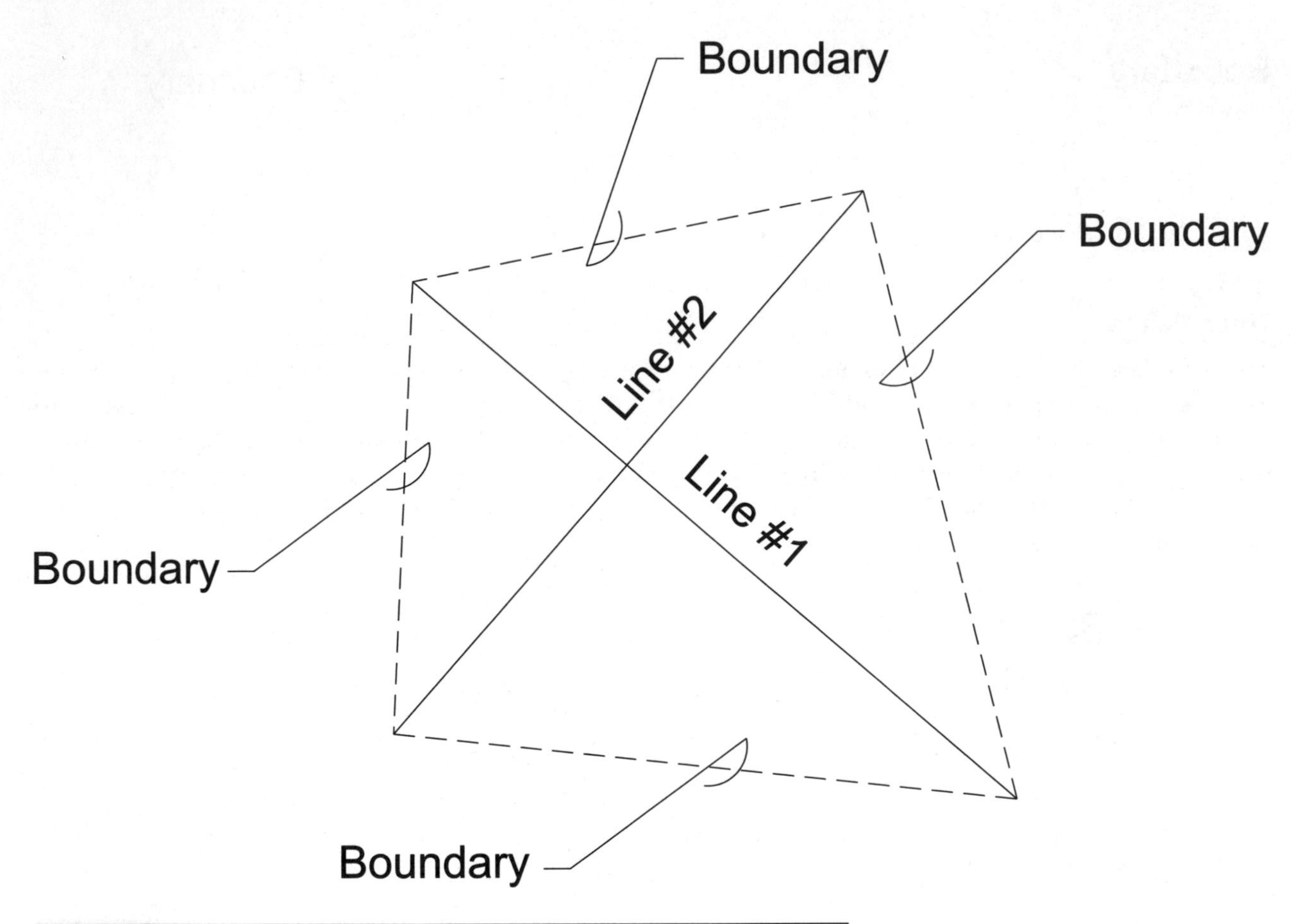


**Figure 4–3** *Plane formed by two line segments intersecting at 90°.*

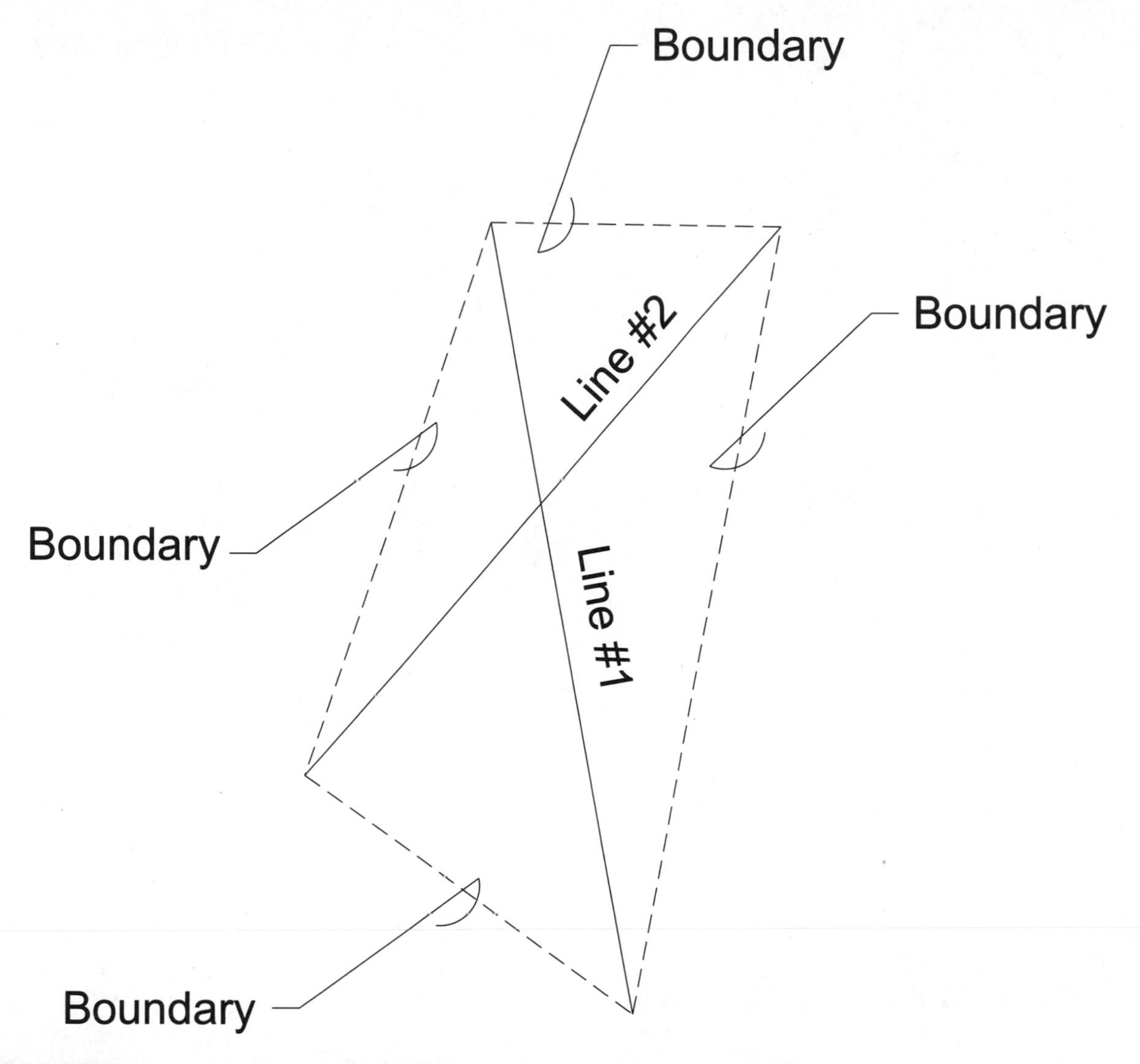


**Figure 4-4** *Plane formed by two line segments intersecting at 51.4368°.*

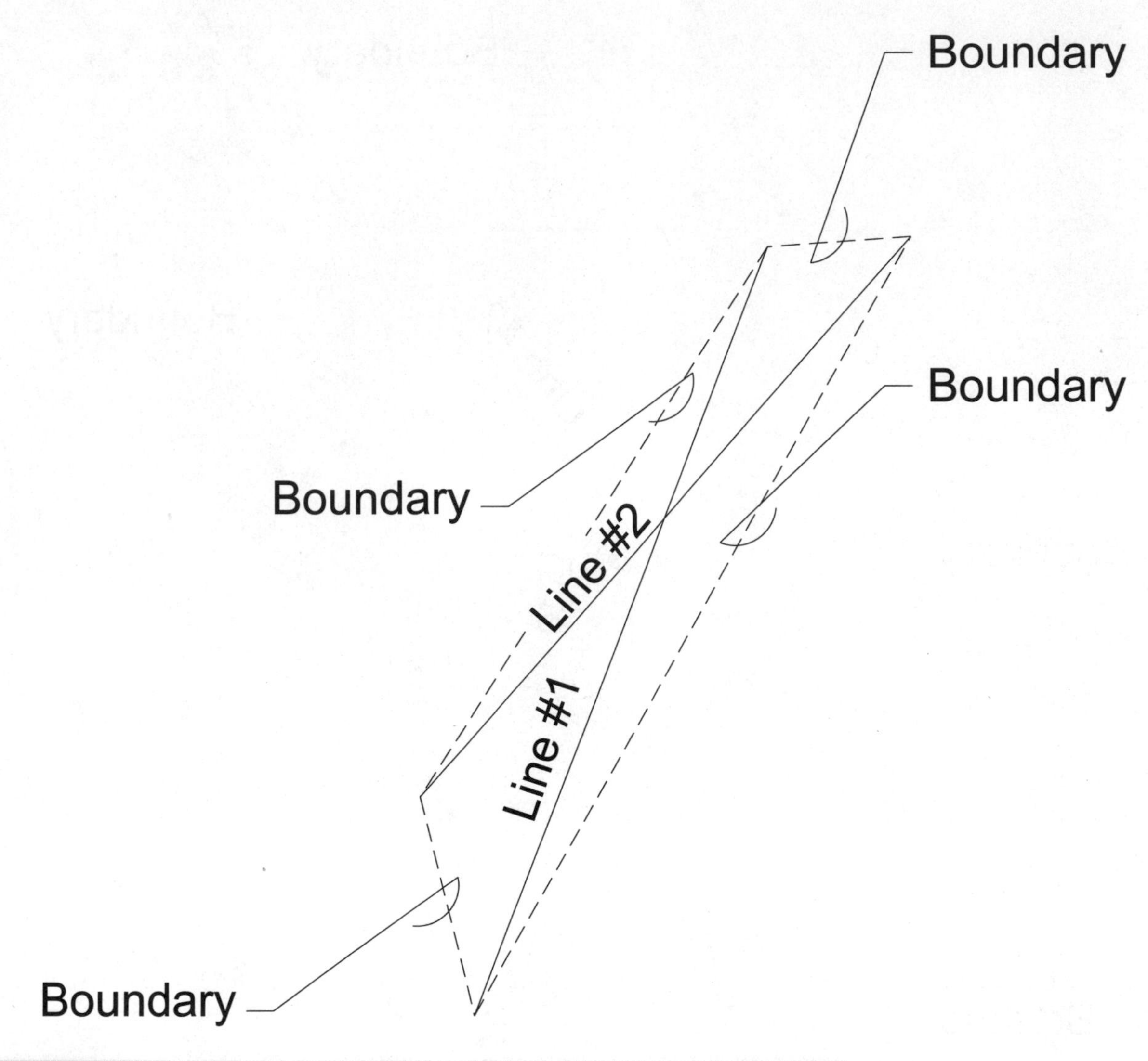


**Figure 4–5** *Plane formed by two line segments intersecting at 20.2355°.*

### Two Parallel Lines

To define a plane using two parallel lines, both line segments must reside on the plane that they are defining. Once this has been determined, the boundaries are created by drawing line segments that connect the endpoints of the parallel lines. This is illustrated in Figure 4–6.

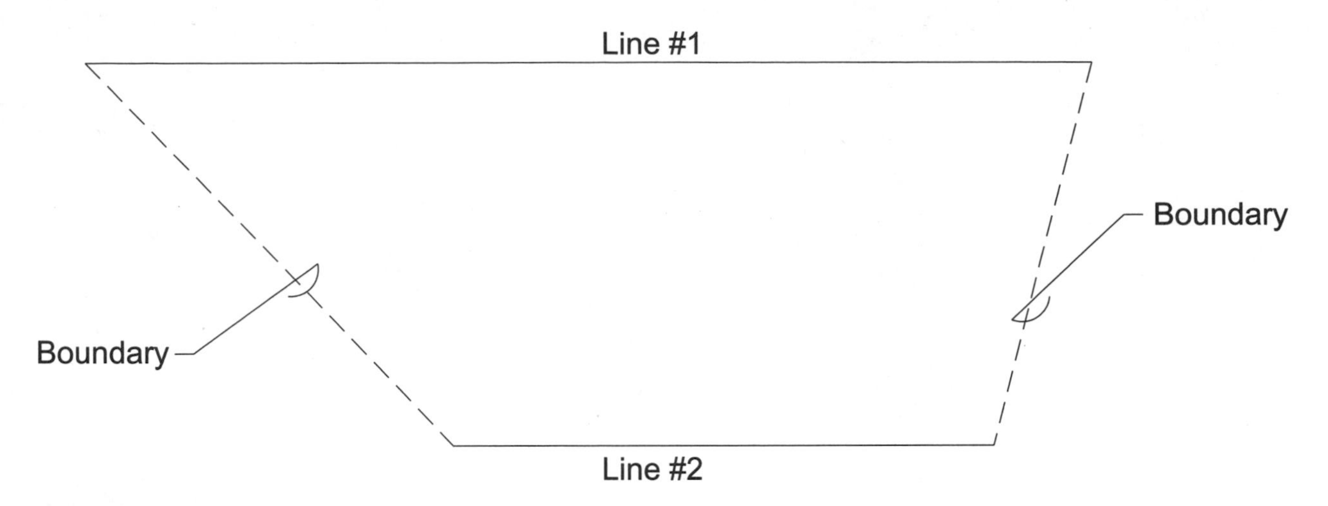


**Figure 4–6** *Plane formed by two parallel lines.*

## Types of Planes

After the boundaries of a plane have been determined, it is necessary to classify the plane. There are three main classifications of planes, with one of the classifications being broken down into three subcategories. By placing the planes into different classifications, a more clear and precise picture of the plane will emerge. The reason for the classification of a plane is similar to the reason for the classification of lines. Recall from Chapter two that if a line is not shown in true length in any of the principal views (horizontal, frontal, profile), then the line is considered to be oblique. To find the true length of this line, an auxiliary view must be constructed in which the line-of-sight is perpendicular to the line. Just by being able to classify the characteristics of an oblique line, it is quickly determined that an auxiliary view must be constructed before the line's true length can be obtained. Likewise, identifying the characteristics of a plane will give a better understanding of how to interact with that plane.

### Normal Planes

If a plane is parallel to one of the three principal planes of projection (frontal, horizontal, profile) then the plane is referred to as a ***normal plane.*** There are three subcategories of normal planes: horizontal, frontal, and profile. A normal plane that is perpendicular to the frontal projection plane and parallel to the horizontal projection plane is known as a horizontal plane. This plane will appear as a line (edge view) in the frontal projection plane, and its true size and shape will be shown in the horizontal projection plane. A normal plane that is perpendicular to the horizontal projection plane and parallel to the frontal projection plane is known as a frontal plane. This plane will appear as a line (edge view) in the horizontal projection plane and its true size and shape will be shown in the frontal projection plane (see Figures 4–7 and 4–8a). A plane that is parallel to the profile projection plane is known as a profile plane. This plane will appear in edge view in both the horizontal and frontal projection planes, and its true size and shape will only be shown in the profile projection plane (see Figure 4–8b).

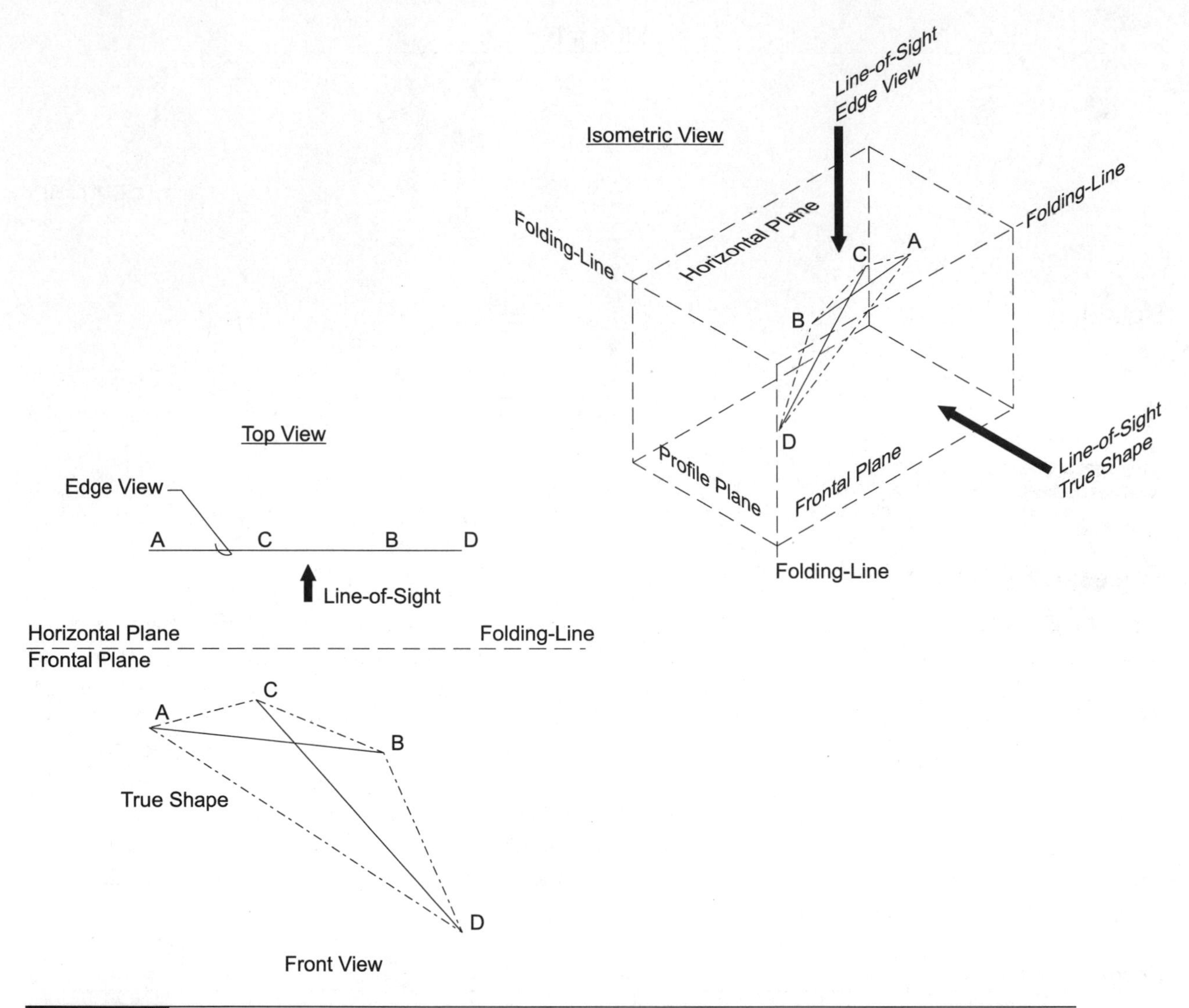


**Figure 4-7** *A normal frontal plane defined by two intersecting line segments. This plane is perpendicular to the horizontal projection plane and parallel to the frontal projection plane. The plane's true shape and size is revealed in the frontal projection plane.*

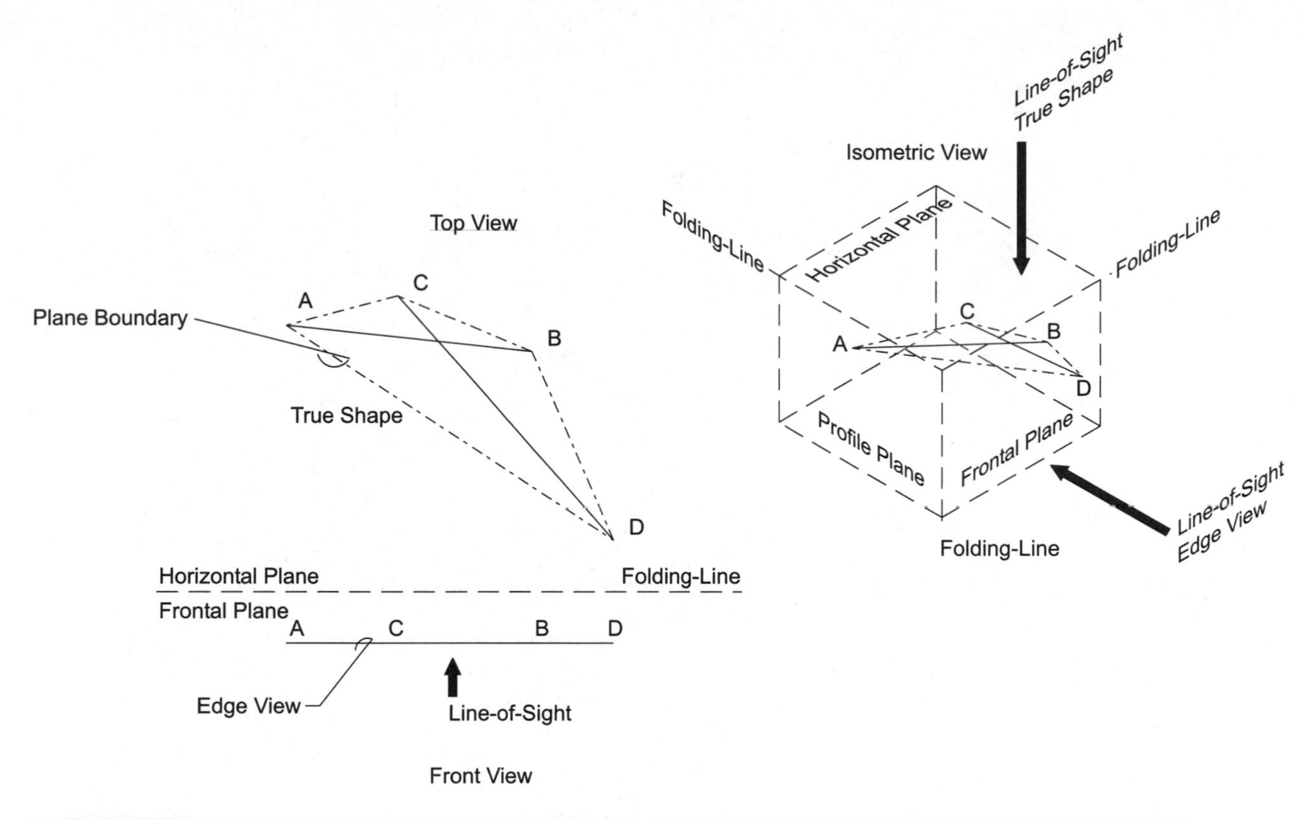


**Figure 4–8a** *A normal horizontal plane defined by two intersecting line segments. This plane is perpendicular to the frontal projection plane and parallel to the horizontal projection plane. The plane's true shape and size is revealed in the horizontal projection plane.*

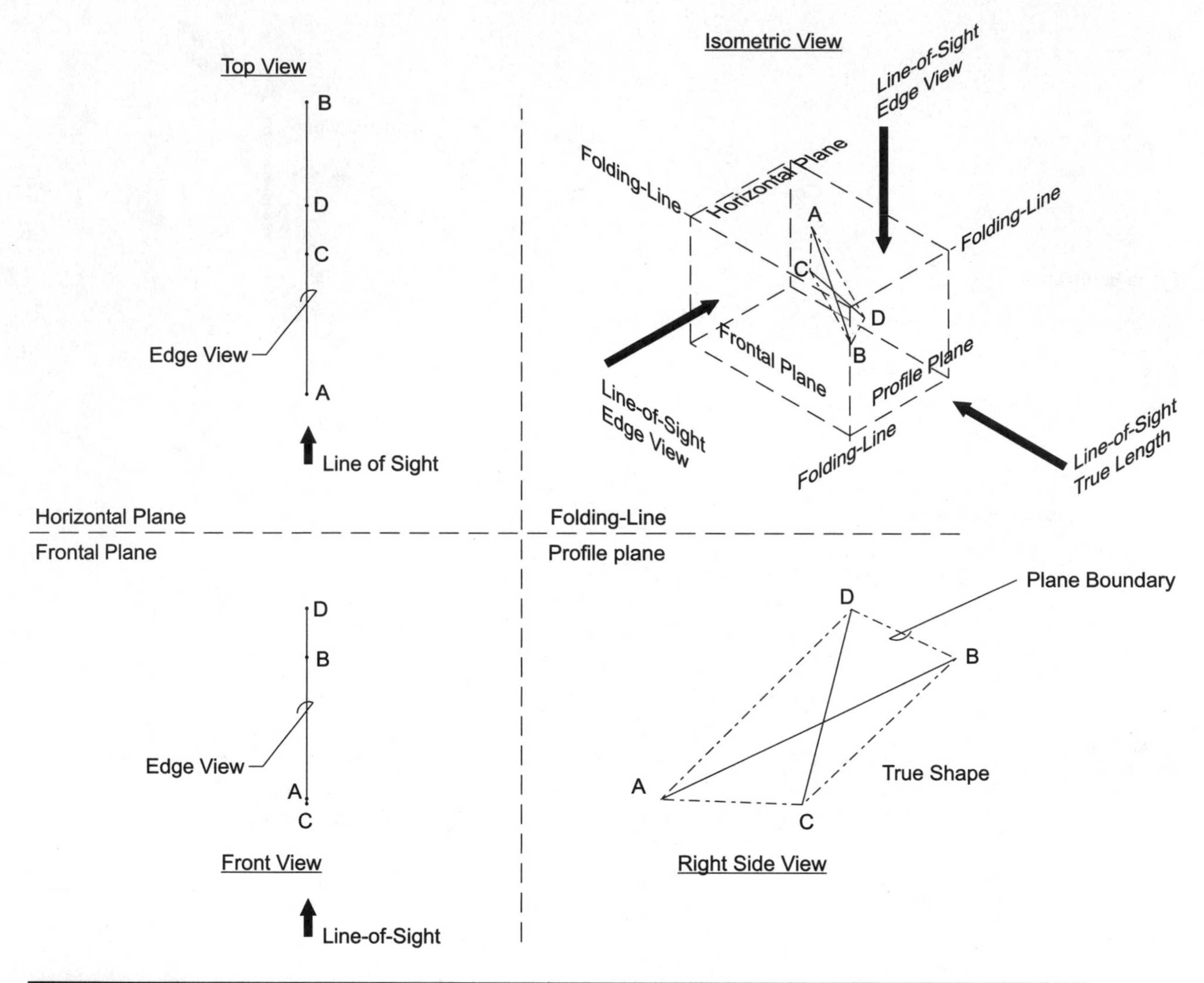


**Figure 4–8b** *A normal profile plane defined by two intersecting line segments. This plane is perpendicular to both the frontal and horizontal planes and parallel to the horizontal projection plane. The plane's true shape and size is revealed in the horizontal projection plane.*

## Inclined Planes

When a plane is not classified as a normal plane but appears as a line (edge view) in one of the principal views and distorted in the remaining two, then this plane is known as an ***inclined plane.*** The true shape and size of an inclined plane can only be found by constructing one or more auxiliary views. This is because the surface of an inclined plane will not be parallel to any of the principal plane's surfaces (see Figure 4–9).

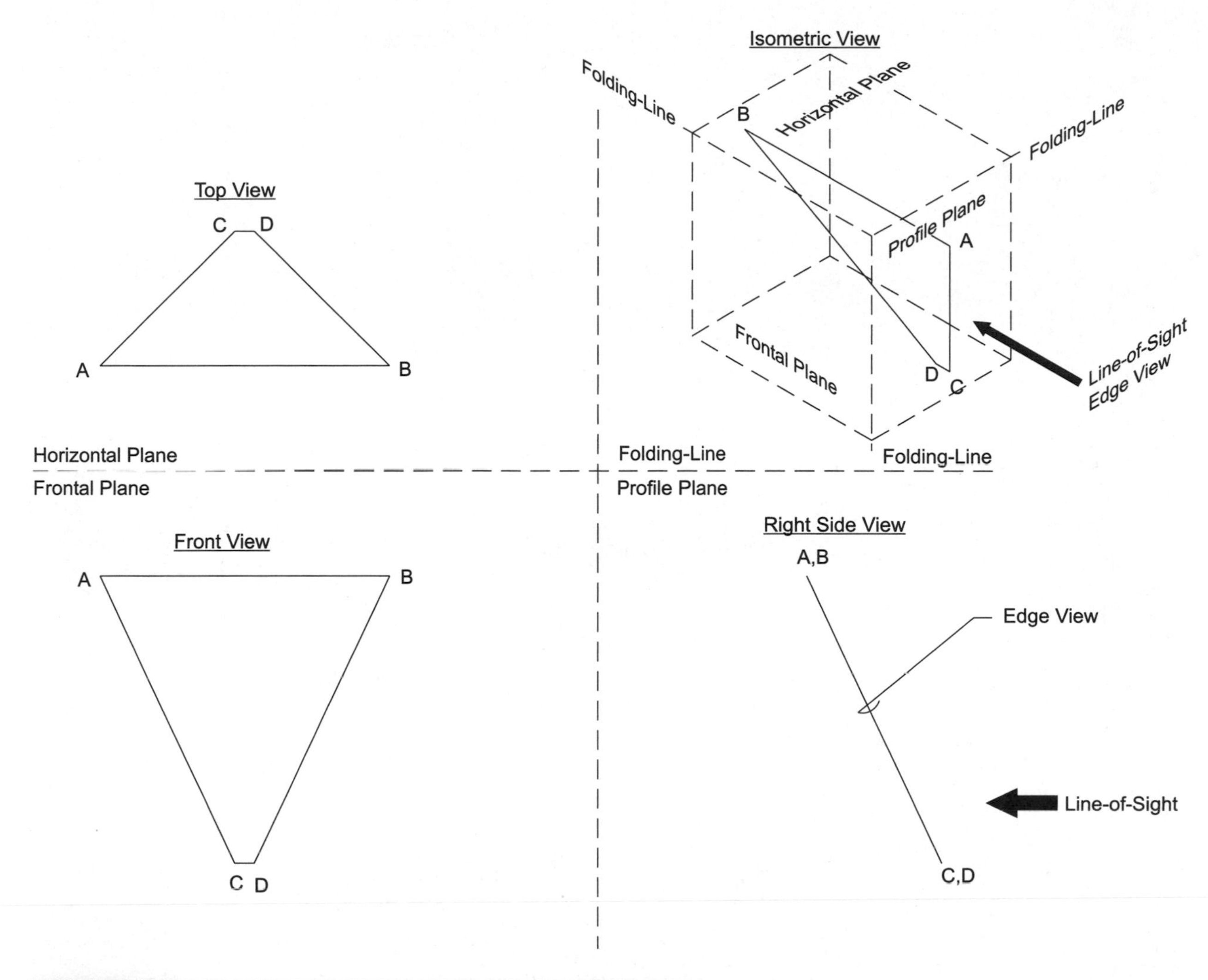


**Figure 4–9** *An inclined plane defined by a line and a point.*

## Oblique Planes

When a plane does not appear in edge view in any of the three principal views then it is known as an ***oblique plane.*** Just like the inclined plane, the true shape and size of an oblique plane can only be determined in one or more auxiliary views (see Figure 4–10).

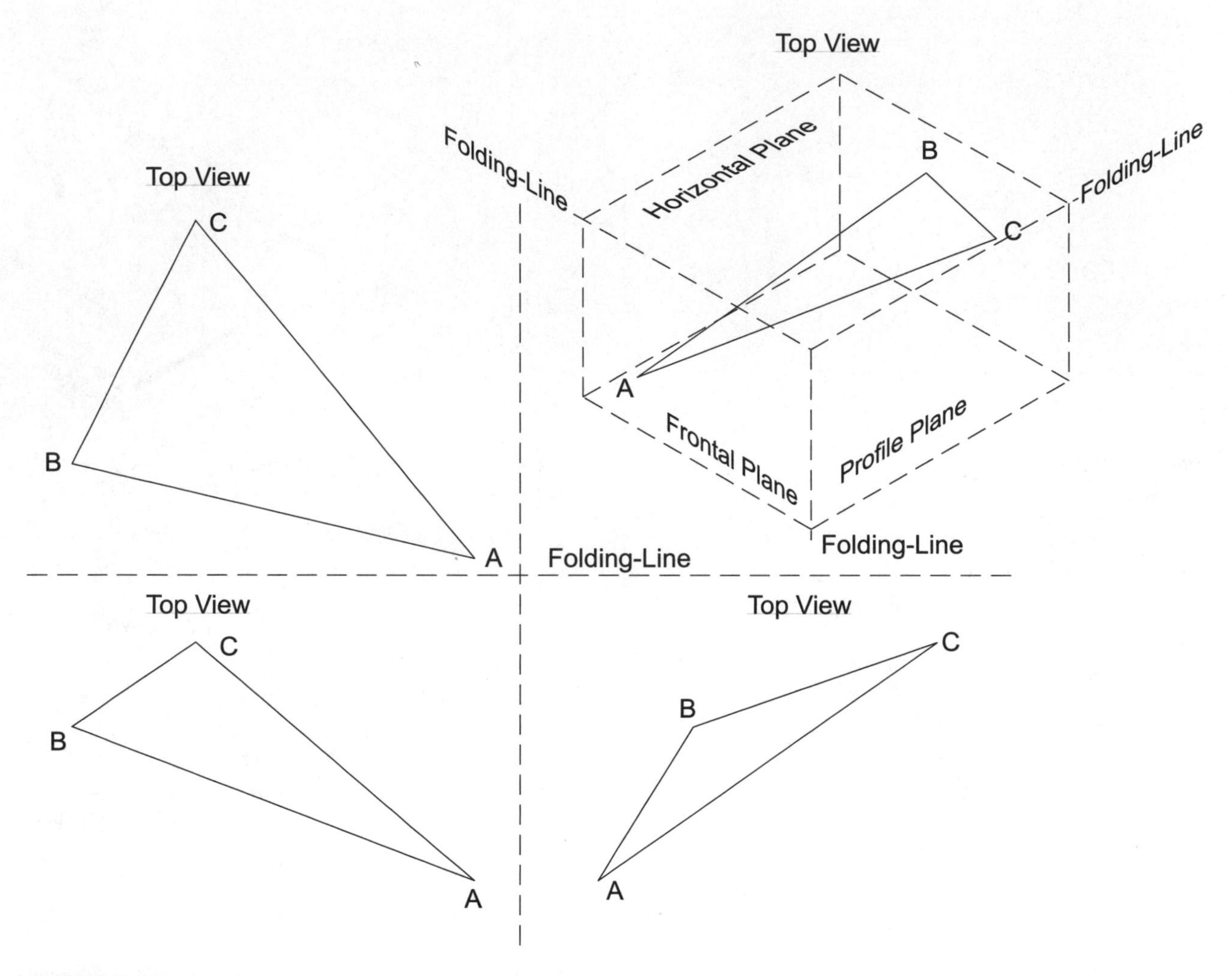


**Figure 4–10** *An oblique plane defined by three points.*

## Relating Planes to AutoCAD

The study and usage of planes has implications that extend far beyond the boundaries of descriptive geometry. In fact, planes and their applications are a subject that the engineering professional is faced with in almost every phase of a project. The makers of AutoCAD recognized this need and incorporated the theories and principles of planes into their software. AutoCAD uses planes to construct, modify, and adjust almost every entity created in it. Even plug-in applications like Mechanical Desktop depend upon the location and positions of planes.

### XY Plane (AutoCAD)

The fact that AutoCAD is fundamentally structured by planes becomes obvious when examining the way entities are created in the software. For example, in AutoCAD both two- and three-dimensional objects are created by using entities such as lines, circles, and arcs to connect a series of points. These points are all comprised of the same three elements: X-, Y-, and Z-coordinates. For three-dimensional objects, the Z-coordinates of the points vary so that not all points share a common value. However, for two-dimensional objects the Z-coordinates will all be the same. In theory, the Z-coordinate is used to describe the position of an object along the Z-axis, but in reality it can also be used to describe the actual plane a surface or feature of an object rests upon. When an object is created in AutoCAD, it is created on the XY-plane. Theoretically, the XY-plane extends indefinitely in both directions along the X- and Y-axes, but does not contain a thickness. Without a thickness an infinite number of XY-planes could be stacked upon one another. When projected into a plan view, this would give the illusion that entities are either resting on the same plane or stacked directly on top of one another. In this instance, the Z-coordinate becomes necessary for clarity. If each object is assigned a separate Z-coordinate, and the line-of-sight is set to an angle other than 90° to the XY-plane, then each object can be seen individually. In Figure 4–11, two rectangles are drawn; the first is constructed with its lower left-hand corner at 4,5,2 and its upper right-hand corner at 8,8,2. The second rectangle is drawn with its lower left-hand corner at 2,5,4 and its upper right-hand corner at 6,8,4. When the objects are completed, and the line-of-sight is set perpendicular to the XY-planes, the objects appear to be drawn on the same plane and on top of one another. By changing the line-of-sight, the true relationship between these objects emerges (see Figure 4–12).

Command: RECTANG (ENTER)
Chamfer/Elevation/Fillet/Thickness/Width/<First corner>: 4,5,2 *(Press ENTER to specify the location of the first corner of the rectangle to be constructed.)*
Other corner: 8,8 *(Press ENTER to specify the location of the second corner of the rectangle to be constructed. This corner is diagonal to the first corner selected.)*
Command: RECTANG (ENTER)
Chamfer/Elevation/Fillet/Thickness/Width/<First corner>: 2,5,4 *(Press ENTER to specify the location of the first corner of the rectangle to be constructed.)*
Other corner: 6,8 *(Press ENTER to specify the location of the second corner of the rectangle to be constructed. This corner is diagonal to the first corner selected.)*

**Note: When selecting the first point of the rectangle all three coordinates were specified. When the second point was entered the Z-coordinate was omitted. This is because the RECTANG command only constructs two-dimensional objects; therefore the Z-coordinate of the second point must be the same as that of the first point selected.**

**Figure 4–11** *Two rectangles appearing to reside on the same plane when viewed with the line-of-sight set perpendicular to the horizontal plane.*

**Figure 4–12** *Same rectangles viewed with the line-of-sight set at 45° to the XY-planes.*

#### Determining That Two or More Objects Reside on the Same Plane in AutoCAD

As mentioned earlier, when the line-of-sight is set perpendicular to the XY-axis, or parallel to the Z-axis, objects contained on separate XY-planes appear to coexist on the same plane. In the previous example it was not until the line-of-sight was positioned either along the X- or Y-axis or set to an angle other than 90° to the XY-plane using the VPOINT command, that the true positions of the objects become apparent. Even though changing the viewpoint in an AutoCAD drawing is a relatively easy task, it is not always the best method for determining whether two objects are indeed on the same plane. In large and complex drawing files, changing the view could cause AutoCAD to regenerate the screen. This can be a very time consuming operation. To circumvent this possibility, the AutoCAD LIST command can be employed to verify the positions of objects within a drawing. This command not only identifies the entity type, layer, and handles of selected objects, it also gives the object's position (X-, Y-, and Z-coordinates) in relation to the current UCS setting. For example, by using the LIST command and selecting the two objects shown in Figure 4–11, the Z-coordinates verify that the objects reside on different planes separated by a distance of two units. This example can be re-created with drawing DG04-11.dwg on the student CD-ROM.

```
Command: LIST (ENTER)
Select objects: (A crossing window can be used to select both objects.)
Specify opposite corner: 2 found
Select objects: (Press ENTER to terminate selection mode.)
                LWPOLYLINE  Layer: "0"
                          Space: Model space
                Handle = 21
         Closed
   Constant width    0.0000
         area   12.0000
     perimeter   14.0000

     at point  X=  2.0000  Y=  5.0000  Z=  4.0000
     at point  X=  6.0000  Y=  5.0000  Z=  4.0000
     at point  X=  6.0000  Y=  8.0000  Z=  4.0000
     at point  X=  2.0000  Y=  8.0000  Z=  4.0000

               LWPOLYLINE  Layer: "0"
                          Space: Model space
                Handle = 20
         Closed
   Constant width    0.0000
         area   12.0000
     perimeter   14.0000

     at point  X=  4.0000  Y=  5.0000  Z=  2.0000
     at point  X=  8.0000  Y=  5.0000  Z=  2.0000
     at point  X=  8.0000  Y=  8.0000  Z=  2.0000
     at point  X=  4.0000  Y=  8.0000  Z=  2.0000
```

**Note: This explains why AutoCAD sometimes reports the error message that the selected objects are non-coplanar (meaning the objects are on different planes).**

**Defining and Editing Planes in AutoCAD with UCS**

A plane can be defined by combining any two of the three axes used in AutoCAD (XY-plane, XZ-plane, and the YZ-plane), but only the XY-plane is used to construct an object (with the Z-plane used to define depth). The XY-plane is repositioned, or rotated, to correspond with the current plane, whether it is the frontal, profile, or horizontal plane. In AutoCAD, the XY-plane is repositioned by using the UCS command. This command allows the user to change the position of the XY-plane by relocating the origin and/or rotating about one of the axes in the User Coordinate System (UCS). After the XY-plane has been repositioned, additional features of the part may be created. For example, to extrude a hole into the profile of the part shown in Figure 4–13, the XY-plane must first be repositioned onto the part's profile plane. This is accomplished by using the 3Point option of the UCS command (see Figure 4–14). Once this has been done, a circle may be drawn in its proper position on the part's profile plane, then extruded to the desired length and subtracted from the cube, to create the hole. This example can be re-created using DG04-13.dwg located on the student CD-ROM.

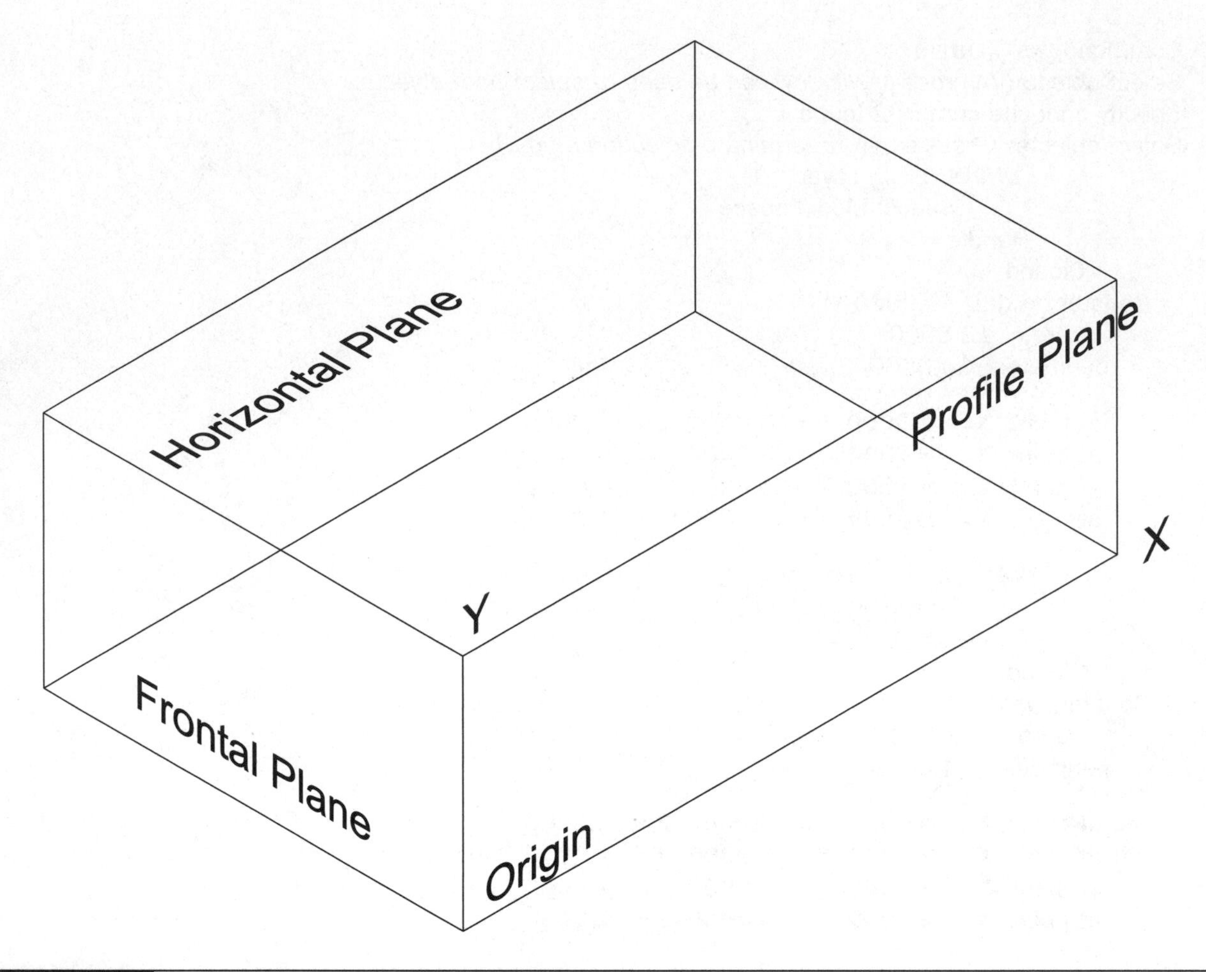


**Figure 4-13** *Three-dimensional cube shown with the line-of-sight set 45° in the XY-plane and 35° from the XY-plane.*

## Step #1

Using the 3Point option of the UCS command, the XY-plane is redefined to coexist with the object's profile plane. The 3Point option allows the user to define the new location of the XY-plane by first picking the XY-plane's origin followed by the extents of the X- and Y-axis. In this example, the origin of the XY-plane has been relocated to the lower left-hand corner of the part's profile plane. The X-axis of the XY-plane is defined as starting from the lower left-hand corner of the part's profile plane and extending through the lower right-hand corner of the part's profile plane. The Y-axis of the XY-plane is defined as starting at the lower left-hand corner of the part's profile plane and extending through the upper left-hand corner of the part's profile plane (see Figure 4–14).

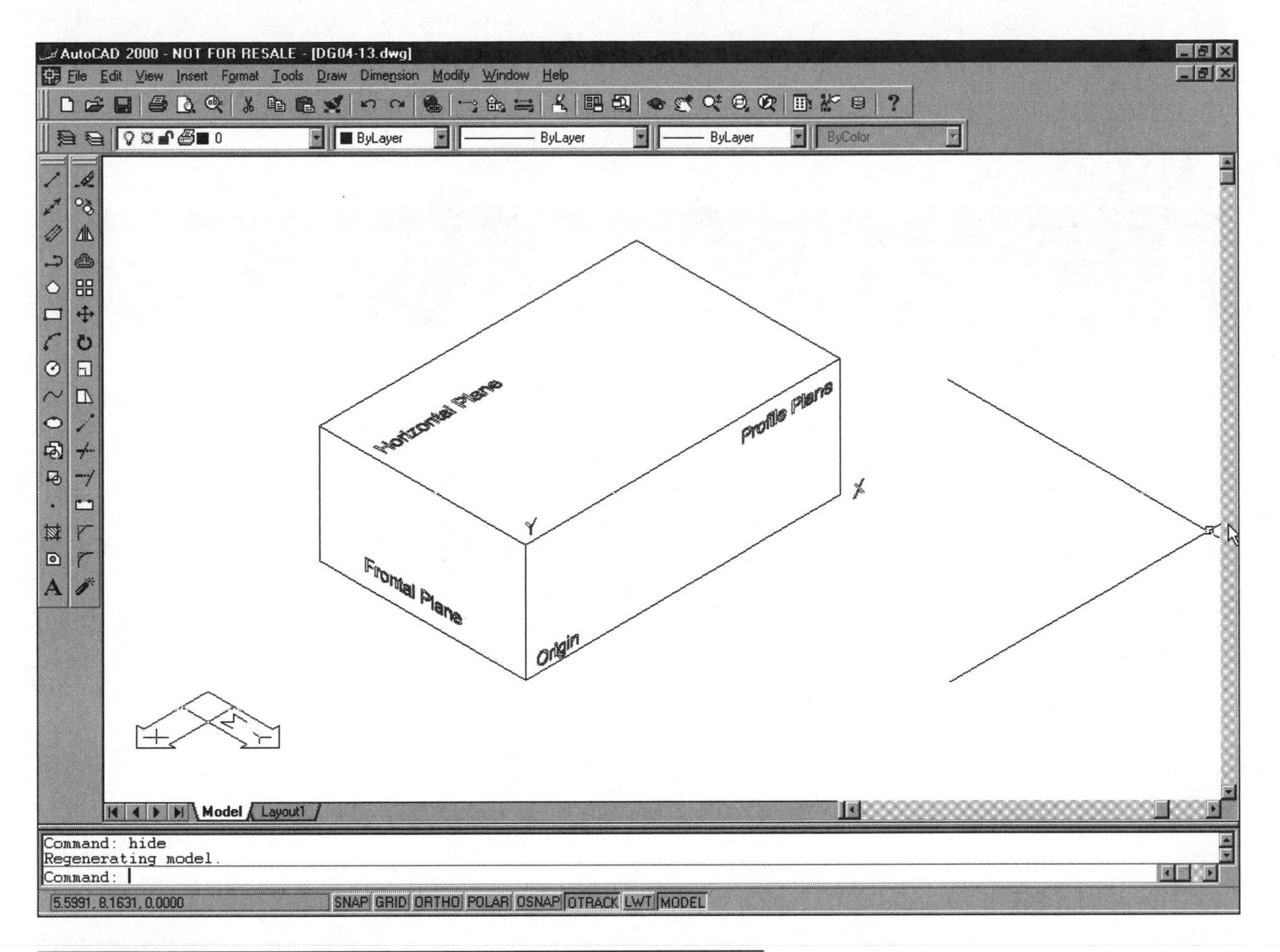


**Figure 4–14** *Position of new XY-plane along the profile plane.*

Command: UCS (ENTER)

Origin/ZAxis/3point/OBject/View/X/Y/Z/Prev/Restore/Save/Del/?/<World>: 3P *(Press* ENTER *to execute the three-point option of the* UCS *command, allowing the user to reposition the XY-plane by selecting three points.)*

Origin point <0,0,0>: END of *(Press* ENTER *to specify the new origin of the XY-plane.)*

Point on positive portion of the X-axis <10.6395,6.3148,0.0000>: END of *(Press* ENTER *to specify the new orientation of the positive X-axis of the XY-plane.)*

Point on positive-Y portion of the UCS XY plane <9.6395,5.3148,0.0000>: END of *(Press* ENTER *to specify the new orientation of the positive Y-axis of the XY-plane.)*

Command:

## Step #2

To construct the circle that defines the circumference of the hole, the CIRCLE command is employed. Once executed, this command prompts the user to select the location of the center of the circle. Since the hole is to be positioned in the center of the part's profile plane, the X-, Y-, and Z-filters can be used. These filters allow the user to retrieve one of more coordinates of a point by entering the filter or a combination of filters that specify the coordinate(s) to be extracted. After determining which coordinate(s) are to be used and that the necessary filter(s) have been supplied, the software prompts the user to select the point containing these coordinate(s). When all three coordinate(s) have been defined (X, Y, and Z), they are combined and the entity is created at that location (see Figure 4–15). This technique is demonstrated in the following steps.

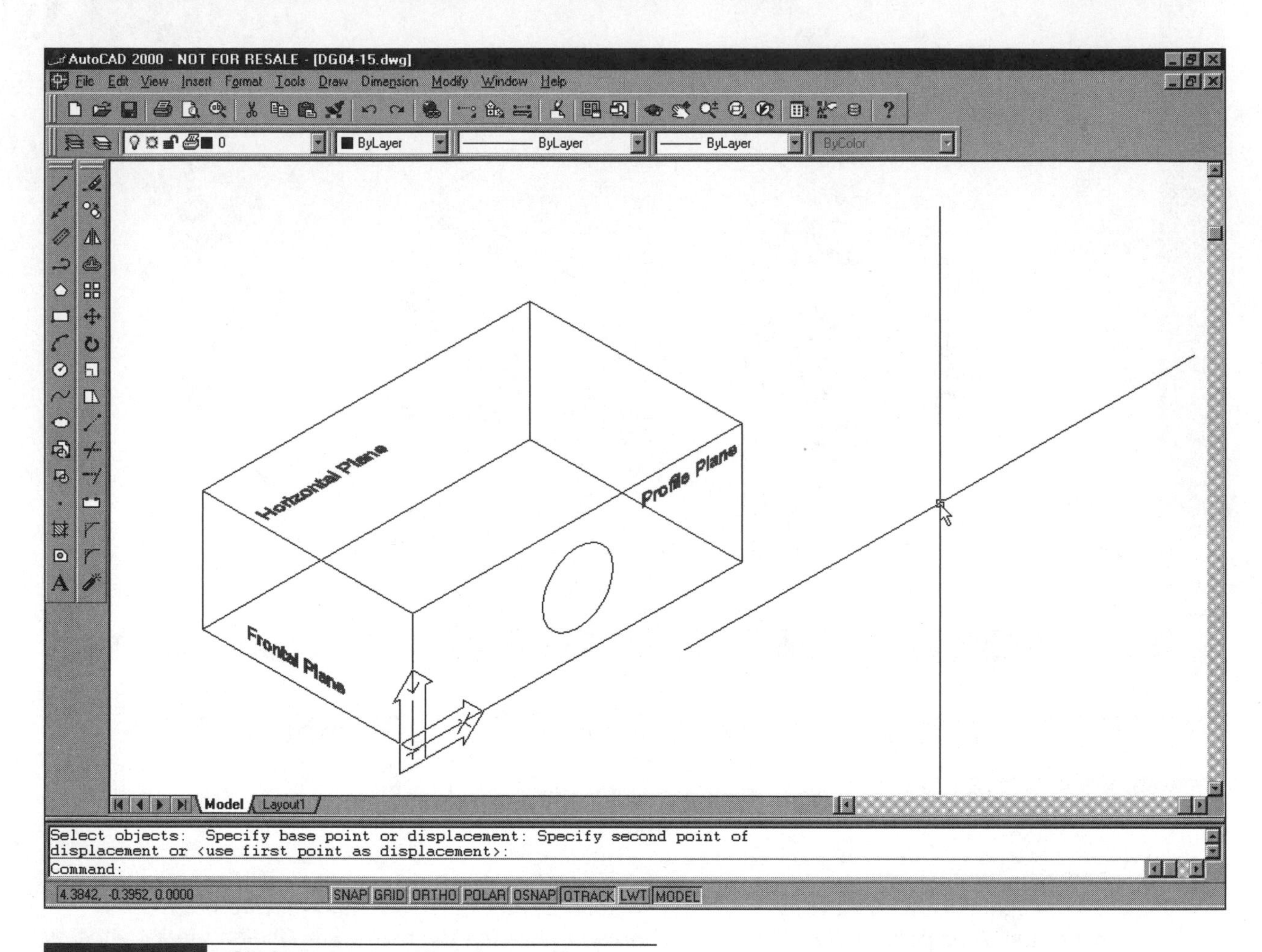


**Figure 4–15** *Circle constructed on the profile plane.*

Command: CIRCLE (ENTER)

CIRCLE 3P/2P/TTR/<Center point>: .X (*Press* ENTER. The .X filter allows the user to specify the X-coordinate for the center of the circle to be drawn.)

of MID *(Press* ENTER. *By using the* MID OSNAP *command in conjunction with the .X filter, the user is able to select the bottom or top edge of the part's profile plane to define the X-coordinate of the circle's center point.)*

of *(Select the bottom or top edge of the profile plane.)*

(need YZ): MID *(Press* ENTER. *Once the X-coordinate has been determined, the user is prompted to specify the Y- and Z-coordinates to be combined with the X-coordinate.)*

of *(Again, using the* MID OSNAP *command allows the user to select either the left or right edges of the part's profile plane, and extract those coordinates to be used with the previously specified X-coordinate. These combined coordinates determine the location of the center of the circle to be drawn.)*

Diameter/<Radius>:0.25 *(Press* ENTER *to specify the radius of the circle to be drawn.)*

Command:

## Step #3

Finally, extruding the circle into a solid cylinder and then subtracting it from the cube completes the hole (see Figure 4–16).

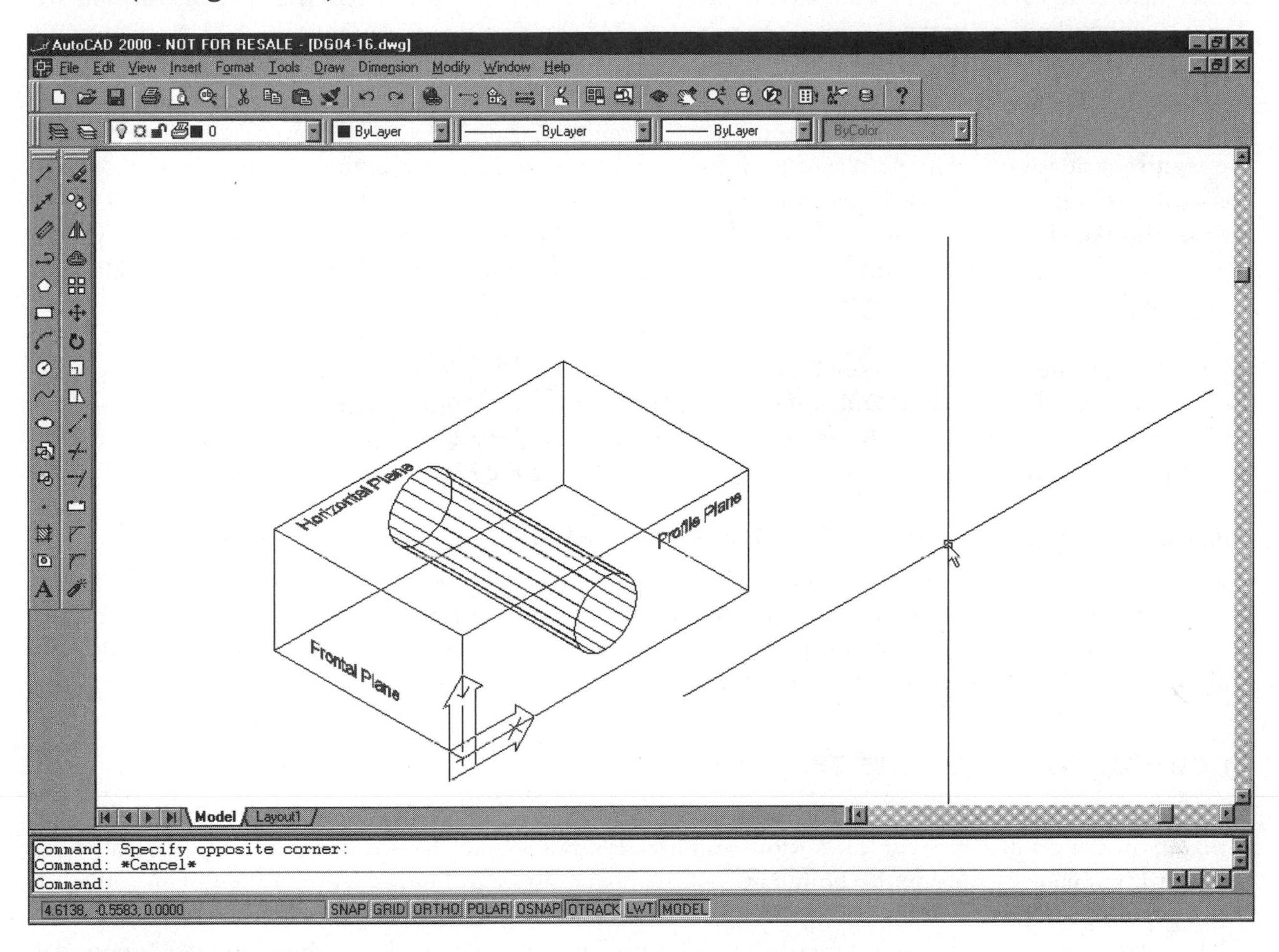


**Figure 4–16** *Completed part.*

Command: EXTRUDE (ENTER)
Select objects: 1 found *(Select the circle to be extruded.)*
Select objects: *(Press* ENTER *to terminate the selection process.)*
Path/<Height of Extrusion>: -5 *(Press* ENTER *to specify the depth of the extruded cylinder.)*
Extrusion taper angle <0>: *(Press* ENTER *to specify a taper angle to apply to the extrusion.)*
Command: SUBTRACT (ENTER)
Select solids and regions to subtract from... *(Select the object(s) the cylinder is to be subtracted from. If more than one object is selected, the* SUBTRACT *command joins these objects together before subtracting from them the object(s) specified in the next step. In this example, the object being subtracted from is the cube.)*
Select objects: 1 found *(Confirms the number of objects selected.)*
Select objects: *(Press* ENTER *to terminate the selection mode.)*
Select solids and regions to subtract... *(Select cylinder.)*
Select objects: 1 found *(Confirms the number of objects selected.)*
Select objects: *(Press* ENTER *to terminate the selection mode and begins the subtraction process.)*
Command:

### Sketch Planes (Mechanical Desktop)

AutoCAD is not the only application that takes full advantage of the use of planes; these principles are also fundamentally ingrained into Mechanical Desktop in the form of ***sketch planes*** and ***work planes***. A sketch plane, in Mechanical Desktop, is one that extends infinitely in two directions and is used to construct the two-dimensional outline of a part's features. The orientation of the X- and Y-axis on the sketch plane determines the position of the Z-axis where the part is extruded. In short, the sketch plane is the equivalent of AutoCAD's XY-plane. It can be repositioned anywhere or along any flat surface. Its position is critical when a part's feature is transposed from an ordinary AutoCAD two-dimensional entity into a Mechanical Desktop profile. If the feature resides on a plane different from the current sketch plane, when the feature is transposed Mechanical Desktop automatically repositions the profile onto the current sketch plane. To avoid this, the position of the sketch plane should be defined before the outline is created, or repositioned to the proper position before the outline is transformed into a Mechanical Desktop profile. To reposition the sketch plane in Mechanical Desktop, the AMSKPLN command is used.

Similar to Mechanical Desktop's sketch planes are the ***work planes***. Work planes are used to define a parametric location for a sketch plane. Only one sketch plane can be set at one time in Mechanical Desktop, but multiple work planes can be defined. This allows the sketch plane to be quickly repositioned by merely selecting an existing work plane. Work planes are defined using the AMWORKPLN command.

#### Defining and Editing Sketch Planes Using Amskpln's UCS Option

One way to redefine the position of a sketch plane is by using the UCS option of the AMSKPLN command. This option repositions the sketch plane to coincide with AutoCAD's current ***U***ser ***C***oordinate ***S***ystem XY-plane position. At times it is easier to first reposition the XY-plane to the desired location in Mechanical Desktop and then reset the sketch plane using the UCS option of the AMSKPLN command.

## Locating Points on Planes

In theory, a point does not contain dimensions, only coordinates (X, Y, and Z) to define its position in space. A plane, on the other hand, can have width and breadth, but will not have a thickness. A point will either coexist with a plane or outside the boundaries of the plane (above or below an established plane). When a plane containing a point is projected into a view where the plane appears as a line, then the point contained on that plane will appear as a point on the line. This can be seen in Figure 4–17, where point "B" appears as a point on line AC in the front view. It can be determined if a point coexists on a plane by constructing any view in which that plane appears as a line (edge view); if the point resides on the line, then it resides on the plane as well. If the point is not located on the line, then the point is not on the plane.

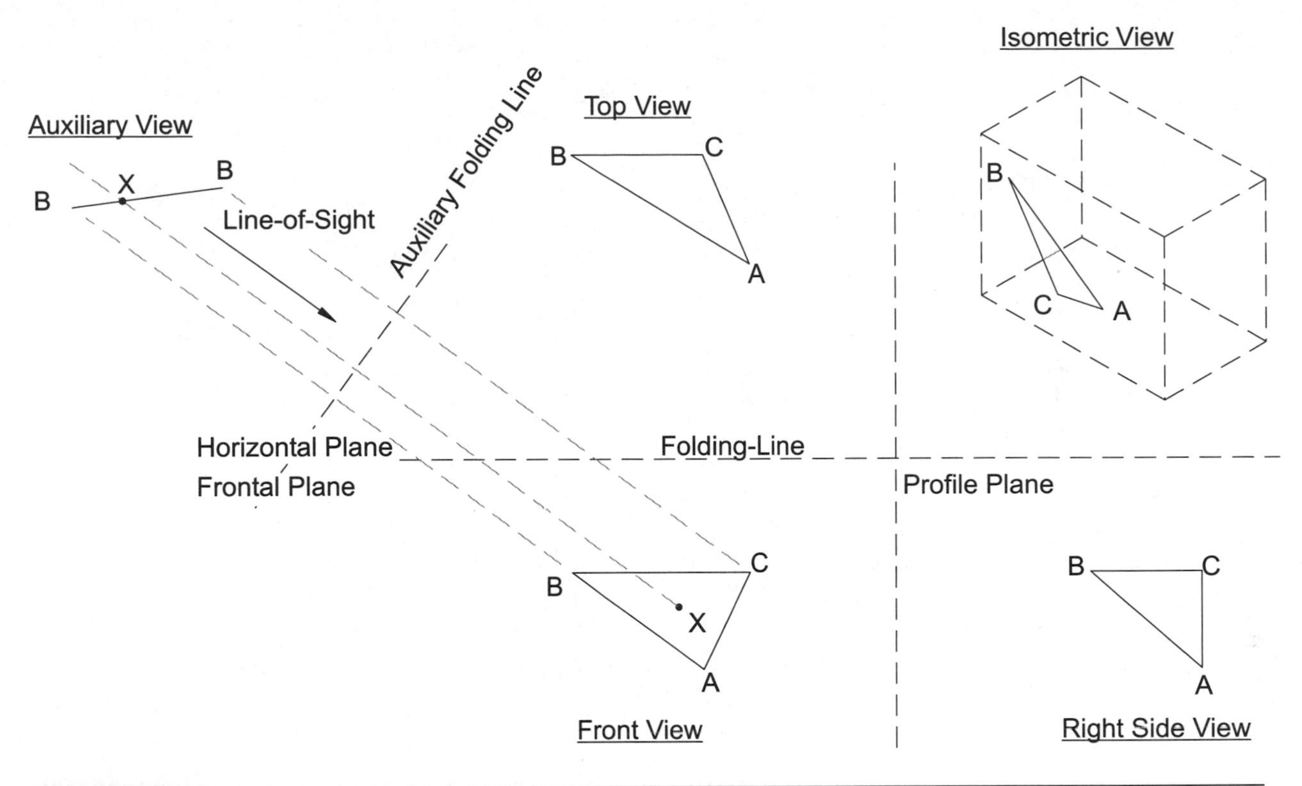


**Figure 4–17** *By projecting the plane into any view where it appears as a line, it can be determined if a point coexists on that plane.*

Once it has been determined that a point does reside on a particular plane, then the location of that point can be defined and transferred to other views using projection lines. In theory, a line coexisting on a plane will contain an infinite number of points that are common to both the plane and the line. Also, a point on a plane can have an infinite number of lines passing through it. By combining these facts, it can be stated that the location of any point contained on a plane can be transferred from one view to another by constructing a line that passes through that point and crosses two plane boundaries. To transfer the location of the point shown in Figure 4–18 from the front view to the profile view, a line is drawn that connects the two boundary points (AA and BB) to the point on the plane (X). Points AA and BB are then transferred from the front view to the profile view using standard orthographic projection techniques, already discussed. Once the points have been transferred, the location of point X in the profile view is determined by connecting AA and BB with a line segment. It is the intersection of this line with the projection line of point X (the line emitting from the point in the front view to the profile view) that marks the point's exact position in the profile view.

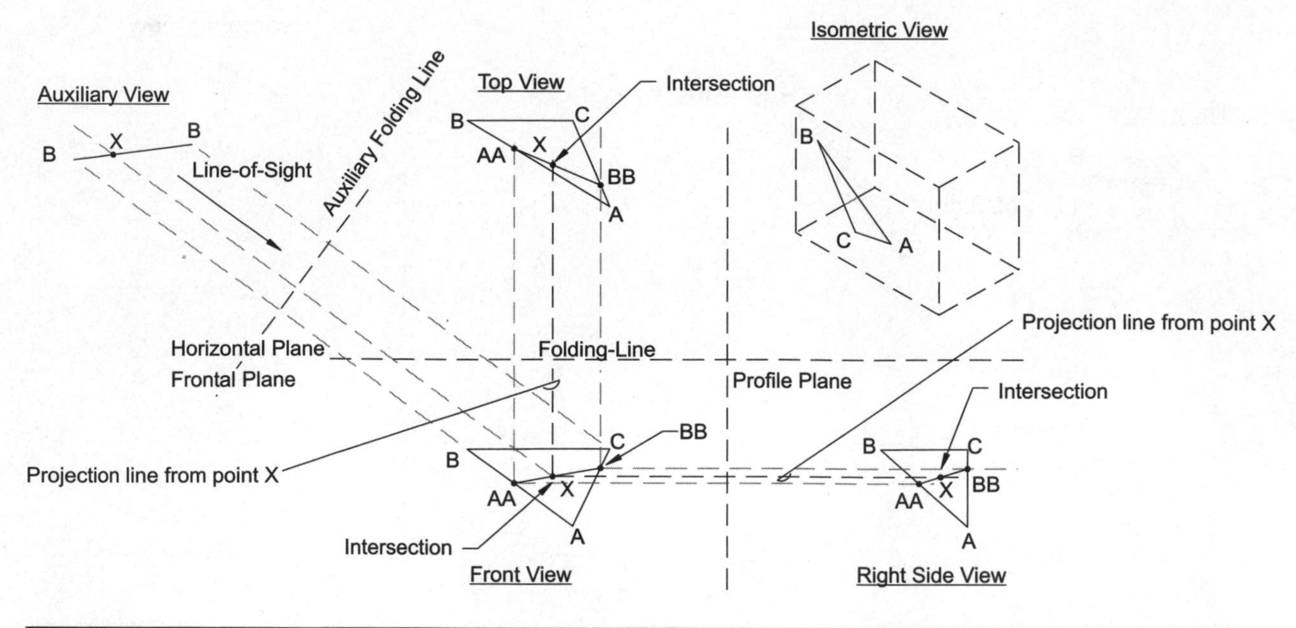


**Figure 4–18** *Extending a line through a point and crossing two boundaries can be used to determine the location of that point in another view.*

## Locating Lines on Planes

The principles, techniques, and theories discussed in the previous section regarding the location of points on a plane can also be applied when transferring the location of a line from one view to another. This is illustrated in the following example, where the location of object MNO is to be transferred from the front view to the plan view (see Figure 4–19). The transfer is accomplished by extending the object lines MN, MO, and NO until they intersect the boundary lines AE, AB, CD, and ED of plane ABCDE, and transferring those points to the plan view using standard orthographic techniques.

**Note: Any of the lines defining the object MNO can be extended in the front view to intersect the boundaries of the plane. These intersection points are then transferred to the top view where a line connecting these points is drawn.**

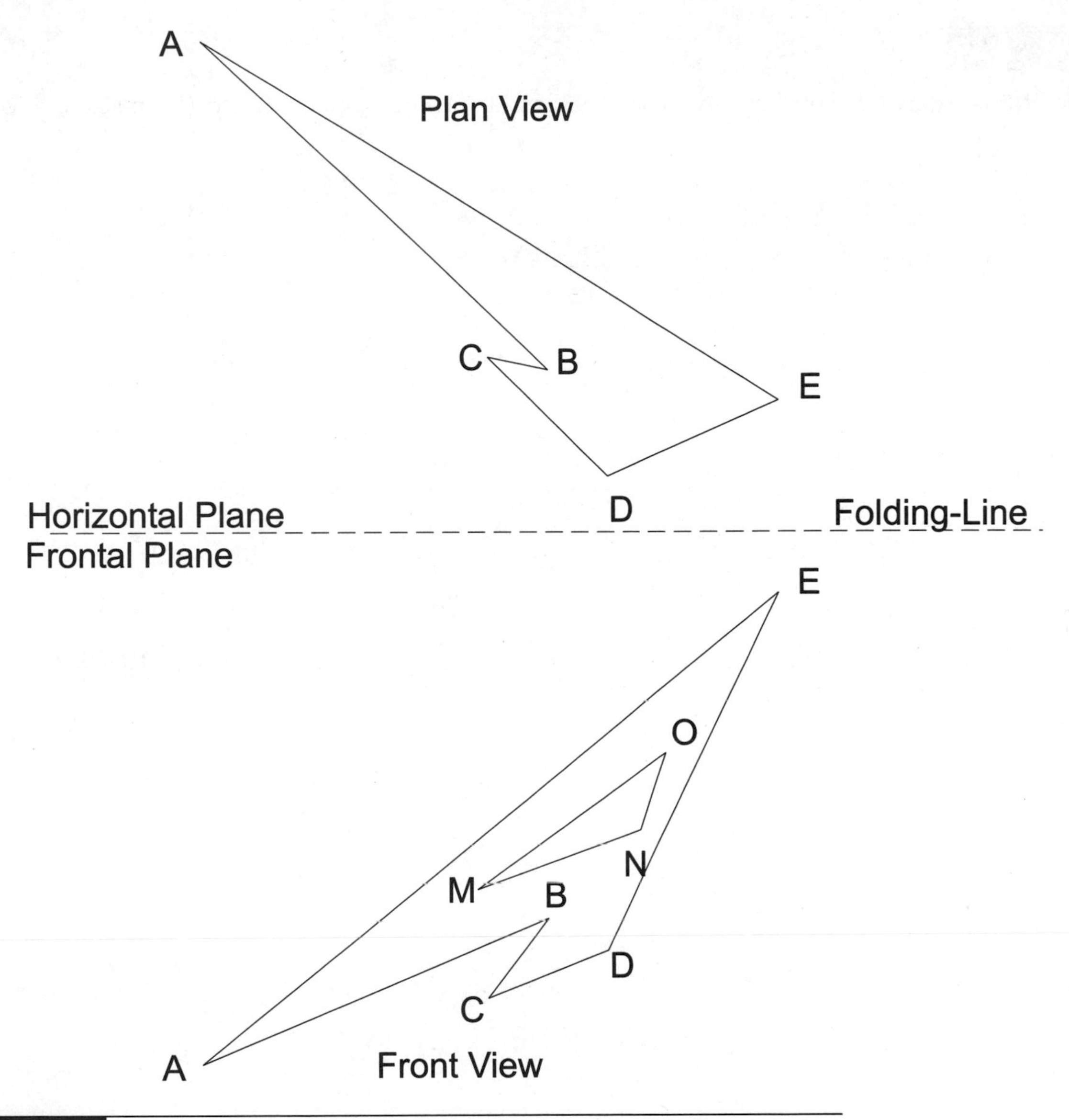


**Figure 4-19** *The position of object MNO is to be determined in the plan view.*

## Step #1

Line NO is extended until it intersects lines AE and ED, producing points S and T. See Figure 4–20.

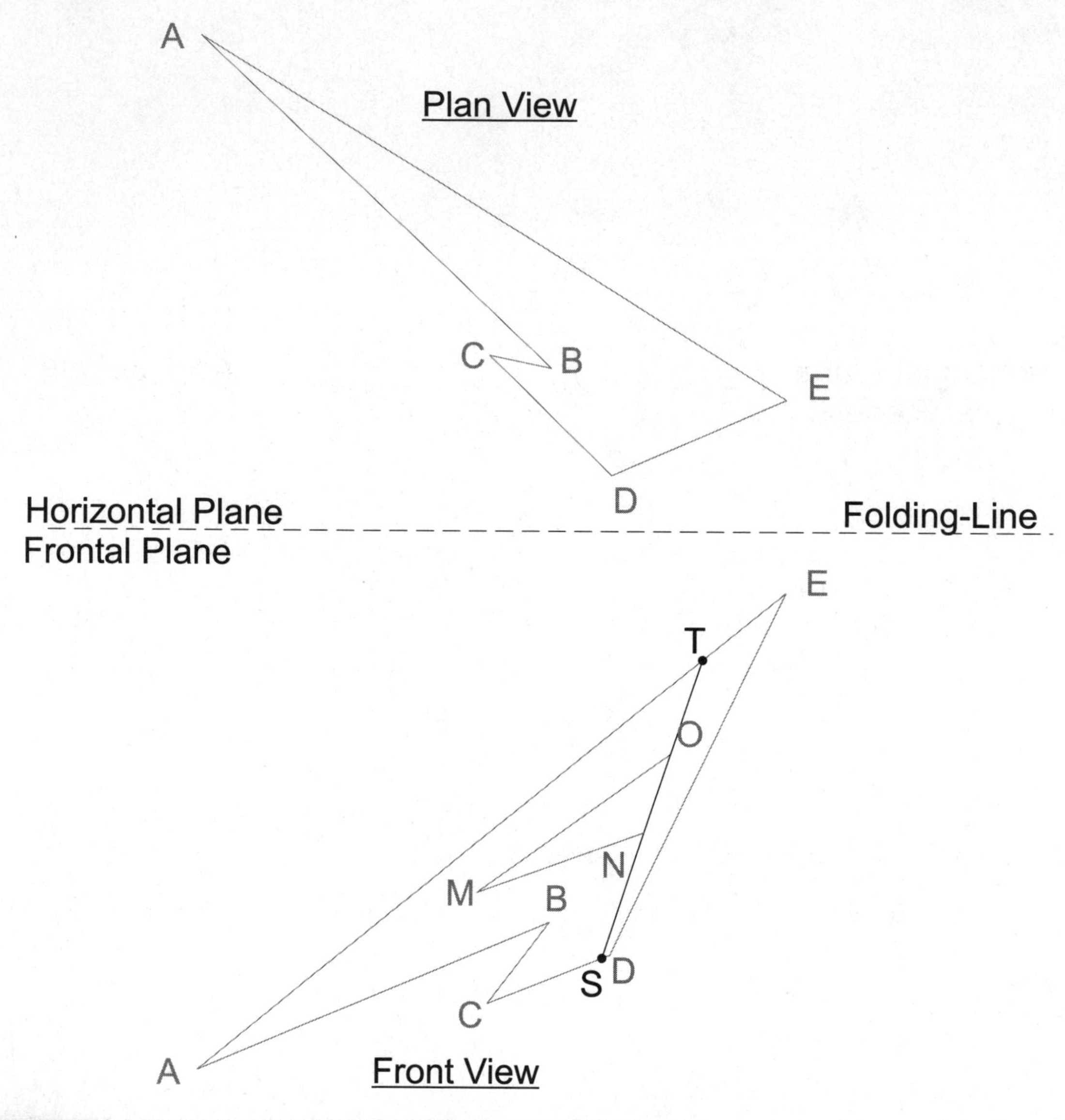


**Figure 4–20** *Extending line NO until it intersects lines AE and ED.*

## Step #2

By using standard orthographic projection techniques the position of points S and T are transferred from the front view to the plan view. Once this has been completed a line segment connecting the two points is drawn (see Figure 4–21).

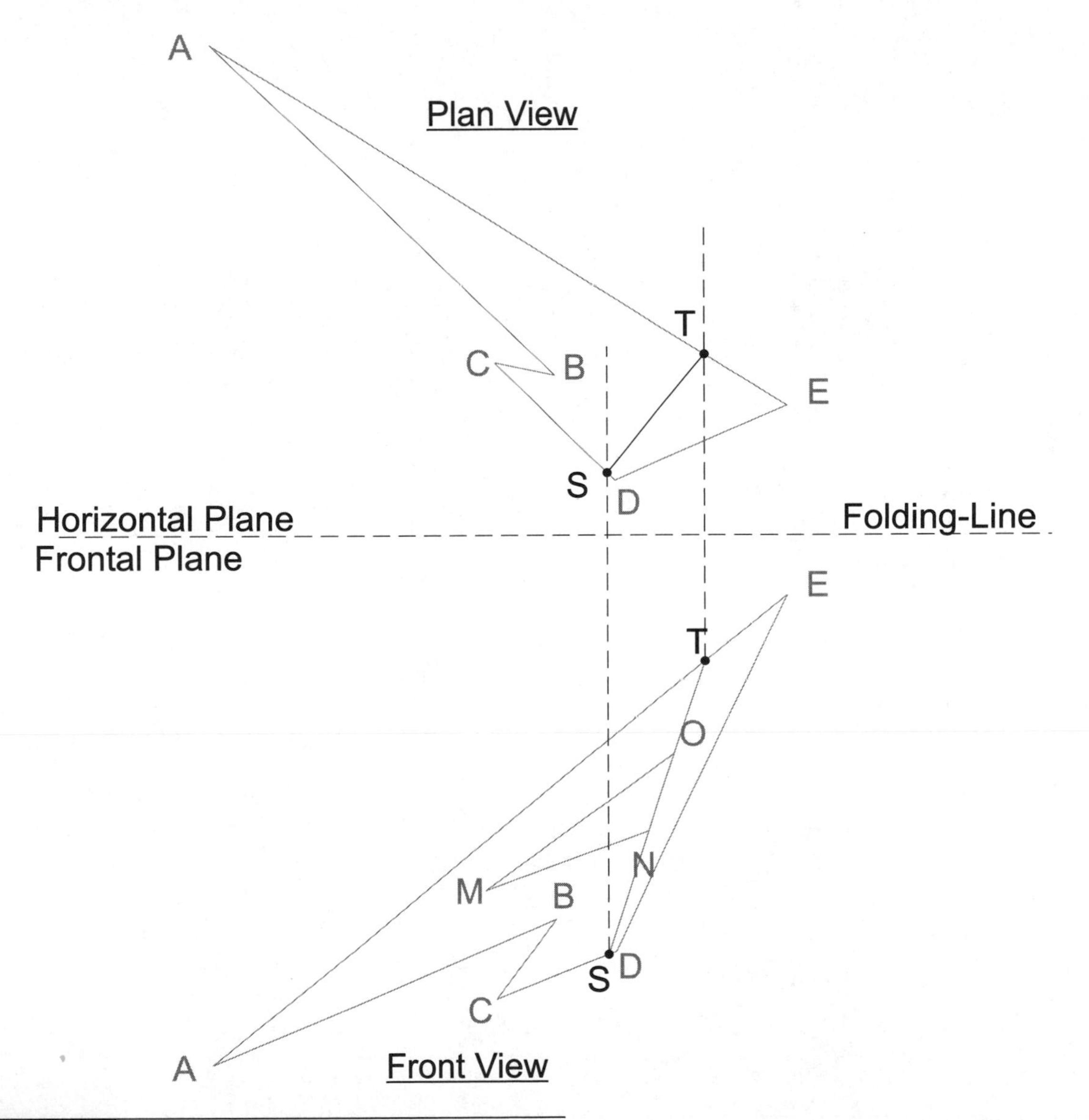


**Figure 4–21** *Line ST is transferred to the plan view.*

## Step #3

Now that line ST has been projected into the plan view, the actual length of line NO can be found in the plan view by projecting lines from points N and O into the plan view and intersecting line ST (see Figure 4–22). Once these points are established in the plan view, lines SN and OT are erased.

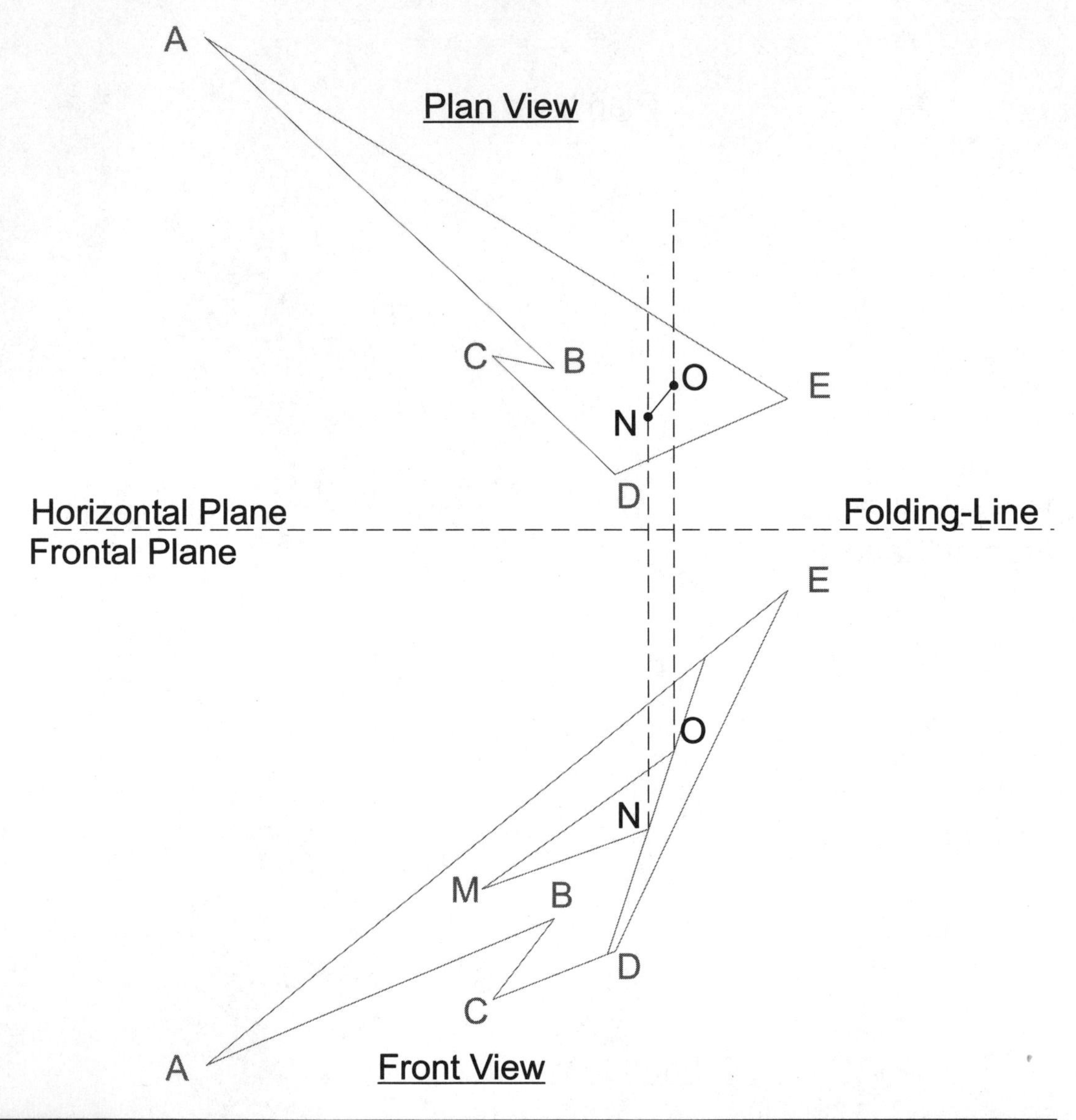


**Figure 4–22** *By projecting points N and O into the plan view, the actual length of line NO can be determined.*

## Step #4

Before line MN can be transferred to the plan view, it must first be extended so that it intersects line AE, producing point U. Point U will be used as an aid in the placement of point M in the plan view (see Figure 4–23).

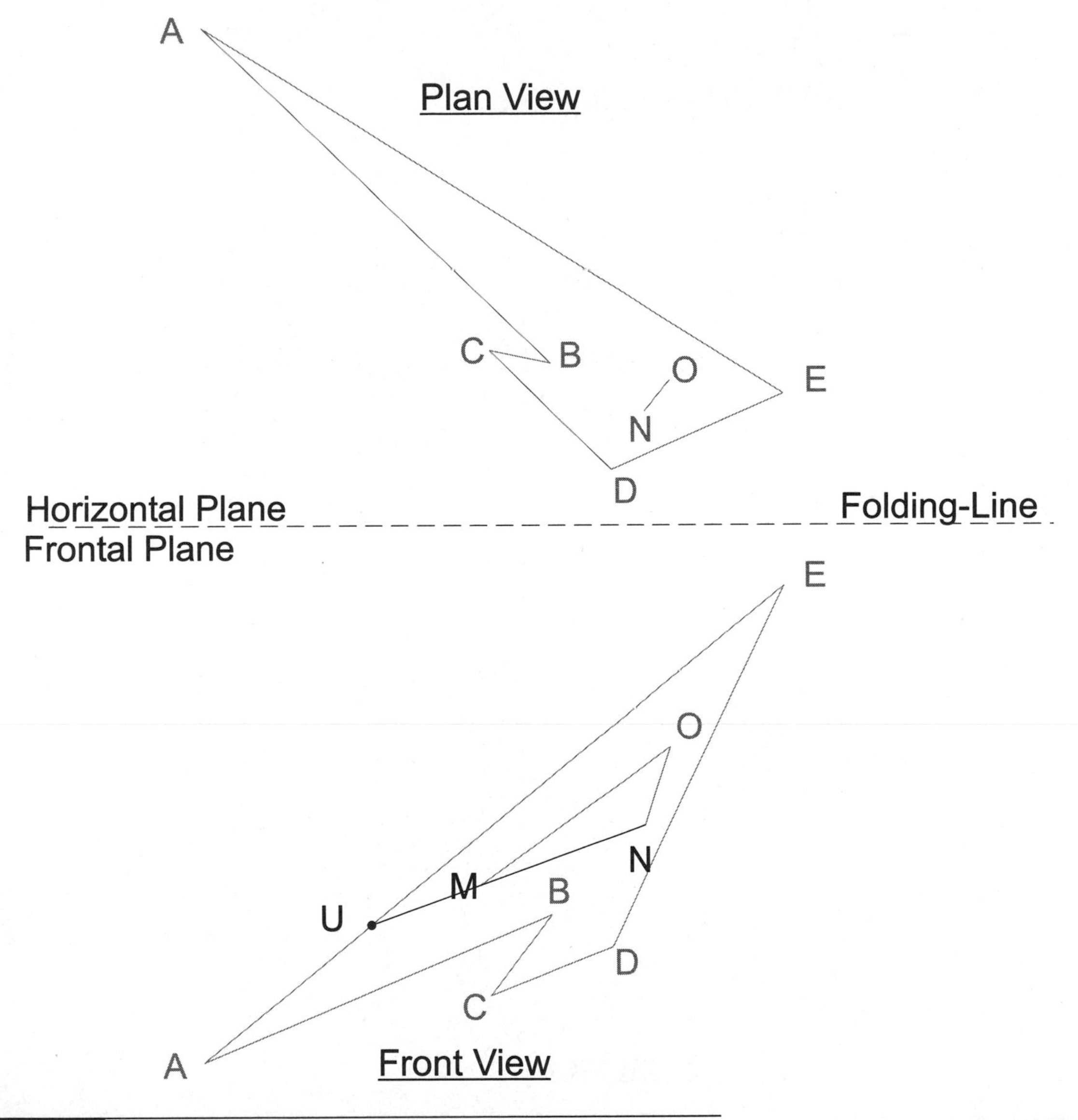


**Figure 4–23** *Extending line MN in the front view produces point U.*

## Step #5

Again, by using standard projection techniques, the location of point U is transferred from the front view to the plan view. A line segment is then drawn connecting point U to point N to produce line UN in the plan view (see Figure 4–24).

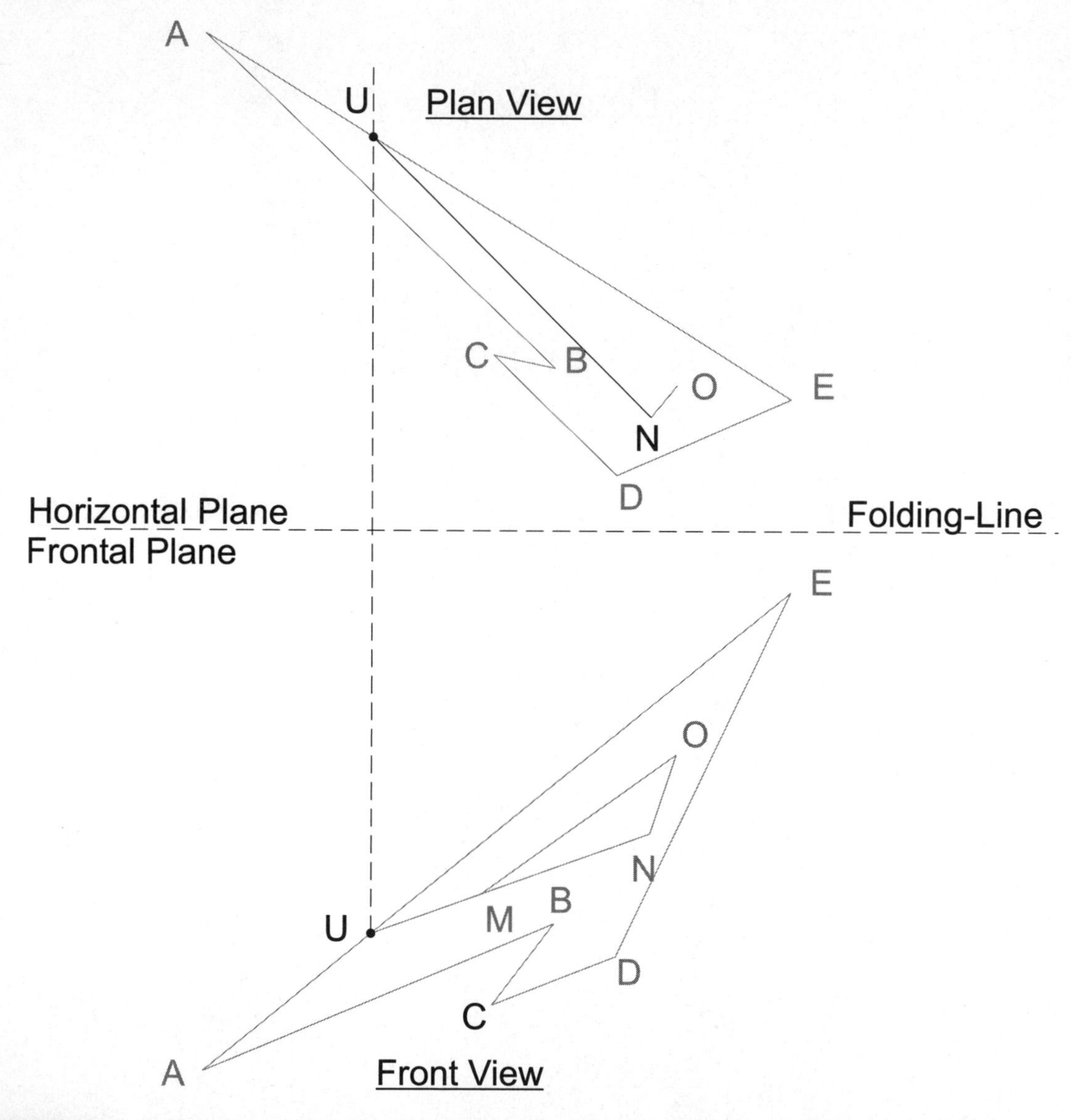


**Figure 4–24** *Line UN is produced by drawing a line from point U to point N in the plan view.*

## Step #6

Just as in step three where the length of line NO was determined by projecting points N and O into the plan view, the length of line MN is determined by projecting point M into the plan view. A line can now be drawn to connect the predetermined points M and O (see Figure 4–25).

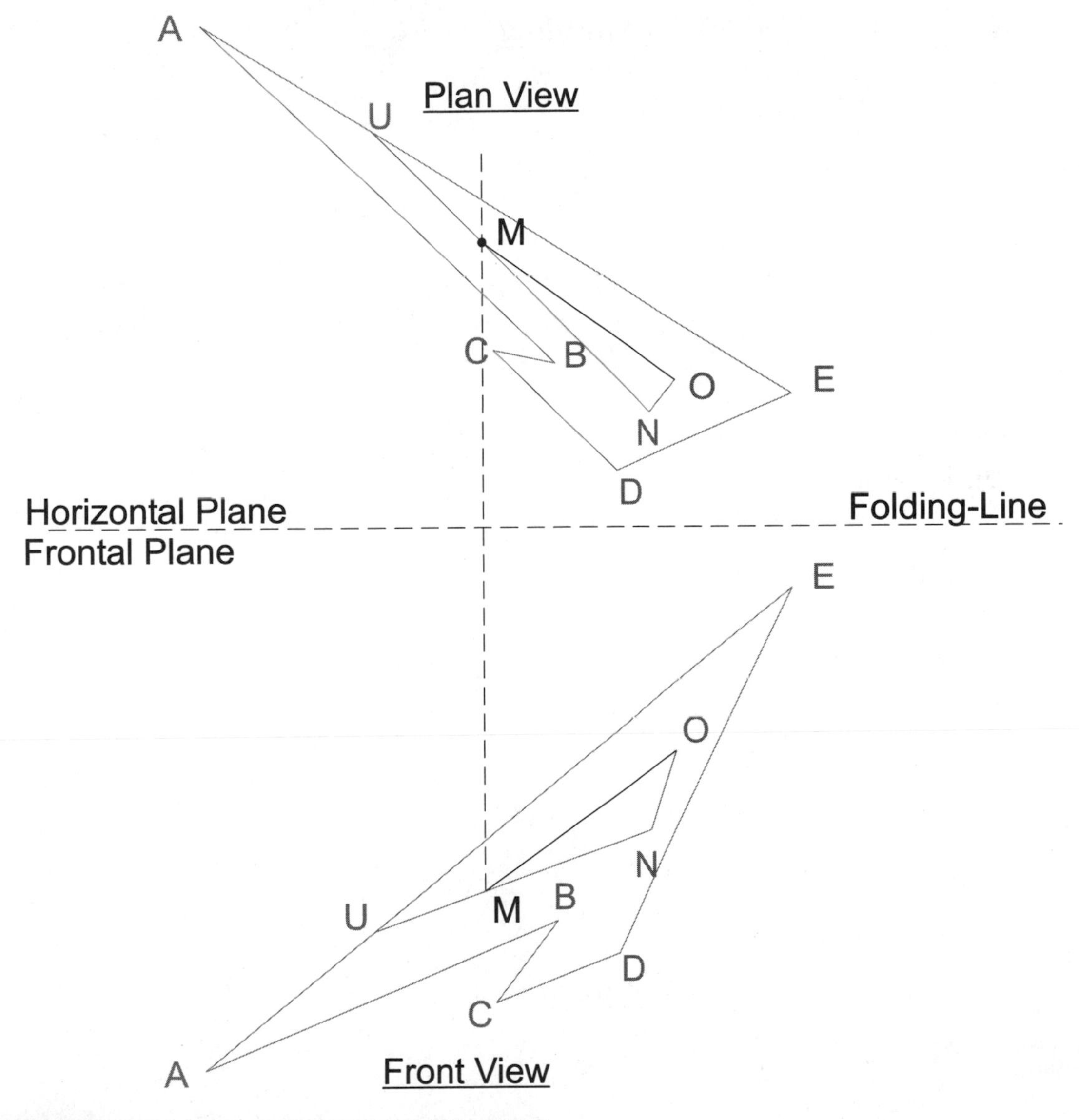


**Figure 4–25** *Projecting point M into the plan view.*

## Step #7

Erasing line UM completes the object (see Figure 4–26).

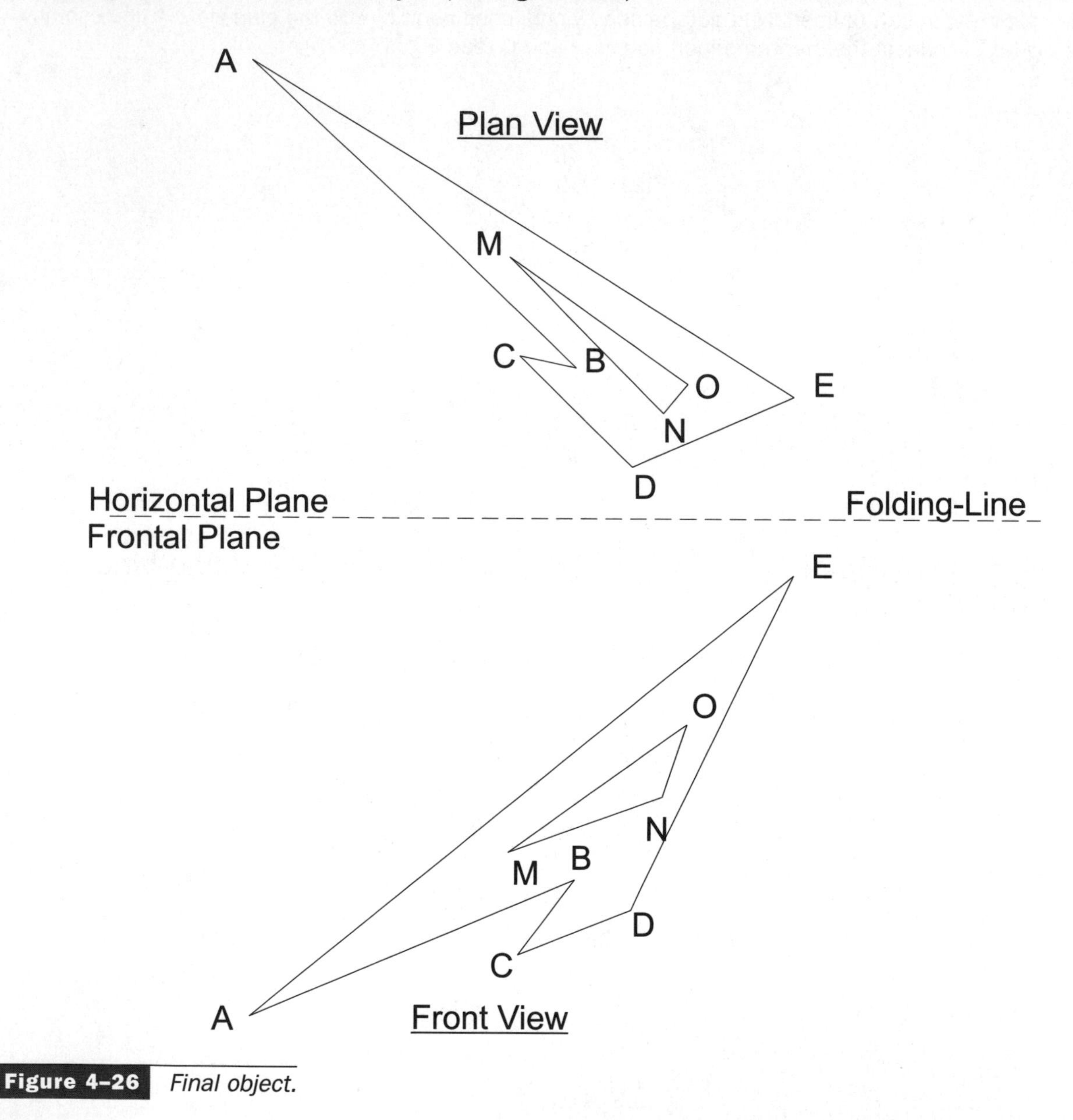


**Figure 4–26** *Final object.*

## Locating the Piercing Point of a Line and a Plane

A point can either be contained within or outside the designated boundaries of a plane. A line, on the other hand, will always be on, parallel to, or intersecting a plane. When a line intersects a plane, the point at which the two entities meet is referred to as the ***piercing point***. This point can either fall within or somewhere outside the established boundaries of the plane. If a line does not coexist on the plane or is parallel to that plane, then a piercing point will exist; because in theory a line extends indefinitely in both directions unless a boundary has been established. When working with line segments it may be necessary to extend the line or the boundaries of the plane so that the piercing point may be found. The location of the piercing point can be determined by using either the ***auxiliary view*** or ***cutting plane methods***.

## Auxiliary View Method

Recall that when a plane is shown as a line (edge view) then any point along that line will be on the plane. The principle for piercing points expands on this concept. When a plane is shown as a line and the intersecting line does not appear as a point, then the entities will intersect at the piercing point. However, if the plane is in edge view and the line appears as a point, then the line is considered parallel to the plane and a piercing point does not exist. At times the piercing point can be determined in one of the principal views, but the odds of this occurring with complex objects is not very likely. If the piercing point cannot be found in any of the principal views, then it can be determined by using one or more auxiliary views. In Figure 4–27, the piercing point of line AB and plane CDE is found by creating a primary auxiliary view.

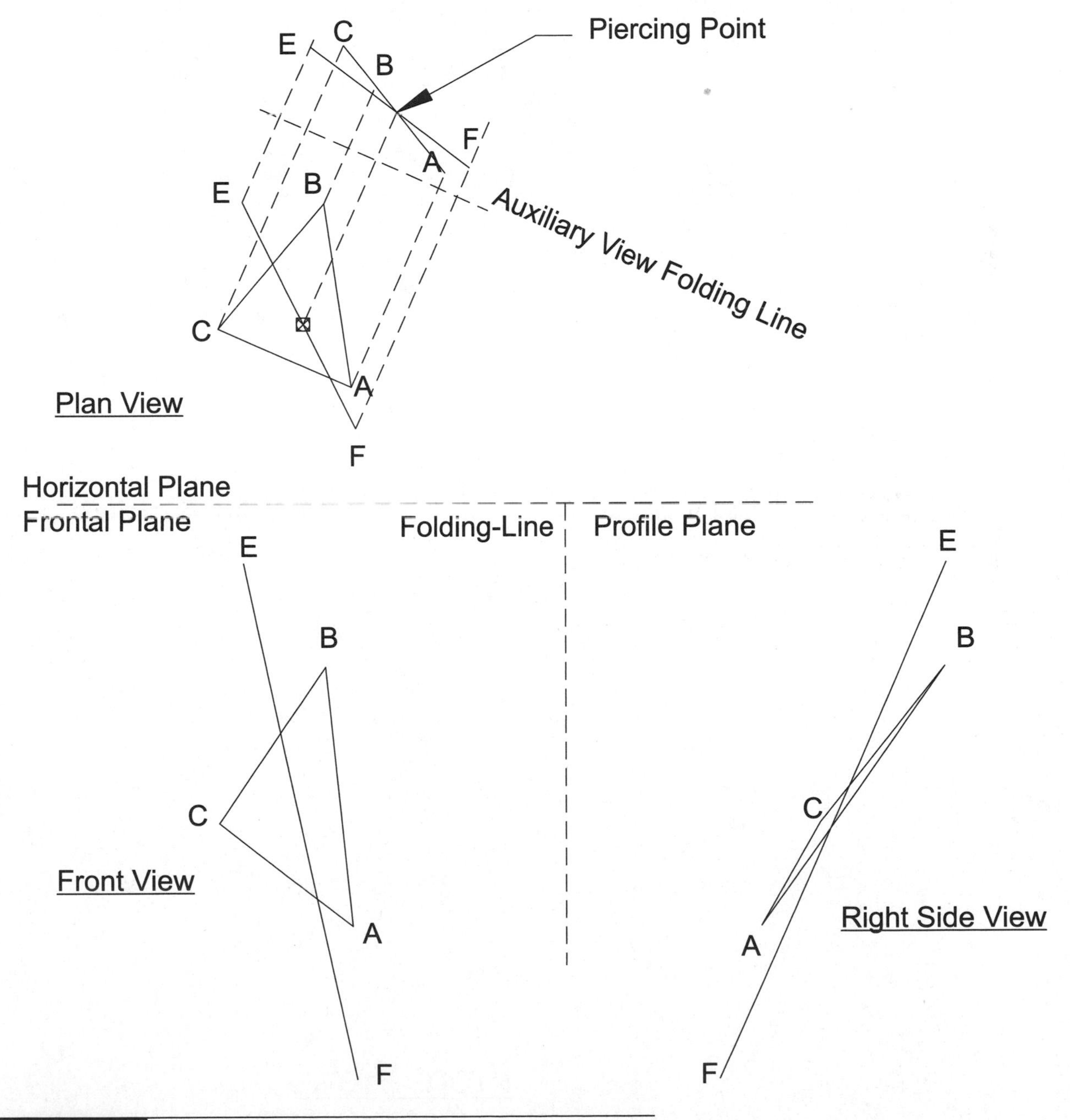


**Figure 4–27** *Auxiliary view method for finding a piercing point.*

## Cutting Plane Method

A second and sometimes much easier approach to finding the piercing point is the ***cutting plane method*** or ***two view method***, as it is often called. In the cutting plane method, the intersecting line is considered to reside on a "cutting plane" which bisects the original plane. The points formed by the intersection of these two planes are then transferred to another view using standard orthographic techniques. The transferred points are connected with a line segment; this line represents the intersection of the planes in that view. It is the junction of this line and the original line that marks the location of the piercing point. Once located, the piercing point can be transferred back to the first view using orthographic projection. This is illustrated in the following example where the piercing point of line XY and plane LMN is found using the cutting plane method (see Figure 4–28).

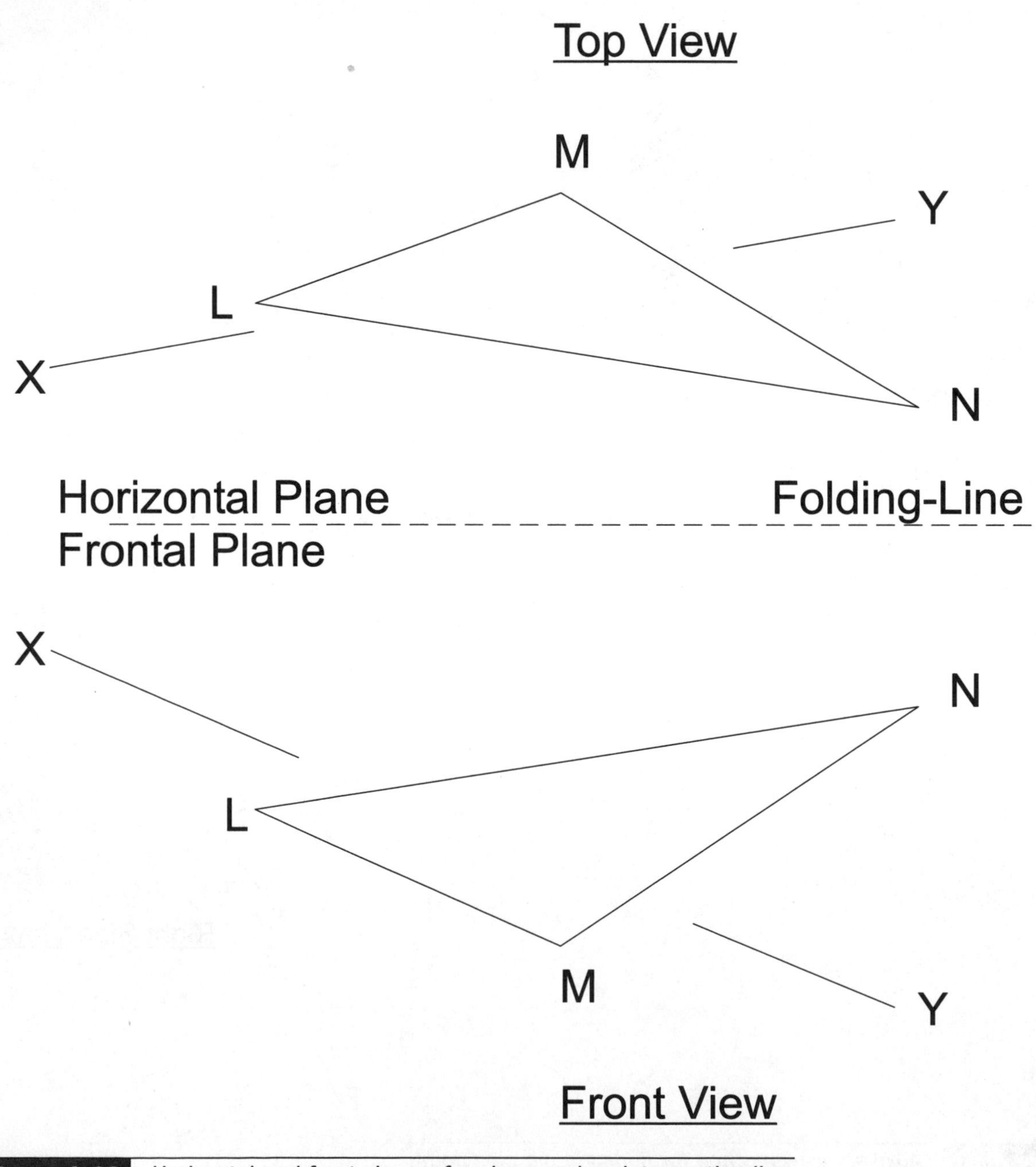


**Figure 4–28** *Horizontal and front views of a plane and an intersecting line.*

**Step #1**

The line segments of line XY are connected in the top view to produce the edge view of an imaginary cutting plane that bisects plane LMN (see Figure 4–29).

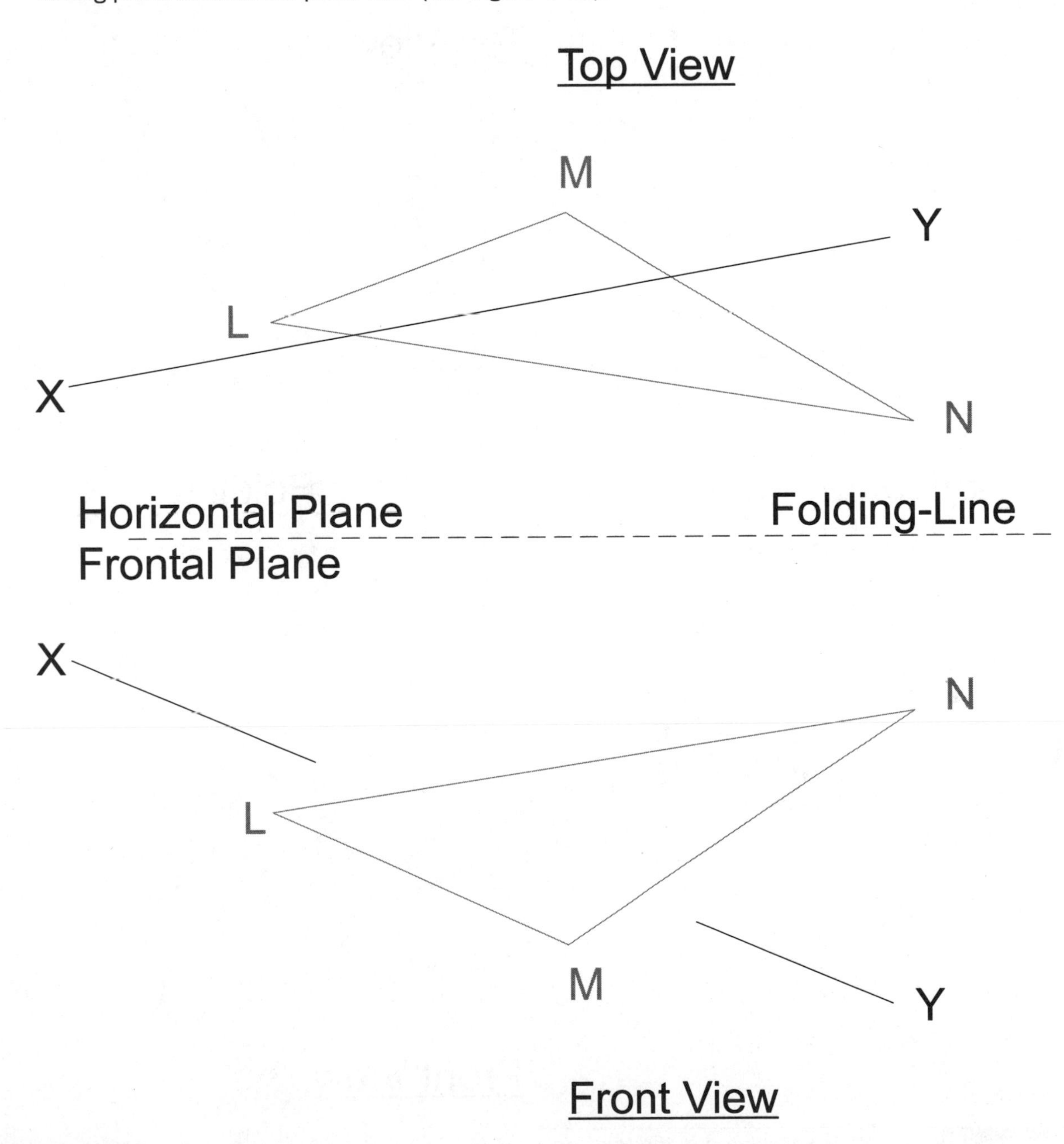


**Figure 4–29** *Line segment XY is transformed into a cutting plane.*

## Step #2

Line XY is defined in the front view by connecting the two line segments (see Figure 4–30).

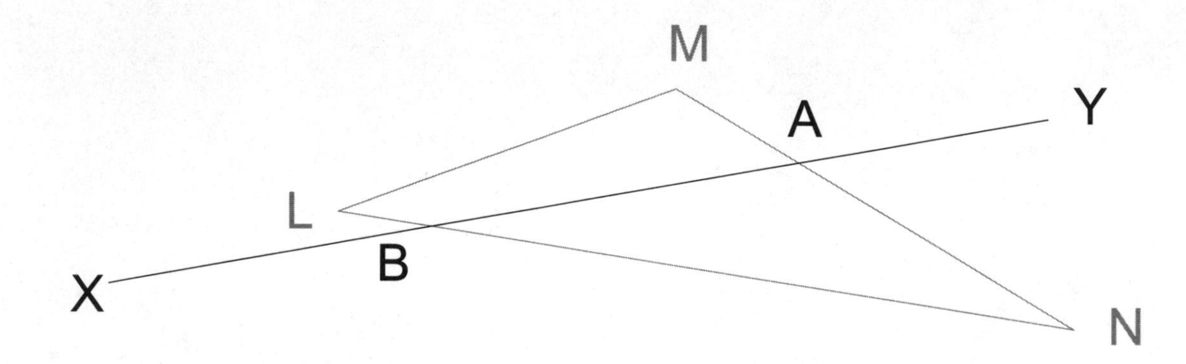


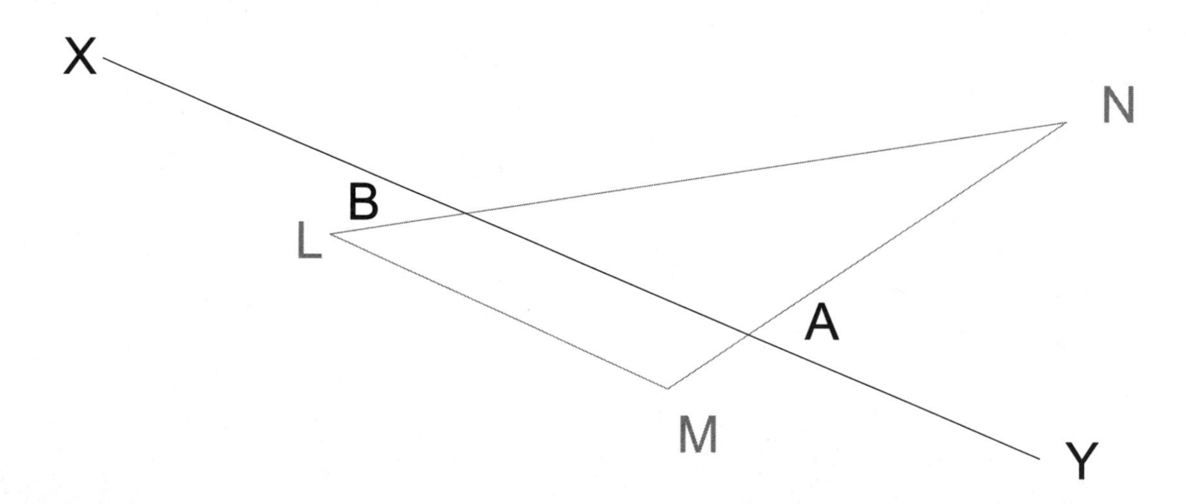


**Figure 4–30** *Line XY is completed in the front view.*

## Step #3

When the line segments were joined in step one, a cutting plane was produced that bisected plane LMN. The points at which the cutting plane crosses the boundary lines of plane LMN (A and B) are transferred down to the frontal plane using standard orthographic techniques (see Figure 4–31).

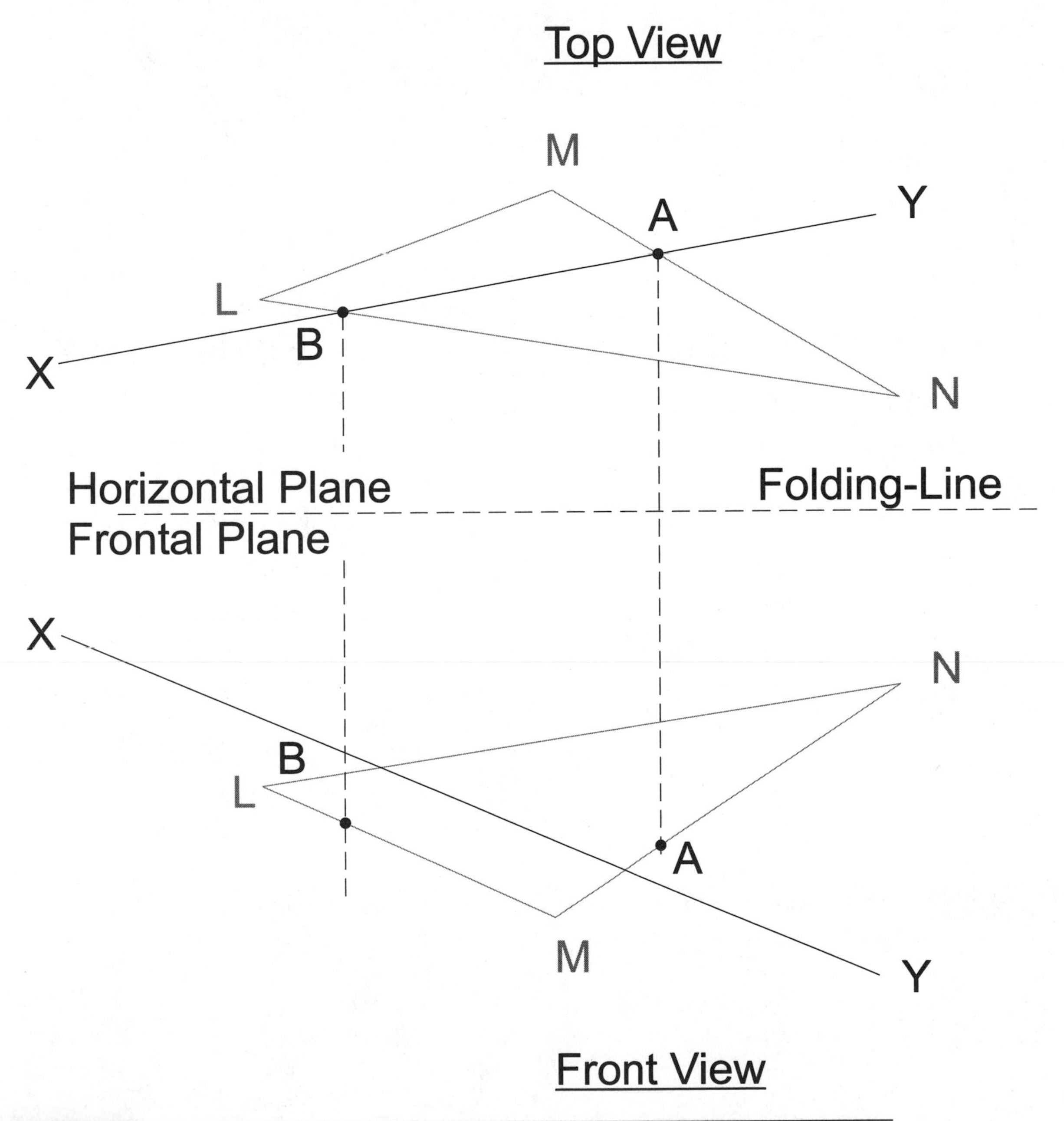


**Figure 4–31** *Points A and B are transferred from the horizontal view to the front view.*

## Step #4

Once points A and B have been transferred to the front view, a line segment is used to connect them. This line represents the intersecting of the two planes in the front view (see Figure 4–32).

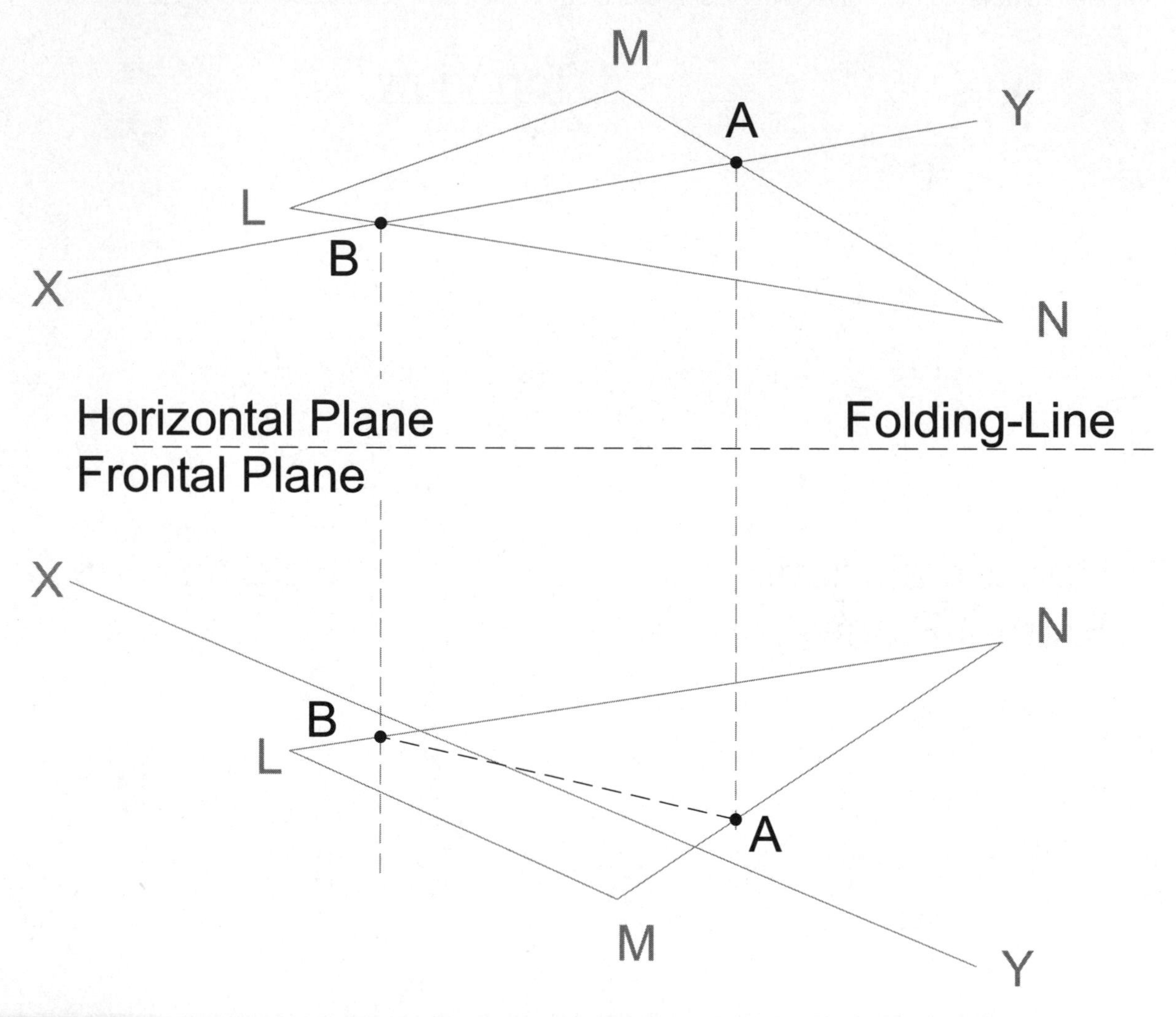


**Figure 4–32** *Line AB represents the intersection of planes LMN and the cutting plane.*

## Step #5

Now that this intersection has been established, the point where this plane crosses line XY is the piercing point of line XY and plane LMN. This point is transferred back to the top view by using standard orthographic techniques (see Figure 4–33).

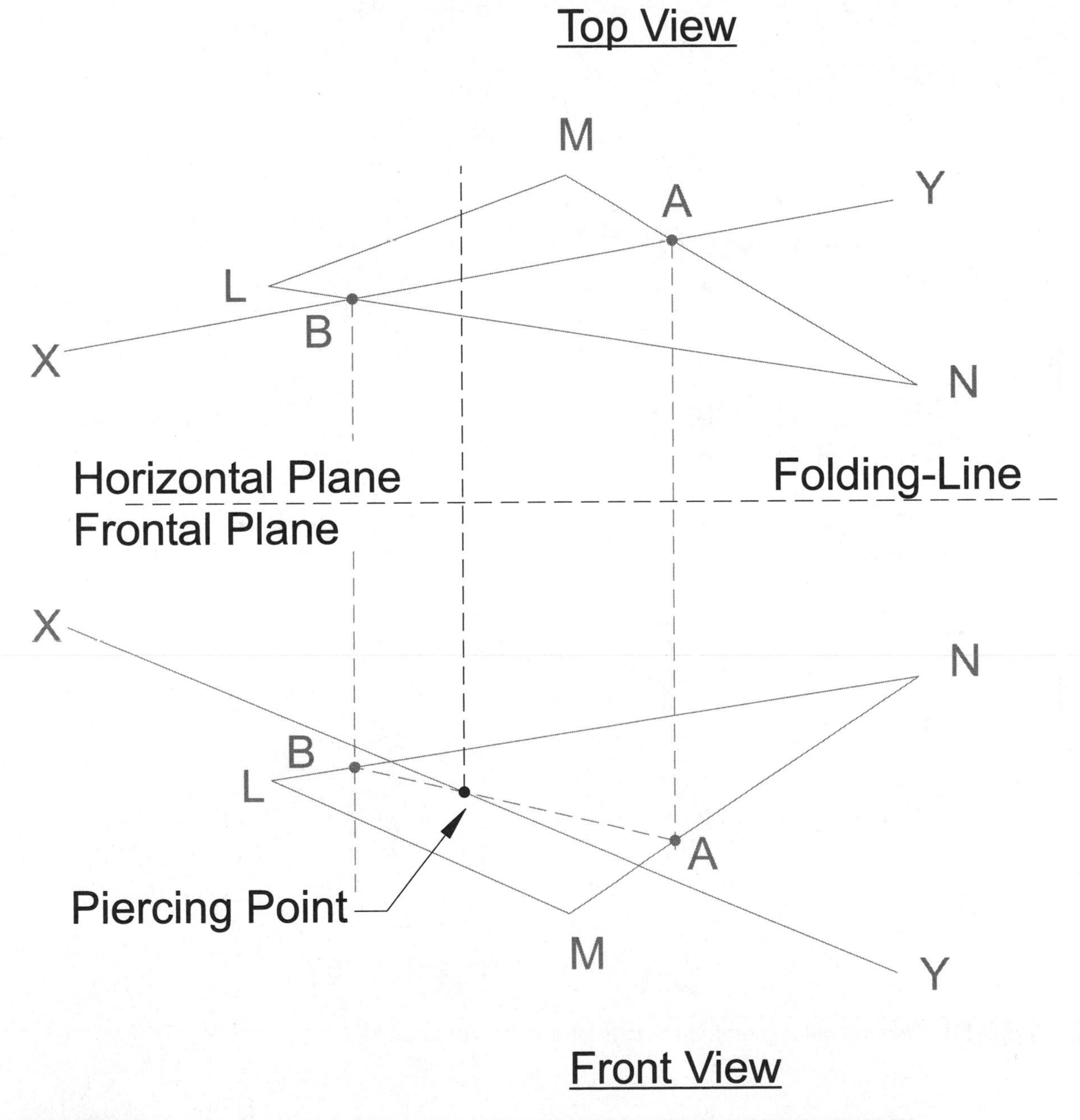


**Figure 4–33** *Piercing point transferred from front view to top view via projection lines.*

## Step #6

It is the intersection of the projection line with line XY that marks the exact location of the piercing point in the top view (see Figure 4–34).

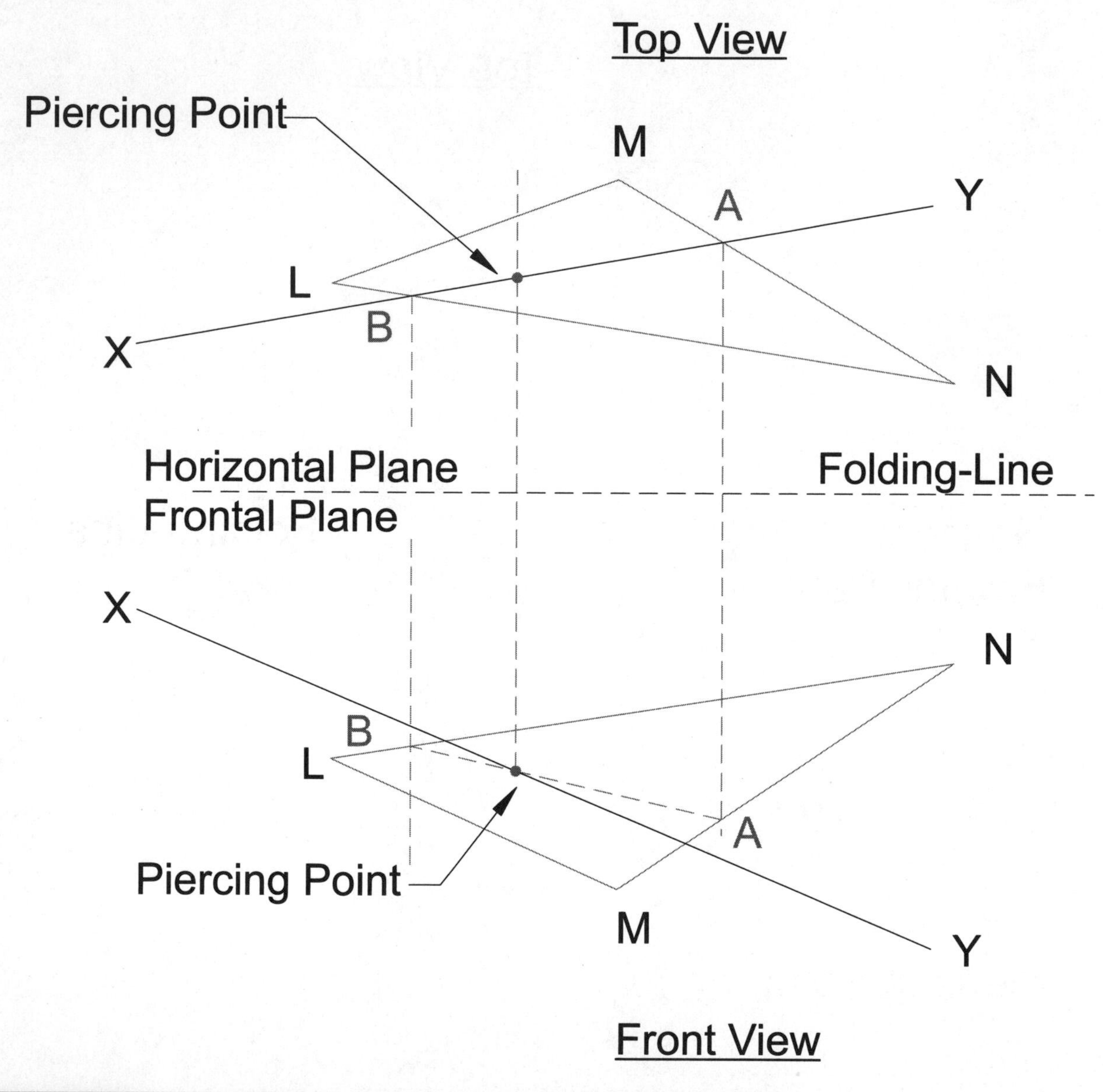


**Figure 4–34** *Piercing point located in both front and horizontal views.*

### Using the Cutting Plane Method in AutoCAD for Locating the Piercing Point of Two Planes

The cutting plane method presented in the previous section can be adapted to AutoCAD without changing the technique. Illustrated in the following example, the piercing point of line XY and plane LMN is determined using AutoCAD (see Figure 4–35). Use drawing DG04-35.dwg on the student CD-ROM to reproduce this example.

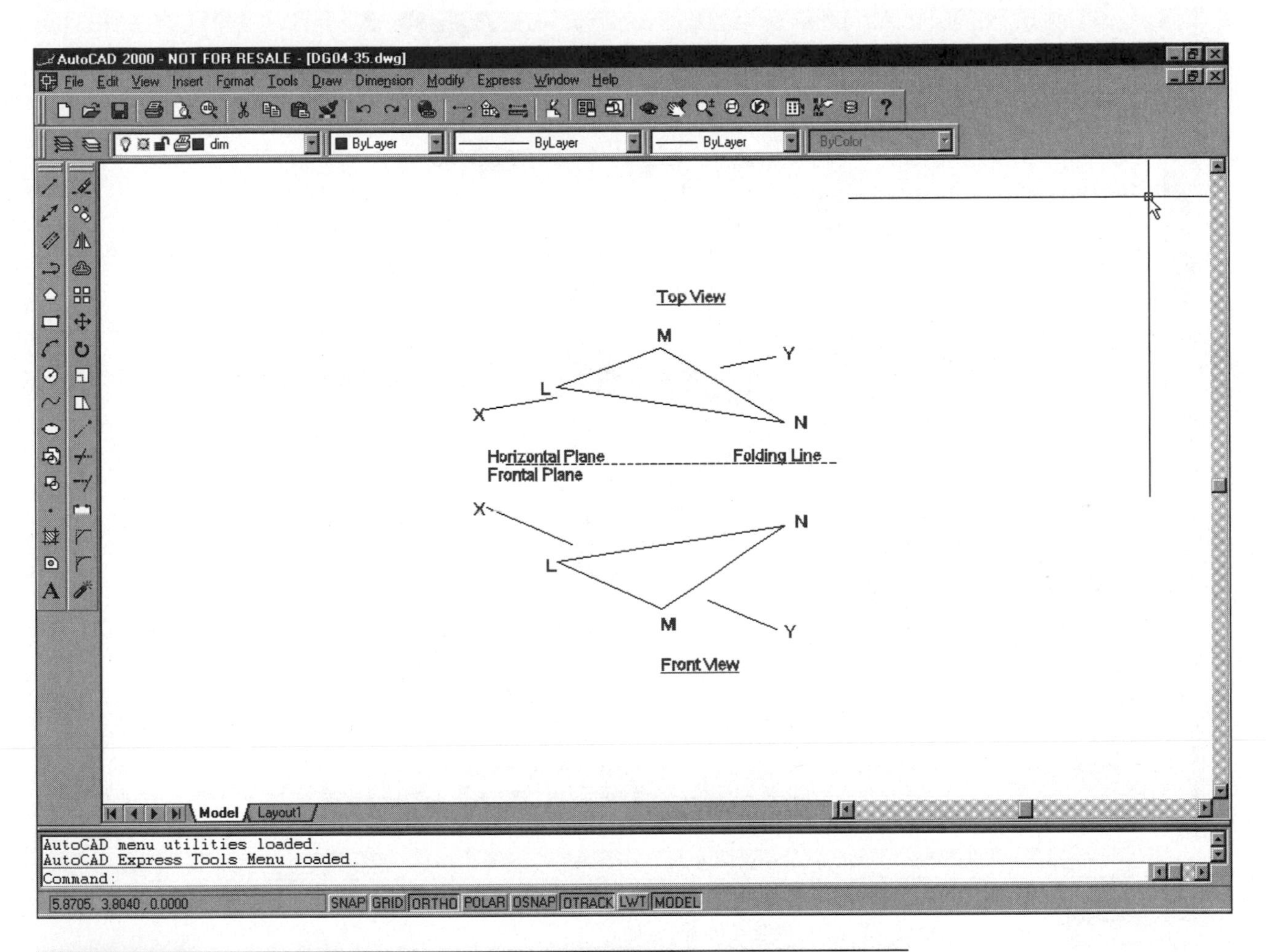


**Figure 4–35** *Horizontal and front views of a plane and an intersecting line.*

## Step #1

Using the LINE command, line XY is completed by connecting in the top view the two segments to produce the edge view of an imaginary cutting plane (see Figure 4–36).

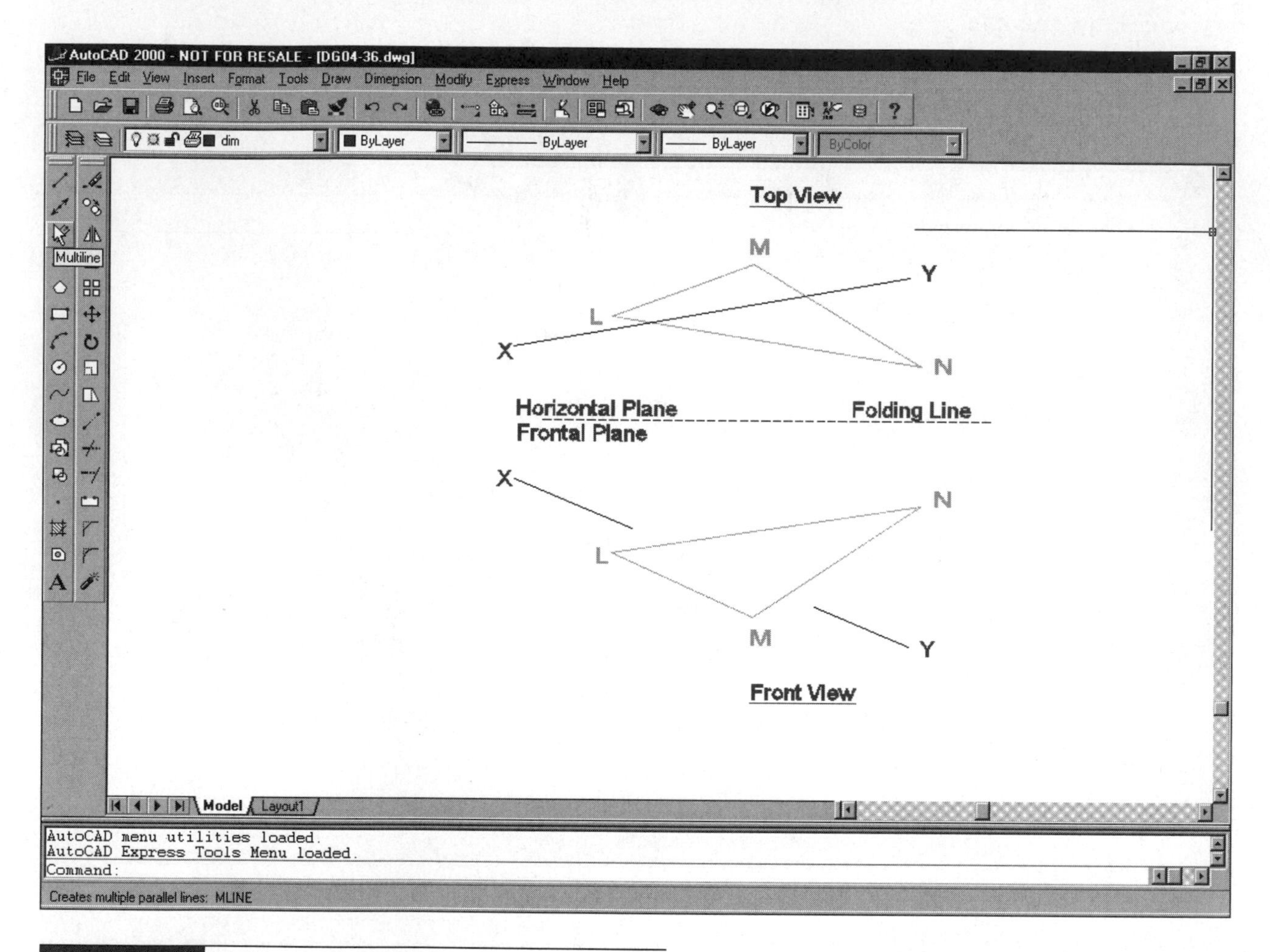


**Figure 4–36** *Line segment XY completed in top view.*

Command: LINE (ENTER)
From point: end of *(Select point X1.)*
To point: end of *(Select point Y1.)*
To point: *(Press ENTER to terminate the LINE command.)*

## Step #2

Using the LINE command, line XY is completed in the front view (see Figure 4–37).

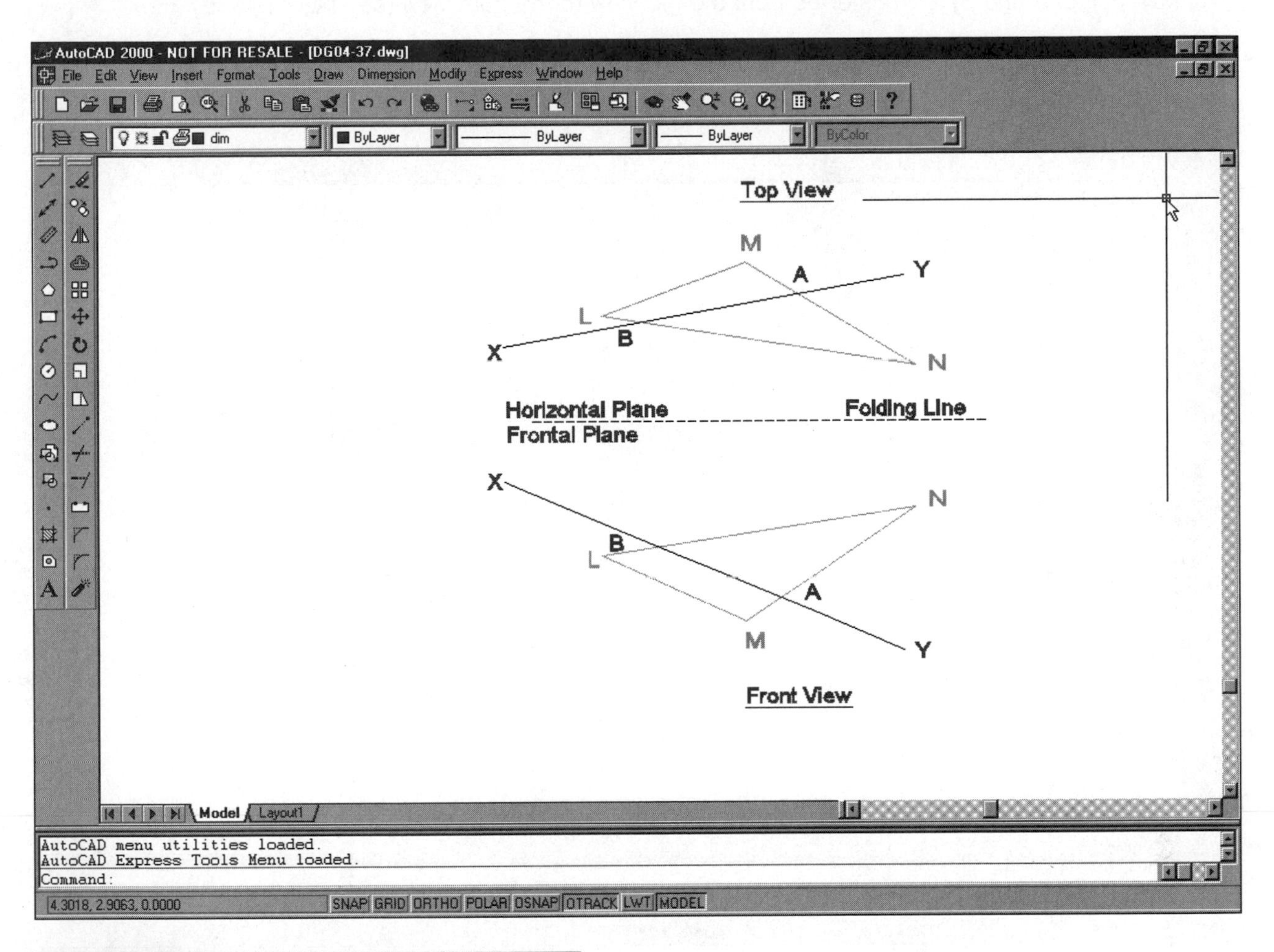


**Figure 4–37** *Line XY completed in front view.*

Command: LINE (ENTER)
From point: end of *(Select point X1.)*
To point: end of *(Select point Y1.)*
To point: *(Press* ENTER *to terminate* LINE *command.)*

## Step #3

Using the LINE command and the INTersection OSNAP option, the intersection of line XY with plane LMN (points A and B) is transferred from the top view to the front view (see Figure 4–38).

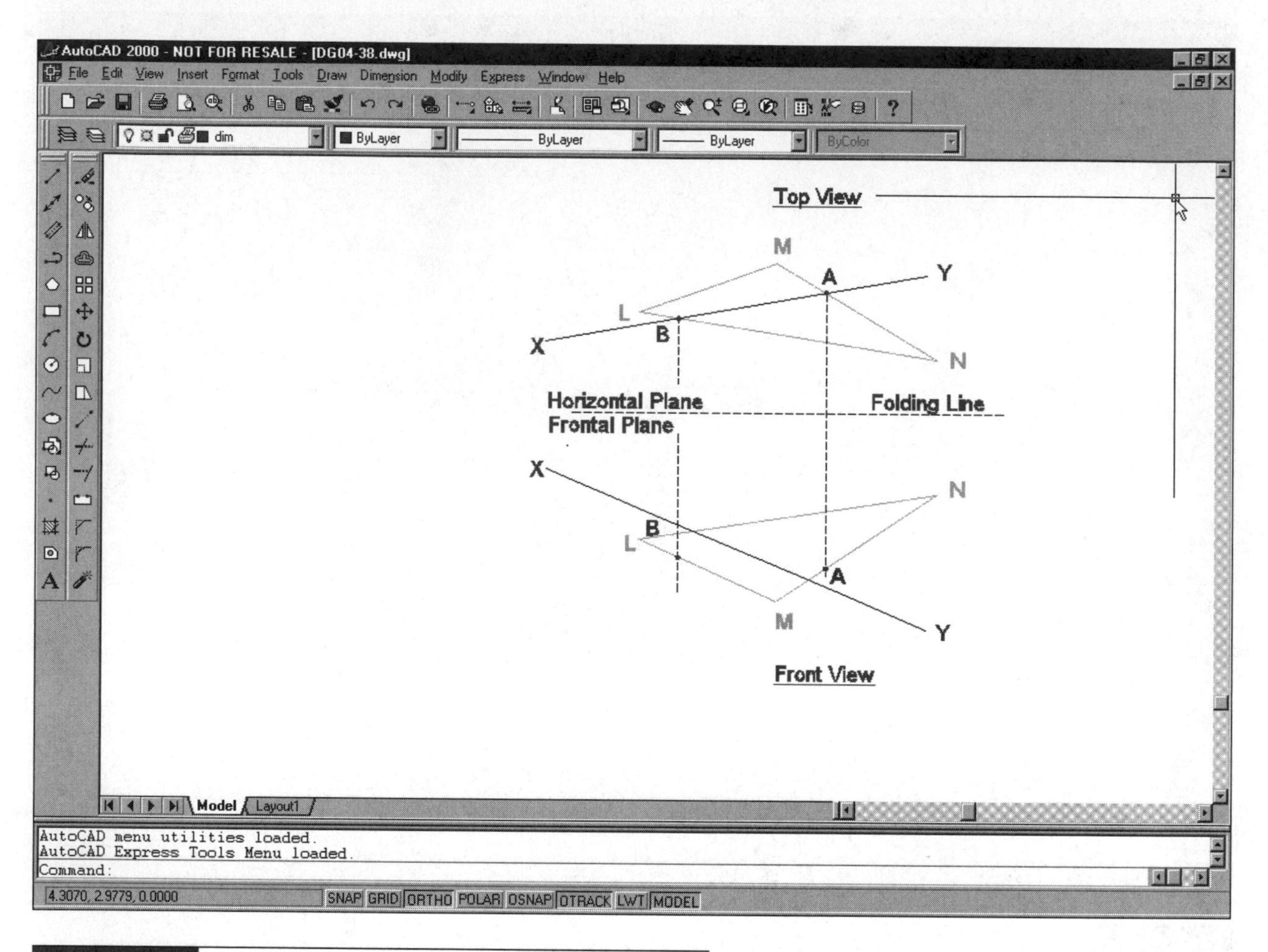


**Figure 4–38** *Points A and B are transferred to front view.*

Command: LINE (ENTER)
From point: int of *(Select point A.)*
To point: *(Select an arbitrary point below the front view.)*
To point: *(Press* ENTER *to terminate the* LINE *command.)*
Command: LINE (ENTER)
From point: int of *(Select point B.)*
To point: *(Select an arbitrary point below the front view.)*
To point: *(Press* ENTER *to terminate the* LINE *command.)*

## Step #4

Once the intersecting points of the two planes have been located in the front view, they are then connected using the LINE command (see Figure 4–39).

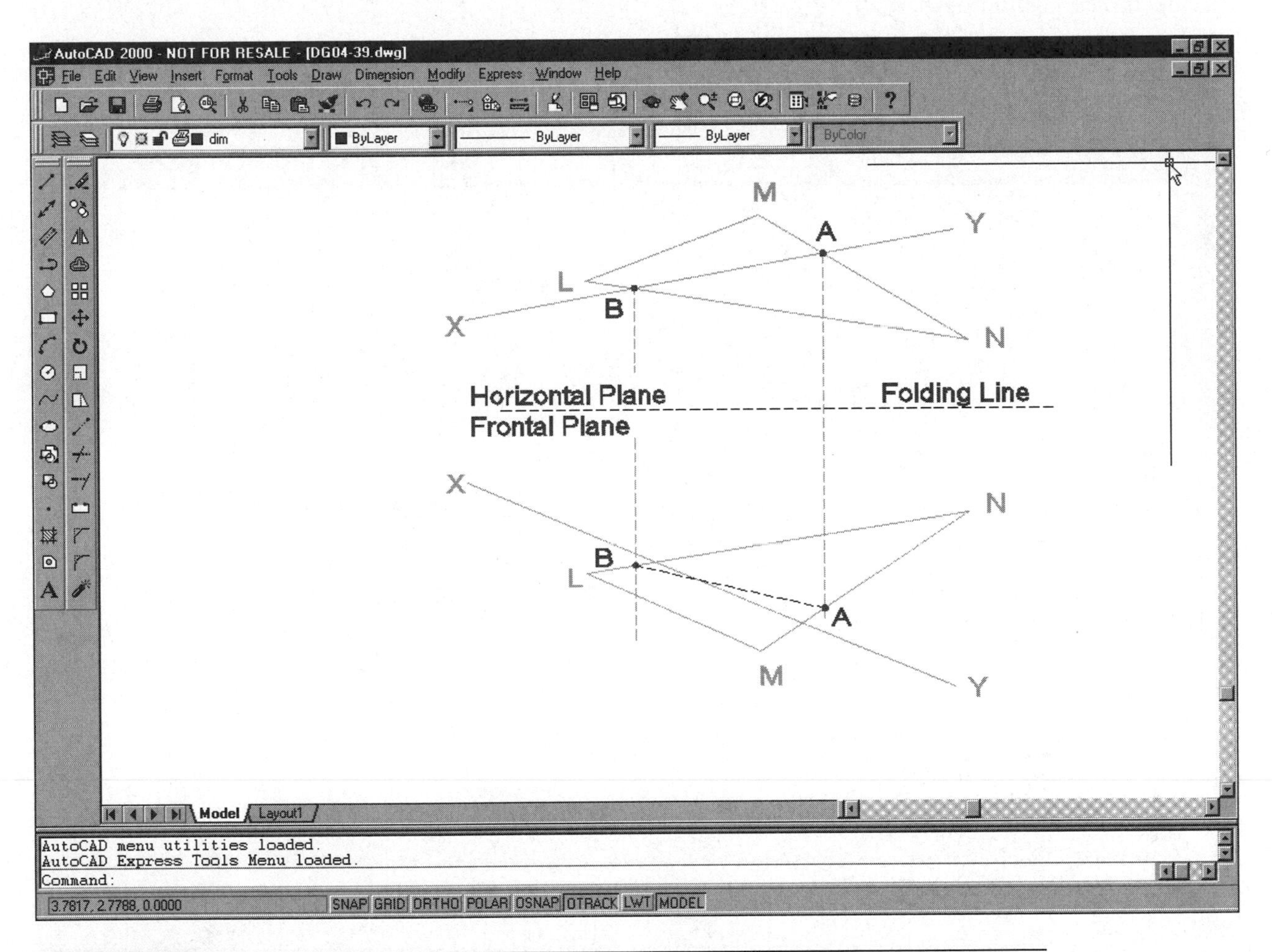


**Figure 4–39** *Line AB represents the intersection of planes LMN and the cutting plane.*

Command: LINE (ENTER)
From point: int of *(Select point A in the front view.)*
To point: int of *(Select point B in the front view.)*
To point: *(Press ENTER to terminate the LINE command.)*

## Step #5

The intersection of line AB with line XY in the front view represents the piercing point of line XY and plane LMN. This point is then transferred from the front view to the top view by using the LINE command (see Figure 4–40).

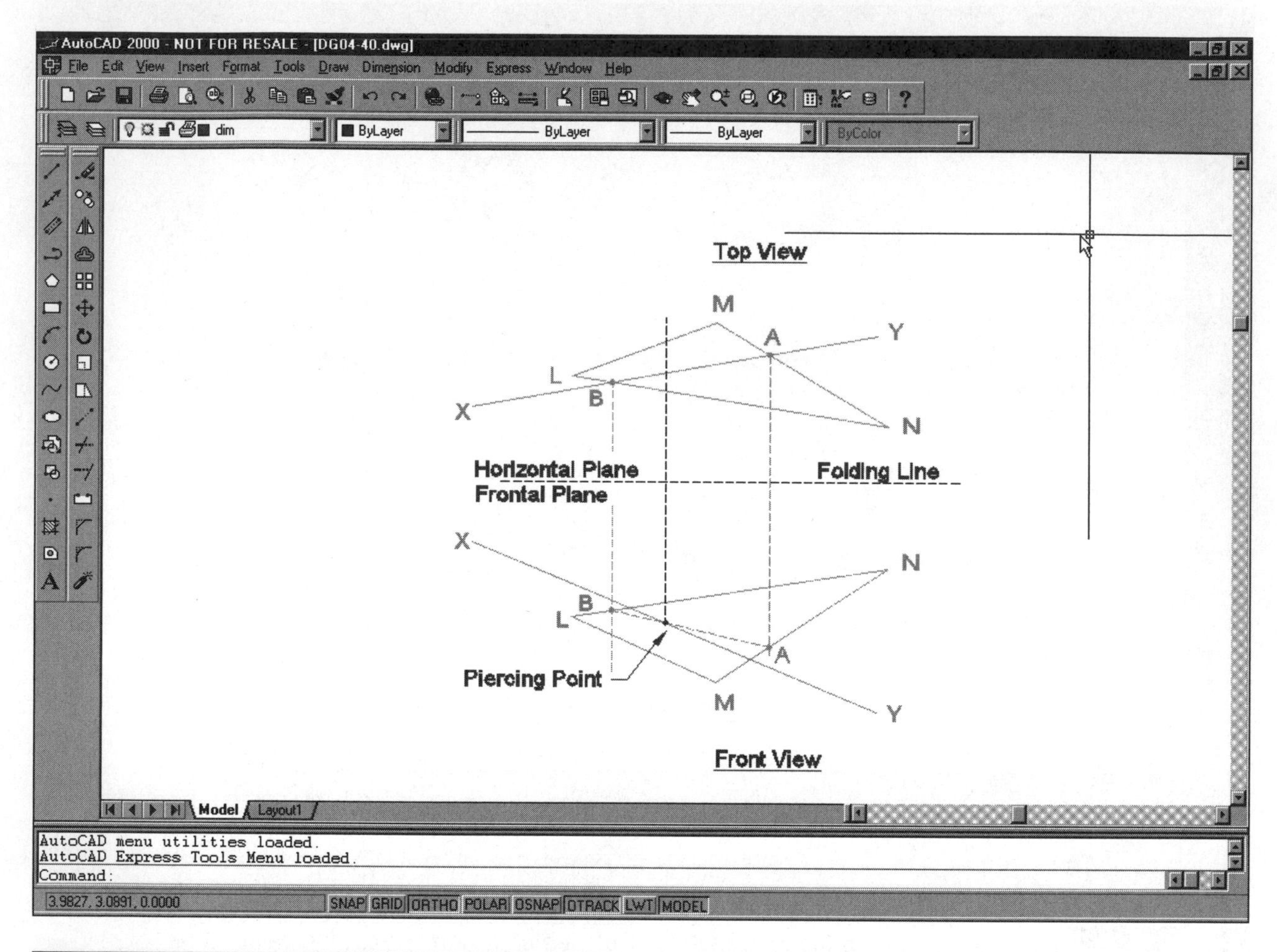


**Figure 4–40** *Transferring piercing point from the front view to the top view.*

Command: LINE (ENTER)
From point: int of *(Select the piercing point in the front view.)*
To point: *(Select an arbitrary point above the plan view.)*
To point: *(Press ENTER to terminate the LINE command.)*

## Step #6

The intersection of the line drawn from the front view in step five and line XY indicate the piercing point of line XY and plane LMN in the top view (see Figure 4–41).

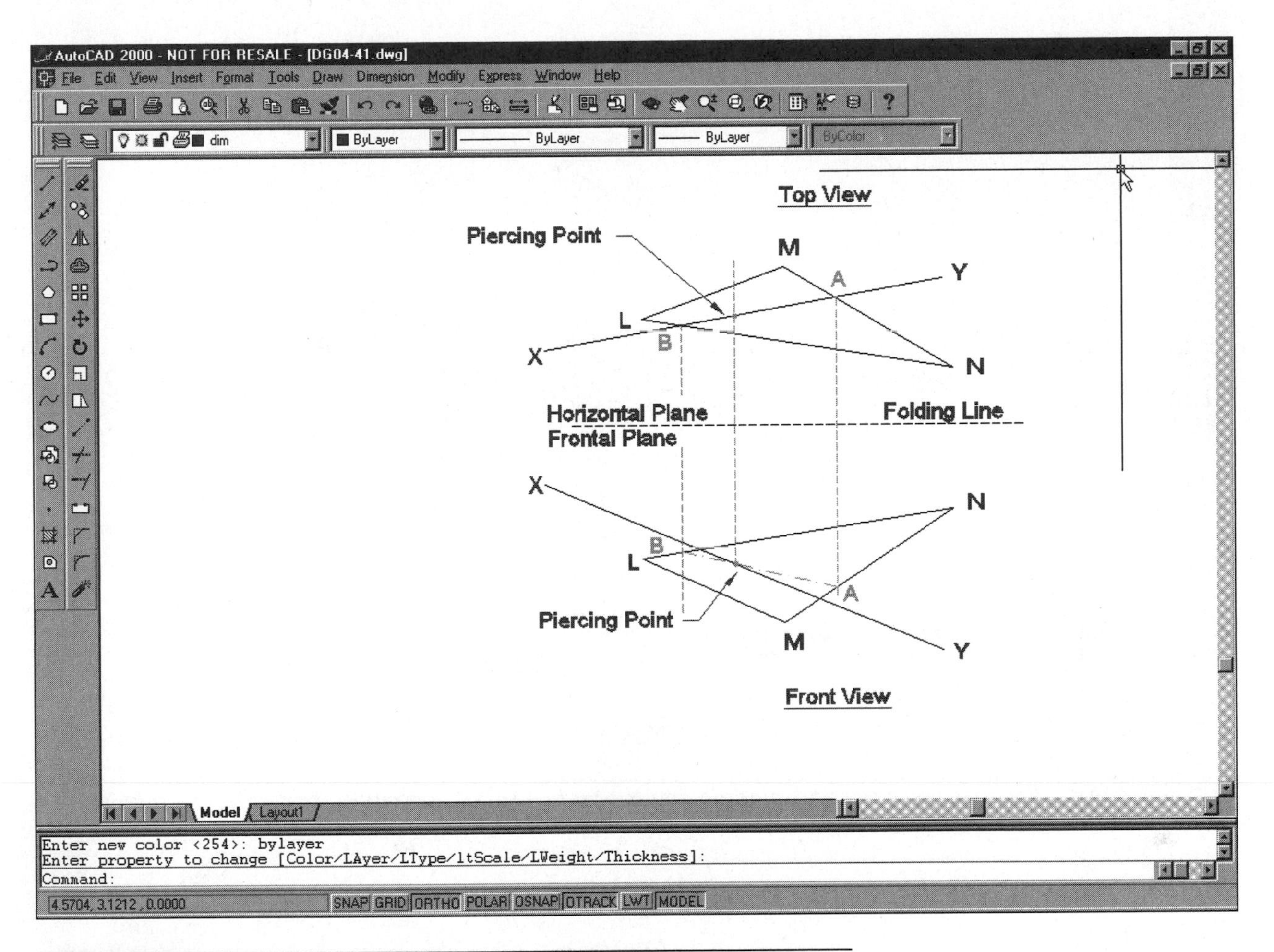


**Figure 4–41** *Piercing point located in both front and horizontal views.*

### Locating the Piercing Point Using the Point Command and App Osnap Option

One of the advantages of using AutoCAD to determine the piercing point of a line and a plane is that AutoCAD has the ability to calculate the exact coordinates of that point. This can be accomplished by modifying the two-view method to incorporate the use of the POINTcommand in conjunction with the APParent intersection osnap option. The APParent intersection option has two modes of operation, apparent and extended apparent intersection. The apparent intersection option is used to snap to the apparent intersection of two objects that appear to intersect on screen, but in reality do not. For example, in Figure 4–42 line XY appears to intersect the boundaries of plane ABCD in two locations, along lines AB and CD. However, when viewed from a plan view, line XY does not appear to cross the boundaries at all (see Figure 4–43). Consequently, to identify the imaginary intersection of these two points, the user would use the ID command along with the APParent intersection osnap option.

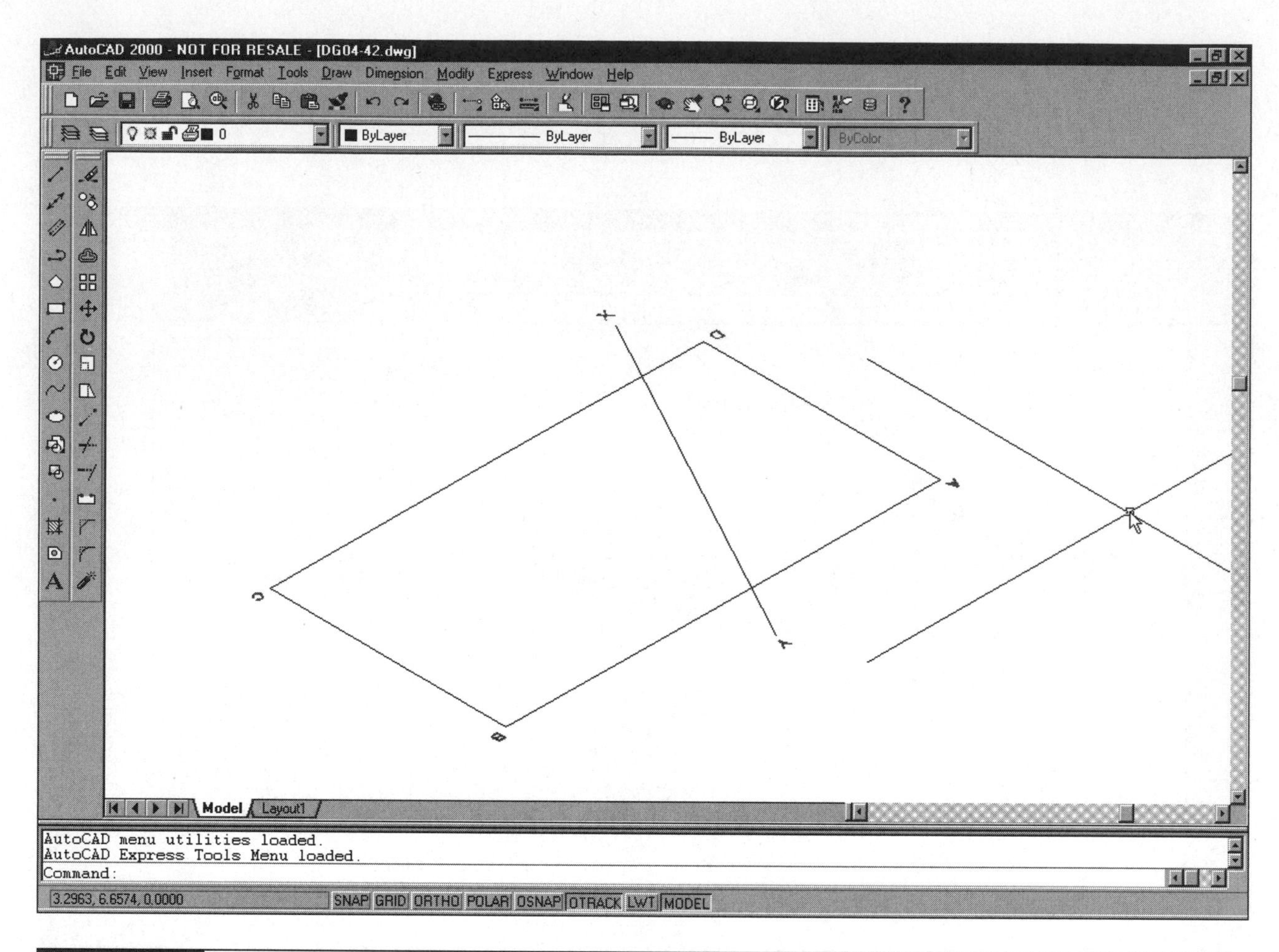


**Figure 4–42** *When viewed from an isometric angle (1,1,1), line XY appears to cross the boundaries of plane ABCD.*

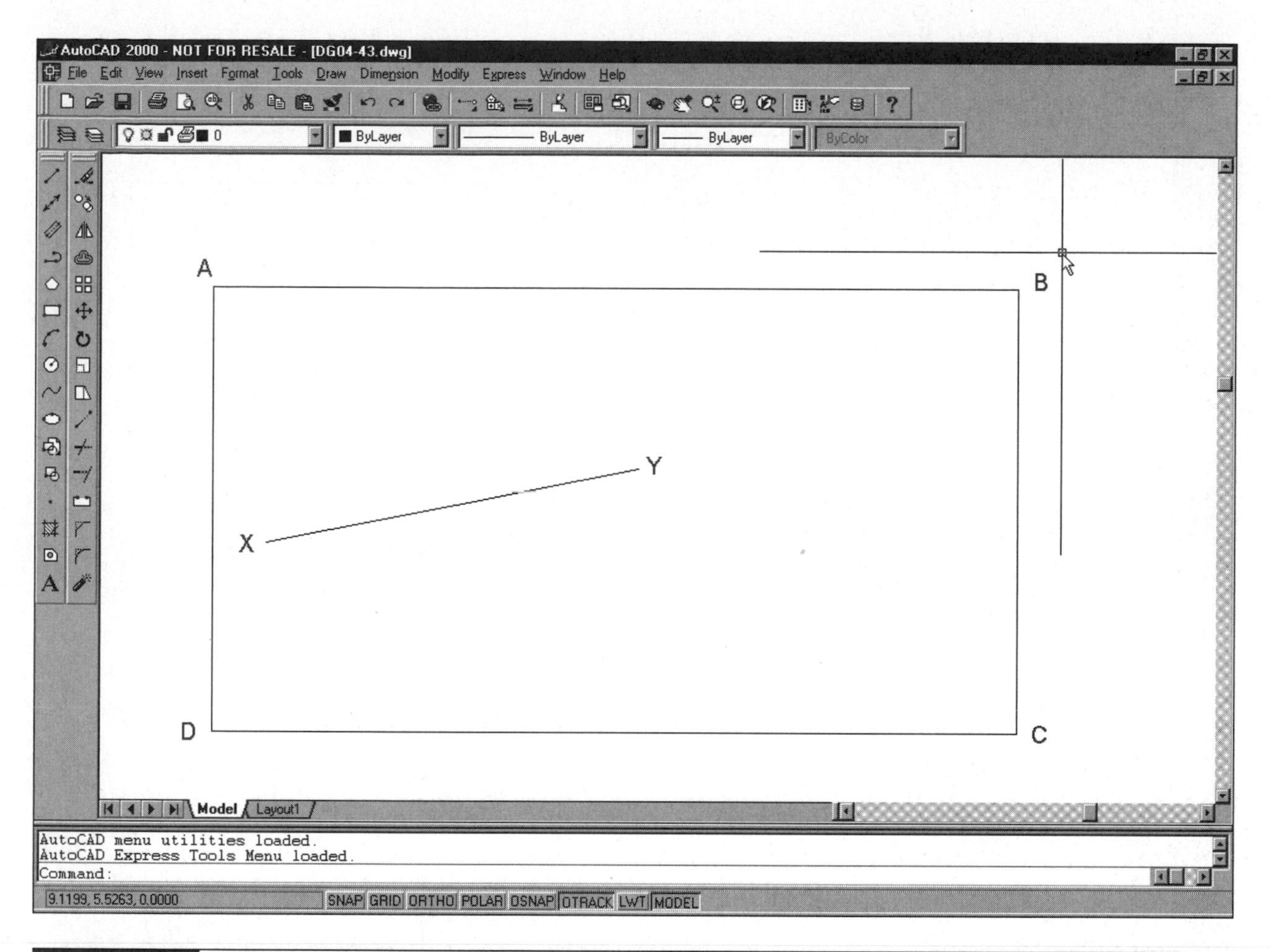


**Figure 4–43** *When viewed from the top view, line XY does not cross the boundaries of plane ABCD.*

Command: ID *(Press* ENTER *to start the* ID *command.)*
Point: APP *(Press* ENTER *to enable the APParent osnap option.)*
of X = 6.0615  Y = 4.1288  Z = 1.1765 *(The coordinates of the apparent intersection of line XY with line AB of plane ABCD.)*
Command: ID *(Press* ENTER *to start the* ID *command.)*
Point: APP *(Press* ENTER *to enable the APParent osnap option.)*
of X = 4.9746  Y = 3.9134  Z = -0.9282 *(The coordinates of the apparent intersection of line XY with line CD of plane ABCD.)*

Not only can the extended apparent intersection option find the coordinates of an apparent intersection, it can also find the coordinates for a projected intersection that would occur if two objects were extended in space. For example, if line XY in Figure 4–44 were extended it would appear to cross lines AD and BC. To determine where these lines would intersect, the ID command along with APParent intersection osnap option would be used.

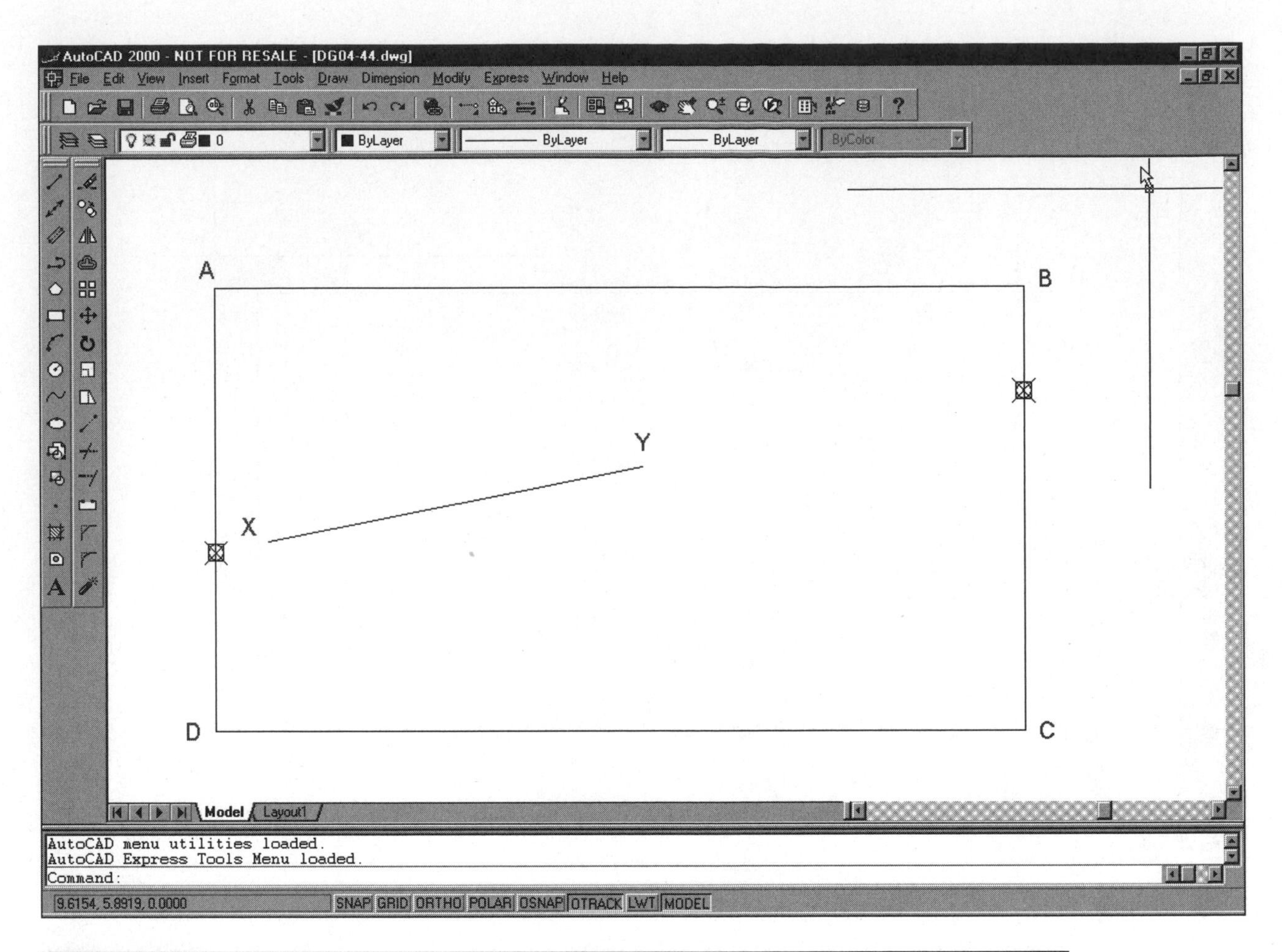


**Figure 4–44** *The extended apparent intersection location between lines XY, AD. and BC.*

Command: ID *(Press* ENTER *to start the* ID *command.)*
Point: APP *(Press* ENTER *to enable the Apparent osnap option.)*
of and X = 4.0465 Y = 3.7295 Z = -2.7254 *(The coordinates of the extended apparent intersection of line XY and line AD of plane ABCD.)*
Command: ID *(Press* ENTER *to start the* ID *command.)*
Point: APP *(Press* ENTER *to enable the Apparent osnap option.)*
of and X = 8.8657 Y = 4.6845 Z = 6.6064 *(The coordinates of the extended apparent intersection of line XY and line BC of plane ABCD.)*

The first step in determining the actual coordinates of the piercing point between a line and a plane in AutoCAD is to reset the observer's line-of-sight so that the line appears to cross the plane in at least two locations. These locations are marked using the AutoCAD POINT command along with the APP osnap option. After this has been completed, the view is switched back to its original settings and the two points are then connected with a line segment. The intersection of this line with the original line marks the location of the piercing point between the line and the plane. The exact coordinates of this piercing point may now be obtained by using the AutoCAD ID command. This procedure is illustrated in the following example where the piercing point of Figure 4–45 will be determined. Drawing DG04-45.dwg on the student CD-ROM can be used to reproduce this example.

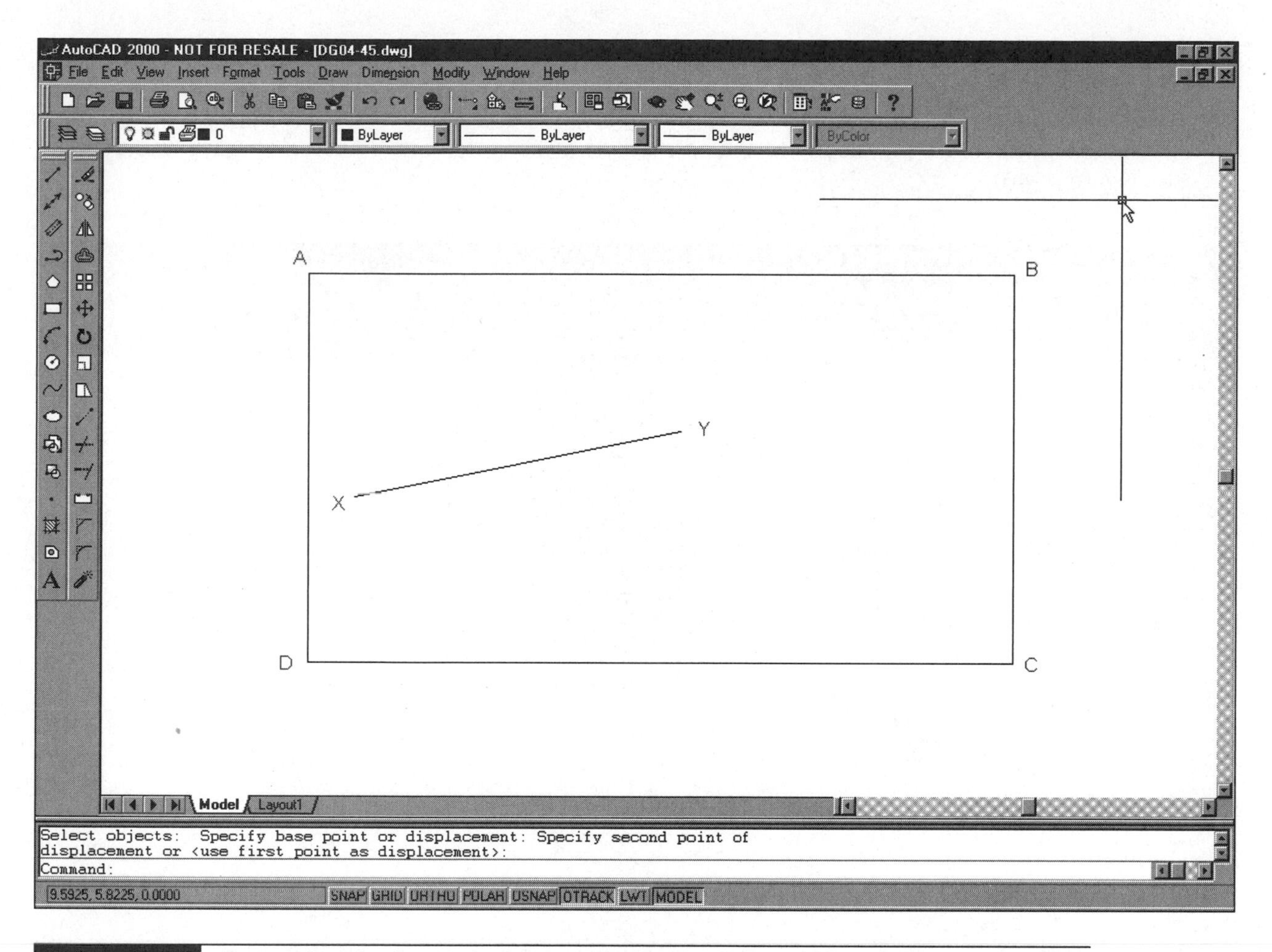


**Figure 4–45** *Determine the location of the piercing point between line XY and plane ABCD.*

## Step #1

Using the VPOINT command, the observer's line-of-sight is changed so that the line appears to cross the boundary in at least two locations. In this example a VPOINT command setting of 1,1,1 was used (see Figure 4–46).

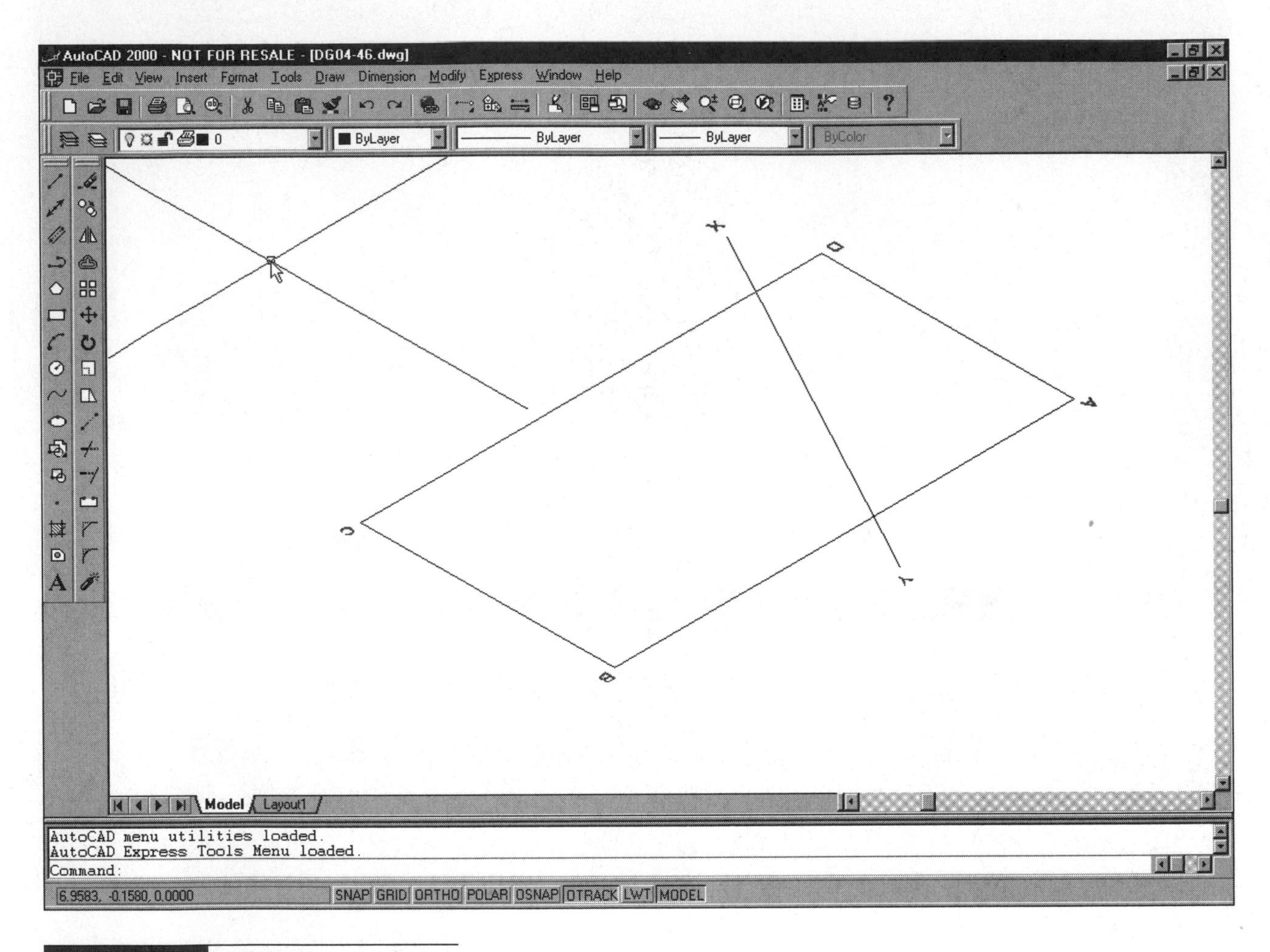


**Figure 4–46** *VPOINT set to 1,1,1.*

Command: VPOINT (ENTER)

Rotate/<View point> <1.0000,1.0000,1.0000>: 1,1,1 *(Press* ENTER *to position the line-of-sight down the positive X, Y, and Z-axes. The line-of-sight is now an angle of 45° in the XY-plane and 35° from the XY-plane.)*

Regenerating drawing.

## Step #2

Using the POINT command, the points at which line XY intersects lines AB and CD are marked (see Figure 4–47).

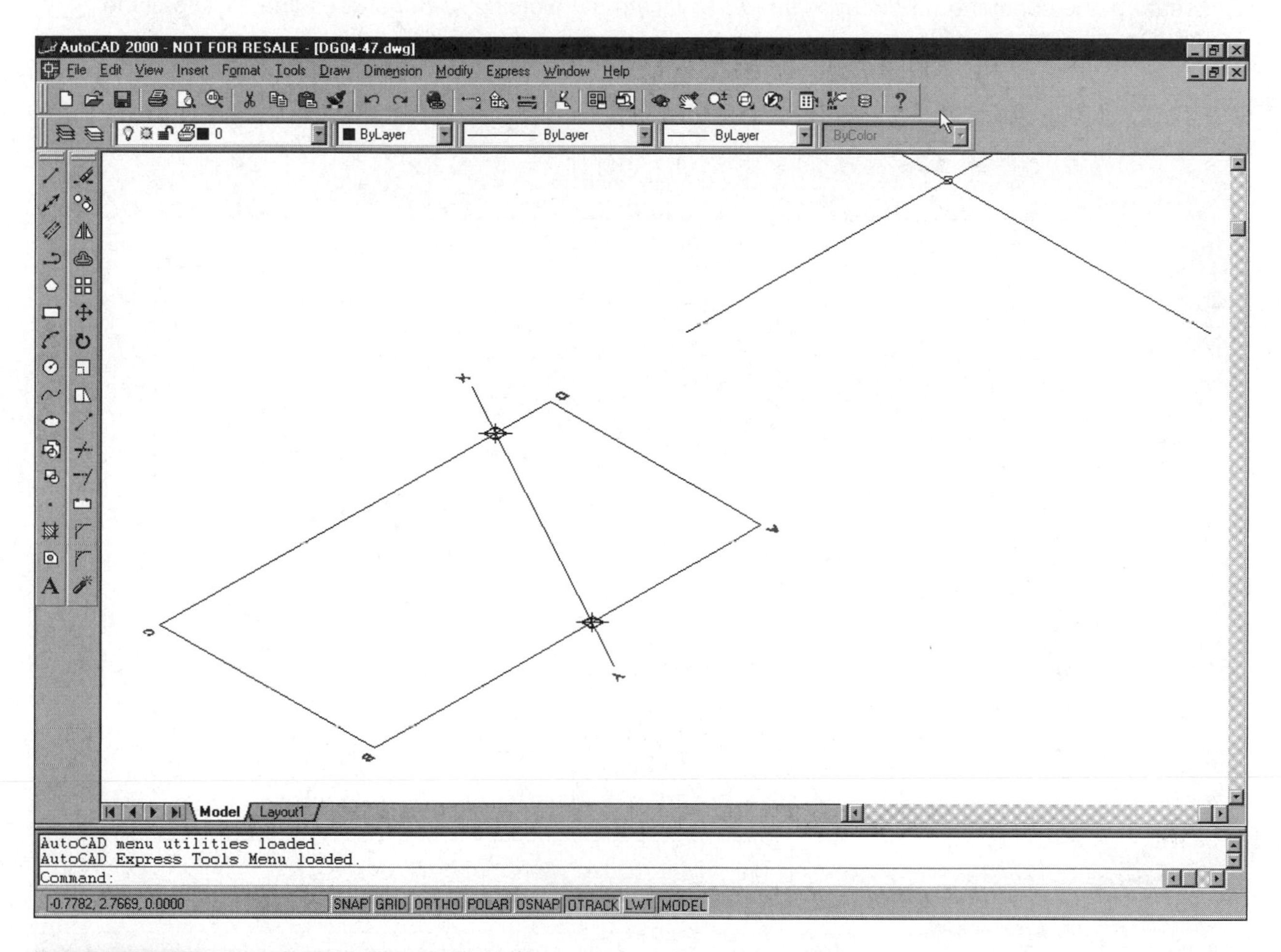


**Figure 4–47** *Intersection points marked.*

Command: POINT (ENTER)
Point: APP of *(Places a node at the apparent intersection of lines XY and DC.)*
Command: POINT (ENTER)
Point: APP of *(Places a node at the apparent intersection of lines XY and AB.)*

## Step #3

Using the VPOINT command to change back to the original viewing angle reveals the two projected intersection points on lines AB and CD. These points, when connected using the LINE command, will create a line segment that crosses line XY to locate the piercing point between line XY and plane ABCD (see Figure 4–48).

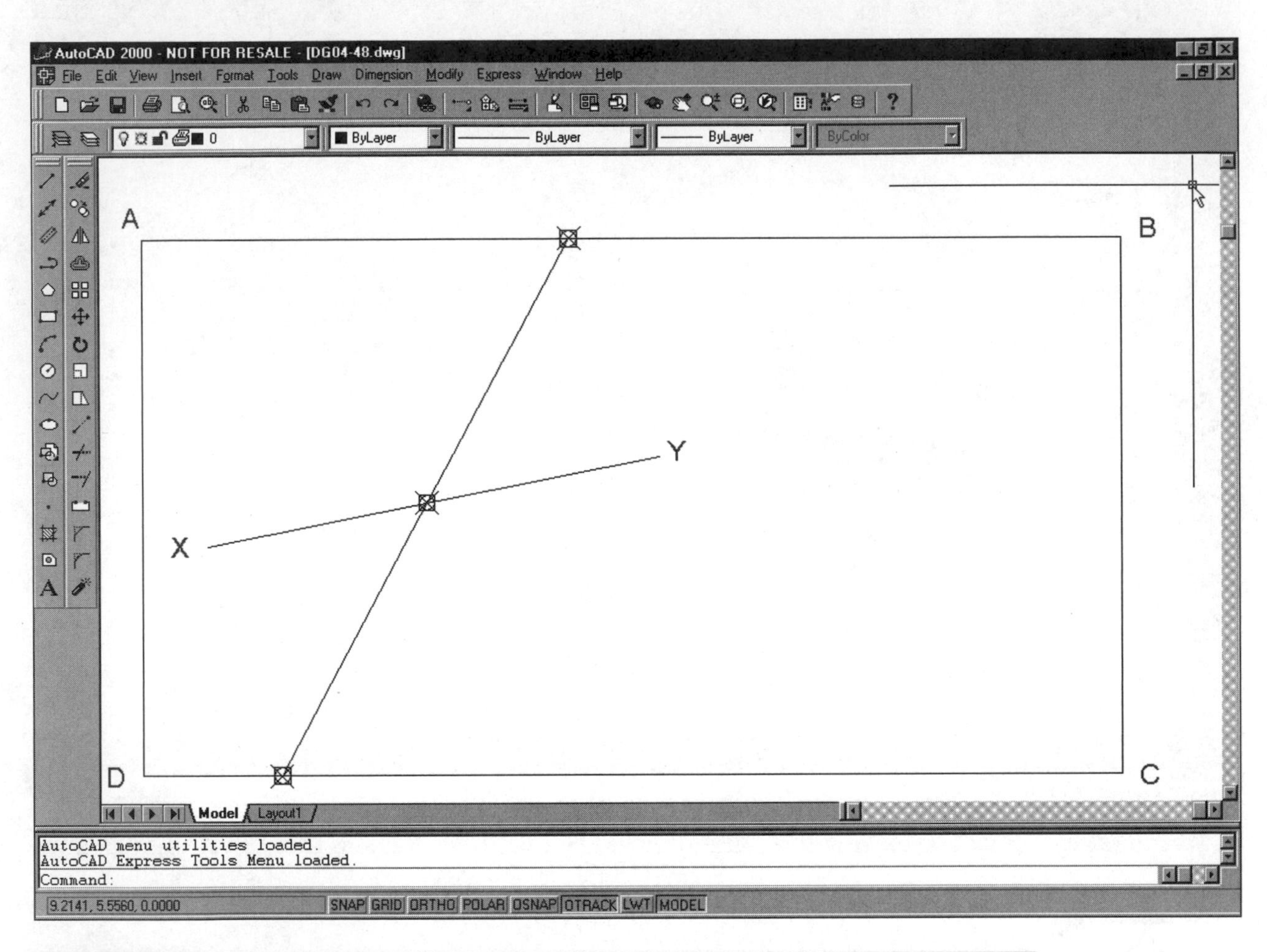


**Figure 4–48** *Piercing point of line XY located by connecting points on line AB and CD.*

Command: VPOINT (ENTER)
Rotate/<View point> <1.0000,1.0000,1.0000>: 0,0,1 *(Press* ENTER *to position the observer's line-of-sight back to its original position.)*
Regenerating drawing.
Command: LINE (ENTER)
From point: NODE *(Press* ENTER *to activate the AutoCAD NODE osnap option.)*
of *(Select the node located on line CD.)*
To point: NODE *(Press* ENTER *to activate the AutoCAD NODE osnap option.)*
of *(Select the node located on line AB to create a line segment between the two points.)*
To point: *(Press* ENTER *to terminate the* LINE *command.)*
Command: POINT (ENTER)
Point: INT *(Press* ENTER *to select the intersection of line XY and the line just created.)*
of

## Locating the Intersection of Two Planes

If a line intersects a plane and that line is contained on another plane, then it stands to reason that the two planes also intersect. In theory, all planes are either parallel or intersecting. When two or more planes intersect, they will share more than one common point. In fact, they will share a series of points that may be connected by a straight-line (***line of intersection***), and every point along that line will be common to both planes. To determine the location of the line of intersection, one of three methods can be used: the auxiliary view method, piercing point method, or cutting plane method.

### Auxiliary View Method

The intersection of two planes forms a straight line. Consequently, if a view is constructed to show one of the planes in edge view, then the edge of the plane will also be the line of intersection between the two planes. Any point along the edge will be common to both planes; by selecting two points and projecting their location to corresponding views, the line of intersection can be determined. In Figure 4–49a the line of intersection is determined by constructing an auxiliary view where plane MNOP appears as line NO. The points at which this line crosses the boundaries of plane ABCD indicate the line of intersection (see Figure 4–49b). These points are then transferred back to the top and front views (see Figure 4–49c), were they are connected with line segments to complete the line of intersection (see Figure 4–49d).

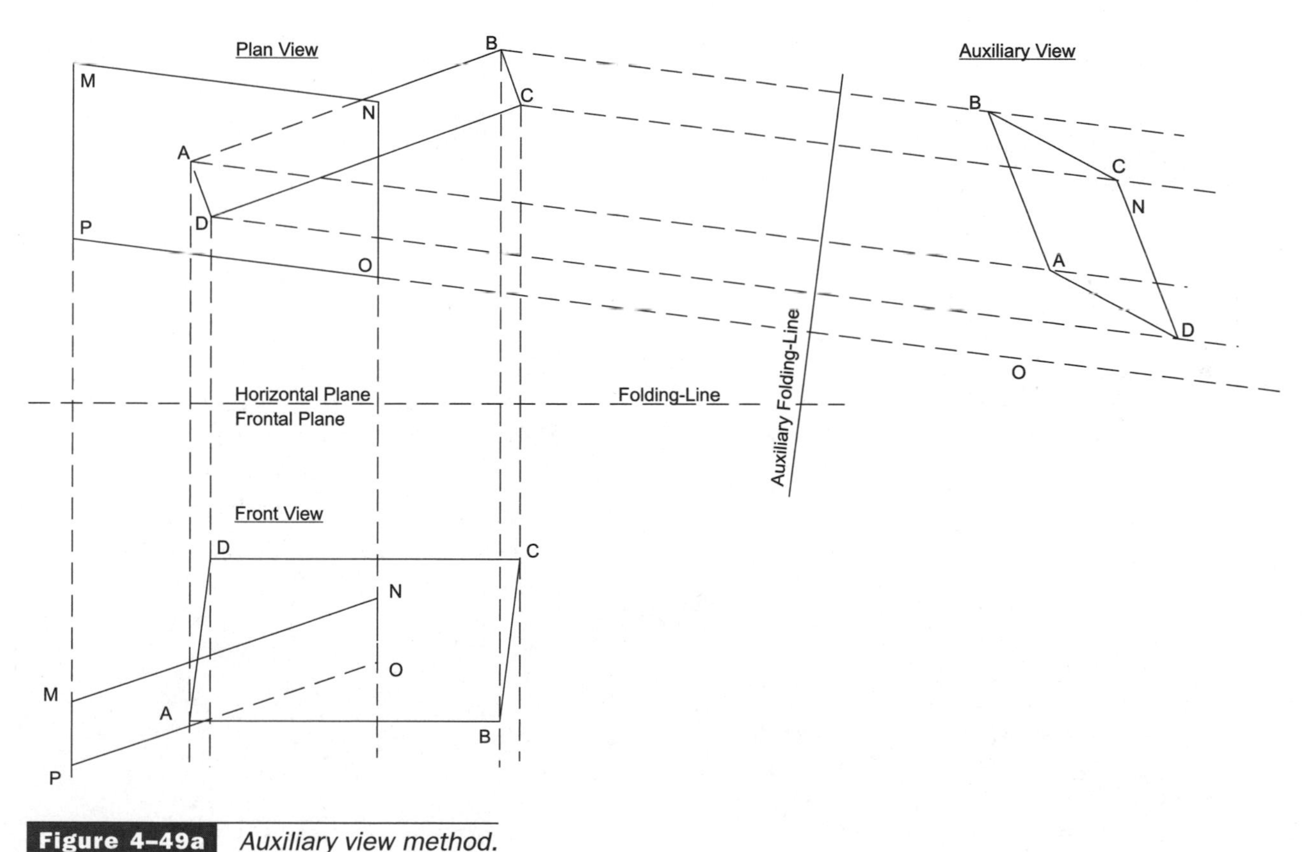


**Figure 4–49a** *Auxiliary view method.*

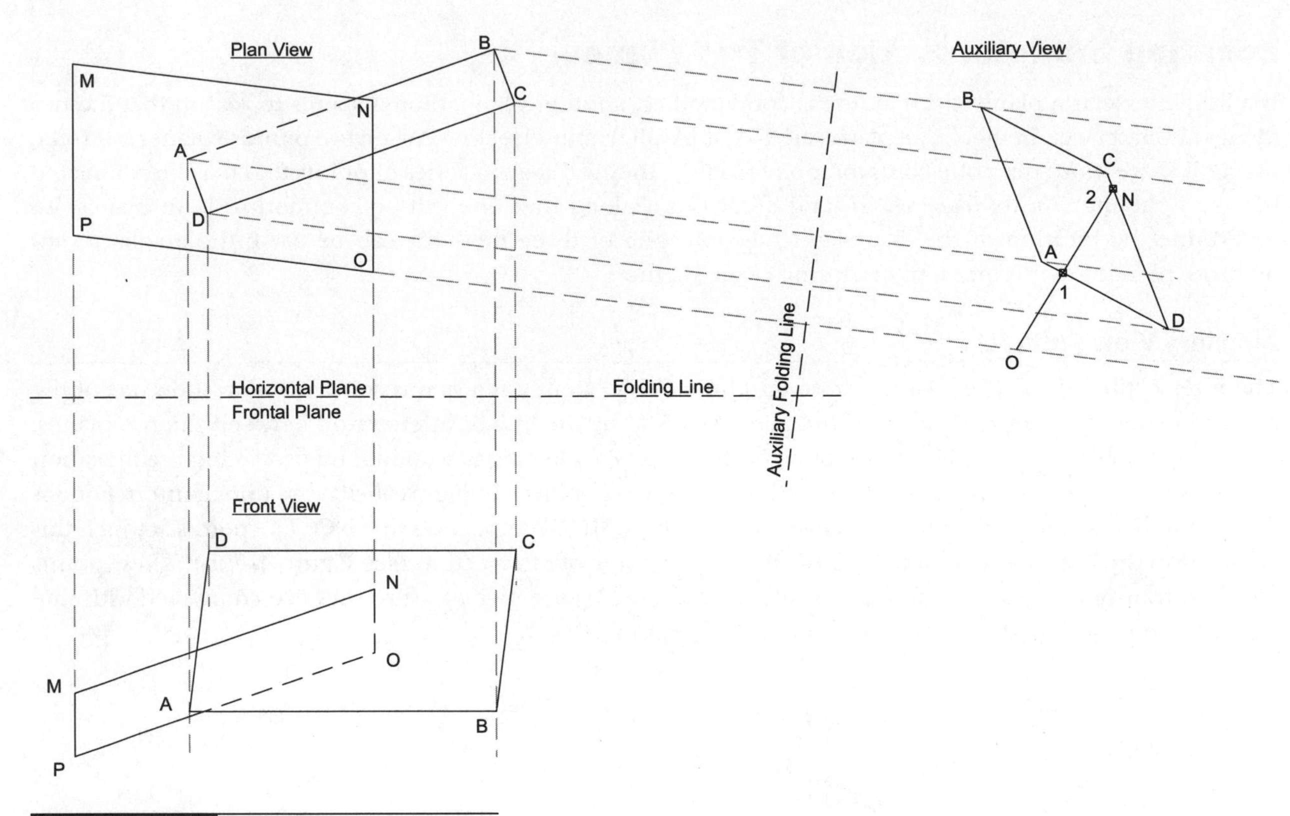


**Figure 4–49b** *Auxiliary view method.*

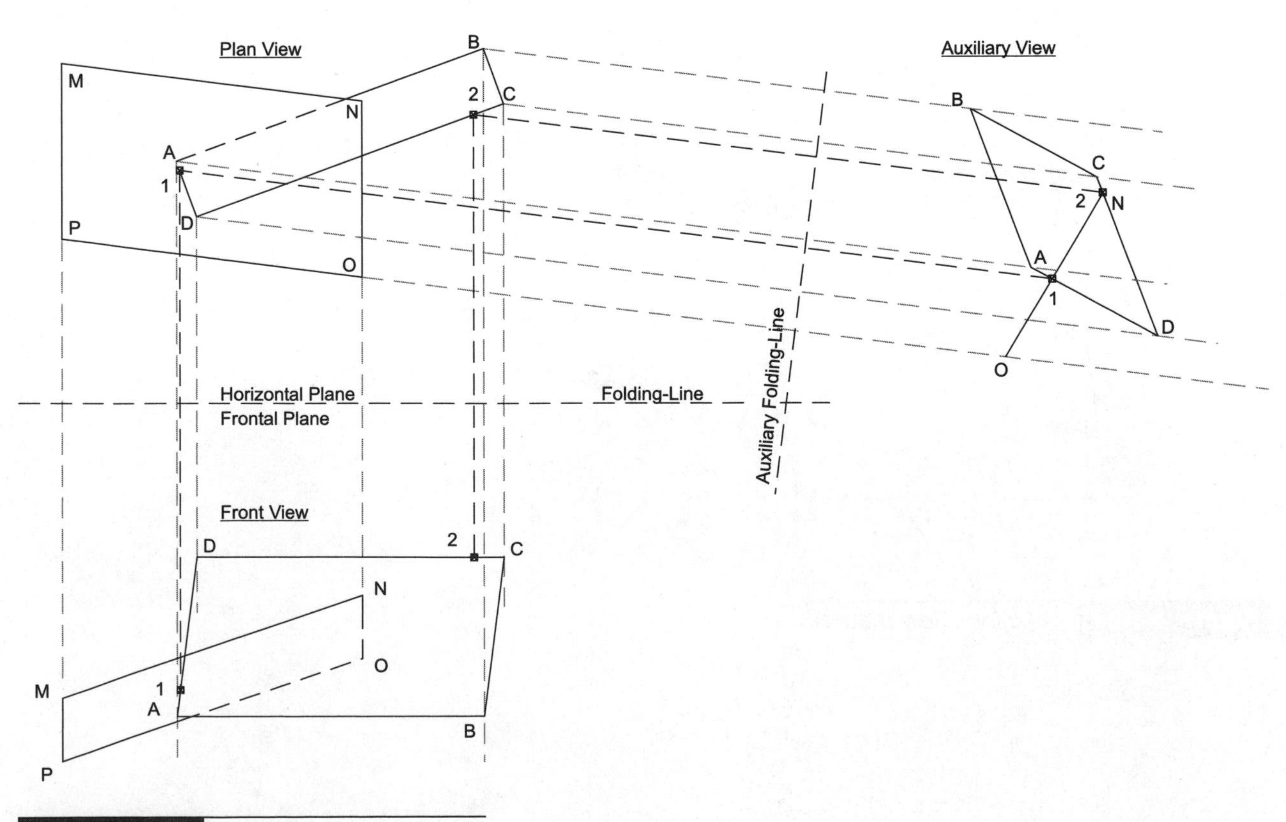


**Figure 4–49c** *Auxiliary view method.*

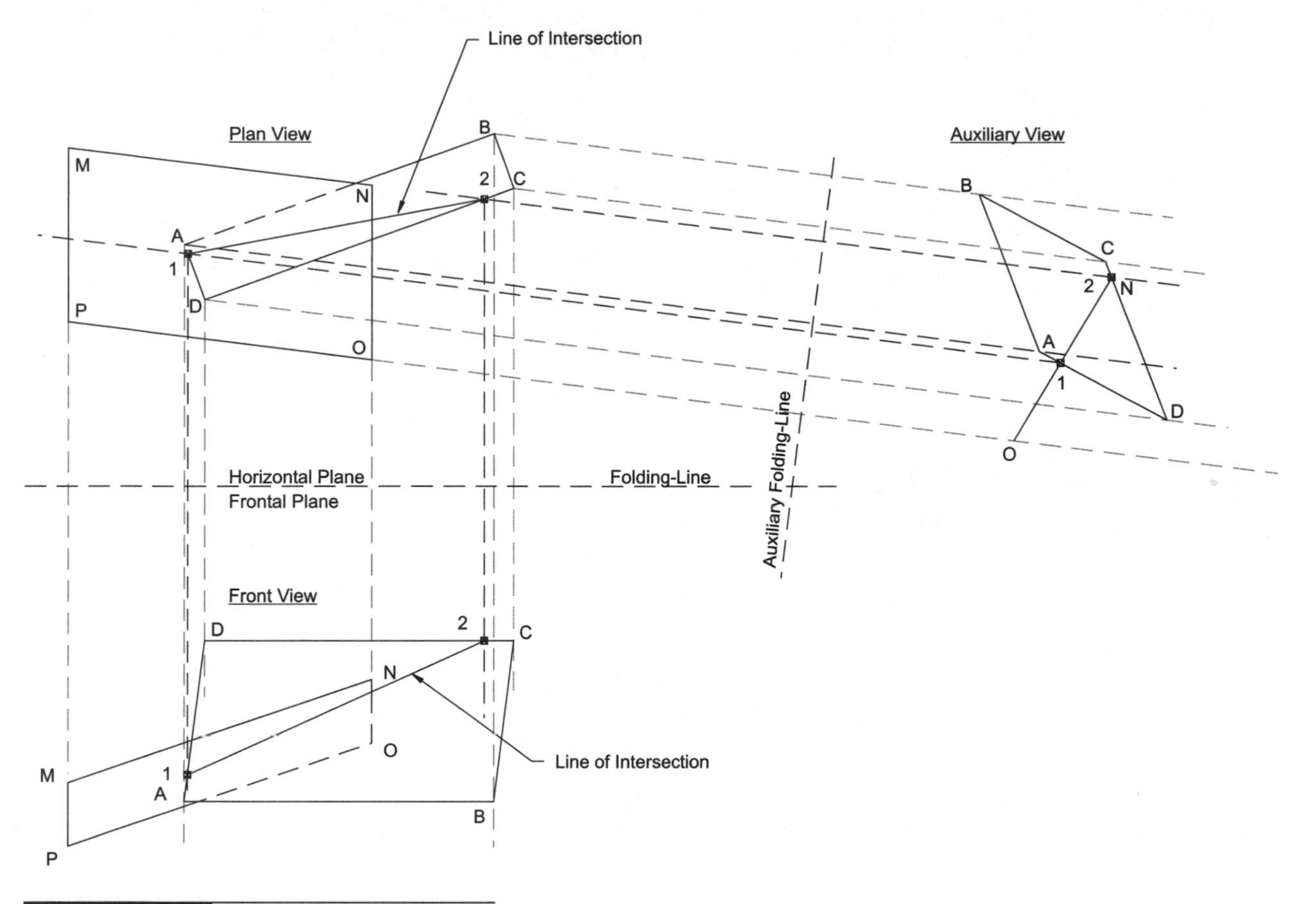


**Figure 4–49d** *Auxiliary view method.*

## Piercing Point Method

The intersection of two planes can also be determined by using the piercing point method. Since the line of intersection between two planes is being determined, this method will have to be employed at least twice to locate at least two points. In some cases the method might have to be repeated a third time if the piercing points happen to be close together.

## Cutting Plane Method

When two planes that are not parallel to one another are crossed by a third plane not parallel to the first two, the cutting plane method can be used to find a point common to all three planes. When this is repeated two or more times, the result is a series of points that can be connected to form the line of intersection between the three planes. For example, using the plan and front views of planes ABCD and MNOP in Figure 4–50, the line of intersection between these two planes can be found by constructing two cutting planes that are perpendicular to the line-of-sight in any of the views given. In this example, the cutting planes are constructed in the front view with the intersecting points of the three planes being transferred from the front view to the plan view using standard orthographic projection techniques. After these points have been transferred, a line segment can be created which extends through these points to mark the line of intersection. Once the line of intersection has been established, its proper location can then be transferred to the front view using orthographic projection techniques. See Figures 4–50 through 4–56.

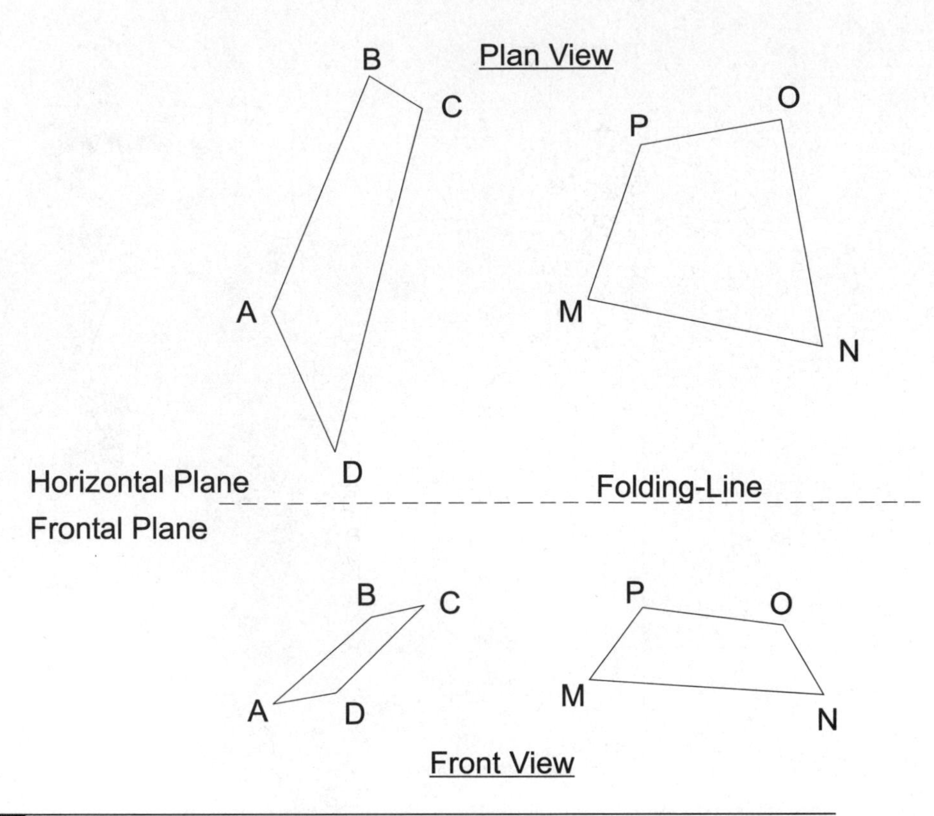


**Figure 4–50** *Determine the line of intersection between planes ABCD and MNOP.*

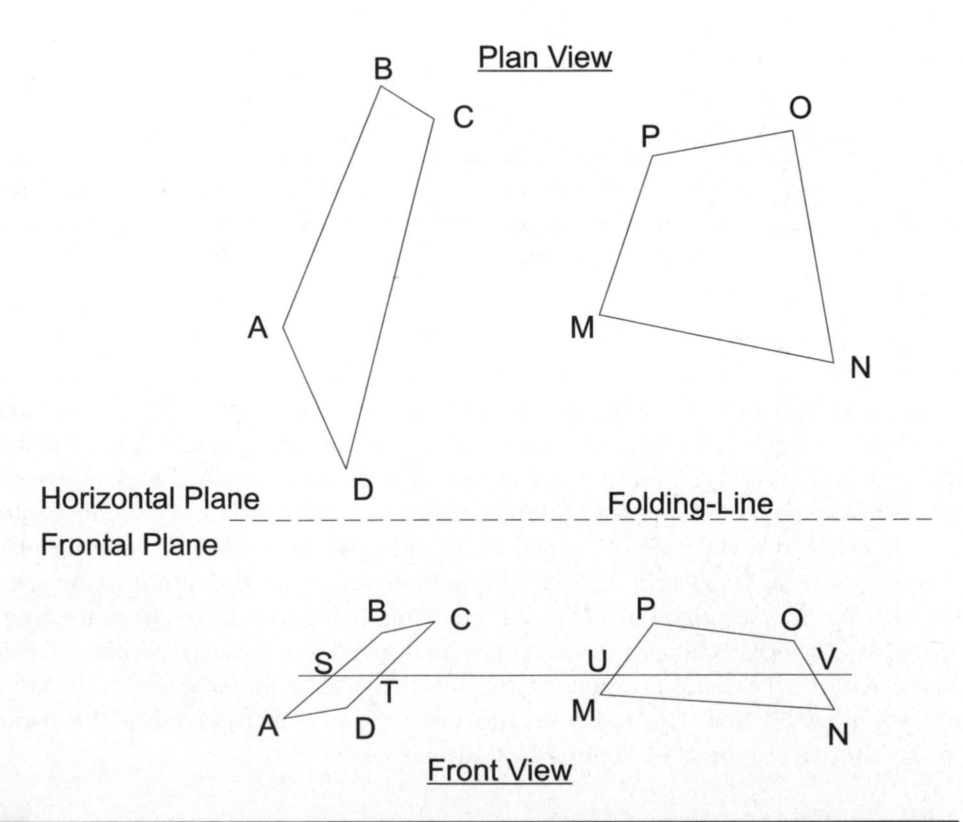


**Figure 4–51** *A cutting plane is first drawn in the front view, producing points S, T, U, and V.*

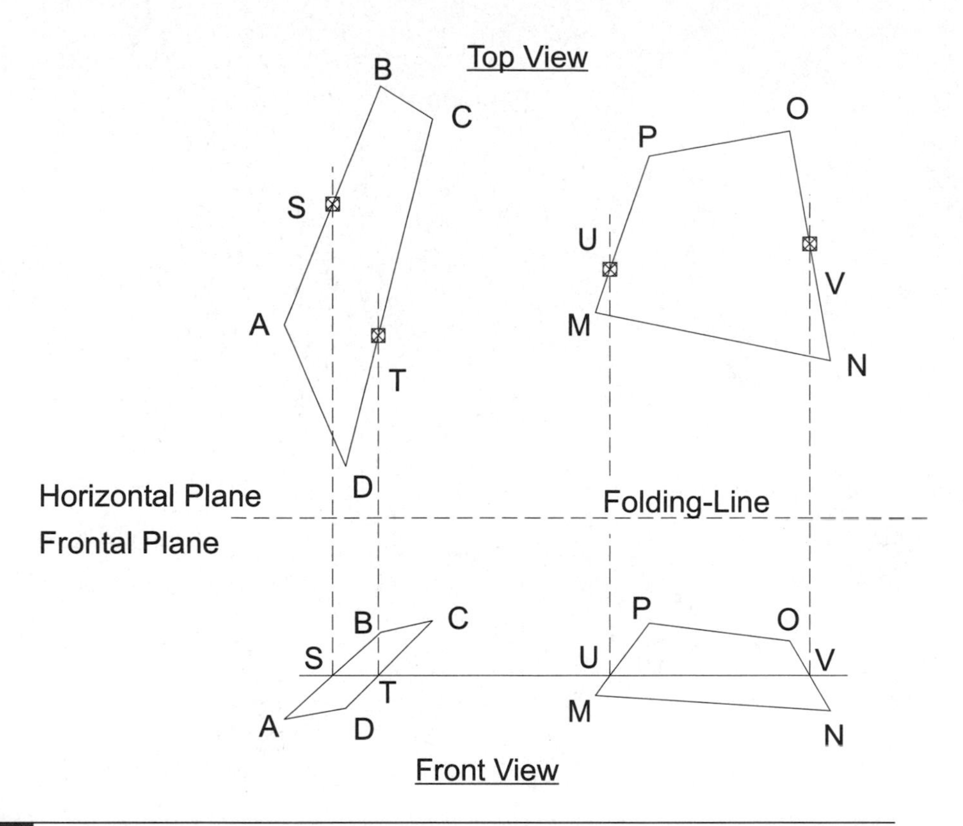


**Figure 4-52** *Points S, T, U, and V are transferred from the front view to the plan view.*

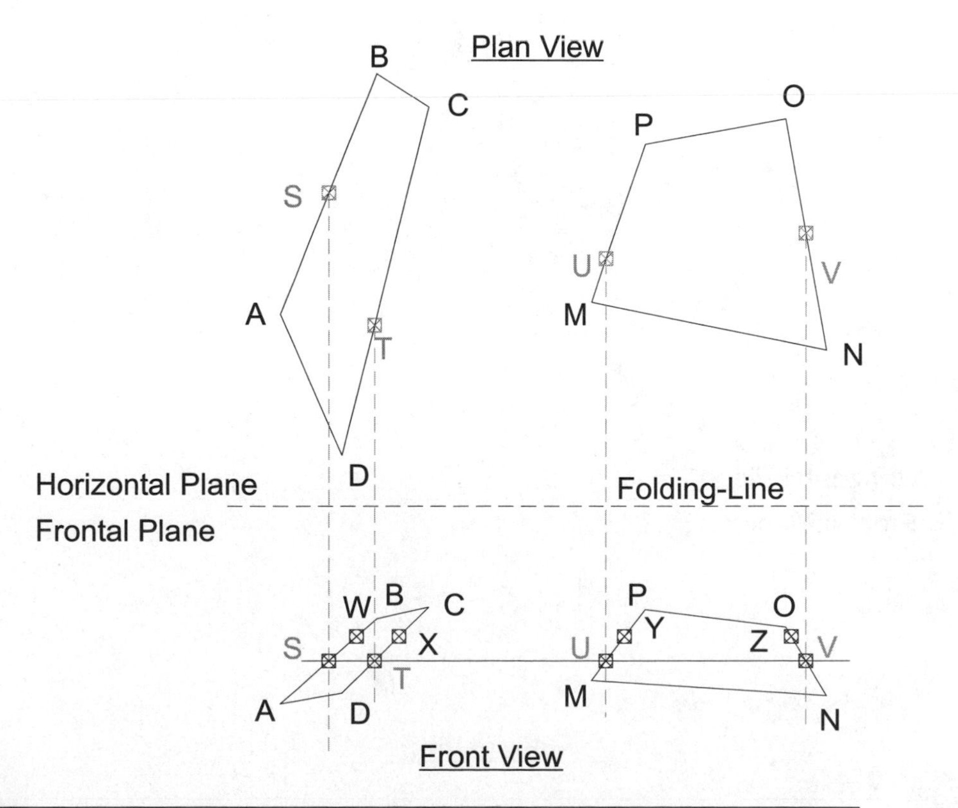


**Figure 4-53** *A second cutting plane is constructed producing points W, X, Y, and Z.*

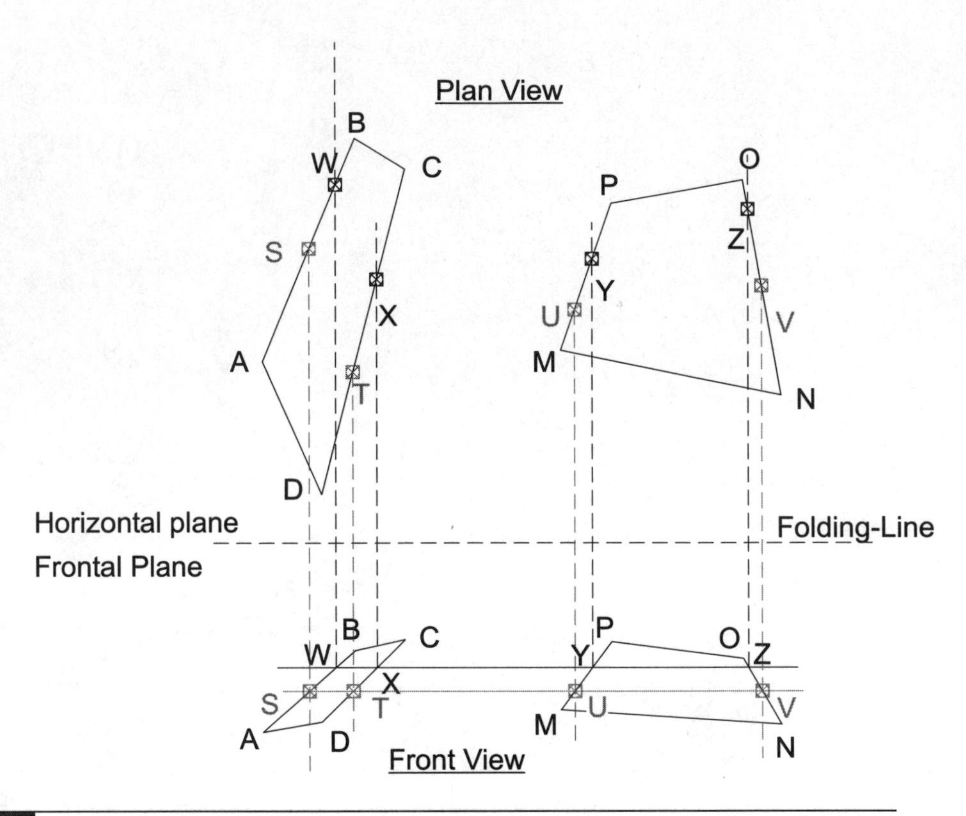


**Figure 4–54** *Points W, X, Y, and Z are transferred from the front view to the plan view.*

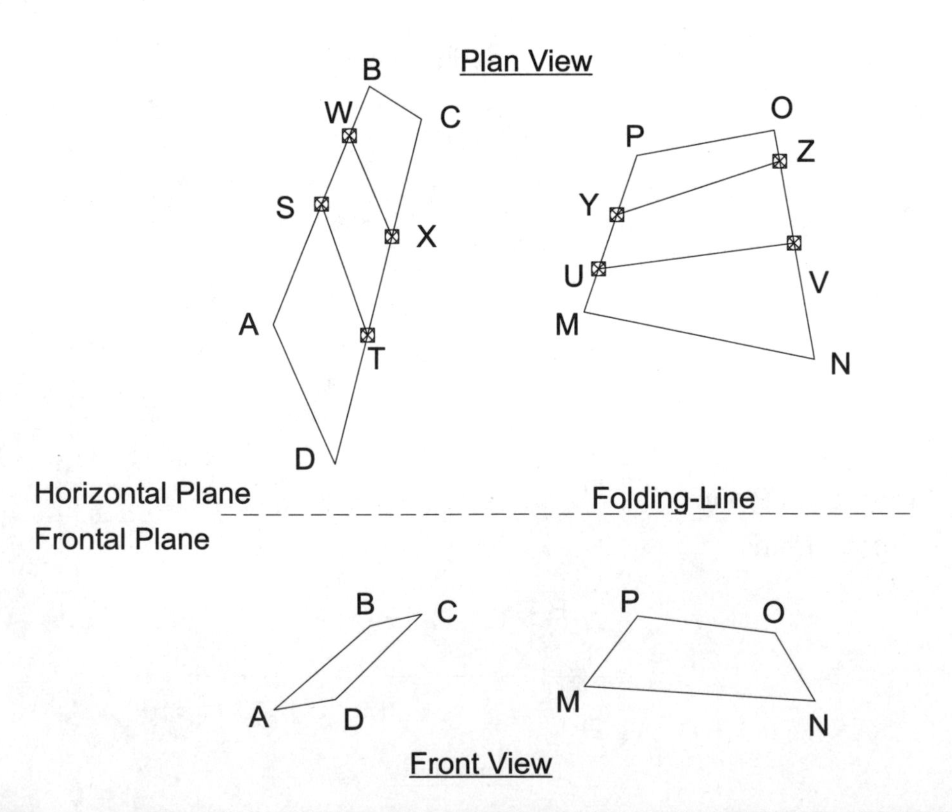


**Figure 4–55** *Points S, T, U, V, W, X, Y, and Z are connected using line segments to produce lines ST, UV, WX, and YZ.*

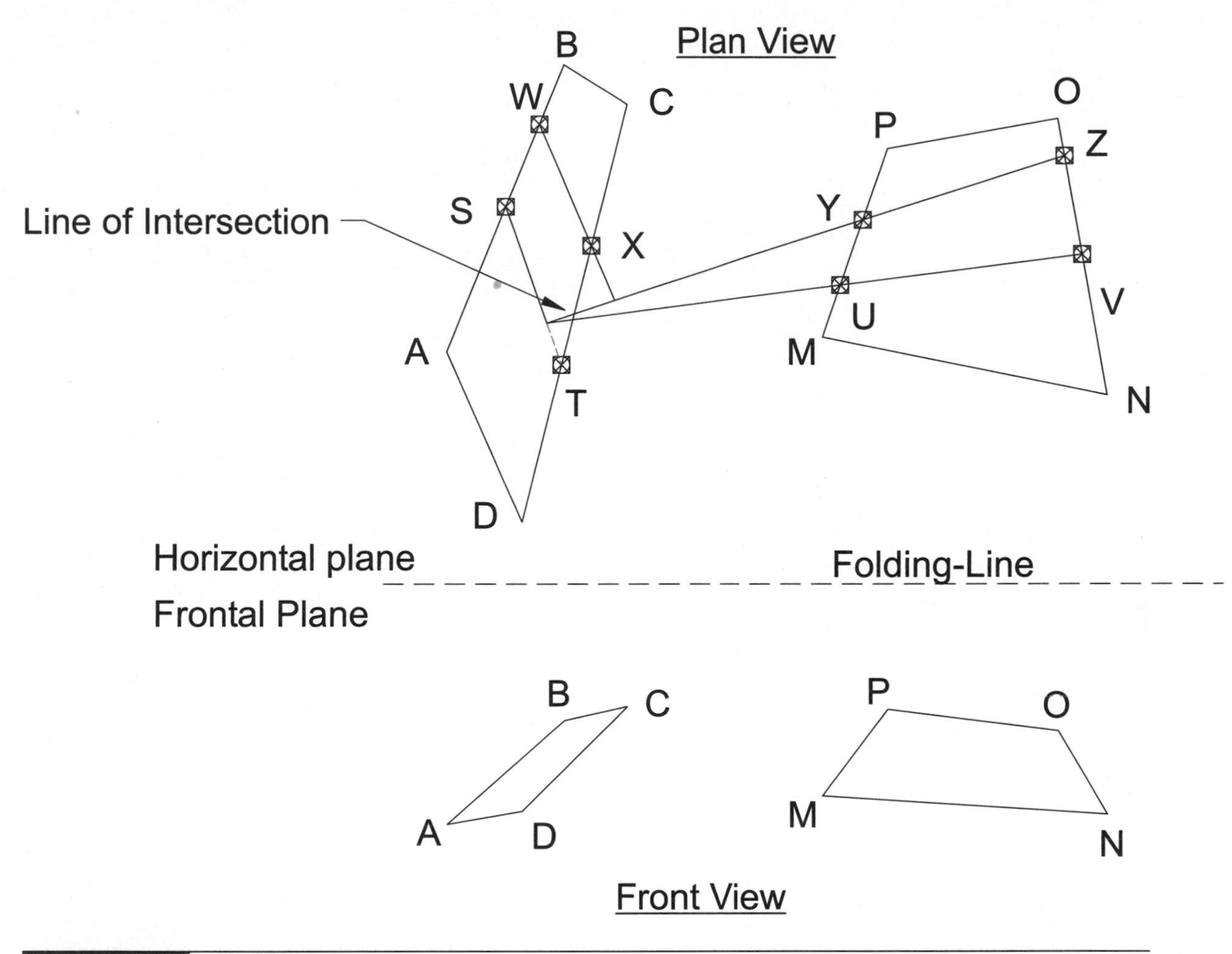


**Figure 4–56** *Lines ST, UV, WX, and YZ are extended to reveal the line of intersection between the three planes.*

### Using AutoCAD to Determine the Line of Intersection

The cutting plane method can easily be adapted for use in AutoCAD without making any changes to the basic technique. However, if the planes are regions or three-dimensional objects then the line of intersection can also be determined by using the piercing point method described earlier with only one slight modification; the procedure must be repeated twice before the true line of intersection can be established.

## Determining the Angle between Two Planes

When two planes intersect, they form an angle known as the ***dihedral angle*** (see Figure 4–57). This angle appears in true size in any view where the two planes appear in edge view, or any view in which the line of intersection appears as a point (see Figure 4–58). In many cases this requires the use of an auxiliary view.

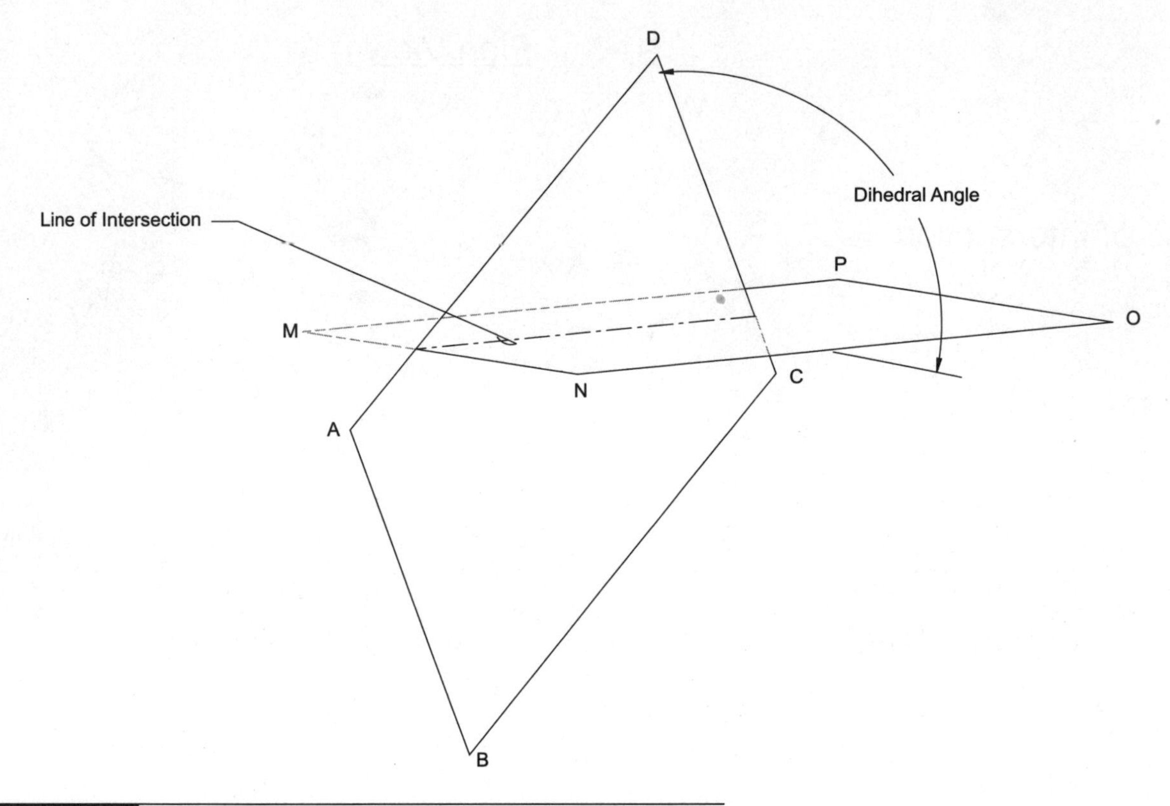


**Figure 4–57** *Dihedral angle between planes ABCD and MNOP.*

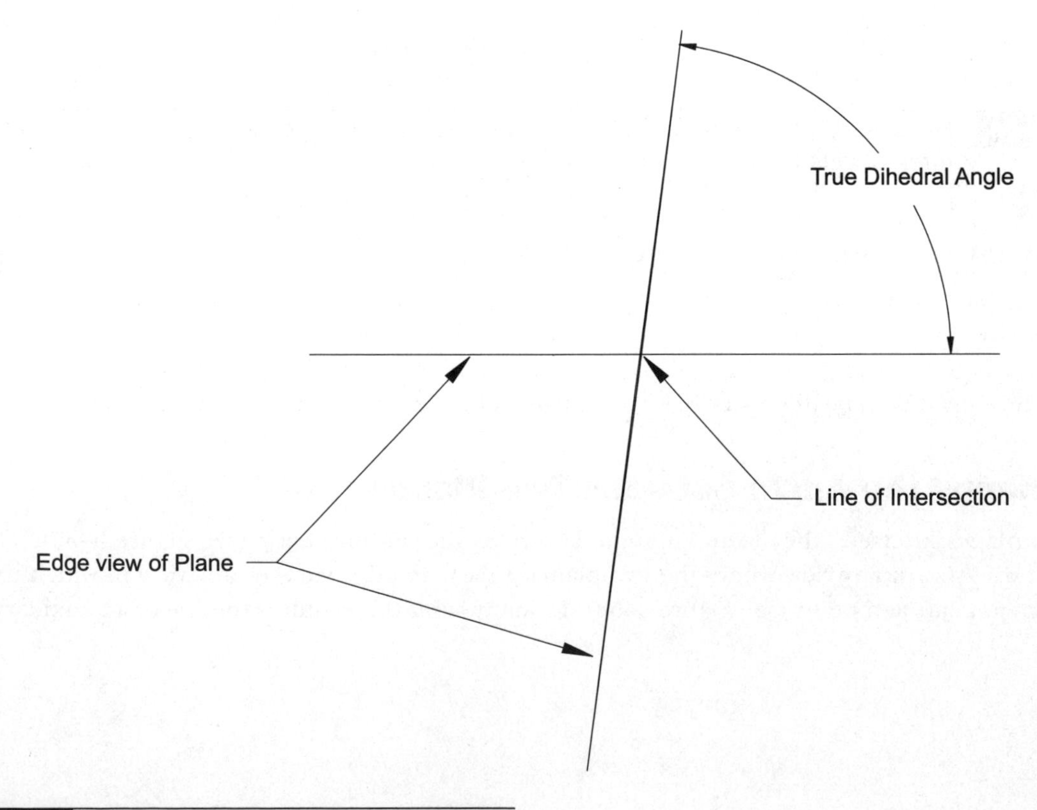


**Figure 4–58** *Both planes appearing in edge view.*

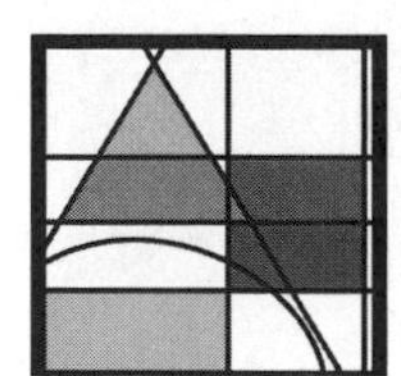

# advanced applications

## Using AutoLISP and DCL to Determine the Piercing Point of a Line and a Plane

In the previous sections, numerous methods were introduced for securing the piercing point(s) of lines and planes. By mastering these methods, the majority of the problems involving piercing points can be solved. These methods are not the only ones that can be used to derive piercing points. Piercing points can also be calculated using ***analytical geometry***. Analytical geometry, sometimes called ***coordinate geometry***, is the combination of algebra and coordinates for solving geometric problems. Although analytical geometry is beyond the scope of this textbook, its theories and principles are often used in the development of computer programs designed to solve geometric problems. To develop computer programs for solving various geometric problems, familiarity with this branch of mathematics is essential. In the following example, AutoLISP and DCL programming languages were used to develop an application that calculates the piercing points of lines and planes. The theory used in the development of this program is based on analytical geometry. The coordinates of a line can be calculated using the formula $(X - X1) / a = (Y - Y1) / b = (Z - Z1) / c$. The coordinates for a plane can be calculated using the formula $AX + BY + CZ + D = 0$. By setting both equations equal to one another, the X-, Y-, and Z-coordinates derived from these equations will be the X, Y, and Z coordinates of the piercing point. Even though both the AutoLISP and DCL programs presented below can be found on the student CD, these programs should be studied carefully to gain an understanding of how they work. The program prompts the user to select three points to define the plane. Once the plane has been defined, the program prompts the user to select two points on the line, then a dialog box is displayed that shows the coordinates of both the plane and the line along with the coordinates of the piercing point (see Figure 4–59).

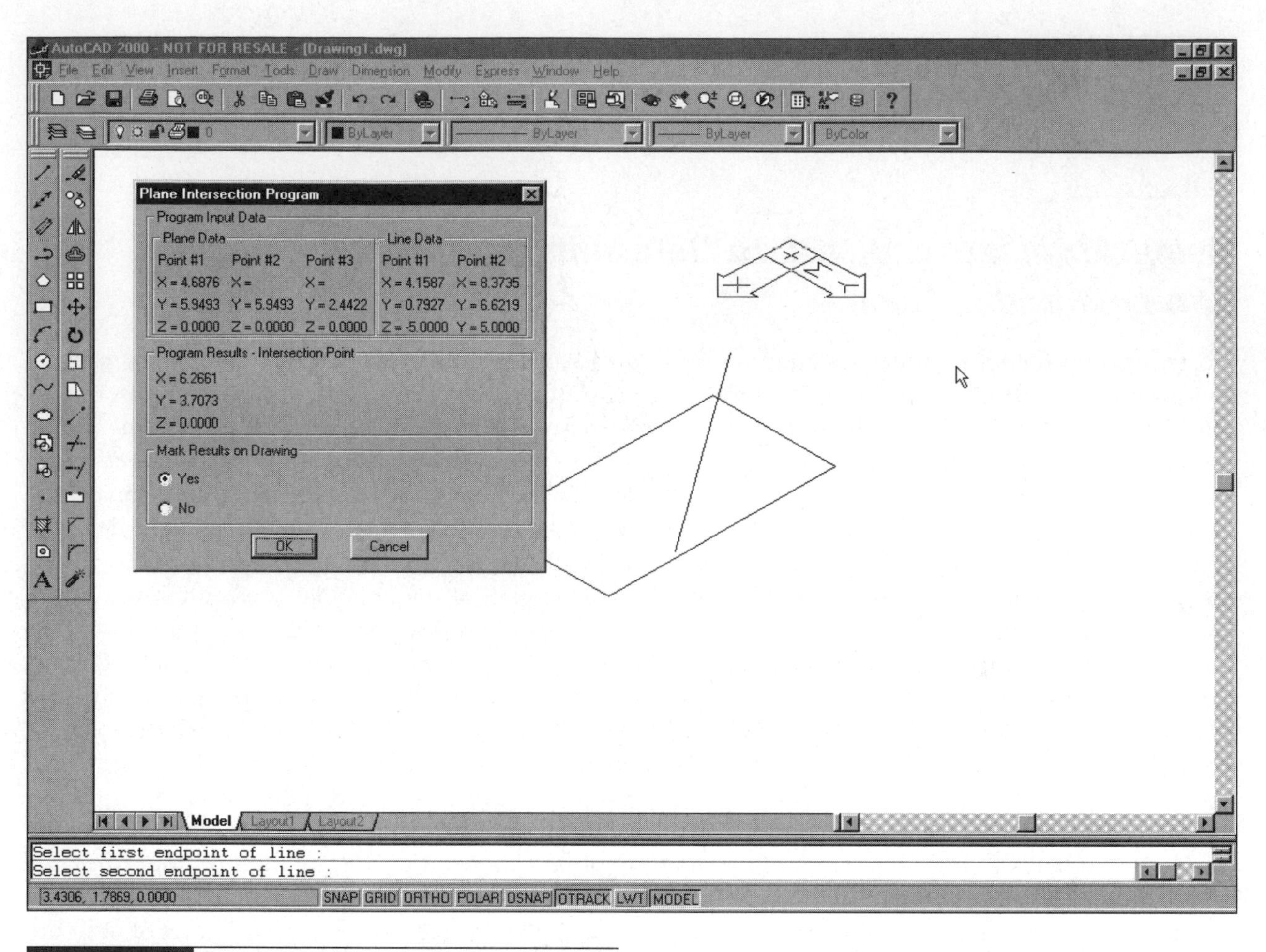


**Figure 4–59** *Dialog box for intersection program.*

## AutoLISP Program

```
;;;*********************************************************************
;;;
;;;  Program Name: Intersection.lsp
;;;
;;;  Program Purpose: To calcualte the intersection of a plane and a line
;;;
;;;  Program Date: 04/21/99
;;;
;;;  Program By: James Kevin Standiford
;;;
;;;*********************************************************************
;;;*********************************************************************
;;;
;;;Main Program
;;;
;;;*********************************************************************
(defun c:intersection ()
  (setqcmd   (getvar "cmdecho")
       osmod (getvar "osmode")
```

```
)
(setvar "cmdecho" 0)
(setvar "osmode" 1)
(initget 1)
(setqpoint_1_plane
     (getpoint "\nSelect first point defining the plane : "
     )
)
(initget 1)
(setq
  point_2_plane
   (getpoint "\nSelect second point defining the plane : "
   )
)
(initget 1)
(setq
  point_3_plane
   (getpoint "\nSelect third point defining the plane : "
   )
)
(initget 1)
(setq
  point_1_line
   (getpoint "\nSelect first endpoint of line : ")
)
(initget 1)
(setq
  point_2_line
   (getpoint "\nSelect second endpoint of line : ")
)
(setvar "osmode" 0)
(setq
  point_1_plane_x    (car point_1_plane)
  point_1_plane_y    (cadr point_1_plane)
  point_1_plane_z    (caddr point_1_plane)
  point_2_plane_x    (car point_2_plane)
  point_2_plane_y    (cadr point_2_plane)
  point_2_plane_z    (caddr point_2_plane)
  point_3_plane_x    (car point_3_plane)
  point_3_plane_y    (cadr point_3_plane)
  point_3_plane_z    (caddr point_3_plane)
  point_1_line_x     (car point_1_line)
  point_1_line_y     (cadr point_1_line)
  point_1_line_z     (caddr point_1_line)
  point_2_line_x     (car point_2_line)
  point_2_line_y     (cadr point_2_line)
  point_2_line_z     (caddr point_2_line)
;;;*****************************************************************
;;;
;;;  Theory behind Calculatons.  If the equation for a plane is defined
;;;  as AX + BY + CZ + D = 0 and the equation for a line can be found by
;;;  using the formula (X - X1)/a = (Y - Y1)/b = (Z - Z1)/c then by setting
;;;  both equation equal to each other the intersection point between the
```

```
;;;  plane and the line can be found.
;;;
;;;****************************************************************
;;;
;;;
;;;  Start Calculaton
;;;
;;;****************************************************************
    iu_plane        (- point_2_plane_x point_1_plane_x)
    ju_plane        (- point_2_plane_y point_1_plane_y)
    ku_plane        (- point_2_plane_z point_1_plane_z)
    iv_plane        (- point_3_plane_x point_1_plane_x)
    jv_plane        (- point_3_plane_y point_1_plane_y)
    kv_plane        (- point_3_plane_z point_1_plane_z)
    a_plane         (- (* kv_plane ju_plane) (* jv_plane ku_plane))
    b_plane         (- (* iu_plane kv_plane) (* iv_plane ku_plane))
    c_plane         (- (* iv_plane ju_plane) (* iu_plane jv_plane))
    d_plane         (+ (* a_plane point_1_plane_x)
                      (* b_plane point_1_plane_y)
                      (* c_plane point_1_plane_z)
                    )
    a_line          (- point_2_line_x point_1_line_x)
    b_line          (- point_2_line_y point_1_line_y)
    c_line          (- point_2_line_z point_1_line_z)
    x_value         (/ (+ (* b_plane b_line point_1_line_x)
                         (* -1 b_plane a_line point_1_line_y)
                         (* c_plane c_line point_1_line_x)
                         (* -1 c_plane a_line point_1_line_z)
                         (* -1 (* d_plane a_line))
                      )
                      (+ (* a_plane a_line)
                         (* b_plane b_line)
                         (* c_plane c_line)
                      )
                    )
    y_value         (/ (+ (* b_line x_value)
                         (* -1 b_line point_1_line_x)
                         (* a_line point_1_line_y)
                      )
                      a_line
                    )
    z_value         (/ (+ (* c_line x_value)
                         (* -1 c_line point_1_line_x)
                         (* a_line point_1_line_z)
                      )
                    )
    intersection_point (list x_value y_value z_value)
    dcl_id4         (load_dialog
                     "c:/windows/desktop/geometry/student/autolisp programs/chap4.dcl"
                    )
  )
;;;****************************************************************
;;;
```

```
;;;
;;;  Provide point and calculation data to dialog box
;;;
;;;
;;;*********************************************************************
  (if (not (new_dialog "chap_4" dcl_id4))
    (exit)
  )
  (set_tile "field_pt_p_1_x"
           (strcat "X = " (rtos point_1_plane_x))
  )
  (set_tile "field_pt_p_1_y"
           (strcat "Y = " (rtos point_1_plane_y))
  )
  (set_tile "field_pt_p_1_z"
           (strcat "Z = " (rtos point_1_plane_z))
  )
  (set_tile "field_pt_p_2_x"
           (strcat "X = " (rtos point_2_plane_x))
  )
  (set_tile "field_pt_p_2_y"
           (strcat "Y = " (rtos point_2_plane_y))
  )
  (set_tile "field_pt_p_2_z"
           (strcat "Z = " (rtos point_2_plane_z))
  )
  (set_tile "field_pt_p_3_x"
           (strcat "X = " (rtos point_3_plane_x))
  )
  (set_tile "field_pt_p_3_y"
           (strcat "Y = " (rtos point_3_plane_y))
  )
  (set_tile "field_pt_p_3_z"
           (strcat "Z = " (rtos point_3_plane_z))
  )
  (set_tile "field_pt_l_1_x"
           (strcat "X = " (rtos point_1_line_x))
  )
  (set_tile "field_pt_l_1_y"
           (strcat "Y = " (rtos point_1_line_y))
  )
  (set_tile "field_pt_l_1_z"
           (strcat "Z = " (rtos point_1_line_z))
  )
  (set_tile "field_pt_l_2_x"
           (strcat "X = " (rtos point_2_line_x))
  )
  (set_tile "field_pt_l_2_y"
           (strcat "Y = " (rtos point_2_line_y))
  )
  (set_tile "field_pt_l_2_z"
           (strcat "Y = " (rtos point_2_line_z))
  )
```

```
  (set_tile "field_pt_i_x" (strcat "X = " (rtos x_value)))
  (set_tile "field_pt_i_y" (strcat "Y = " (rtos y_value)))
  (set_tile "field_pt_i_z" (strcat "Z = " (rtos z_value)))
  (set_tile "plot_yes" "1")
  (action_tile "accept" "(inquire) (done_dialog)")
  (start_dialog)                        ;  Start Dialog Box
  (if (= plot_yes "1")
    (command "point" intersection_point)
  )
  (setvar "cmdecho" cmd)
  (setvar "osmode" osmod)
)
(defun inquire ()
  (setqplot_yes (get_tile "plot_yes")
      plot_no (get_tile "plot_no")
  )
)
//%%%%%%%%%%%%%%%%%%%%%%%%%%%%%%%%%%%%%%%%%%%%%%%%%%%%%%%%%%%%
//
//     Dialog Box for Intersection Program.
//
//     03-13-98
//
//     James Kevin Standiford
//
//%%%%%%%%%%%%%%%%%%%%%%%%%%%%%%%%%%%%%%%%%%%%%%%%%%%%%%%%%%%%
//%%%%%%%%%%%%%%%%%%%%%%%%%%%%%%%%%%%%%%%%%%%%%%%%%%%%%%%%%%%%
//
//            Main Program
//
//%%%%%%%%%%%%%%%%%%%%%%%%%%%%%%%%%%%%%%%%%%%%%%%%%%%%%%%%%%%%
chap_4 : dialog {
  label = "Plane Intersection Program";
        : boxed_row {
         fixed_width = true;
         children_fixed_width = true;
          label = "Program Input Data";
          : row {
          label = "Plane Data";
          :column {
          : text {
          label = "Point #1";
          }
          : text {
          label = "X = ";
          key = "field_pt_p_1_x";
          }
          : text {
          label = "Y = ";
          key = "field_pt_p_1_y";
          }
          : text {
          label = "Z = ";
```

```
key = "field_pt_p_1_z";
}
}
:column {
: text {
label = "Point #2";
}
: text {
label = "X = ";
key = "field_pt_p_2_x";
}
: text {
label = "Y = ";
key = "field_pt_p_2_y";
}
: text {
label = "Z = ";
key = "field_pt_p_2_z";
}
}
:column {
: text {
label = "Point #3";
}
: text {
label = "X = ";
key = "field_pt_p_3_x";
}
: text {
label = "Y = ";
key = "field_pt_p_3_y";
}
: text {
label = "Z = ";
key = "field_pt_p_3_z";
}
}
}
: row {
label = "Line Data";
:column {
: text {
label = "Point #1";
}
: text {
label = "X = ";
key = "field_pt_l_1_x";
}
: text {
label = "Y = ";
key = "field_pt_l_1_y";
}
: text {
```

```
label = "Z = ";
key = "field_pt_l_1_z";
}
}
:column {
: text {
label = "Point #2";
}
: text {
label = "X = ";
key = "field_pt_l_2_x";
}
: text {
label = "Y = ";
key = "field_pt_l_2_y";
}
: text {
label = "Z = ";
key = "field_pt_l_2_z";
}
}
}
}
: boxed_column {
label = "Program Results - Intersection Point";
: text {
label = "X = ";
key = "field_pt_i_x";
}
: text {
label = "Y = ";
key = "field_pt_i_y";
}
: text {
label = "Z = ";
key = "field_pt_i_z";
}
}
: boxed_column {
label = "Mark Results on Drawing";
: radio_button {
label = "Yes";
key = "plot_yes";
}
: radio_button {
label = "No";
key = "plot_no";
}
}

ok_cancel;
}
```

# Review Questions

Answer the following questions on a separate sheet of paper. Your answers should be as complete as possible.

1. List the three categories of planes and give an example of each.
2. List the four ways that the boundaries of a plane may be defined and give an example of each.
3. How do normal planes differ from inclined planes?
4. On which plane does AutoCAD construct entities?
5. How do surfaces differ from planes?
6. Explain in detail the differences between the auxiliary view method and the cutting plane method.
7. What is a piercing point and how does it differ from a line of intersection?
8. List the methods used to identify the piercing point of a line and a plane.
9. List the methods used to identify the line of intersection between two planes.
10. How are the coordinates of the piercing point of a line and a plane calculated in AutoCAD?
11. What AutoCAD command can be used to determine if two of more entities reside on the same plane?

# Review Exercises

## Determining if Two or More Entities Coexist on the Same Plane Using AutoCAD

Using AutoCAD, determine if the following objects coexist on the same XY-plane. If they do not, then by what distance are they separated? All drawings with an * can be found on the student CD-ROM.

***#1**

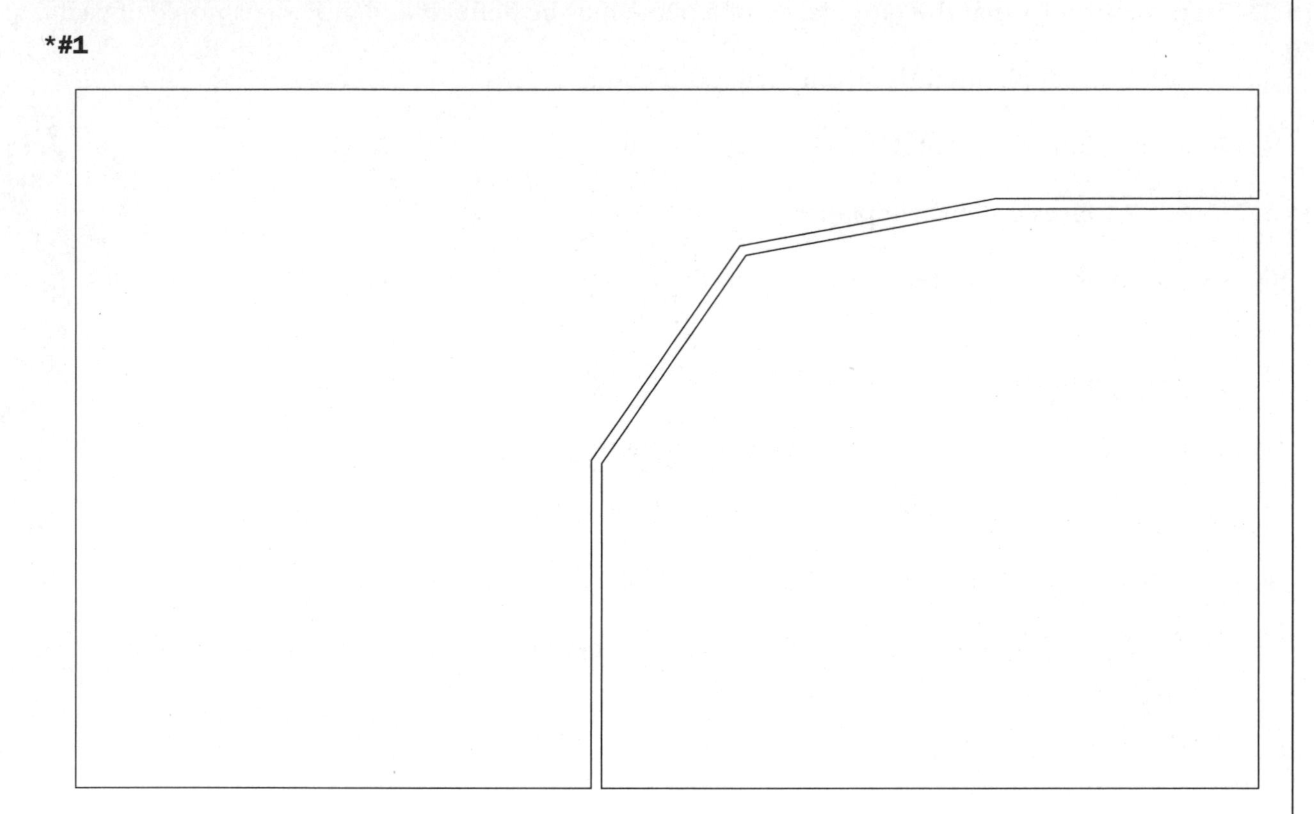

*#2

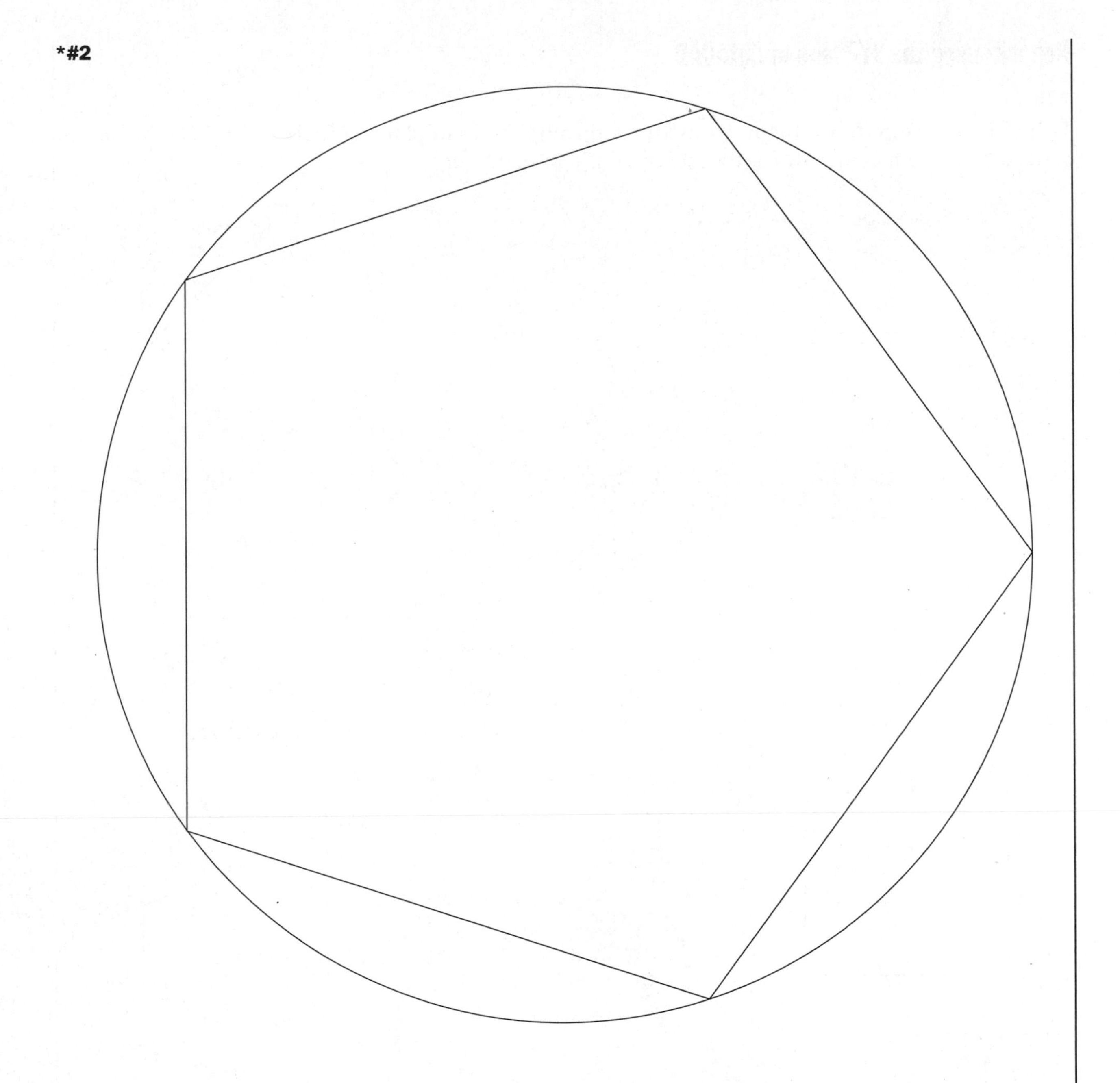

## Repositioning the XY-Plane in AutoCAD

***#3**

Using the drawing shown below, construct 4 through holes in planes ABCD and EFGH that are spaced 3/4" from center to center and are 0.25" in diameter.

Plan View

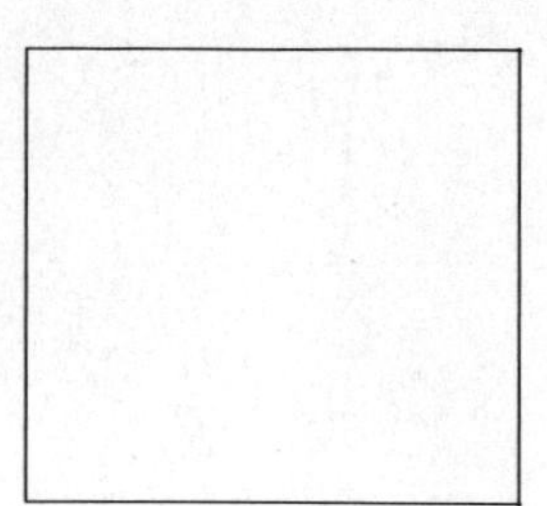

Isometric

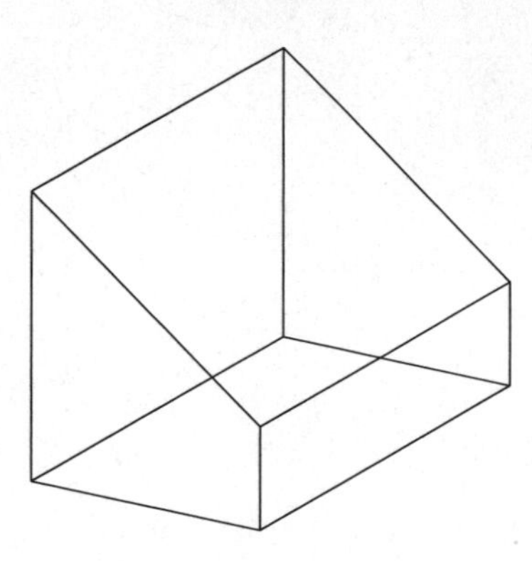

Front View

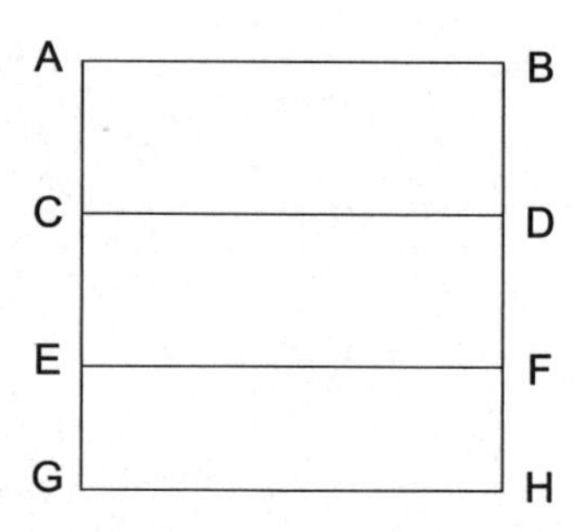


Right Side View

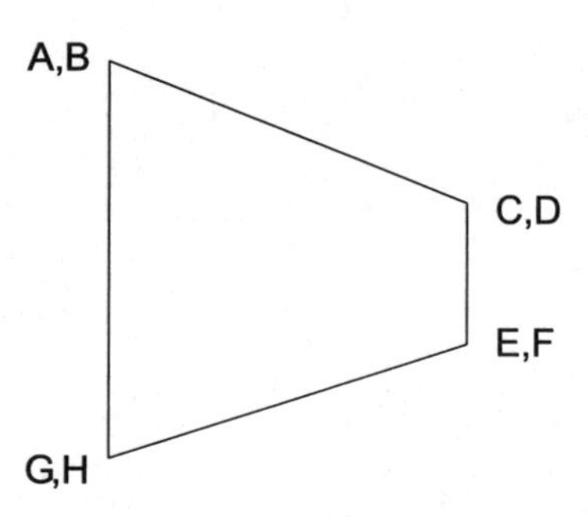

***#4**

Using the drawing shown below, create a hole in the center of plane ABCD that is 0.25" in diameter.

## Plan View

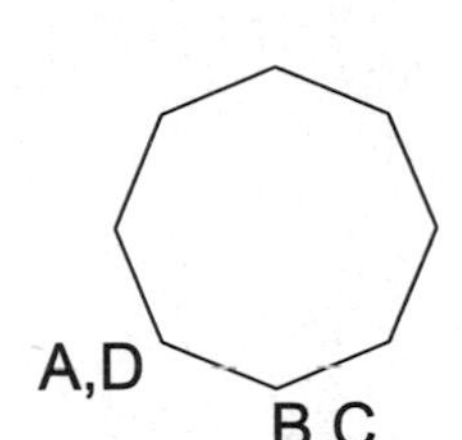


## Isometric

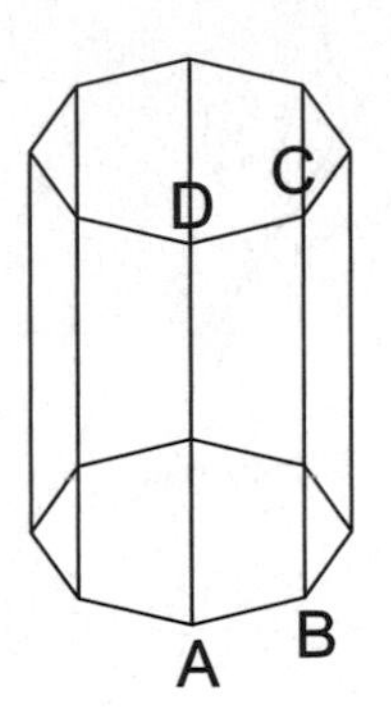


## Front View

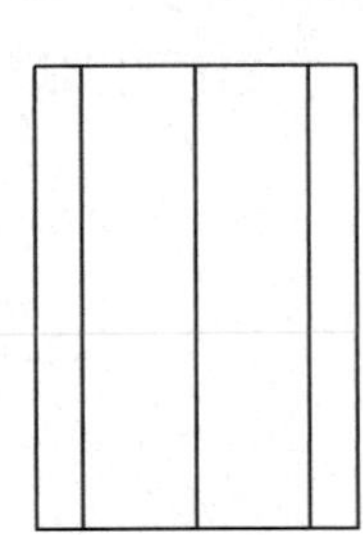

## Right Side View

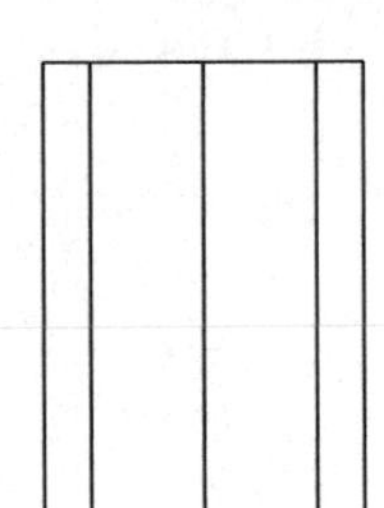

## Piercing Point of a Line and a Plane

Determine the piercing points in the following problems.

*#5

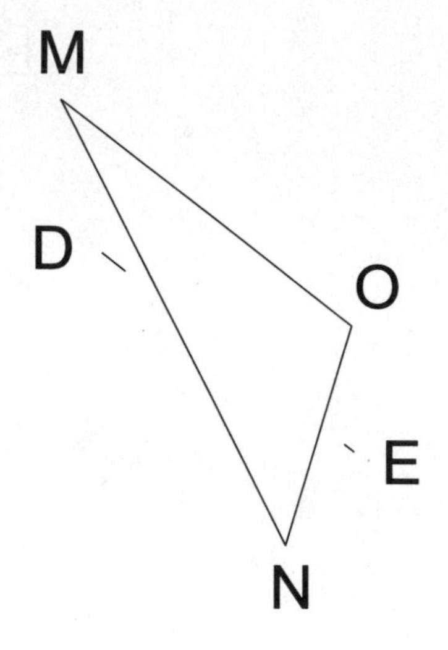


Horizontal Plane Folding-Line

Frontal Plane

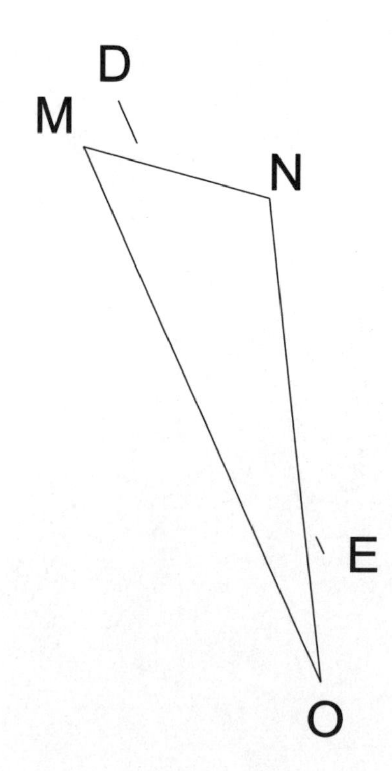

*#6

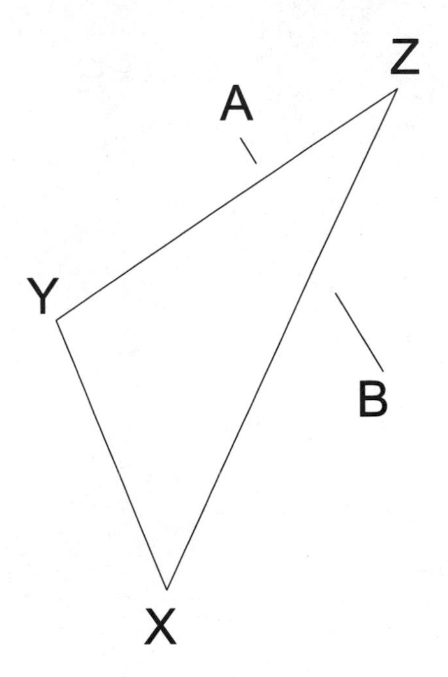


Horizontal Plane

Folding-Line

Frontal Plane

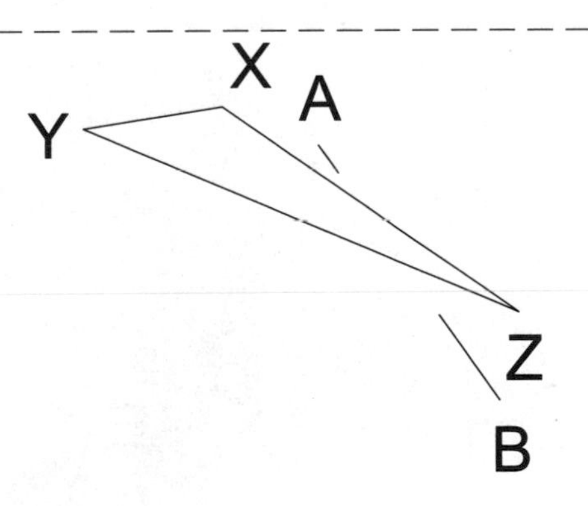

*#7

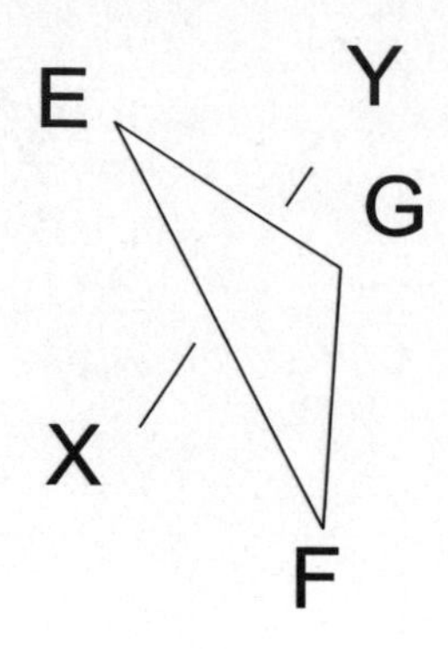


Horizontal Plane

Folding-Line

Frontal Plane

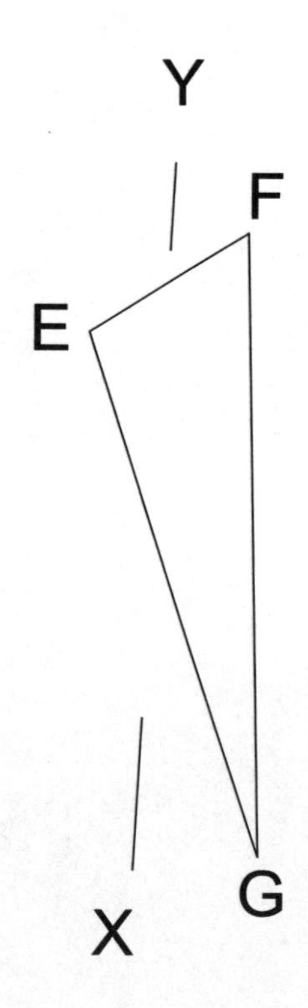

*#8

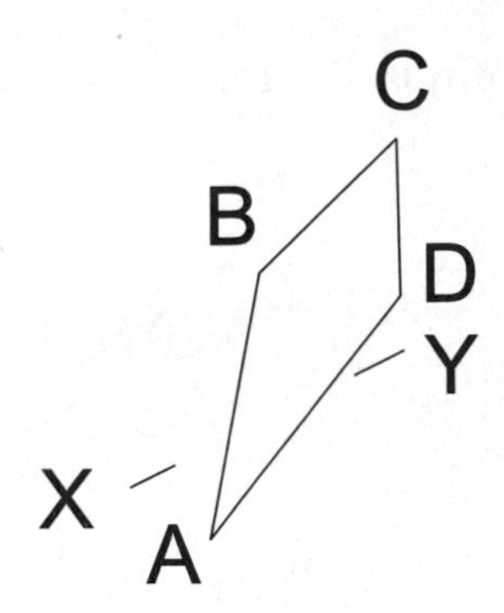


Horizontal Plane Folding-Line

Frontal Plane

X

A

D

B C

Y

## Intersection of Two Planes

Determine the intersection of the planes shown in the following exercises.

***#9**

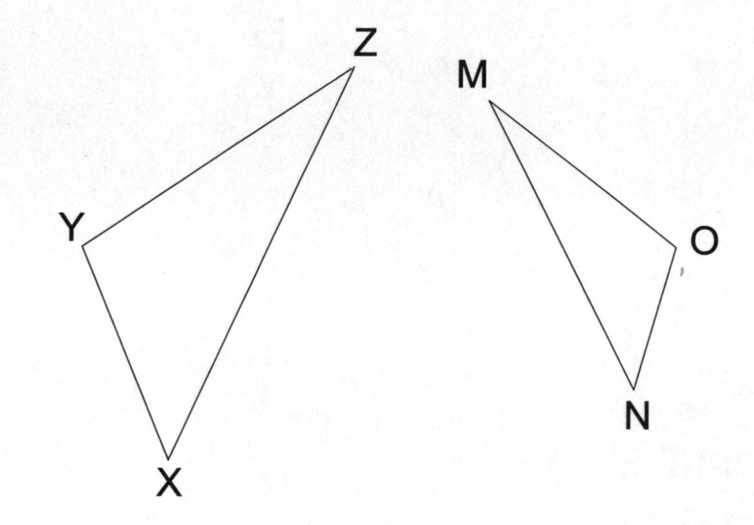


Horizontal Plane

Folding-Line

Frontal Plane

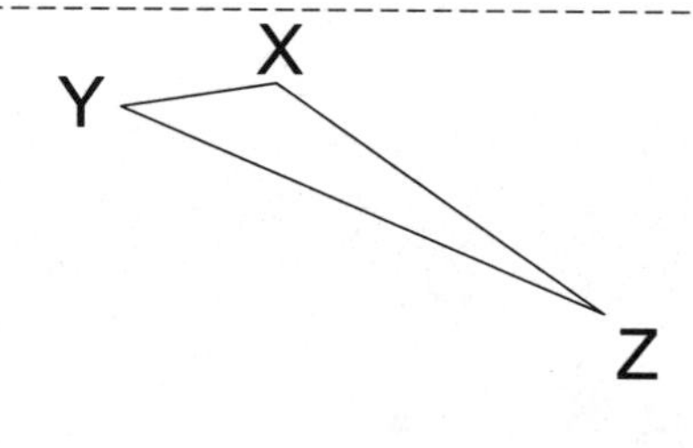


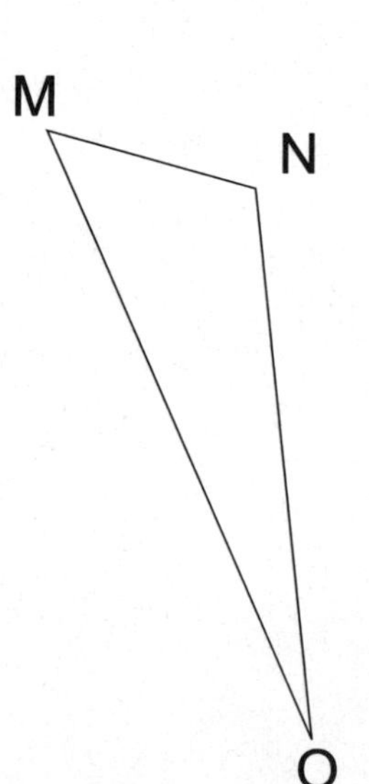

*#10

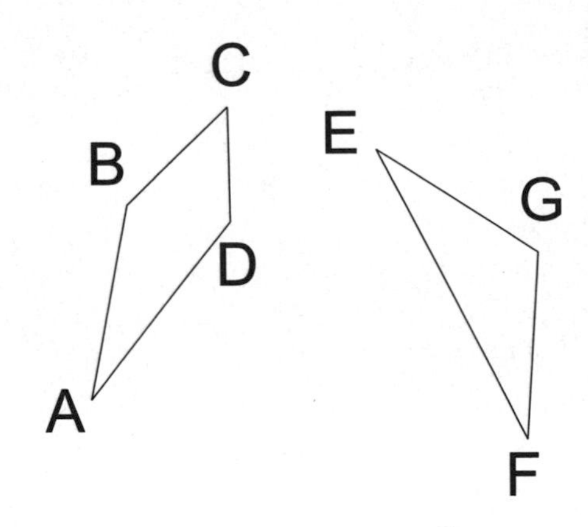


Horizontal Plane Folding-Line

Frontal Plane

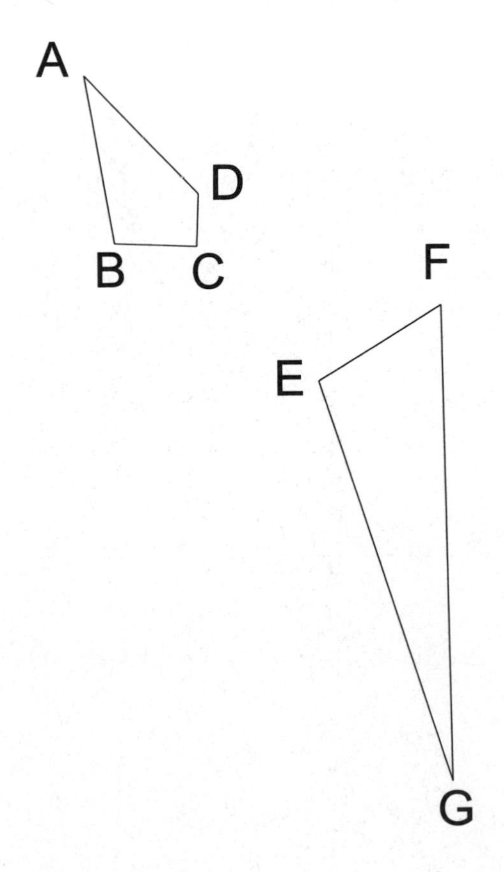

***#11**

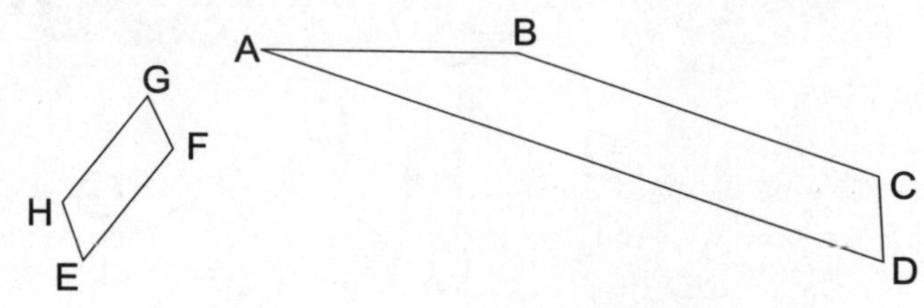


Horizontal Plane

Folding-Line

Frontal Plane

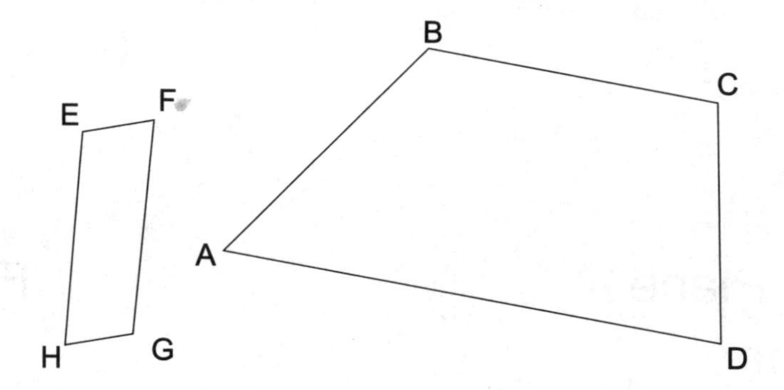


***#12**

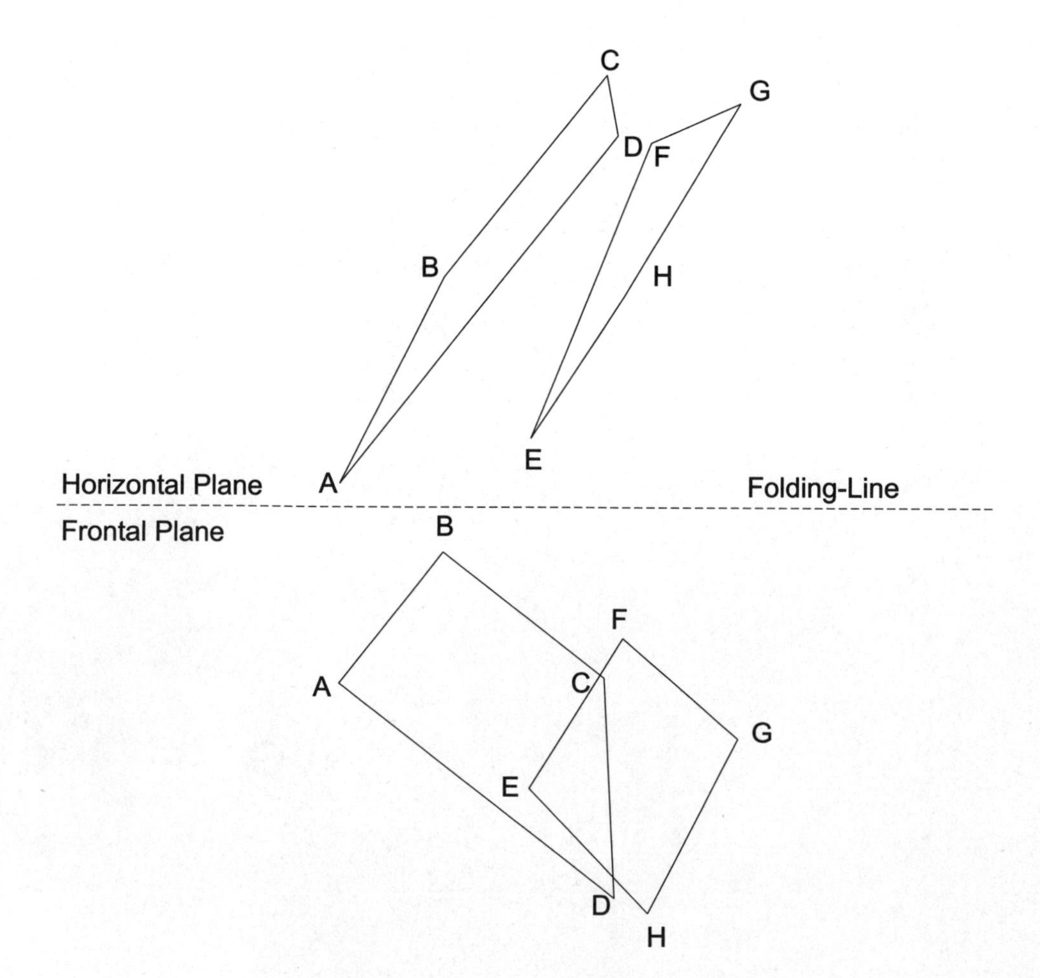

## Locating Lines in Different Views

Complete the following views.

***#13**

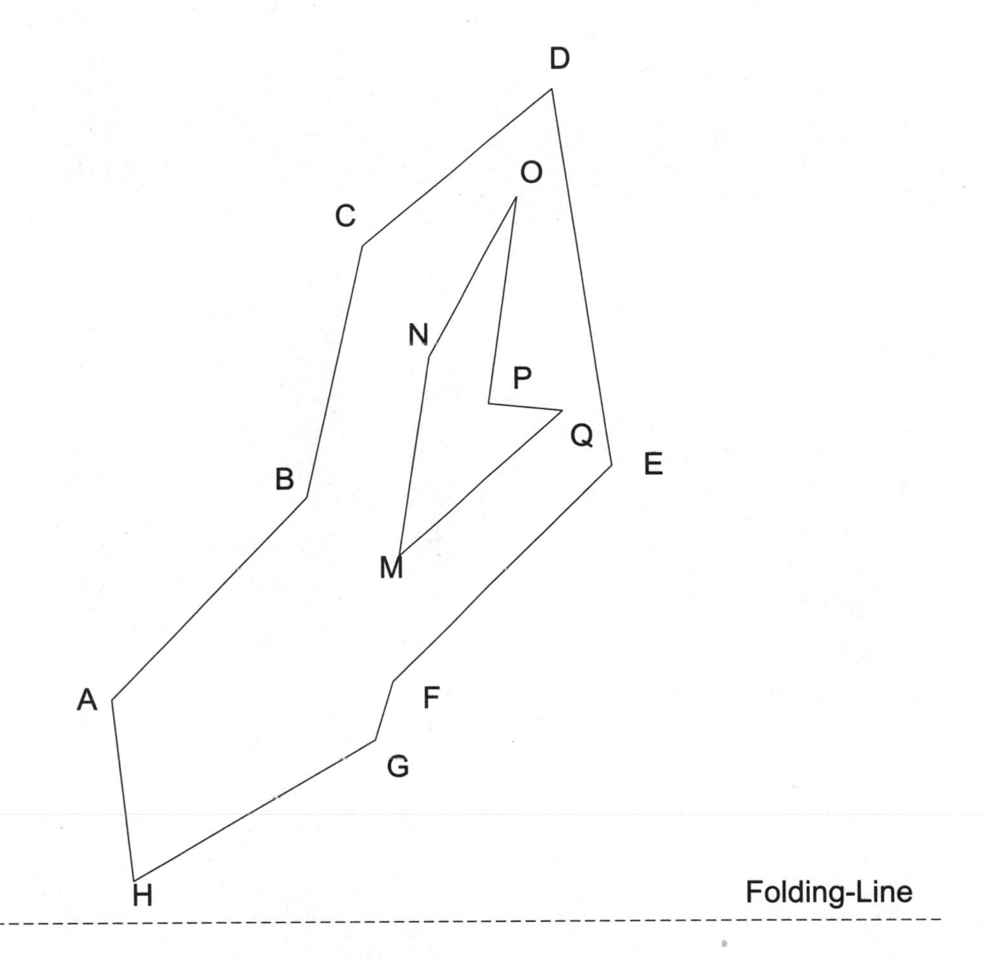


Horizontal Folding-Line

Frontal

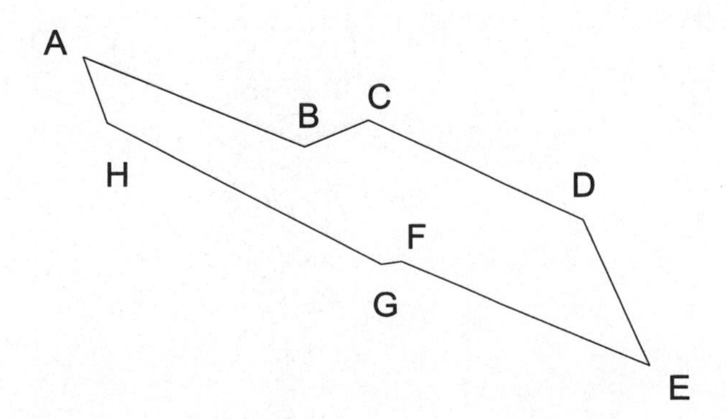

*#14

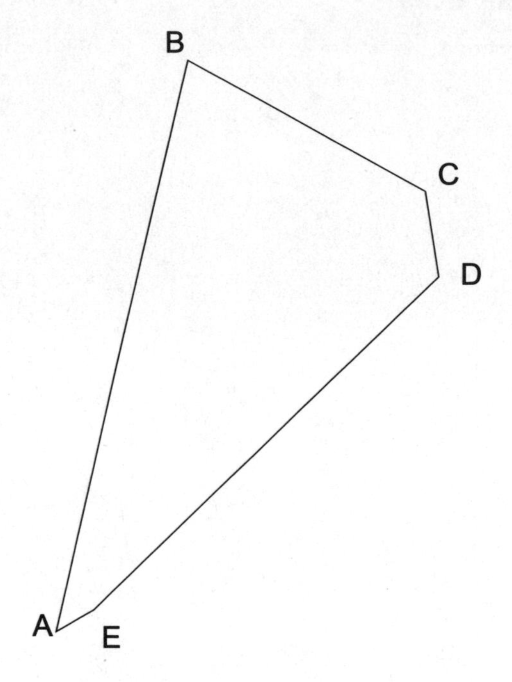


Horizontal                                                                Folding-Line

Frontal

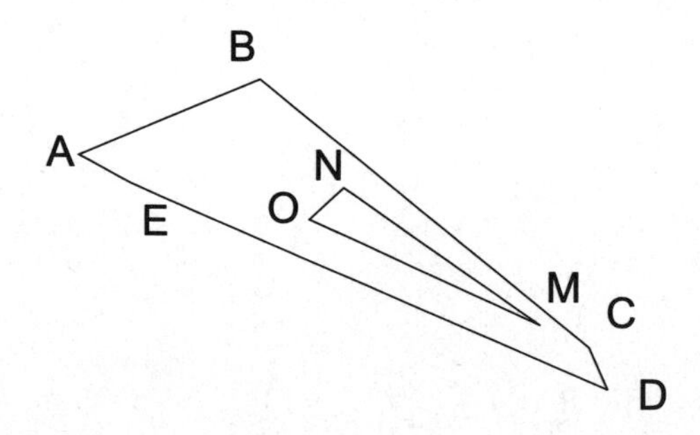

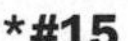

***#15**

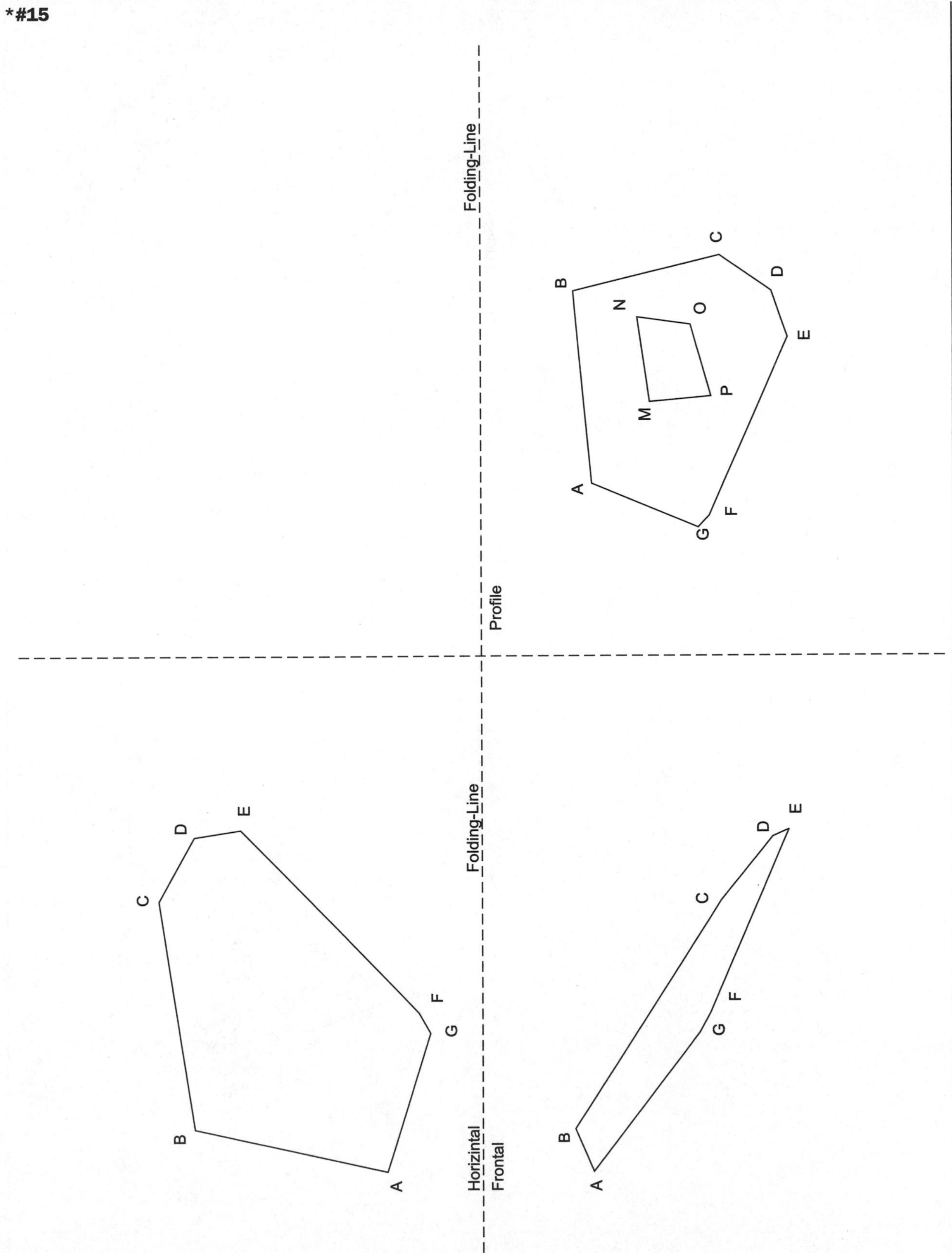
Folding-Line
Profile
Folding-Line
Horizintal
Frontal
A
B
C
D
E
F
G
M
N
O
P

*#16

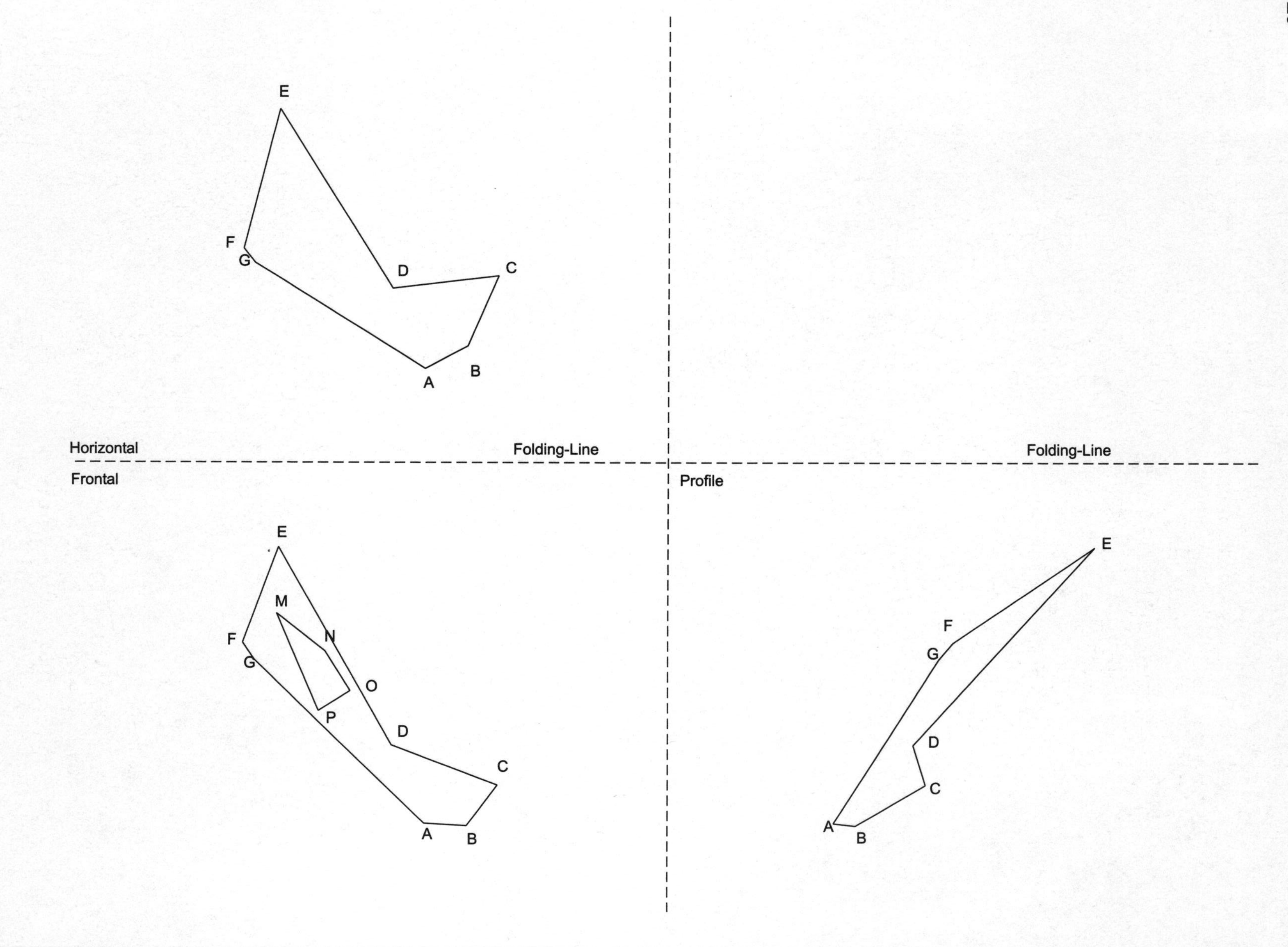

Horizontal
Frontal
Folding-Line
Profile
Folding-Line
A
B
C
D
E
F
G
M
N
O
P

# Revolutions

## Objectives:

**Upon completion of this chapter the student will be able to do the following:**

- *Describe the difference between the change-of-position method and the revolution method.*
- *Define the terms axis of revolution, revolution plane, and successive revolution.*
- *List the principles regarding the revolution of a point, a line, a plane, and a solid.*
- *Revolve the following: points, lines, planes, and solids.*
- *Determine the true length of a line using the revolution method.*
- *Determine the true shape and size of a plane using the revolution method.*
- *Use the revolution method in AutoCAD to determine the true length of a two-dimensional line.*
- *Revolve an AutoCAD three-dimensional solid.*
- *Understand why a mechanical desktop solid should not be edited using standard AutoCAD commands.*

## Key Words and Terms

**Change-of-position**
**Path**
**Revolution axis**
**Revolution method**
**Revolution plane**

## Introduction to Revolutions

In descriptive geometry there are two methods associated with the construction of views, the change-of-position and the revolution methods. All of the various projection techniques are derived from these two methods. The ***change-of-position method*** involves repositioning the observer so that the line-of-sight is set perpendicular to the projection plane where the object's contours are currently being projected. Recall from Chapter one that when an orthographic drawing is created, the object is placed inside of a glass cube and its contours are projected onto the surface of the cube.

After the contours have been projected, the cube is unfolded to reveal the six principal views. In reality the object is not actually placed inside of a glass cube; instead, the observer's line-of-sight is repositioned each time for each view (keeping the line-of-sight perpendicular to the current projection plane). This same technique is also employed in the production of an auxiliary view. The line-of-sight is repositioned so that it is perpendicular to the inclined surface of the object. Once the line-of-sight has been relocated to the proper position, the contours are projected onto the auxiliary plane, and the auxiliary view is created. The part remains stationary while the observer moves to the different locations required for the development of the various views (see Figure 5–1). Passing an imaginary axis through the center of an object and revolving the object about that axis is tantamount to rotating the line-of-sight around an object. Making the observer stationary and revolving the part into various positions is known as a ***revolution*** or ***revolution method*** (see Figure 5–2).

Both methods produce the same results, but there are advantages and disadvantages specific to each method. The revolution method has a tendency to produce a drawing with crowded and overlapping views. This drawing would be difficult to interpret, and increases the possibility of costly mistakes. However, the revolution method often requires less work and can be easier to execute. The change-of-position method produces a drawing that is less congested and easier to read, but usually requires more time to complete. When deciding which method should be used, the decision should be based upon the amount of clarity needed to convey the necessary information. When a view is created from an AutoCAD two-dimensional drawing, either method may be used effectively. The decisive factors in this instance would be the level of accuracy needed and the complexity of the object being illustrated. Greater accuracy and intricate parts require illustrations that are uncluttered and easy to comprehend. Using the revolution method on an AutoCAD three-dimensional or solid model is not as easy as using the change of position method. When different views of a solid are produced in layout space, AutoCAD uses the change-of-position method to revolve the line-of-sight to different positions, not the model. When a solid is rotated in one view then it will be rotated in the same direction and by the same amount in the remaining views.

**Warning: Solid models created using mechanical desktop should not be rotated using conventional AutoCAD commands because unexpected results can occur.**

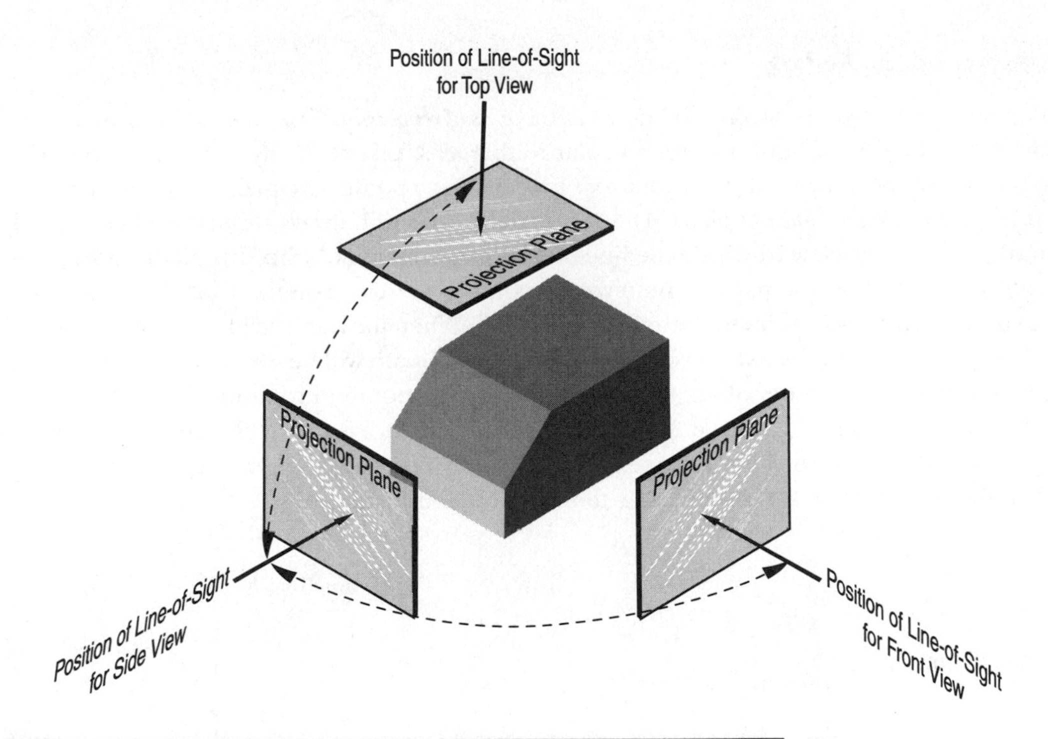


**Figure 5-1** *Rotating the line-of-sight to achieve different views of a part.*

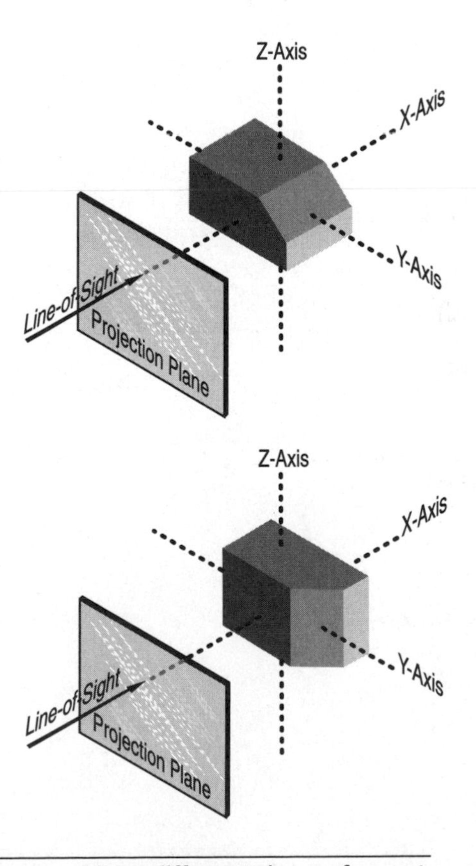


**Figure 5-2** *Rotating the object to achieve different views of a part.*

## Revolution of a Point

The axis that a point is revolved around is known as the ***axis of revolution.*** This axis will appear as a straight line in any view where the line-of-sight is perpendicular to that axis. Likewise, any view where the line-of-sight is parallel to the axis of revolution will cause that axis to appear as a point. The plane containing the point being revolved is known as the ***revolution plane.*** The revolution plane will always be perpendicular to the axis of revolution; therefore, any view in which the line-of-sight is perpendicular to this plane (and parallel to the axis of revolution) will show the path of the revolved point as a circle. The radius of the circle (path) will be equal to the distance from the point to the axis of revolution. When the line-of-sight is parallel to the revolution plane (and perpendicular to the axis of revolution), the point's path will be a straight line (see Figure 5–3). A point can be revolved any number of degrees, but typically the amount of rotation will be greater than 0° and less than 360°. When a point is revolved 360° its path will appear as a full circle in any view where the line-of-sight is perpendicular to the revolution plane. When a point is revolved less than 360° the path will appear as an arc where the line-of-sight is perpendicular to the revolution plane.

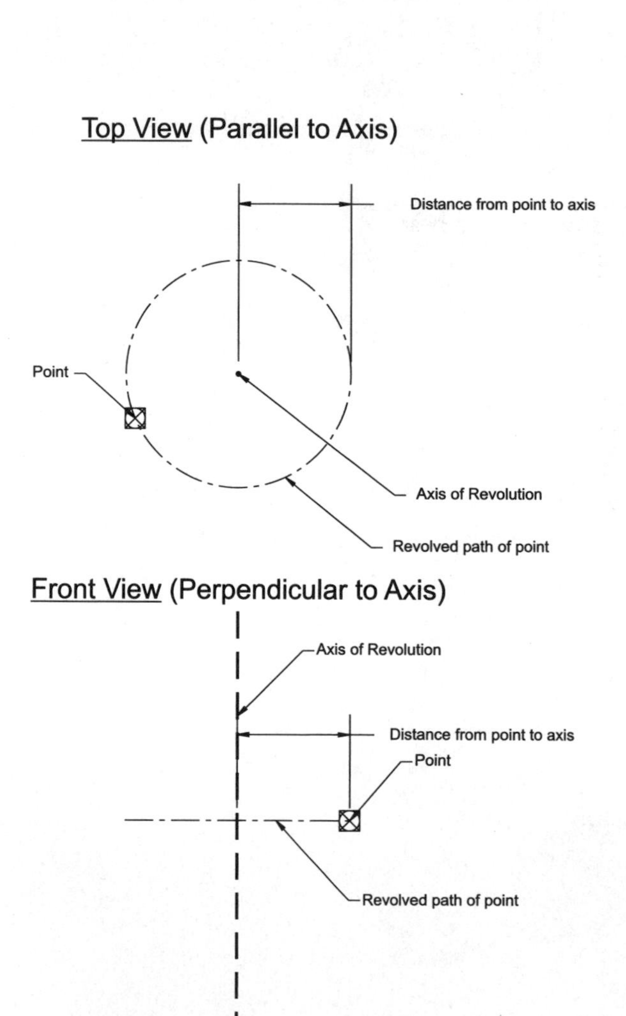


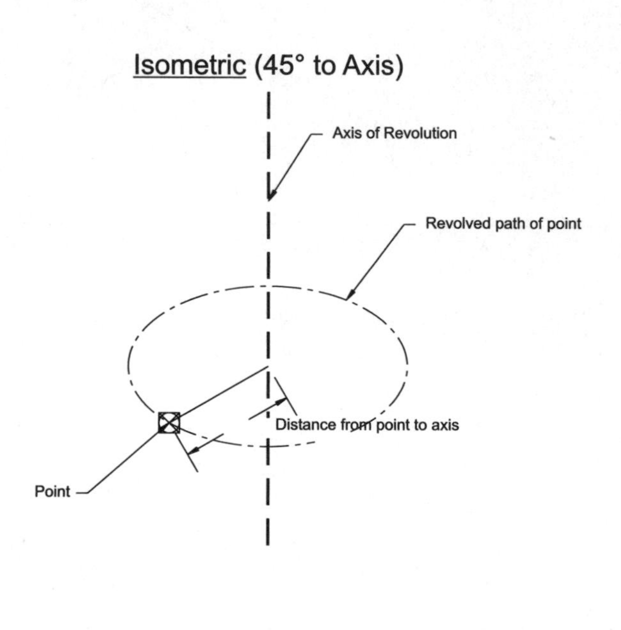


**Figure 5–3** *Revolving a point.*

## Revolution of a Line

The principles that govern the behavior of a revolved point prepare the foundation for performing revolutions, but three additional principles must be introduced for the revolution of a line. First, when a line segment is revolved, one end will be affixed to the axis of revolution while the other end is free to rotate about that axis. The same principles that dictate the behavior of a revolved point also apply to the non-stationary end of the line segment. Second, the true length of a revolved line will remain constant regardless of its position in space. When a line is revolved, its true length will not be affected—only the line's coordinates will change. If the true length of a line does change, then the new length will be an attribute of a new line and not the original. Third, a line will appear in true length when the line is parallel to the projection plane, or any view where the adjacent view illustrates the line as a point (see Figure 5–4). In this example, line AB appears as a point when the line-of-sight is positioned perpendicular to the XY-axis. However, by establishing an axis of revolution and affixing one end of line AB to that axis, the line may be rotated 90° about either the X- or Y-axis to produce a view that reveals the true length.

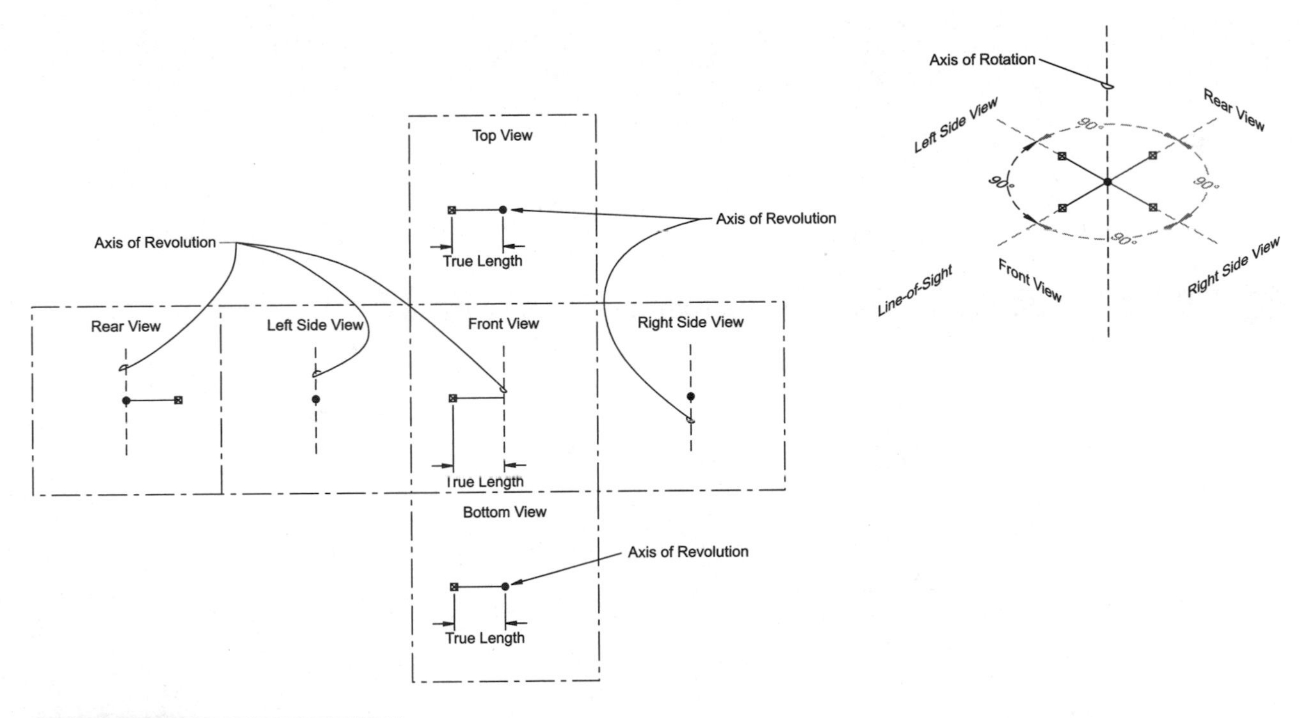


**Figure 5–4** *Revolving a line.*

### Determining the True Length of a Line

As stated earlier, the true length of any line can be obtained from a view where the line appears as either a point in an adjacent view or is parallel to the current projection plane. When a line appears foreshortened in two or more principal views, like in the case of a pyramid, the true length of the line can still be determined by revolving the line into a horizontal position in either the top or front views. This is accomplished by placing a temporary XY-axis at one end of the line (the stationary point) and rotating the line until its slope is equal to zero. For a pyramid, the stationary point is typically positioned at the apex (the point at which the vertices extending from the base connect). Line AB in Figure 5–5 is shown foreshortened in both the top and front views. The true length of the line can be found by locating a temporary XY-axis at the apex of the pyramid (point A in Figure 5–6), and rotating the line counterclockwise until its slope is equal to zero (see Figure 5–7). Finally, projecting the new end point of line AB into the adjacent view and connecting that point to the apex reveals the true length of line AB (see Figures 5–8, 5—9, and 5–10).

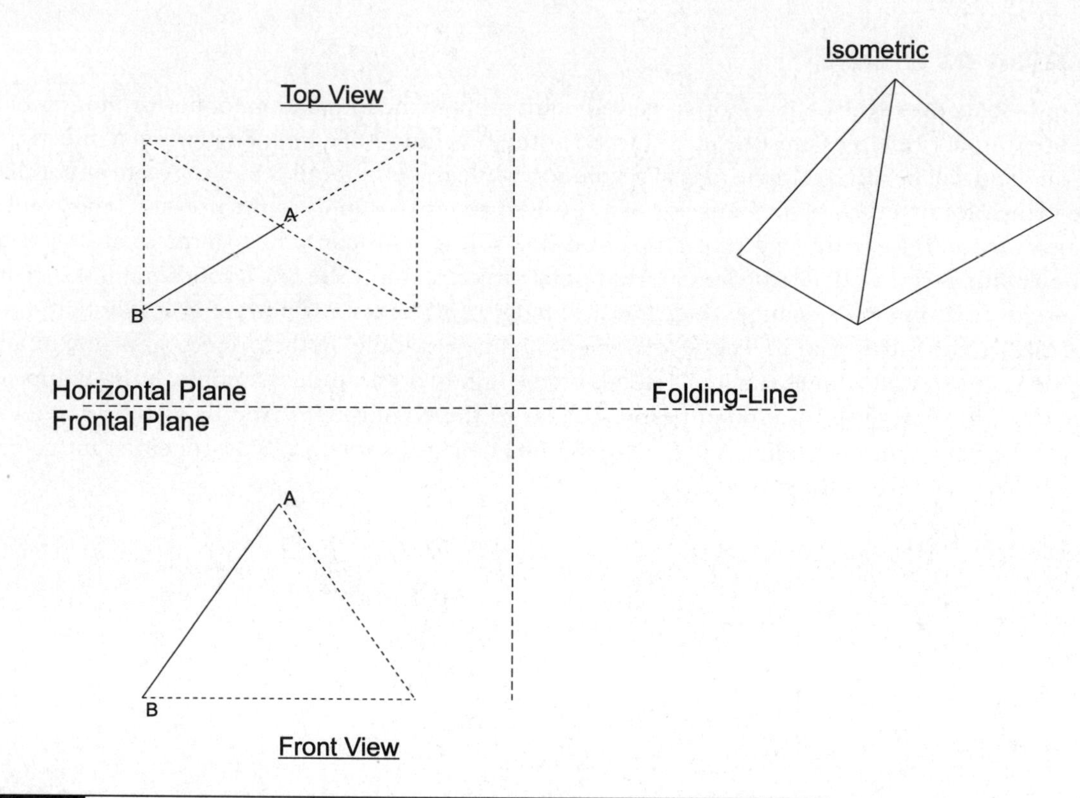


**Figure 5–5** *Line AB appears forshortened in both the top and front views.*

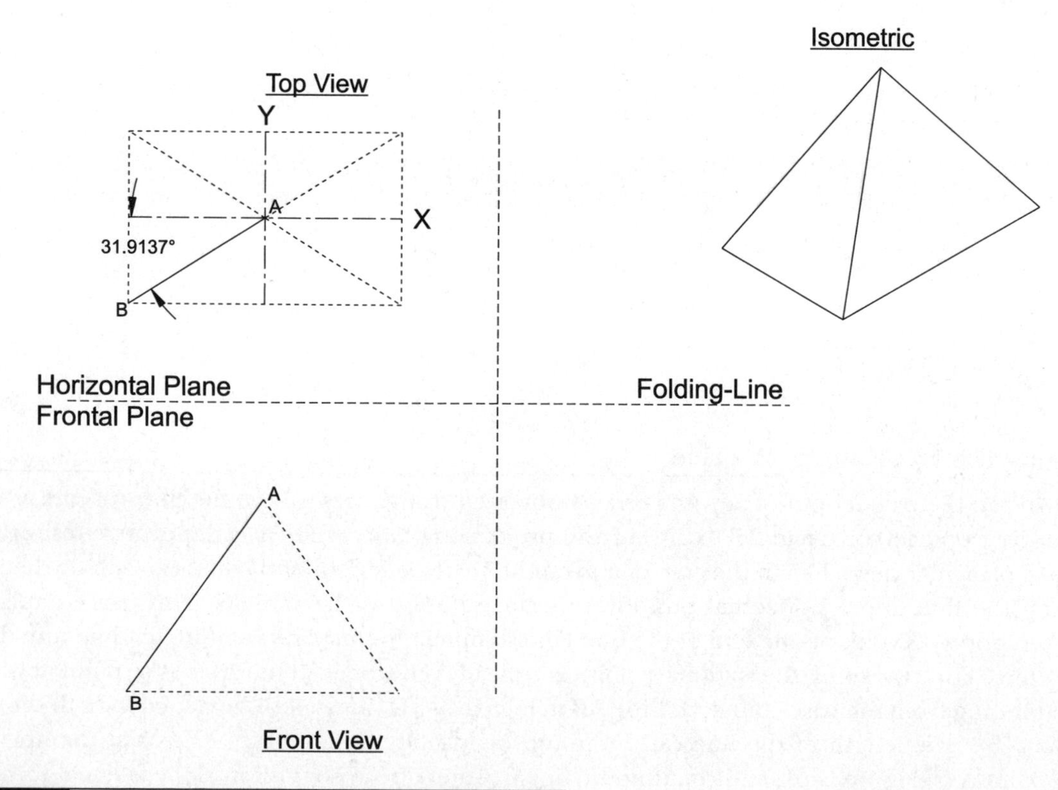


**Figure 5–6** *Axis of rotation is established at point A.*

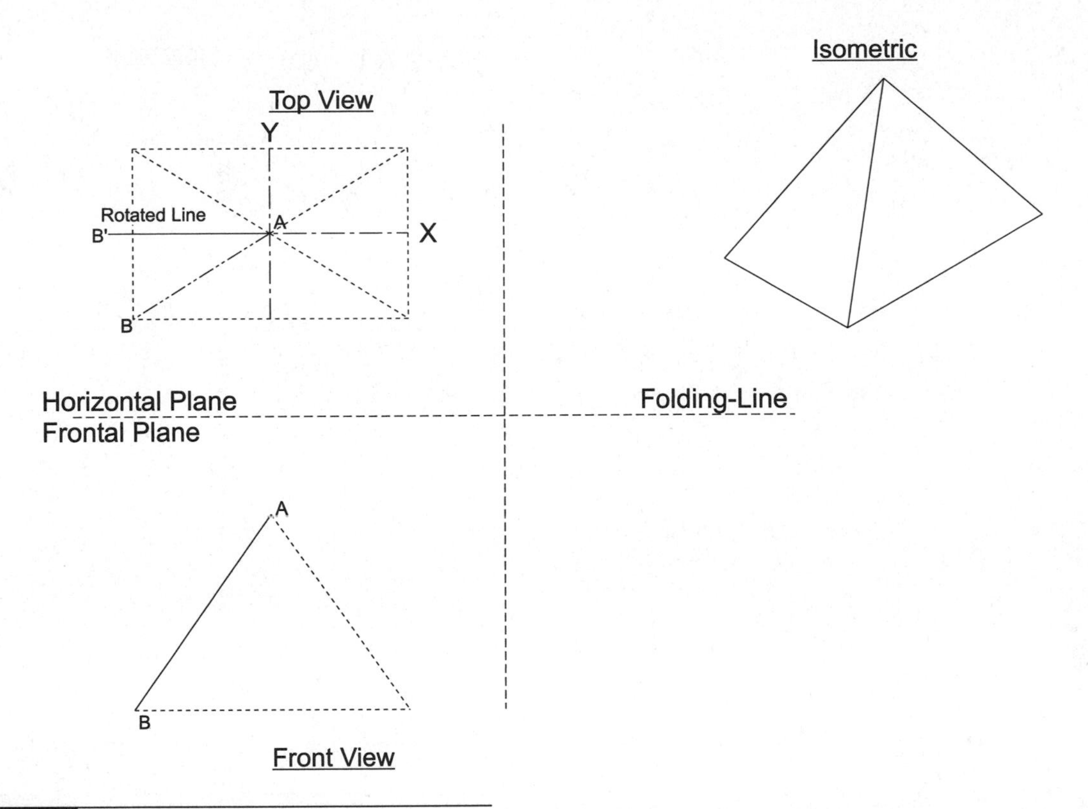


**Figure 5–7** *Line AB is rotated in the top view.*

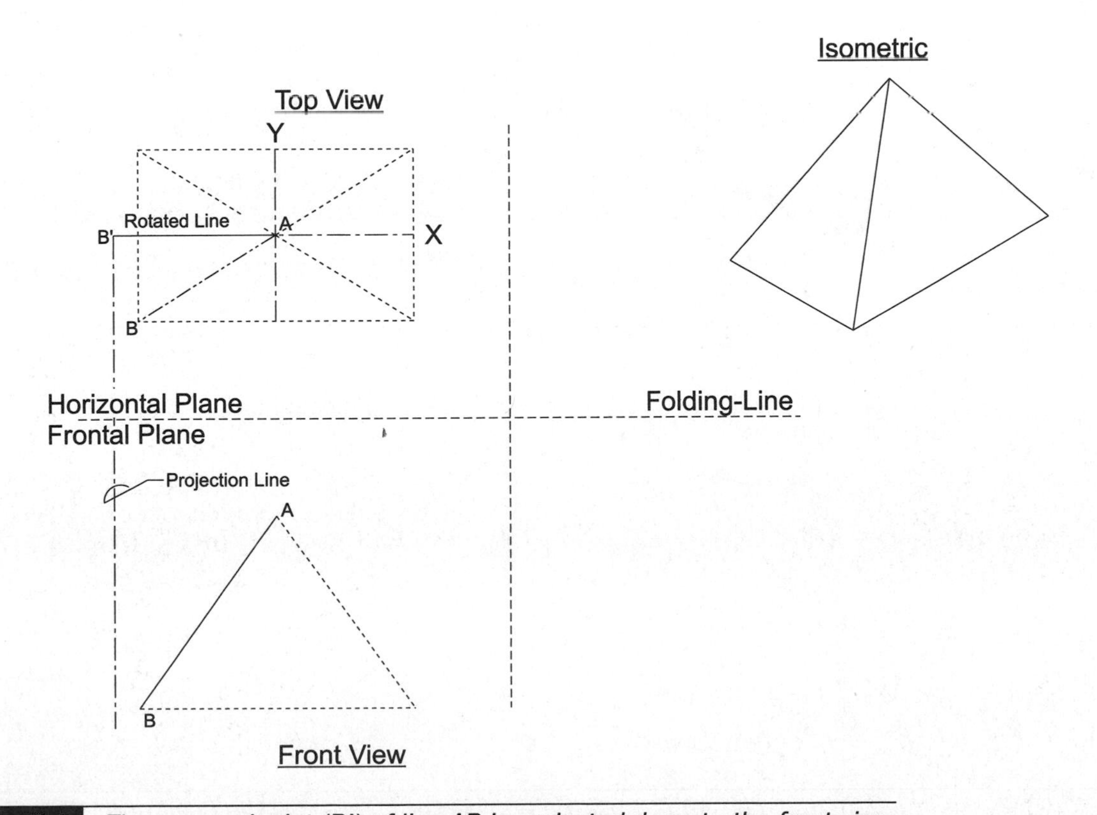


**Figure 5–8** *The new endpoint (B') of line AB is projected down to the front view.*

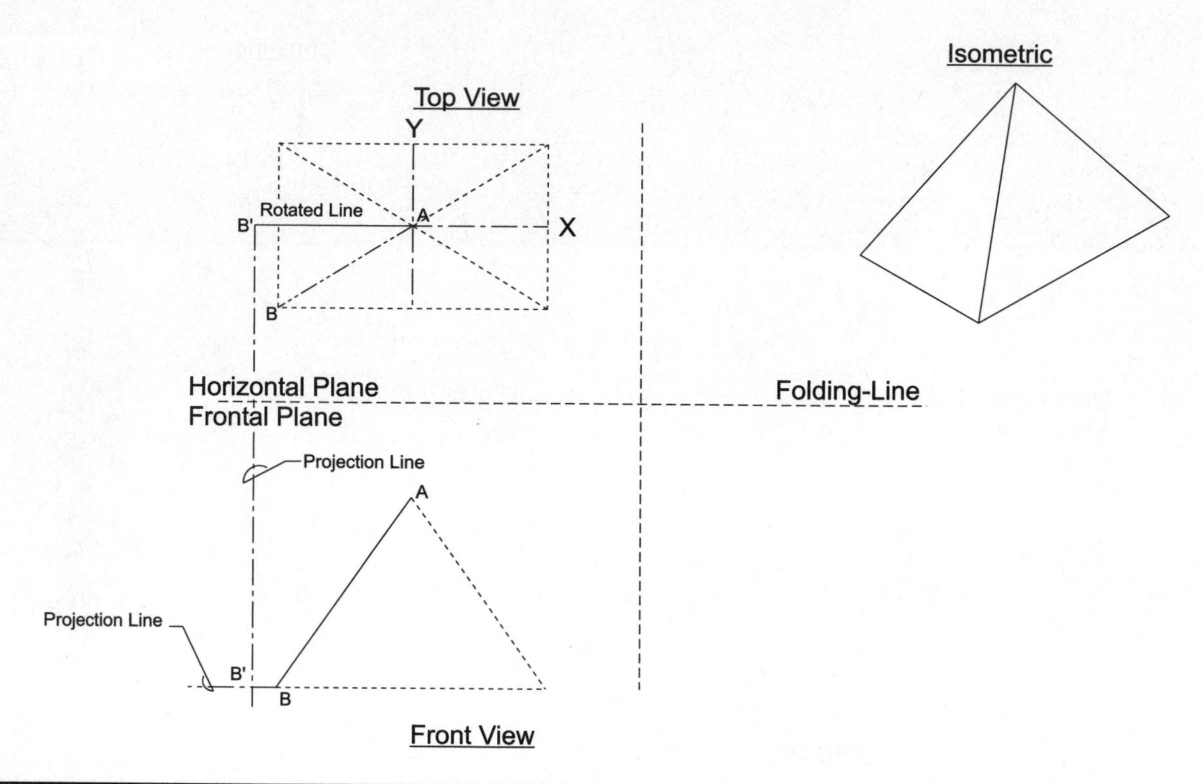


**Figure 5-9** *Point B' is located in the front view by projecting a line from the base of the pyramid until it intersects the projection line extended from the top view.*

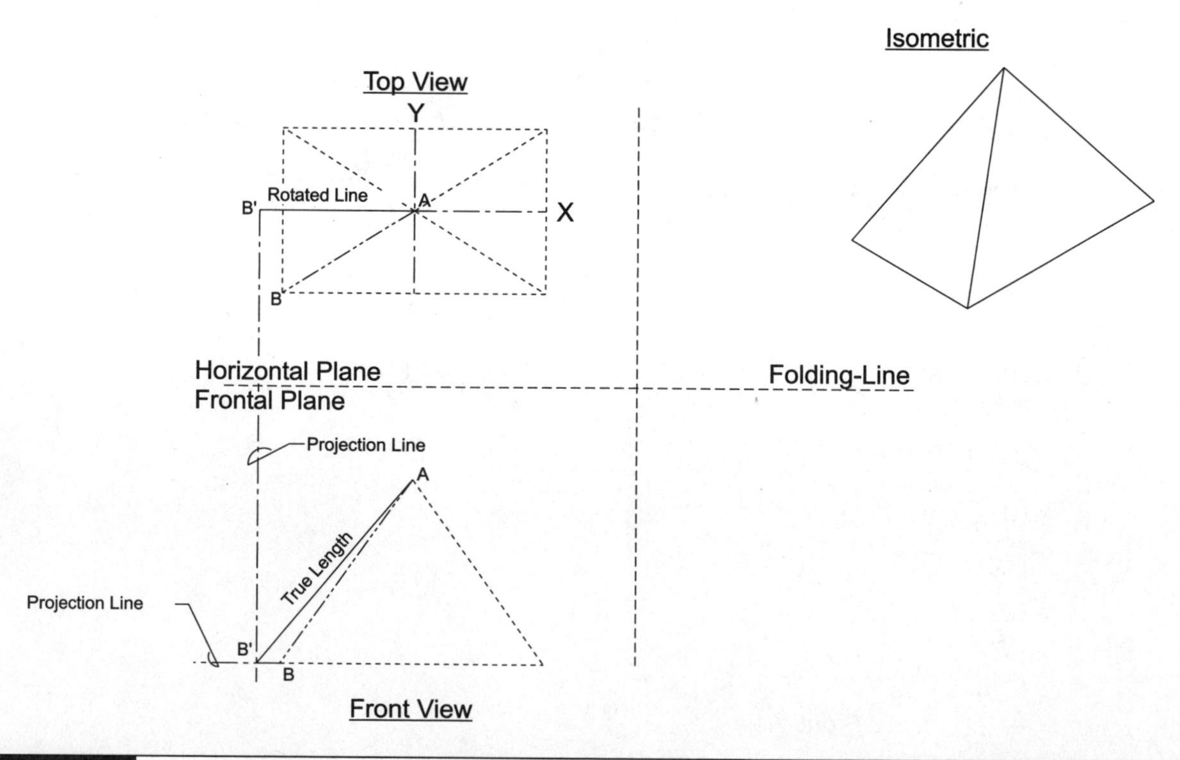


**Figure 5-10** *The true length of line AB is finally determined by connecting the point at the intersection of the projection lines to the apex (A) in the front view.*

## Determining the True Length of a Two-Dimensional Line

When the true length of a two-dimensional oblique line is to be determined using the revolution method in AutoCAD, the basic technique is the same as that used for manual drafting. Starting with either the top or front view, a temporary revolution axis is established at one end of the oblique line. It is not necessary to physically draw this axis, but at times it provides a handy visual reference when working with complex two-dimensional drawings. Once the rotation axis has been established, the ROTATE command is used to revolve the oblique line into a horizontal position. Next, using the LINE command, the new position of the revolved line is projected into the adjacent view (which could be either the top or front view depending upon which view the line was rotated in). Again, using the LINE command, the exact location of the new end point is determined by projecting a line from the old endpoint perpendicular to the previous projection line. The intersection of these two lines marks the location of the new endpoint of the line. Drawing a line from this intersection to the opposite end of the oblique line reveals the true length. This is illustrated in figure 5–11.

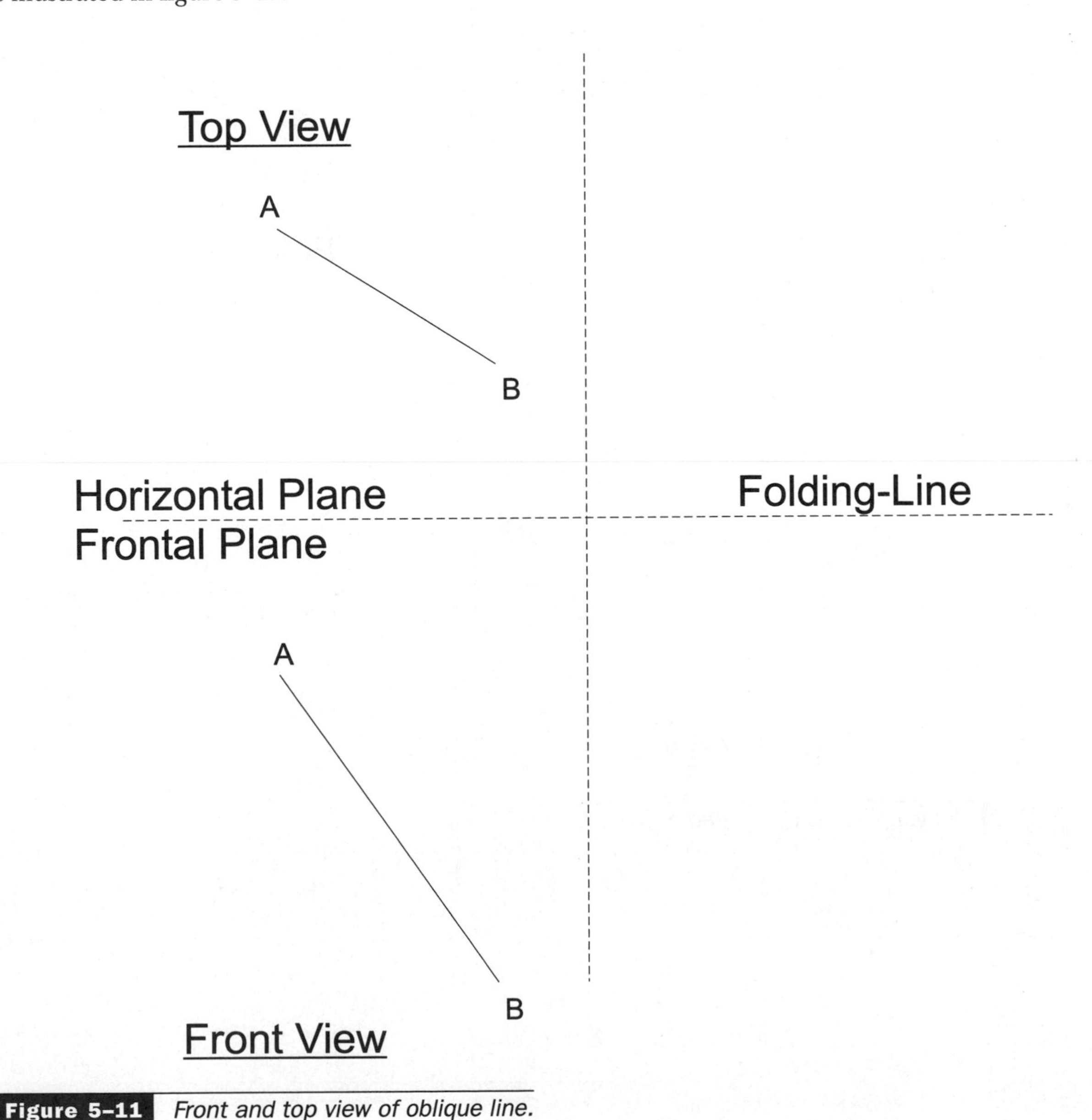


**Figure 5–11** *Front and top view of oblique line.*

## Step #1

Select a view where the revolution will take place. Determine the point through which the revolution axis will be located. In this example, the view the revolution will take place in is the front view and the point the revolution axis will act through is point A (see Figure 5–12).

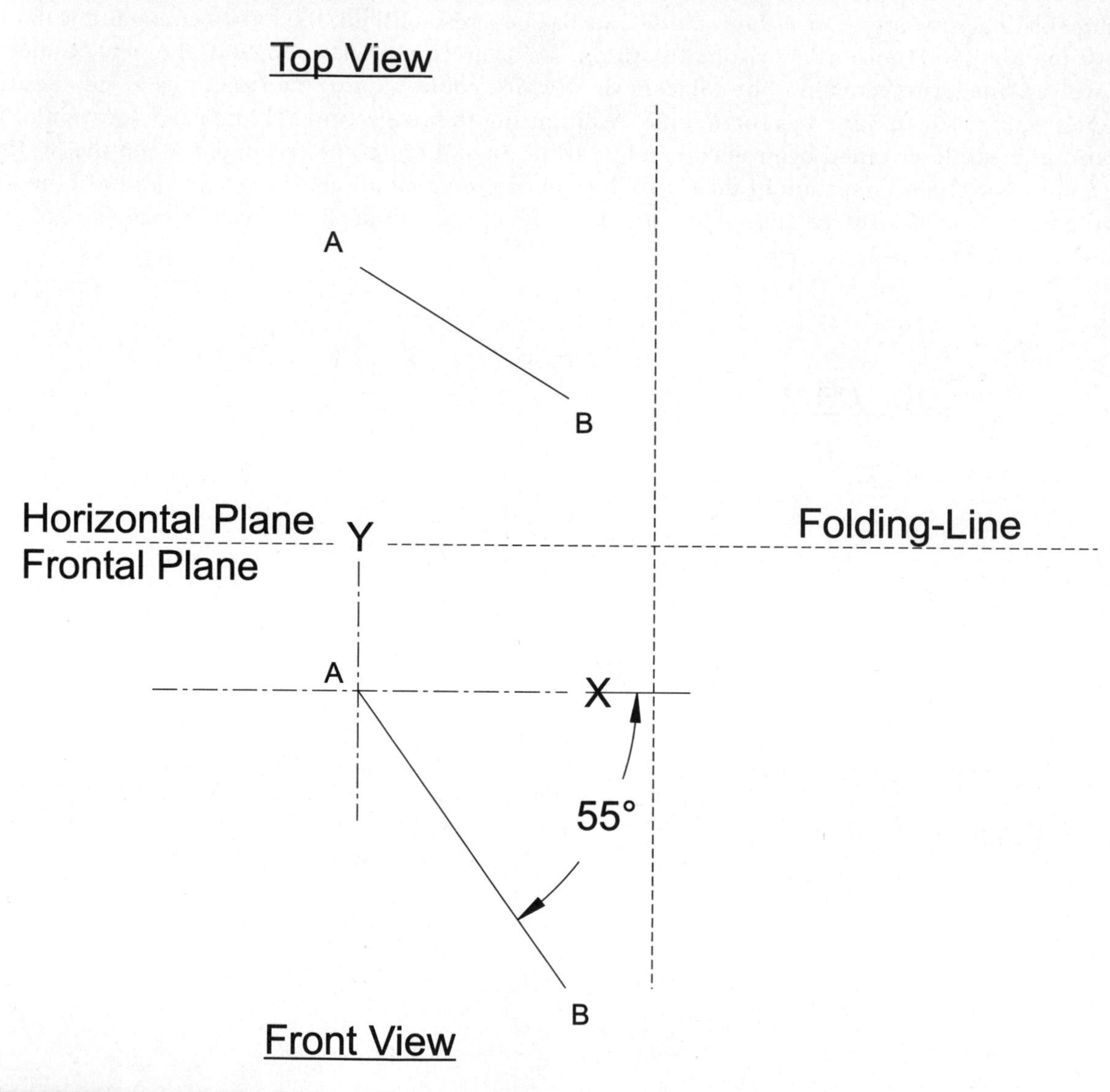


**Figure 5–12** *Imaginary XY-axis located at point A.*

## Step #2

Using the AutoCAD ROTATE command, the oblique line is rotated into a horizontal position using point A as the rotation base point (see Figure 5–13).

Command: ROTATE (ENTER)
Select objects: 1 found *(Select Line AB in the front view.)*
Select objects: *(Press* ENTER *to terminate* ROTATE *command selection mode.)*
Base point: end of *(Select point A as rotation base point.)*
<Rotation angle>/Reference: R *(Press* ENTER*. This option allows the user to specify the absolute rotation and the new rotation angle.)*
Reference angle <0>: end of *(Select point A in the front view.)*
Second point: end of *(Select point B in the front view.)*
New angle: *(With* ORTHO *enabled,* F8*, move cross hairs until line appears in the horizontal position or enter 0.)*

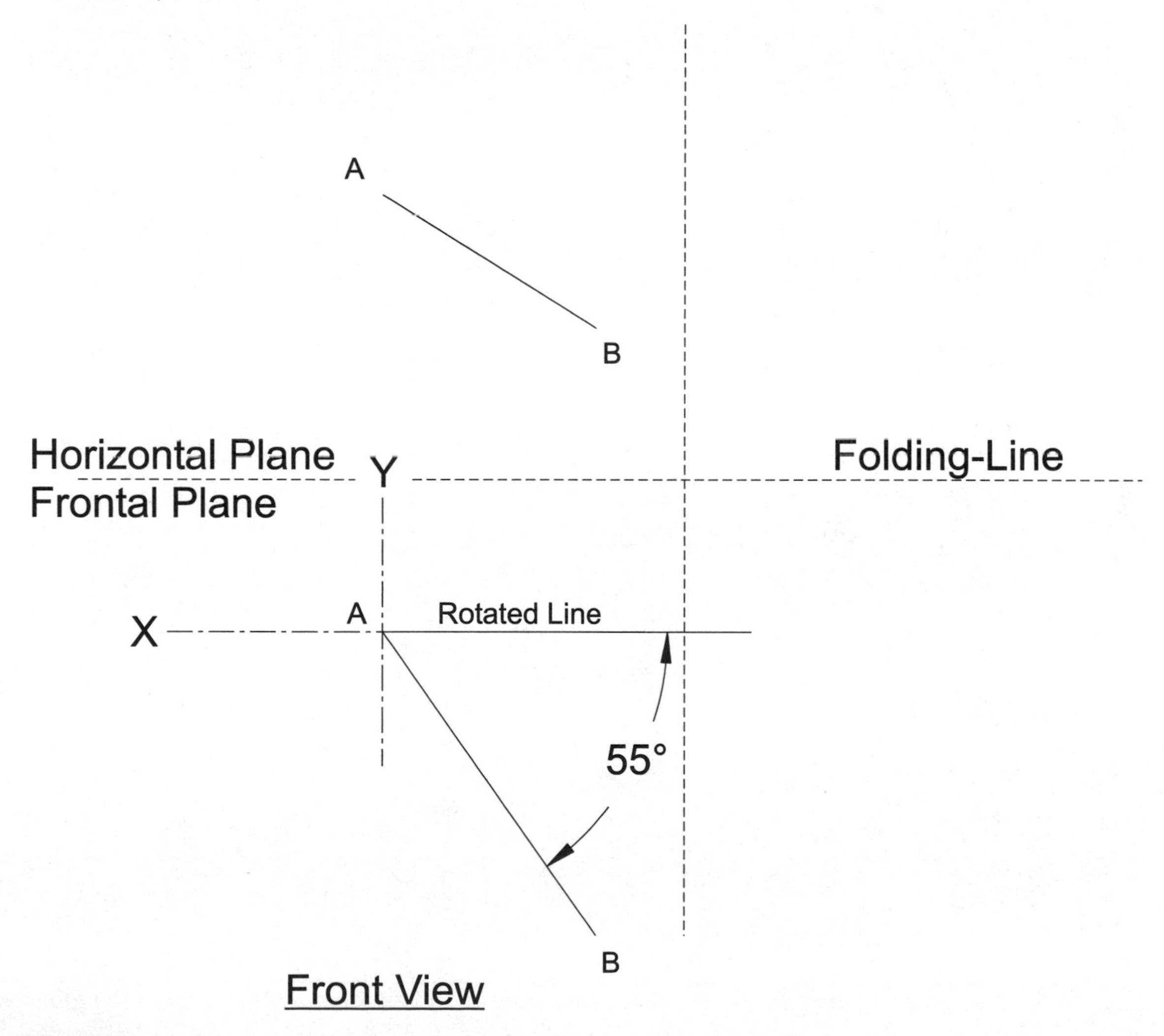


**Figure 5–13** *Line AB rotated into a horizontal position in the front view.*

**Step #3**

Using the LINE command, the location along the X-axis of the new endpoint of line AB is projected up from the front view into the top view (see Figure 5–14).

Command: LINE (ENTER)
From point: end of *(Select point B in the front view.)*
To point: *(Select arbitrary point above the object in the top view.)*
To point: *(Press ENTER to terminate LINE command.)*

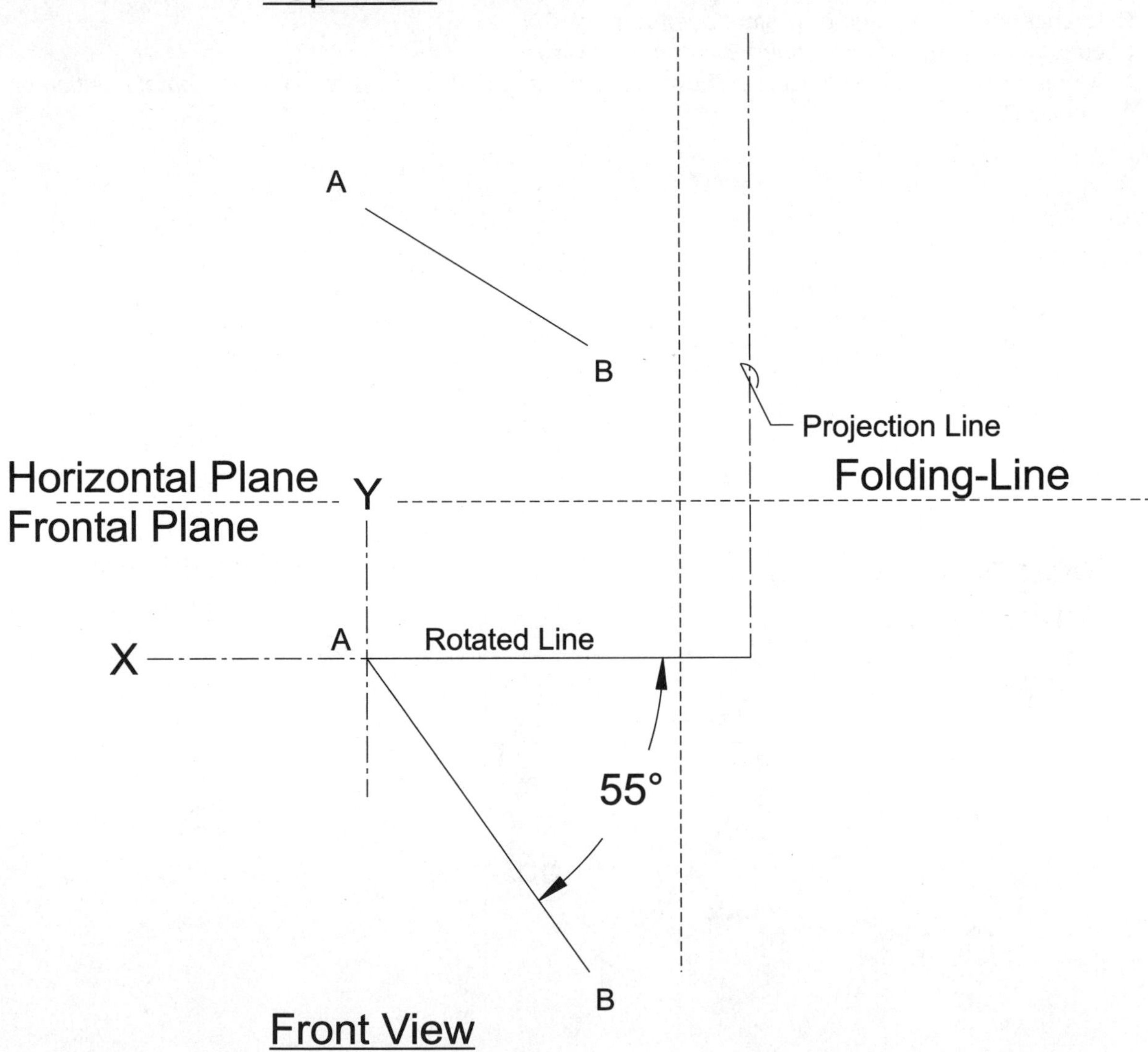


**Figure 5–14** *The location along the X-axis of the new endpoint of line AB is projected up to the top view.*

## Step #4

Using the LINE command, the exact location of the new endpoint of line AB is established in the top view by projecting a line from the old endpoint of line AB (in the top view) perpendicular to the established projection line (see Figure 5–15).

Command: LINE (ENTER)
From point: end of *(Select point B in the top view.)*
To point: *(Select arbitrary point to the right of the projection line in the top view.)*
To point: *(Press ENTER to terminate LINE command.)*

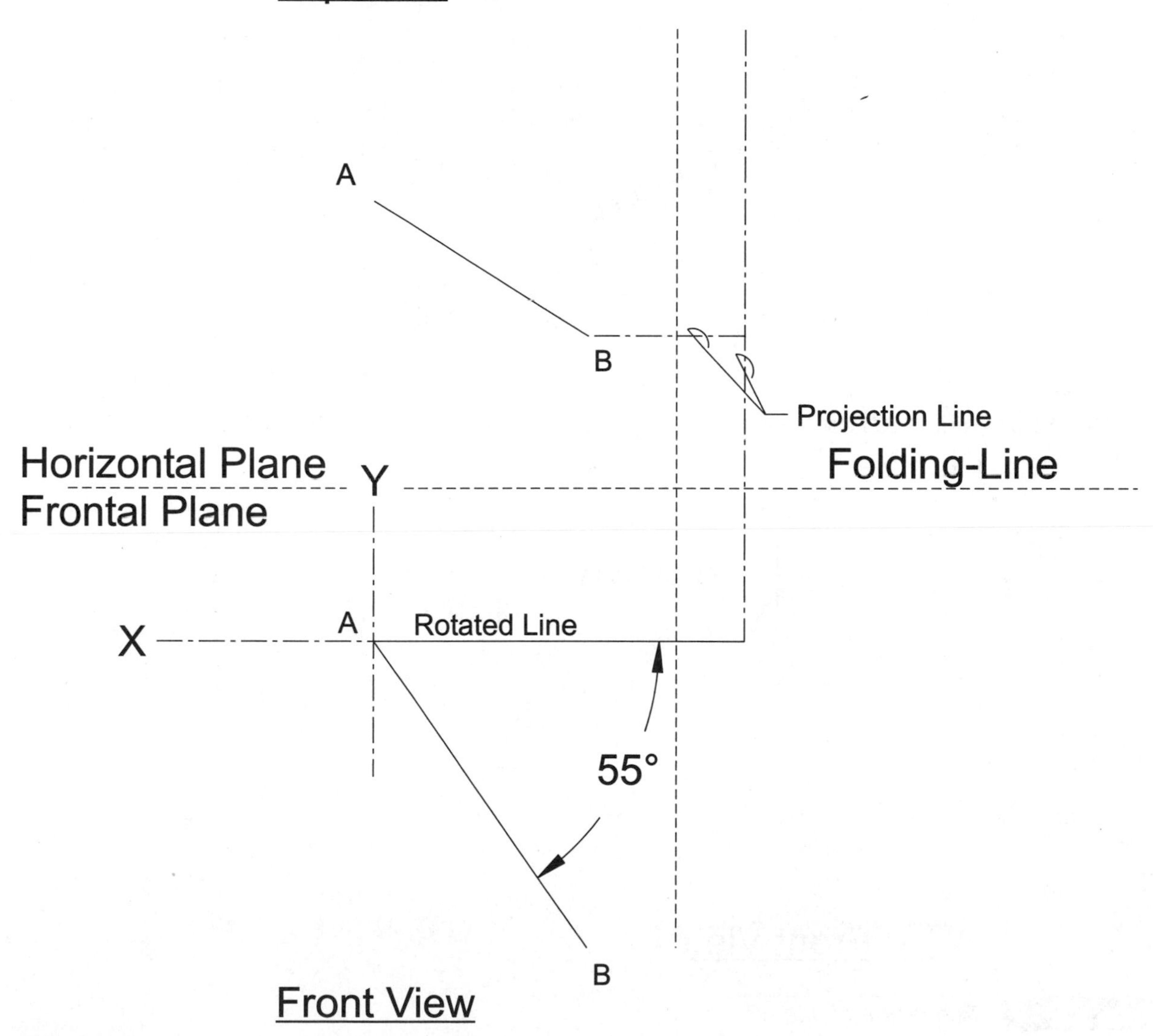


**Figure 5–15** *Point B is projected perpendicular to the previous projection line in the top view. The intersection is the new endpoint of line AB.*

## Step #5

Using the LINE command, the true length of line AB is illustrated in the top view by drawing a line from point A to the intersection of the two projection lines (see Figure 5–16).

Command: LINE (ENTER)
From point: end of *(Select point A in the top view.)*
To point: int of *(Select point B in the top view.)*
To point: *(Press* ENTER *to terminate* LINE *command.)*

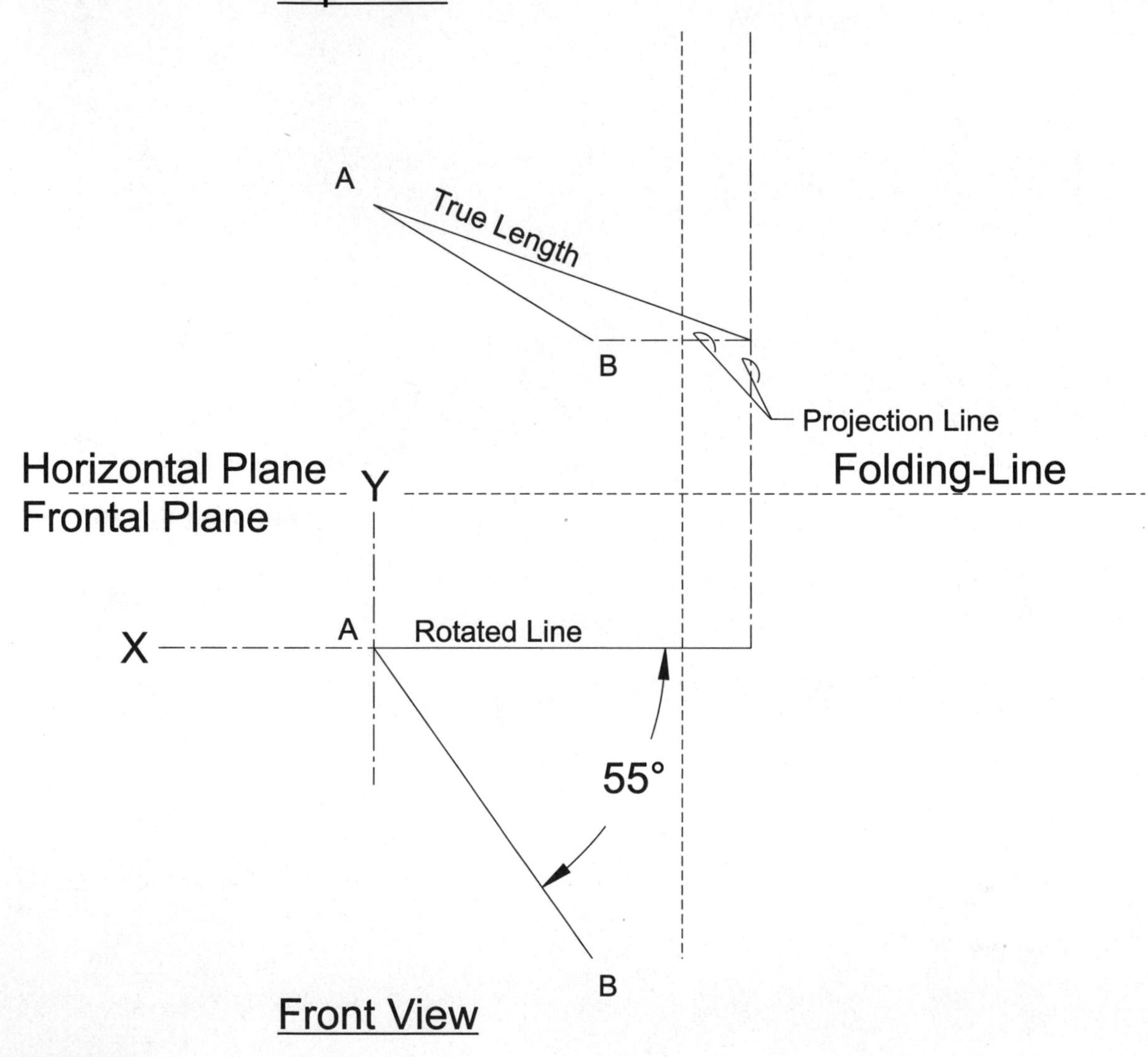


**Figure 5–16** *True length of line AB.*

## Determining the True Shape of a Plane

Chapter four demonstrated that when a plane is shown in true shape and size in one view, then it must appear in edge view in all other adjacent views. This simplifies the task of finding the true shape when the plane can be illustrated in one of the principal views. When a plane cannot be shown in true shape and size in one of the principal views, then an auxiliary view or a revolution must be employed. The axis of revolution for a plane typically lies on that plane. As a result, the axis of revolution will appear as a point in any view where the plane appears in edge view and in true length in any view that shows the plane in true shape and size. Determining the true shape of a plane using the revolution method requires that an auxiliary view be created that will establish the edge view of the plane. In this view, the plane is revolved about the axis of revolution until it is perpendicular to the line-of-sight, and then the new endpoints are projected back to the original view. The exact location of the new endpoints in the original view will be the intersection of a line that extends through the original point and is perpendicular to the projection line. To demonstrate this method, the true shape and size of plane ABC in Figure 5–17 will be determined in the following steps.

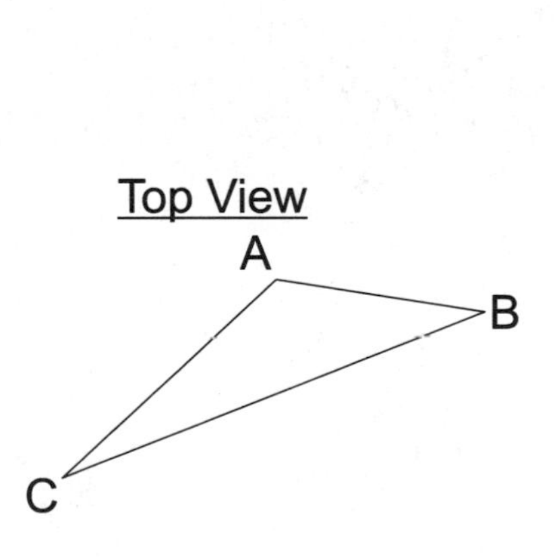

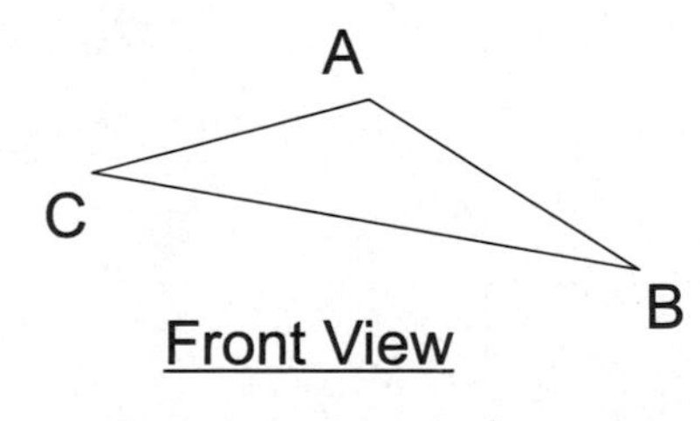


**Figure 5–17** *Top and front view of plane ABC.*

## Step #1

The axis of revolution will reside on plane ABC and will be established by constructing a horizontal line that connects two boundary lines of the plane in the front view. The axis of revolution should be drawn exactly in a horizontal position so that once it has been transferred to the adjacent view it will be shown in true length (see Figure 5–18). Also, an auxiliary view is created to show plane ABC in edge view (see Figure 5–19).

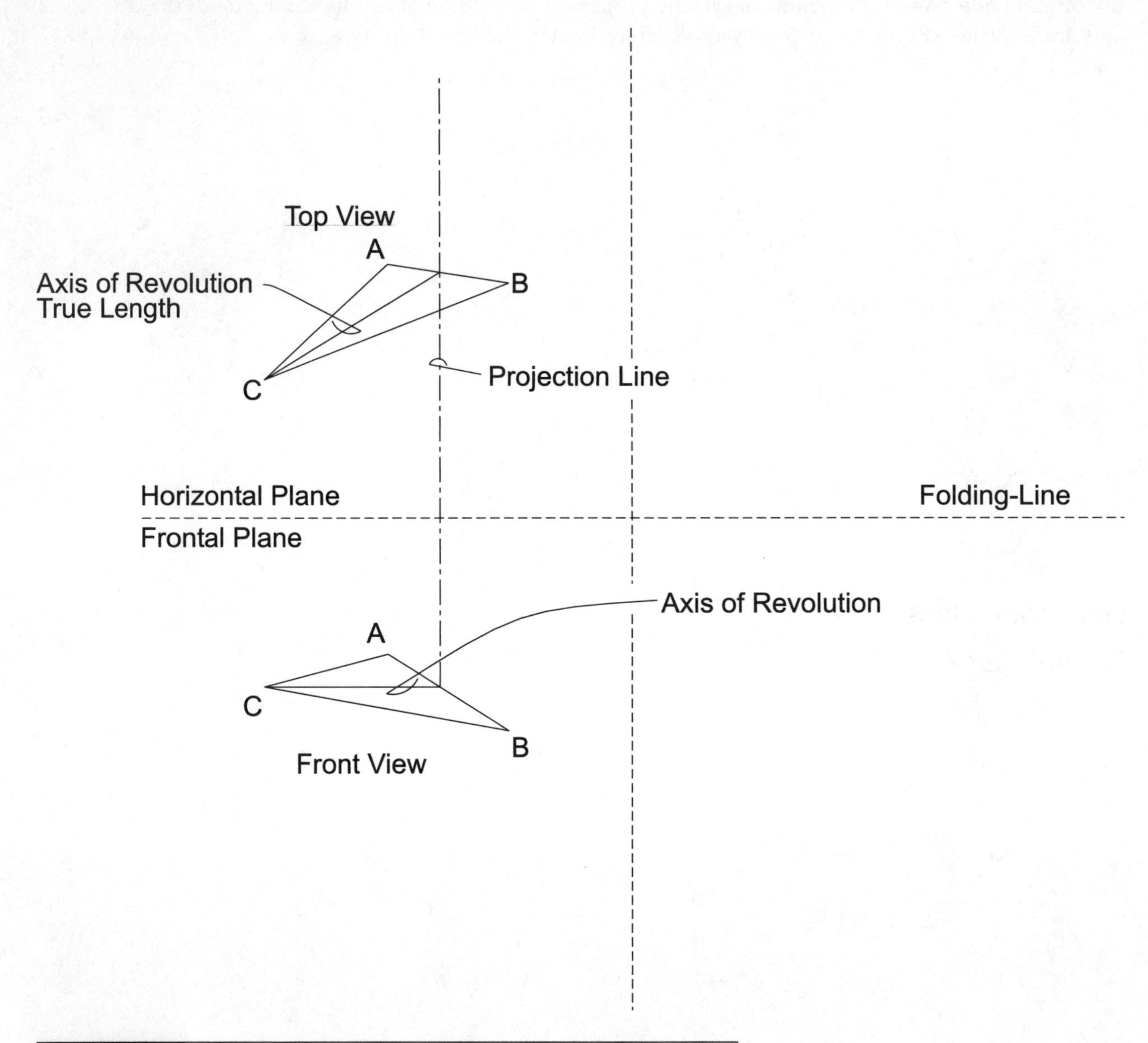


**Figure 5–18** *Establish the axis of revolution in top and front view.*

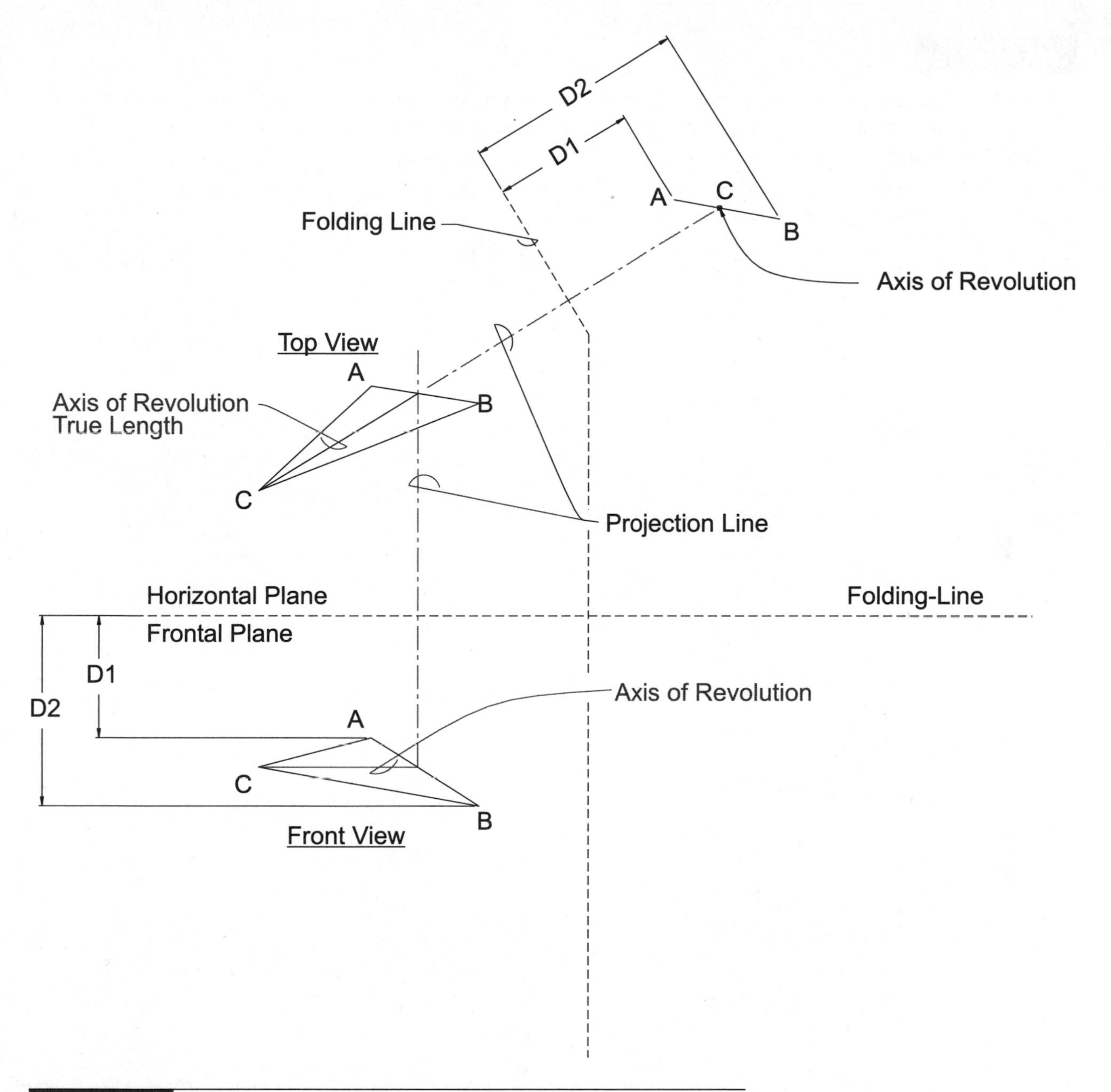


**Figure 5-19** *Auxiliary view is created to show plane ABC in edge view.*

## Step #2

Once the auxiliary view has been established, then plane ABC is revolved about the axis of revolution until it is perpendicular to the line-of-sight for the auxiliary view (parallel to the folding line), as shown in Figure 5–20.

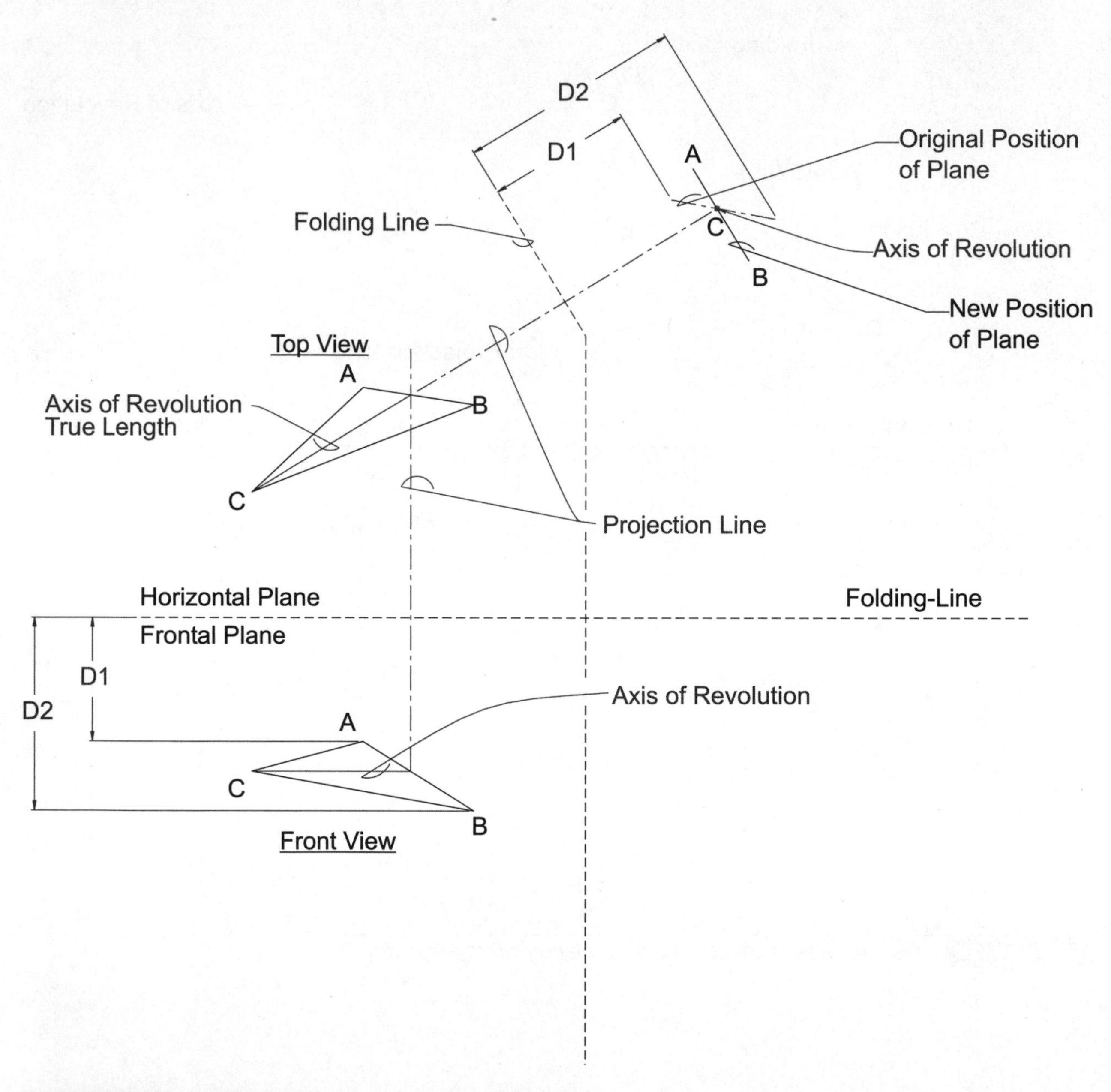


**Figure 5–20** *Plane ABC revolved in the auxiliary view.*

## Step #3

The new location of points B and C are transferred to the top view via projection lines extended from the rotated endpoints of plane ABC in the auxiliary view (see Figure 5–21). The exact location of these points can be determined by constructing projection lines in the top view that pass through the original points and are perpendicular to the projection lines extended from the auxiliary view (see Figure 5–22).

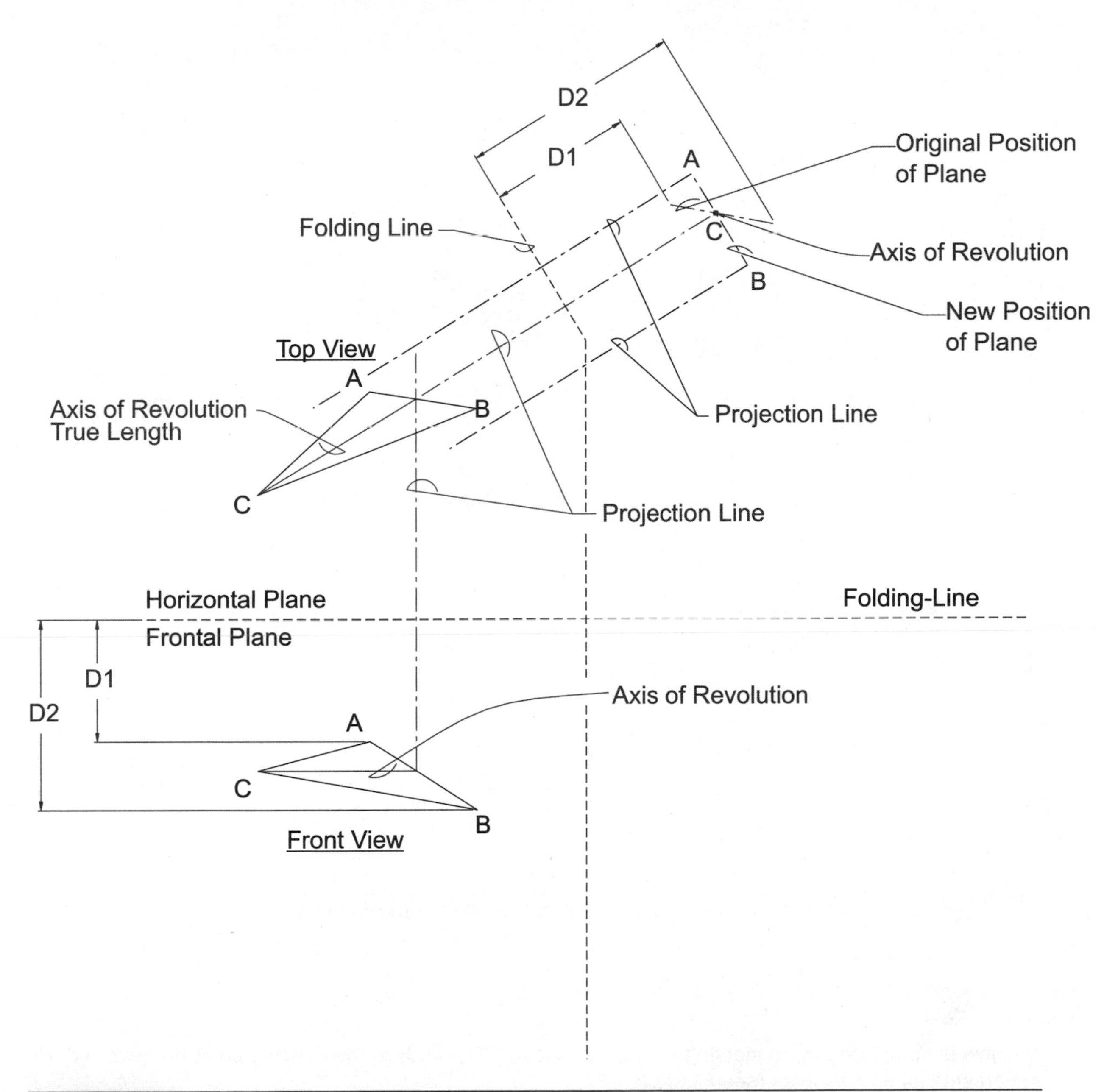


**Figure 5–21** *New locations of point B and C are transferred from the auxiliary view to the top view.*

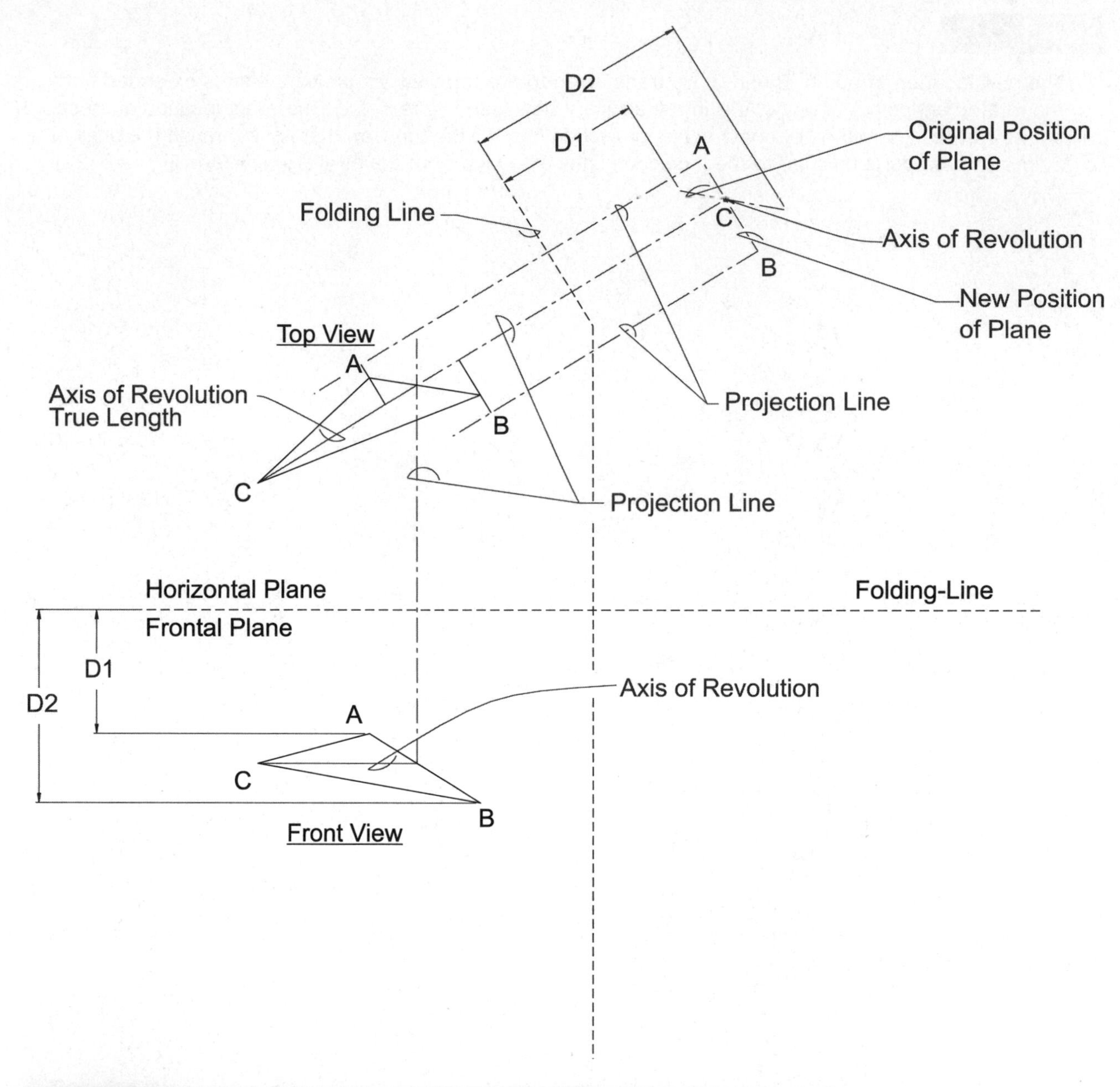


**Figure 5–22** *The exact location of points ABC are established in the top view.*

## Step #4

The view is completed by connecting the new endpoints (B and C) to the existing point not revolved (A) and erasing all excess lines (see Figure 5–23).

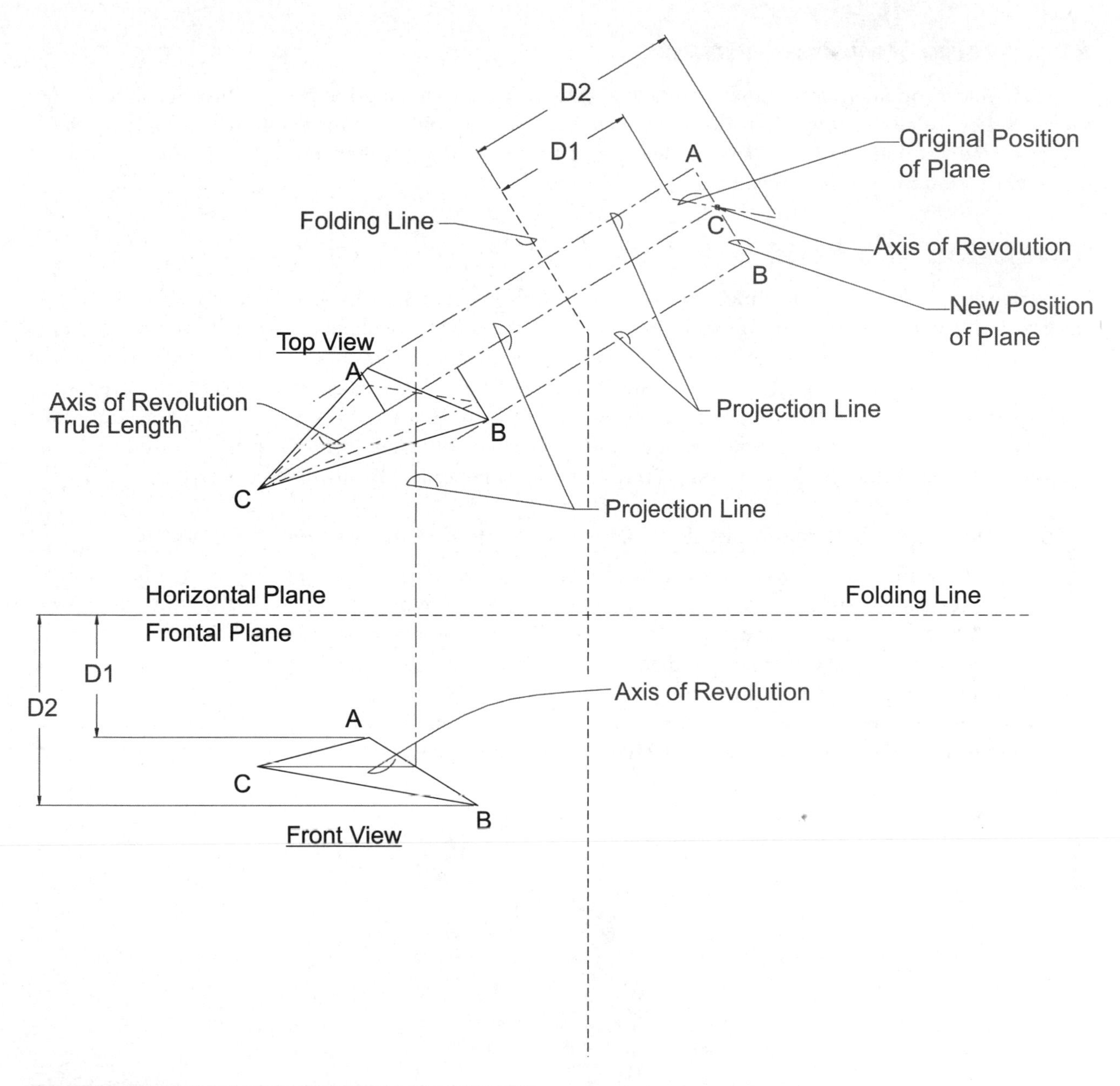


**Figure 5–23** *True shape and size of plane ABC.*

## Determining the True Angle between a Line and a Plane

Chapter four demonstrated that the true angle between two planes can be found in any view in which the two planes appear in edge view. The same can be said about finding the true angle between a line and a plane; any view where the line appears in true length and the plane appears in edge view will show the true angle between the two entities.

## Successive Revolved Views

When an object contains features that cannot be shown clearly in either a primary or secondary auxiliary view, then a successive auxiliary view must be constructed. The same holds true for revolutions. When an object contains features that cannot be shown clearly in a single revolution, then multiple revolutions must be made, each from the previous revolution.

## Revolution of a Solid

Understanding how points and lines are revolved is an important step in being able to solve real world applications using the revolution method. In the real world, the professional is not limited to just working with points and lines; they must also work with solid objects. Since all solid object are comprised of the same basic elements (points and lines), the techniques introduced in the previous sections are also employed when working with solid objects. However, there is one additional principle that is necessary for the revolution of a solid. When a solid object is revolved, all attributes associated with that object must be revolved the same amount and their distances from the axis of revolution must be uniformly maintained.

### Rotating an AutoCAD Three-Dimensional Object instead of Using the VPoint Command

When working with a three-dimensional object in AutoCAD, views can either be created by changing the observer's line-of-sight (using the VPOINT command) or rotating the object into the desired position (using either the ROTATE or ROTATE3D commands). Remember that when a three-dimensional object is rotated in AutoCAD, then that object will appear rotated in any other view that features the object. If the object is to be rotated in one view and not in others, then the object should be copied onto a different layer before it is rotated. Once the object has been copied, then the layer containing the original part may be frozen and the duplicate object rotated about the X-, Y-, or Z-axis, using either the ROTATE3D or ROTATE commands.

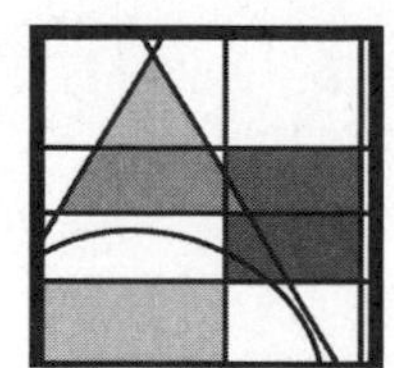

# advanced applications

## Using AutoLISP and DCL to Determine the True Angle Between a Line and a Plane

Chapter four featured a program that calculates the piercing point of a line and a plane using formulas derived from analytical geometry theories and principles. By making a few minor modifications, the program can be expanded to include calculating the true angle between a line and plane. Combining three principles discussed earlier with a few basic trigonometry functions determines the true angle. The three principles are as follows: first, the true angle between a line and a plane will only appear in a view that shows the line in true length and the plane in edge view. Second, if a point is contained on a plane, when that plane appears in line view, then the point will reside on that line. Third, a point contained on a plane will have an infinite number of lines passing through it. Dividing the distance between the endpoint of the line perpendicular to the plane by the distance between the point on the plane and the piercing point provides a quotient that can be used with the tangent trigonometry function to find the true angle. Once the modifications have been completed, the program interface should look like Figure 5–24.

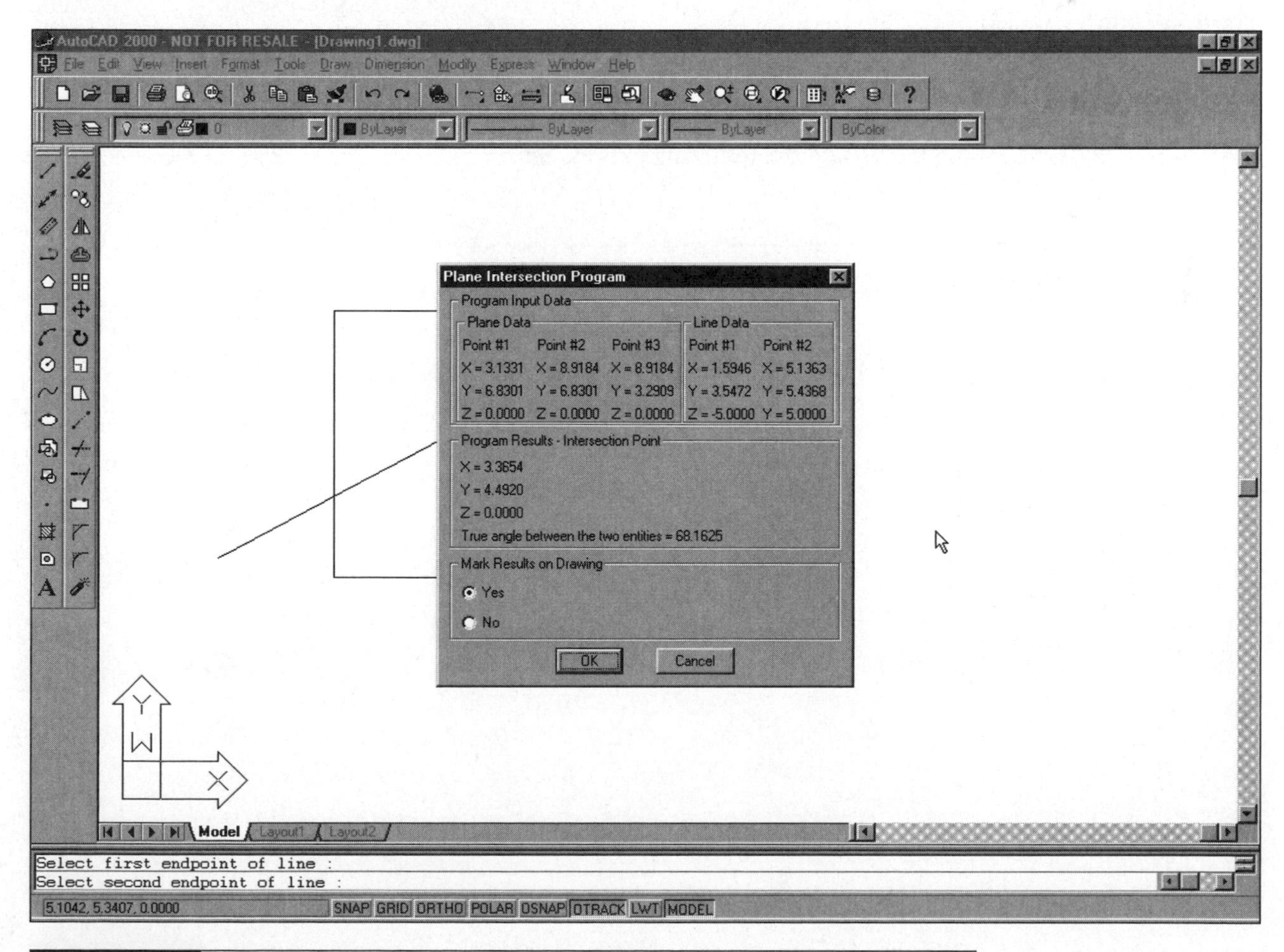


**Figure 5–24** *Dialog box for intersection program with modifications completed.*

## AutoLISP Program Supplement

First open the DG04.lsp file and locate the expressions:

```
dcl_id4  (load_dialog "/windows/desktop/geometry/
                            student/Dialog_Box_Source_Files/chap_4.dcl")
)
```

Immediately after this expression, insert the following code:

```
    pt2         (list point_1_line_x point_1_line_y z_value)
    dis1               (distance point_1_line pt2)
    dis2               (distance pt2 intersection_point)
    angle1           (* (atan (/ dis1 dis2)) (/ 180 3.14))
```

Second, locate the expression:

```
(set_tile "field_pt_i_z" (strcat "Z = " (rtos z_value)))
```

Immediately after this expression, insert the following code:

```
(set_tile "angle1" (strcat "True angle between the two entities = " (rtos angle1)))
```

Third, save the changes.

## DCL Supplement

First, open the DG04.dcl file and locate the following expression:

```
: text {
          label = "Z = ";
          key = "field_pt_i_z";
     }
```

Immediately after this expression, insert the following code:

```
: text {
          label = "Angle between the two entities";
          key = "angle1";
          }
```

Second, save the file.

# Review Questions

Answer the following questions on a separate sheet of paper. Your answers should be as complete as possible.

1. Describe the difference between the change-of-position method and the revolution method.
2. Define the terms axis of revolution, revolution plane, and successive revolution.
3. Discuss the basic principles regarding the revolution of a point, a line, a plane, and a solid.
4. List and describe the steps involved in determining the true length of a two-dimensional line using the revolution method in AutoCAD.
5. List and describe the steps involved in determining the true shape of a plane using the revolution method.
6. When are successive revolutions used and how are they created?
7. What method does AutoCAD use to generate a view of a three-dimensional object?
8. When working with a solid object in AutoCAD, why is it not a good idea to rotate the solid?
9. What AutoCAD commands are typically used to rotate a solid?
10. Why should a mechanical desktop solid not be edited using standard AutoCAD commands?

# Review Exercises

## Revolving

All problems marked with an * can be found on the student CD-ROM.

***#1**

Using the revolution method, find the true length of line AB.

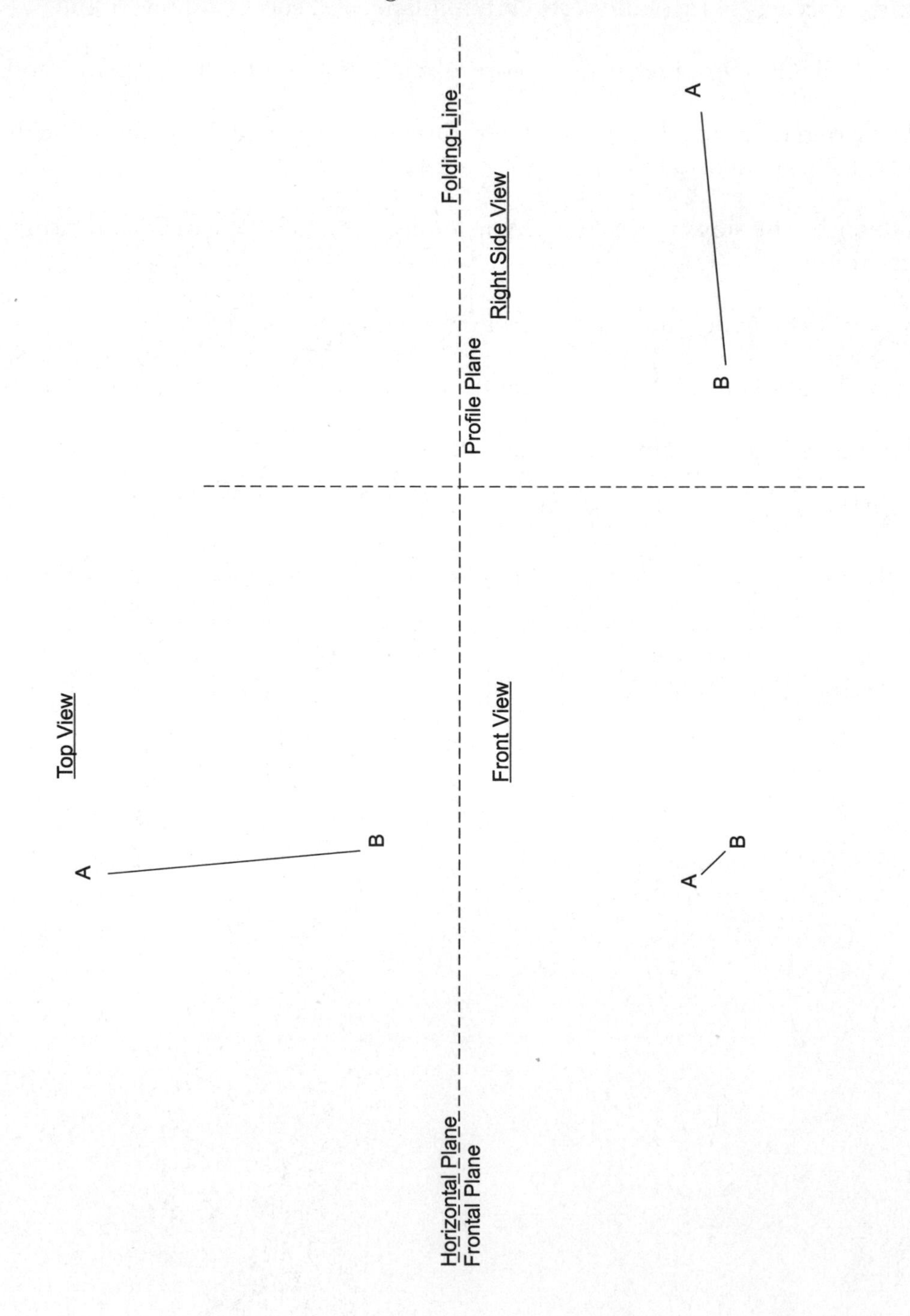

***#2**

Using the revolution method, find the true length of line AB.

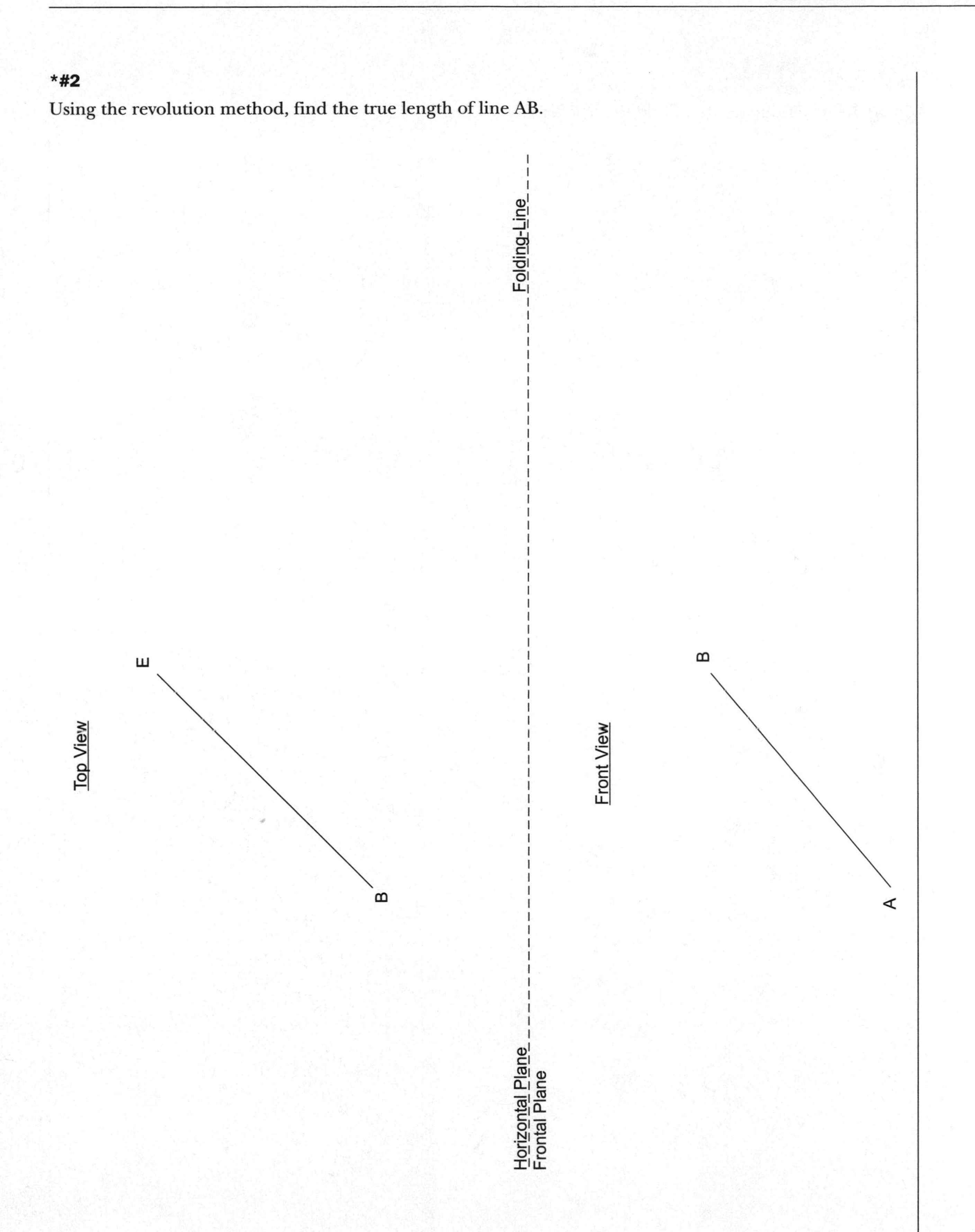

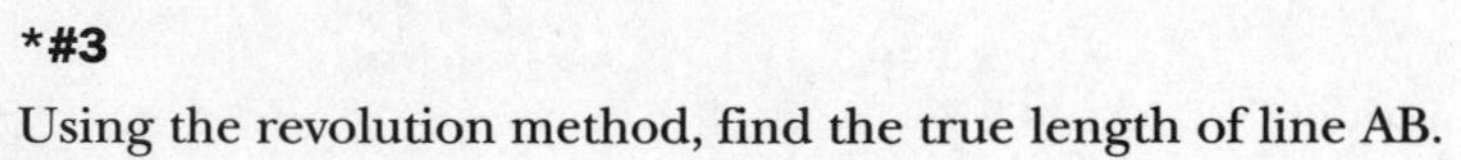

***#3**

Using the revolution method, find the true length of line AB.

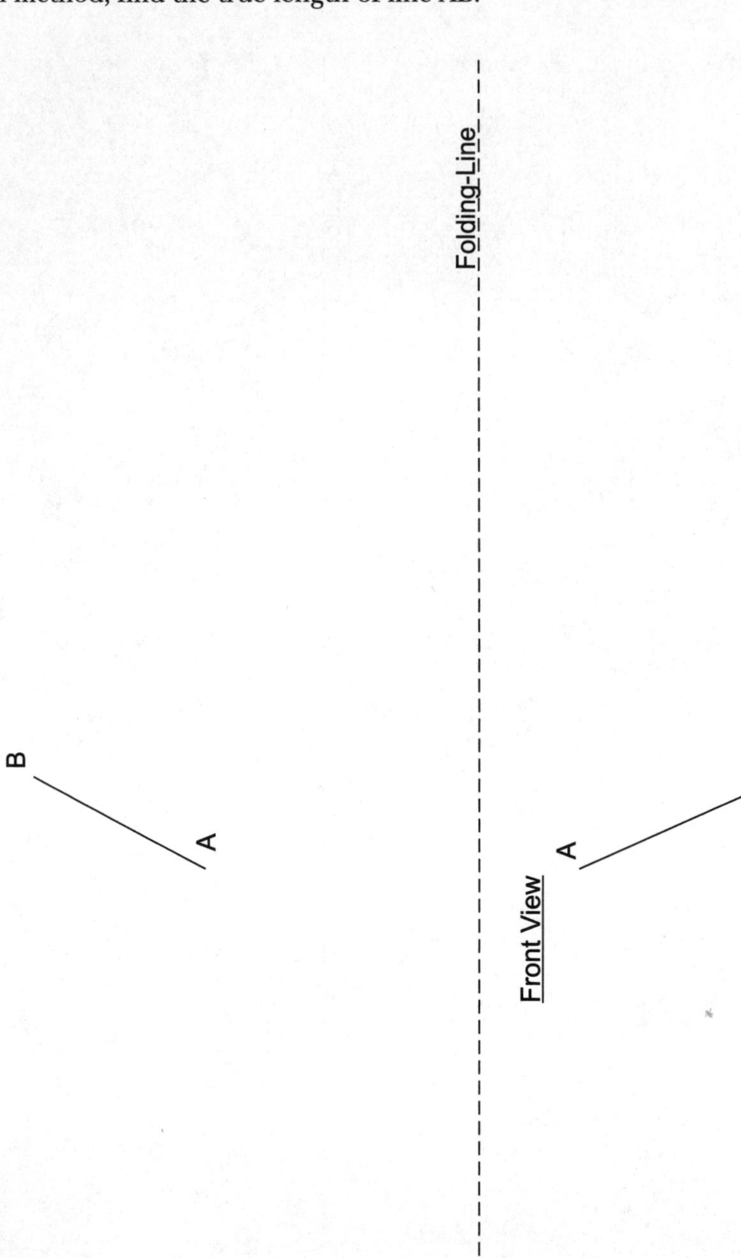

***#4**

Using the revolution method, find the true length of line BE.

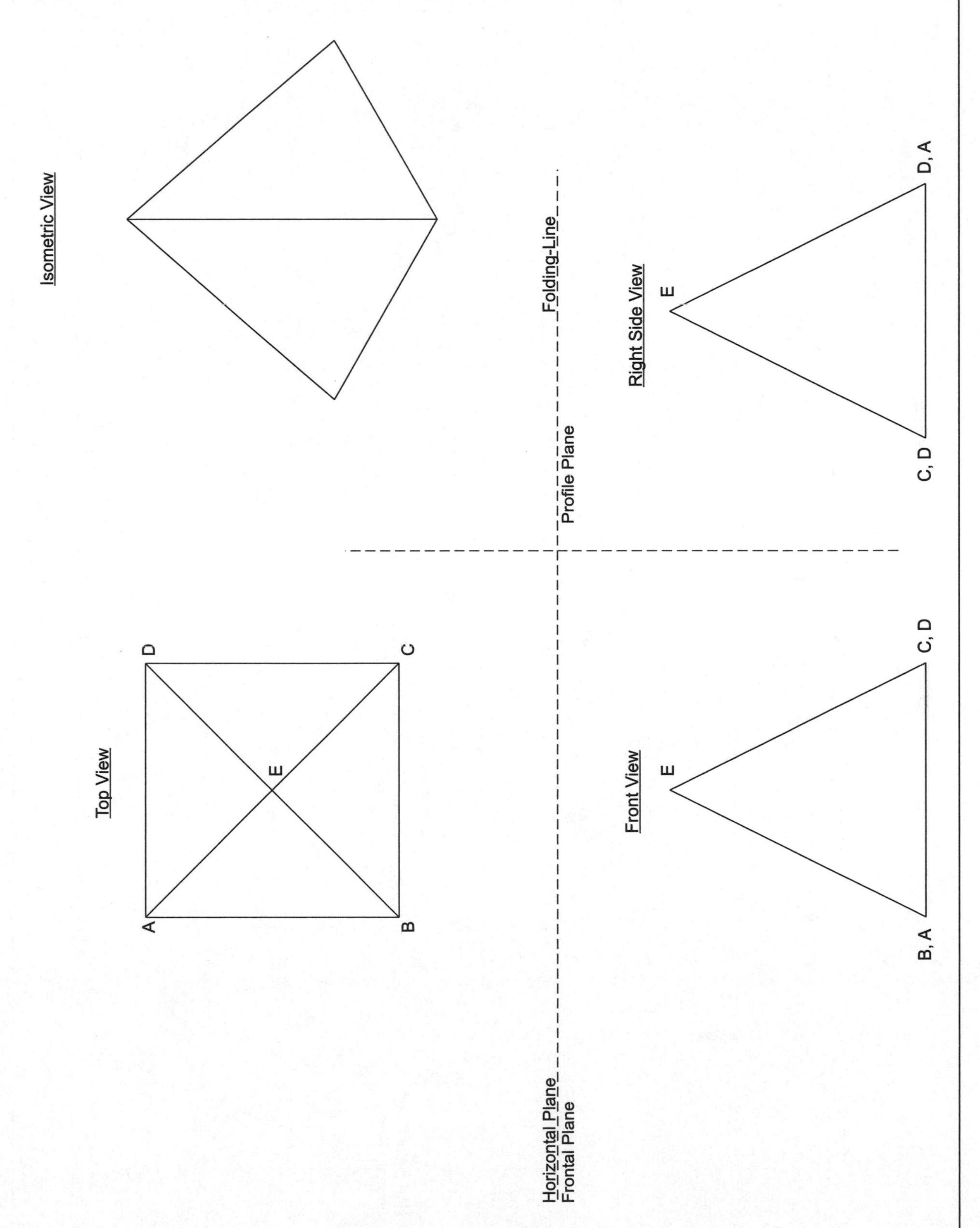

***#5**

Using the revolution method, find the true length of line CD.

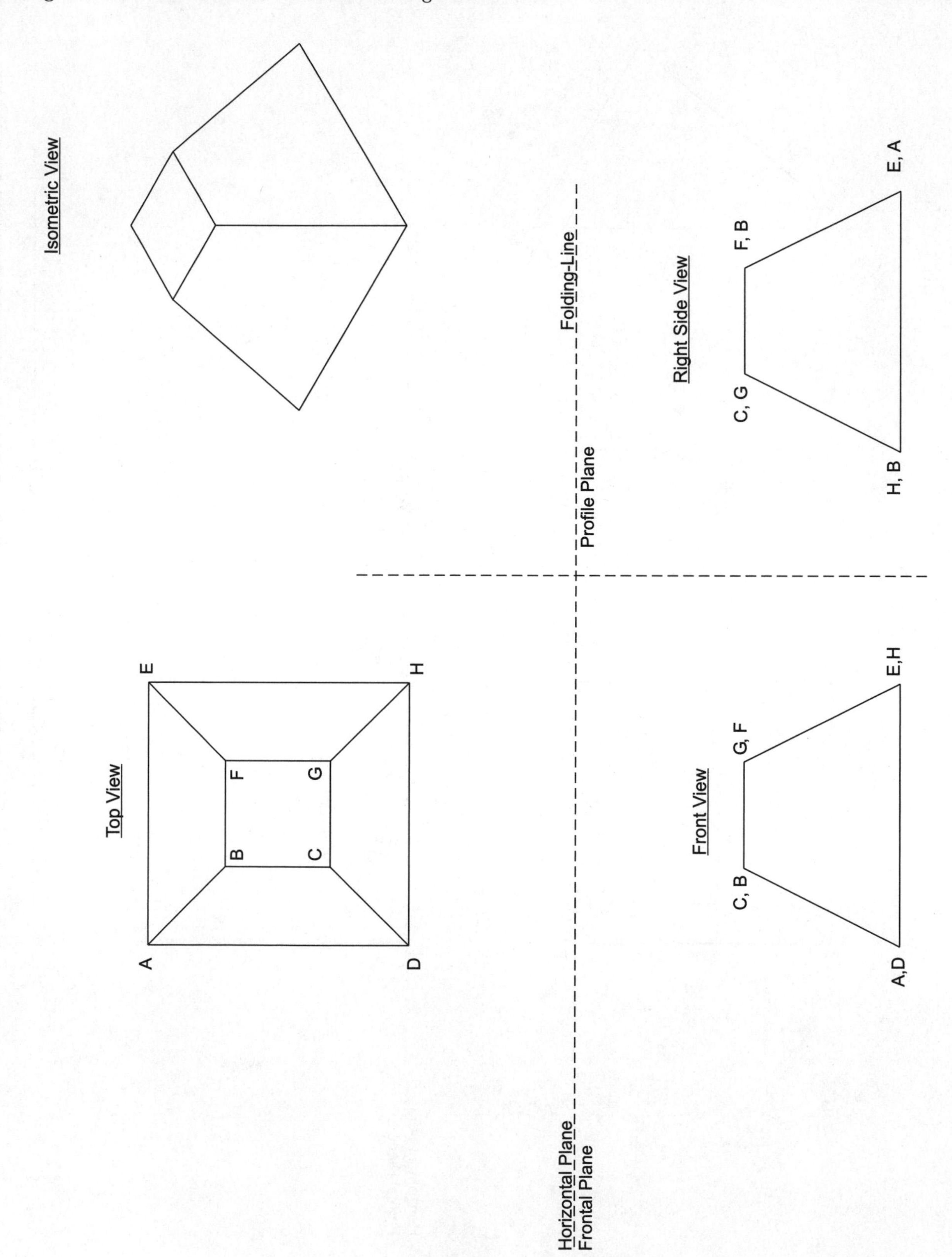

***#6**

Using the revolution method, find the true length of line BC.

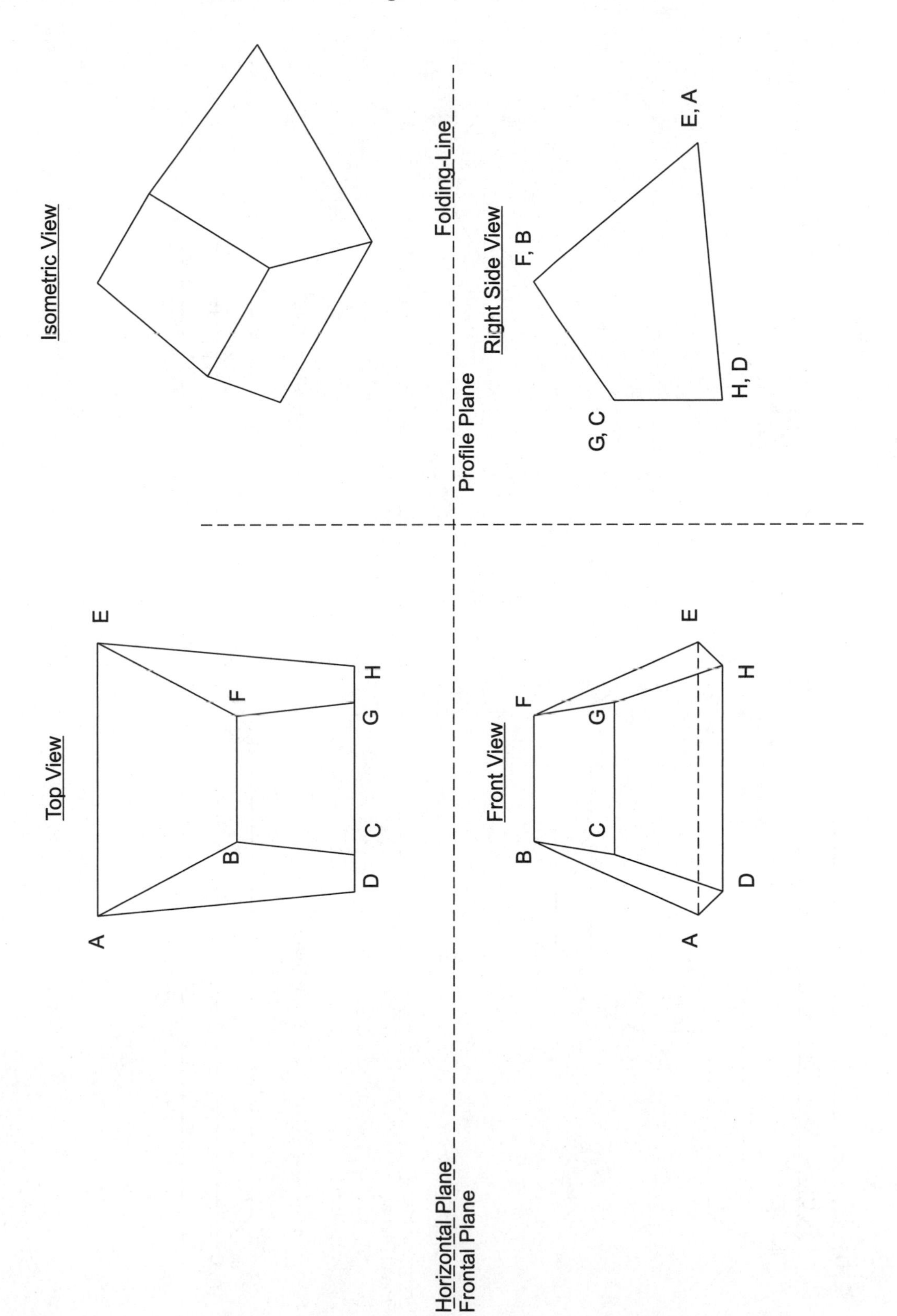

***#7**

Using the revolution method, find the true shape of plane ABCD.

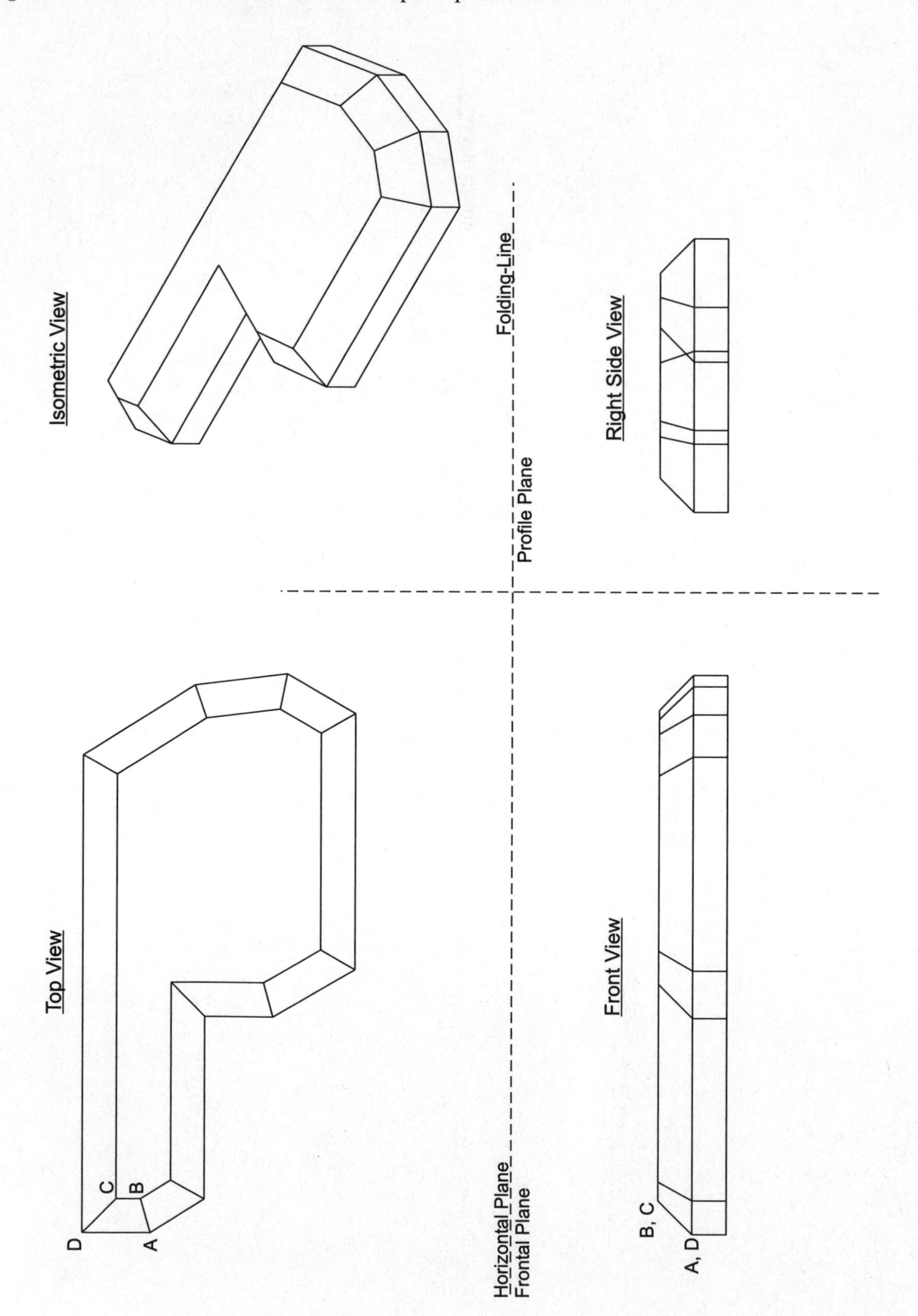

***#8**

Using the revolution method, find the true length of plane ABCD.

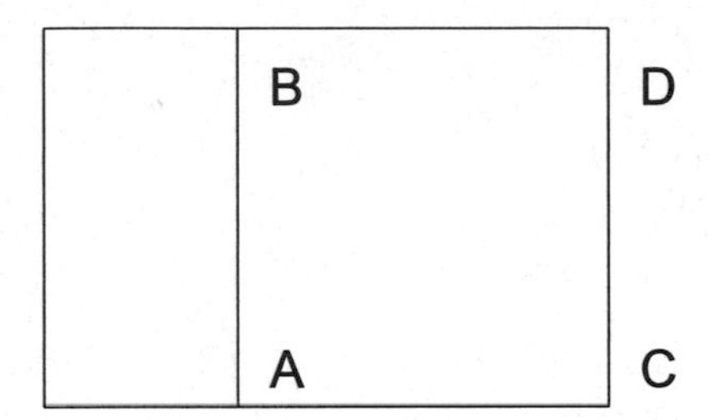


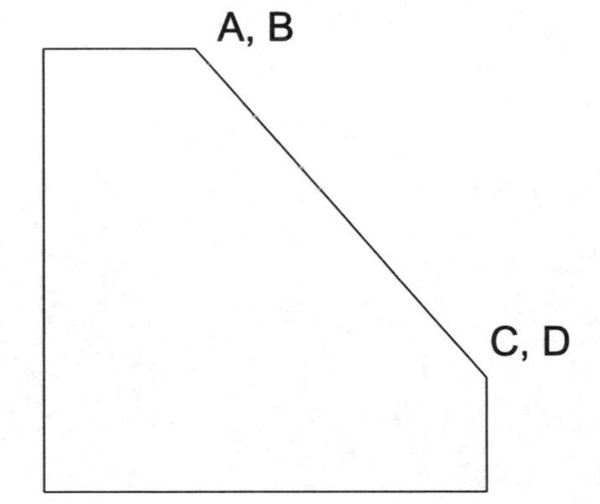


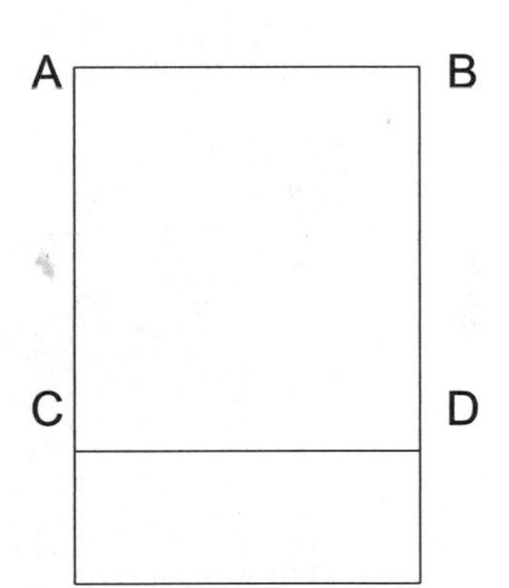


**#9**

Using the revolution method, find the true length of plane BCFG for problem #6.

**#10**

Using the revolution method, find the true length of plane EFGH for problem #6.

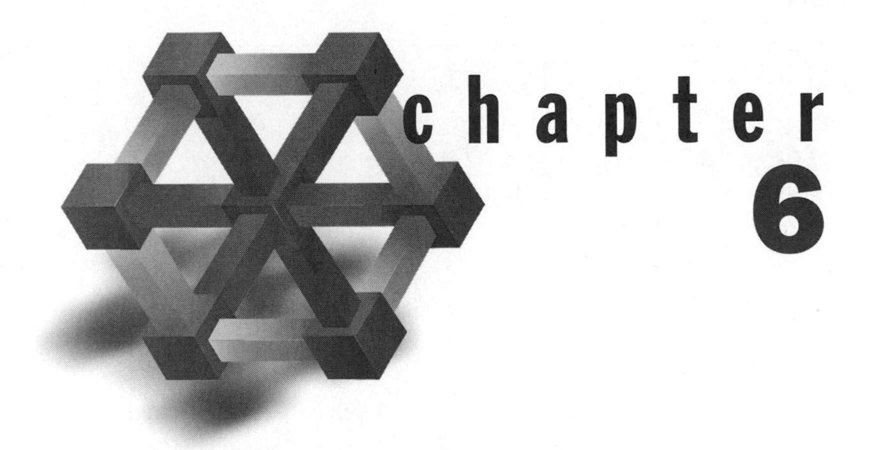


# Developments

## Objectives

**Upon completion of this chapter the student will be able to do the following:**

- *Define the term development and describe what it is used for.*
- *List the five categories of materials.*
- *Define the term bend allowance and calculate the bend allowance for soft copper, soft brass, aluminum, medium-hard copper, medium-hard brass, soft steel, bronze, hard copper, cold-rolled steel, and spring steel.*
- *Identify the common methods of seaming and hemming sheet metal.*
- *Calculate the stock length required to construct a seam or hem.*
- *Define the following terms: prism, right prism, right square prism, oblique prism, cylinder, right cylinder, oblique cylinder, truncated prism, and truncated cylinder.*
- *Use the parallel-line technique to develop a right prism, oblique prism, truncated prism, right circular cylinder, oblique cylinder, and truncated cylinder.*
- *Use the radial-line technique to develop a right cone, oblique cone, truncated cone, right pyramid, oblique pyramid, and truncated pyramid.*
- *Use the triangulation technique to develop a transitional part.*
- *Define the following terms: row, column, key, table, database, field, and record.*
- *Define the terms ASI and ASE.*
- *Define and use the command* DBCONNECT.
- *Set up a basic formula using the math features of a database.*

## KEY WORDS AND TERMS

Apex
Bend allowance
Columns
Cone
Cylinder
Database
Database connectivity manager
Database management system
Development
Environment
Fields
Flat pattern layout
Hem
Link
Link template
Oblique circular cone
Oblique cylinder
Oblique prism
Oblique regular pyramid
Parallel-line technique
Prism
Pyramid
Radial-line technique
Record
Right circular cone
Right cone
Right cylinder
Right prism
Right regular pyramid
Right square prism
Rows
Seaming
Slant height
Stock length
Table
Transitional fittings
Triangulation method
Truncated cylinder
Truncated prism

*Photo courtesy of Smith Fiberglass Products Company*

## Introduction to Developments

Products today are manufactured from a combination of different material types and thicknesses. These materials can be divided into five categories: **plastics** (thermoplastics, thermoset plastics, and elastomers), **ceramics** (oxides, nitrides, carbides, glass, graphites), **composites** (reinforced plastics, metal-matrix, ceramic-matrix, laminates), **pulp** (wood and paper), and **metals** (ferrous and nonferrous). The products made of metals, in particular sheet metal, are the focus of this chapter.

Once designed, metal products must have a ***development*** created before they can be manufactured (see Figures 6–1 and 6–2). A ***development*** is the pattern produced when a three-dimensional geometric shape is unfolded onto a flat surface. Developments can also be called ***flat patterns*** or ***flat pattern layouts***. Their purpose is to show the true shape and size of each area of an object before it is folded into its final shape. Products manufactured from sheet metal are not the only ones that require the use of developments; any product that receives its shape from a bending operation needs a development. Heating and air-conditioning, aerospace, automobile manufacturing, and packaging are just a few of the industries that require the extensive use of developments. For this reason, it is easy to see why conceptualizing and producing the development of a part is an integral part of the engineering process.

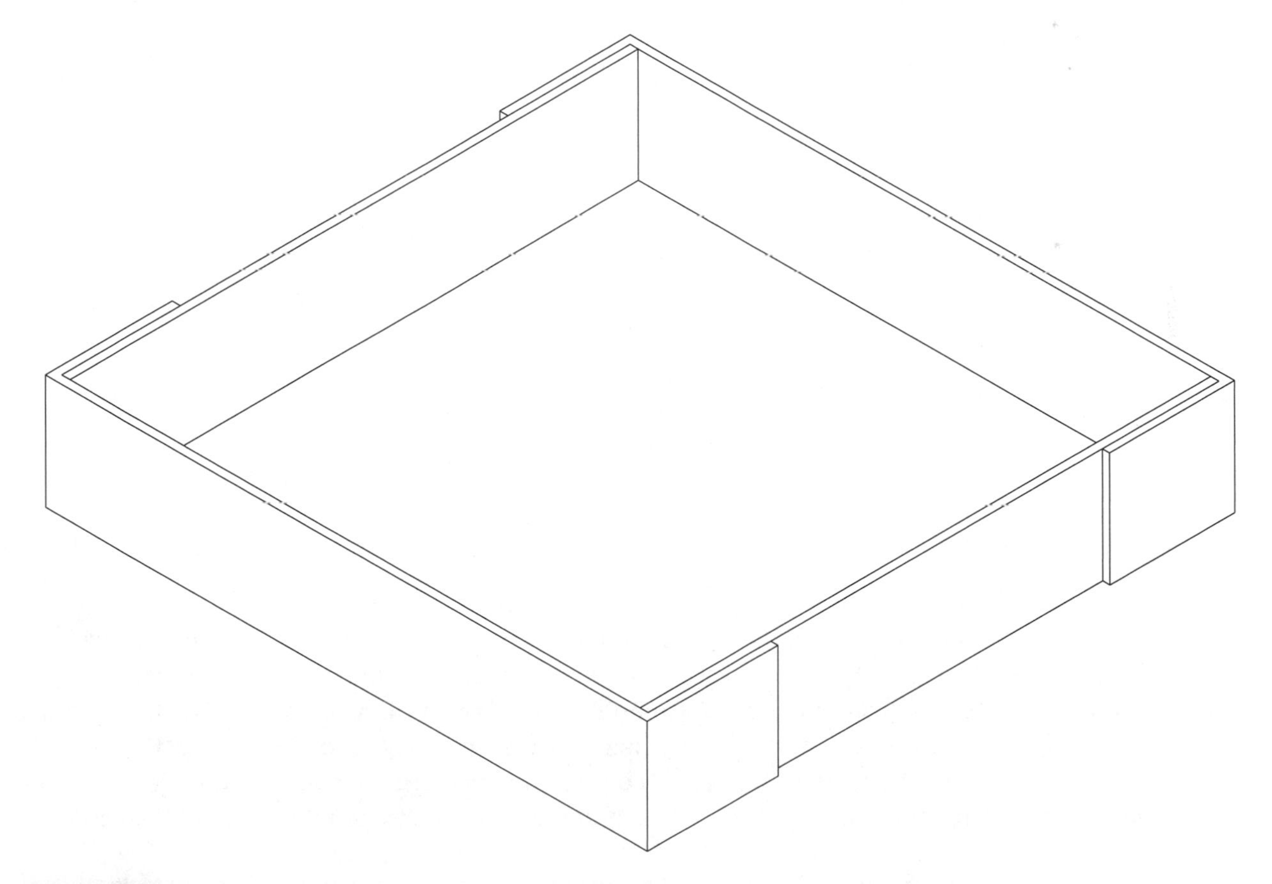

**Figure 6–1** *Isometric view of a design for a metal box.*

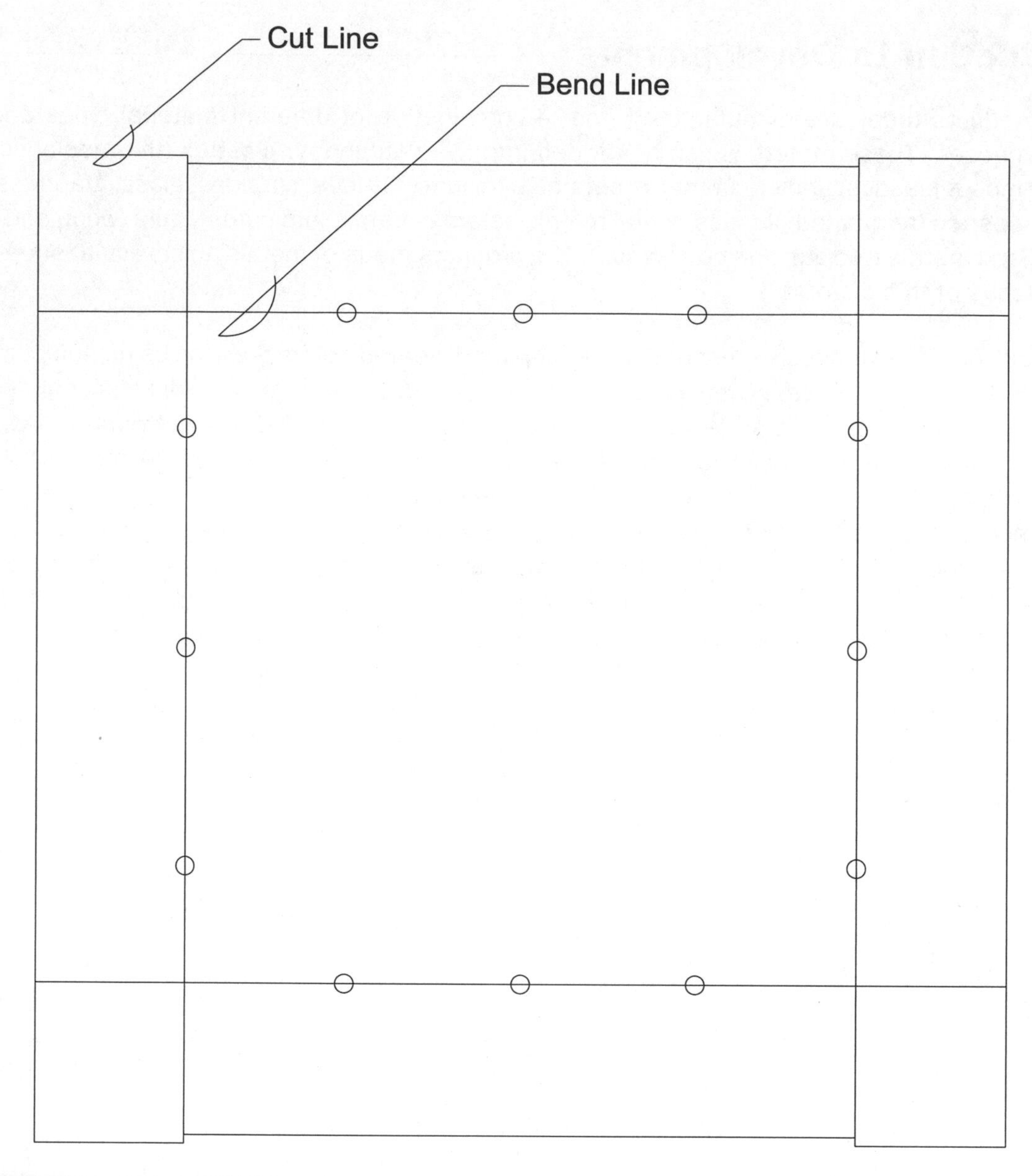


**Figure 6–2** *The development, or flat pattern of the metal box shown in Figure 6–1.*

When sheet metal products are manufactured, special consideration must be given to the thickness of the material and the angle at which it is bent. This consideration is important because the material, when subjected to a bending operation, will experience tension on one side causing that side to stretch, while the other side is exposed to compressive forces causing it to shrink (see Figure 6–3). The result is an angle (or a bend) that does not contain sharp edges or corners. Additional length, called ***bend allowance***, must be added to the stock material to compensate for these rounded corners.

For materials thinner than 0.65 mm (0.025") the allowance can be omitted, but for any material thicker than 0.65 mm (0.025") the allowance must be calculated to ensure that the part is produced within design tolerances. Bend allowances can be found using the allowance tables in the *Machinery's Handbook* or *ASME Handbook*. If neither of these reference books are available, then the allowance can be calculated using the following formulas. For soft copper or soft brass the formula is: Allowance = (0.55 × Thickness) + (0.5 × 3.14 × Radius of Bend). For aluminum, medium-hard copper, medium-hard brass, and soft steel, the formula is: Allowance = (0.64 x Thickness) + (0.5 × 3.14 × Radius of Bend). For bronze, hard copper, cold-rolled steel, and spring steel, the formula is: Allowance = (0.71 × Thickness) + (0.5 × 3.14 × Radius of Bend). In all three of these formulas, the calculations are based on a bend angle of 90°; when an angle other than 90° is to be used, the modifier (angle°/90°) must be applied to these formulas. For example, using soft steel and a bend angle of 56°, the formula now becomes: Allowance = [(0.64 × Thickness) + (0.5 x 3.14 × Radius of Bend)] × (56°/90°). If the same sheet of soft metal has a thickness of 0.125 and a bend radius of 0.25 at an angle of 56°, the complete equation would be: Allowance = [(0.64 × 0.125) + (0.5 × 3.14 x 0.25)] × (56°/90°). After plugging the information into a calculator, a bend allowance of 0.2871 is computed.

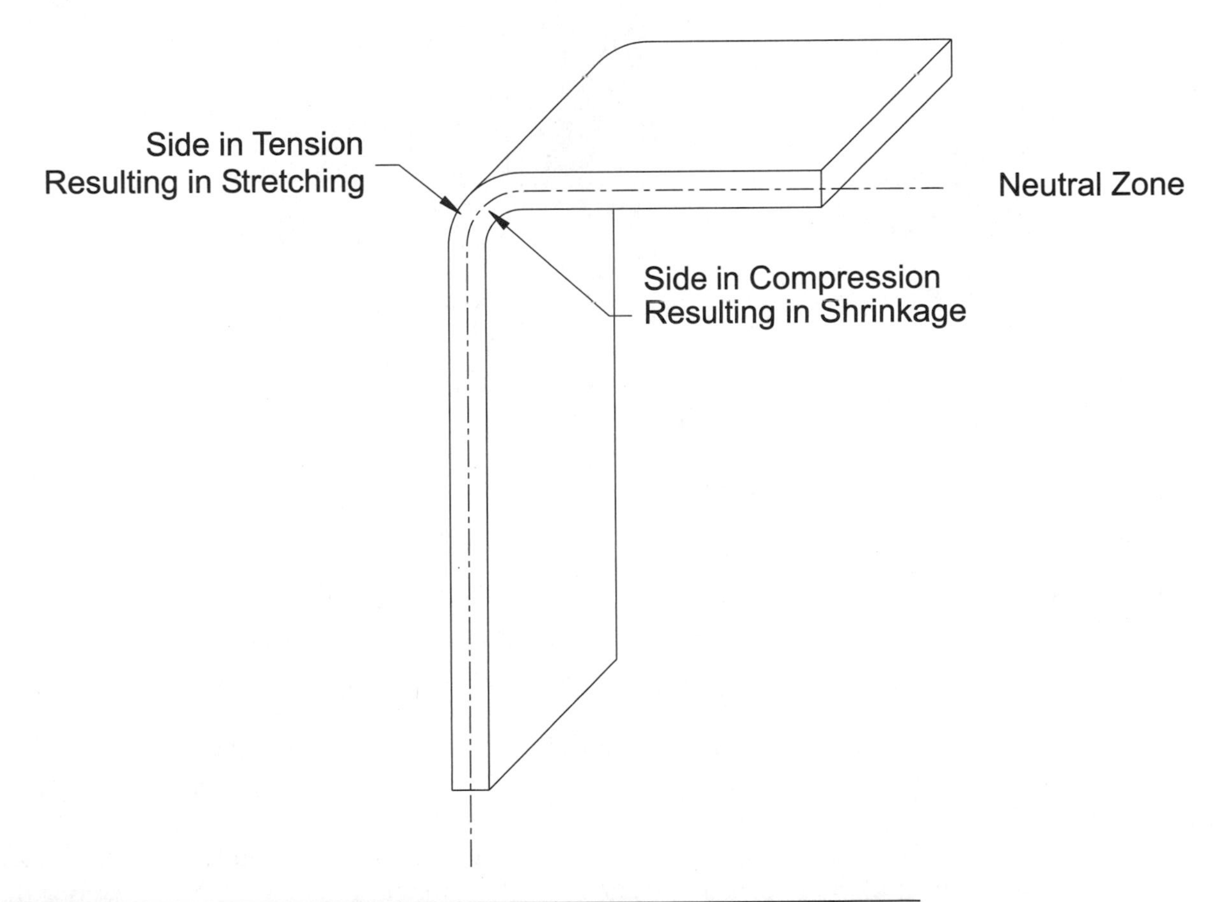


**Figure 6–3** *Forces experienced by a piece of steel subjected to a 90° bend.*

Once the bend allowance is found, it is then used to calculate the overall stock length needed to manufacture the part. This is accomplished by adding the inside distance from either allowance to allowance ($Distance_1$ in Figure 6–4), or from edge-of-part to allowance ($Distance_2$ in Figure 6–4). This particular part would yield the formula: Stock length = $Distance_1$ + $Distance_2$ + $Distance_3$ + $Allowance_1$ + $Allowance_2$. Naturally, the complexity of the part determines the number of distances and allowances in the formula. For this equation to work, the distances supplied must be minus the bends. If the distances supplied do contain the bends, the equation would become: ($Distance_{1a}$ – Radius – Thickness) + ($Distance_{2a}$ – Radius – Thickness) + ($Distance_{3a}$ – Radius – Thickness) + $Allowance_1$ + $Allowance_2$.

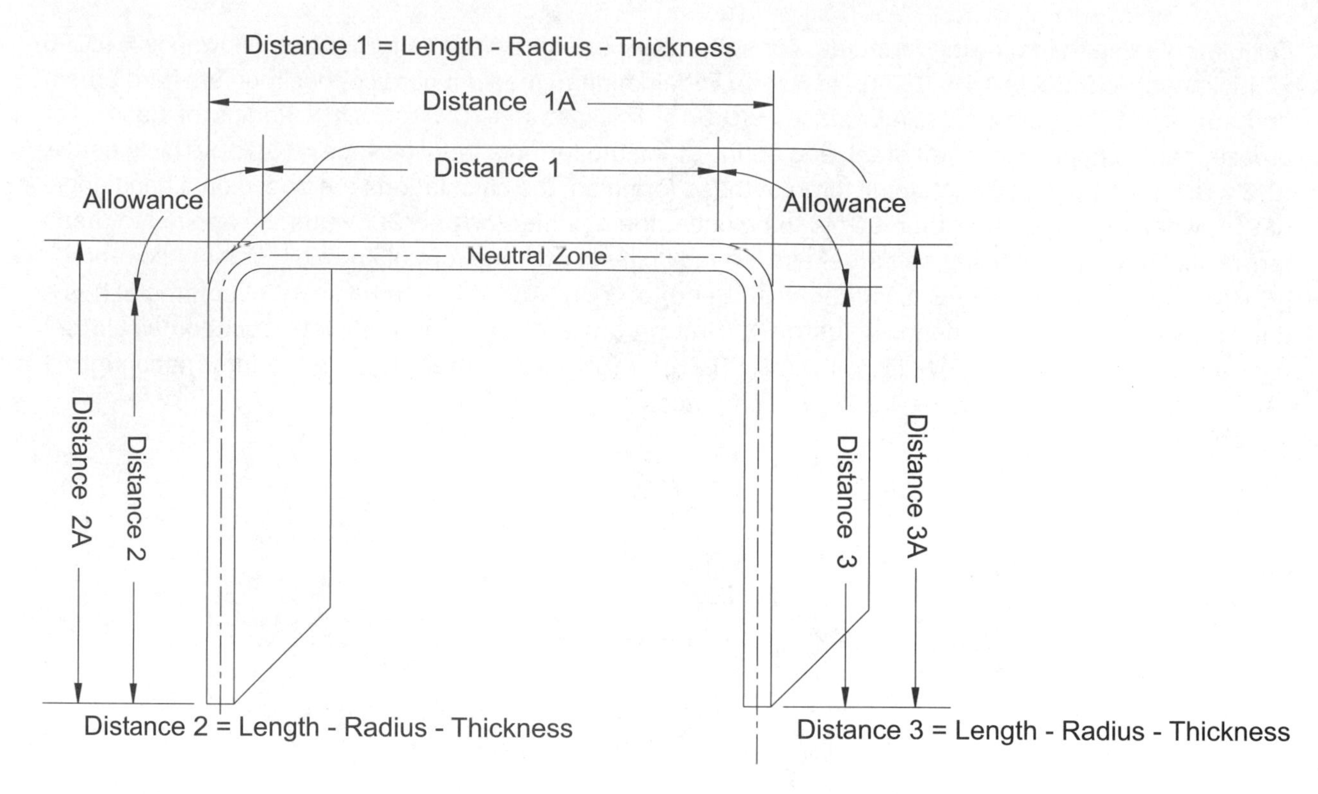


**Figure 6–4** *Boundaries used to measure the distances and allowances for stock length formulas.* *$Distance_1$ + $Distance_2$ + $Diatance_3$ + $allowance_1$ + $Allowance_2$*
*or*
*Stock Length = [($Distance_{2a}$ – Radius – Thickness) + ($Distance_{1a}$ – Radius – Thickness) + $allowance_1$ + $allowance_2$]*

In the above example, the formula could have been written as Distance + Distance + Distance + allowance – (2R + 2T). This practice is not recommended for technicians with little experience, because it can lead to confusion and possible mistakes when calculating large problems.

Another variable that must be taken into consideration when measuring for stock length is the method used to fasten sheets together. The most common methods used are: ***welding***, ***brazing***, ***soldering***, ***adhesive bonds***, and ***mechanical joining***. For each of these methods a different amount of material must be allowed to make an adequate connection. The method used is determined by the application of the final product, strength of material, and the forces to which the part will be subjected. Among the most common mechanical joining methods used today are seaming and hemming (examples of this method are illustrated in Figure 6–5). When calculating the length of stock required to construct a seam, the formulas discussed earlier must be used for each bend or fold in a particular seam. This is shown in Figure 6–6.

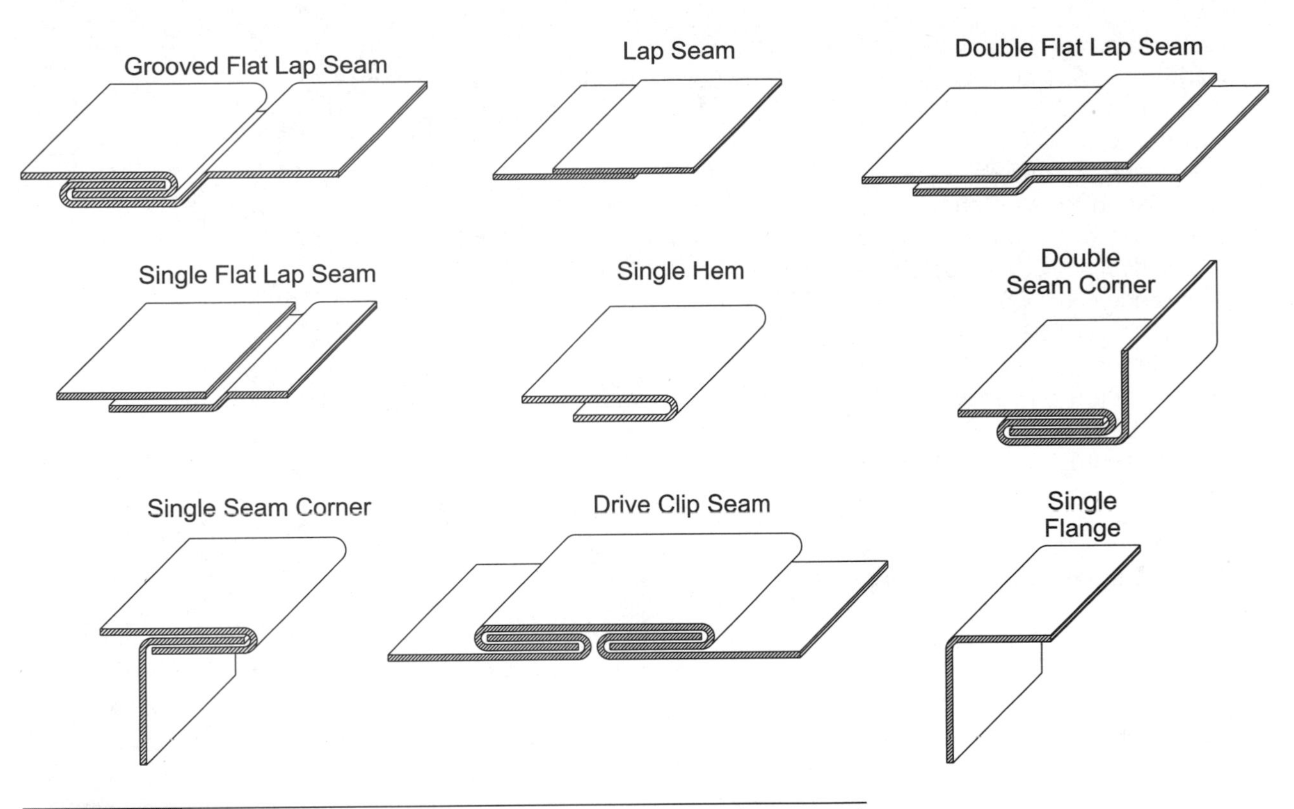


**Figure 6–5** *Common seams and hems used to connect sheet metal.*

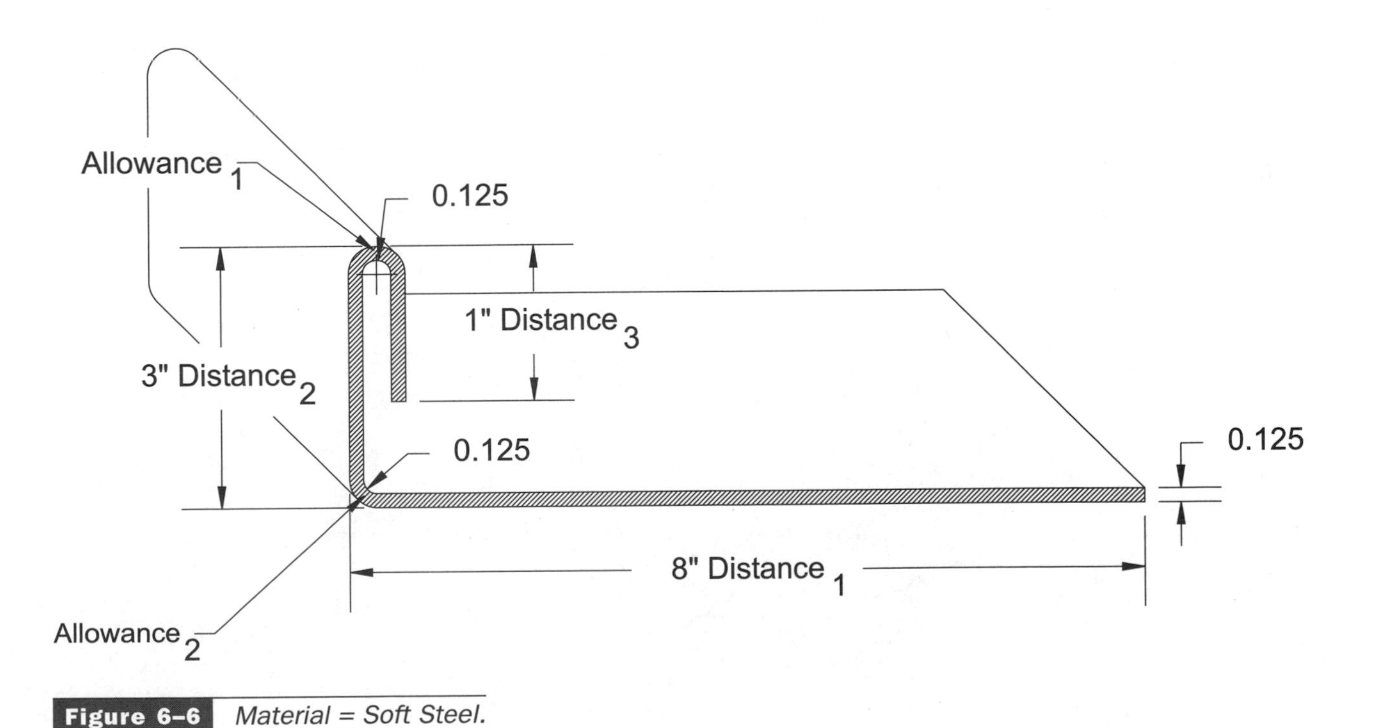


**Figure 6–6** *Material = Soft Steel.*

Calculation for $\text{Allowance}_1$

$\text{Allowance}_1 = [(0.64 \times \text{Thickness}) + (0.5 \times 3.14 \times \text{Radius of Bend})] \times \text{Angle}°/90°$

$\text{Allowance}_1 = [(0.64 \times 0.125) + (0.5 \times 3.14 \times 0.125)] \times 180°/90°$

$\text{Allowance}_1 = [0.08 + 0.19625] \times 2$

$\text{Allowance}_1 = 0.5525$

Calculation for $\text{Allowance}_2$

$\text{Allowance}_2 = [(0.64 \times \text{Thickness}) + (0.5 \times 3.14 \times \text{Radius of Bend})] \times \text{Angle}°/90°$

$\text{Allowance}_2 = [(0.64 \times 0.125) + (0.5 \times 3.14 \times 0.125)] \times 90°/90°$

$\text{Allowance}_2 = [0.08 + 0.19625] \times 1$

$\text{Allowance}_2 = 0.27625$

Calculation for Stock Length

Stock Length = (Distance – (0.125 + 0.125)) + (Distance – (0.125 + 0.125)) + (Distance – (0.125 + 0.125)) + Allowance + Allowance

Stock Length = 7.75 + 4.25 + 0.75 + 0.5525 + 0.27625

Stock Length = 12.82875

## Parallel-Line Developments

Flat patterns of solids containing parallel lateral edges and elements, primarily prisms and cylinders, are produced using the ***parallel-line development technique.*** This technique positions the object with one of its faces onto the development plane (see Figures 6–7a and 6–7b). The object is then unrolled or unfolded to create a flat pattern. All parallel attributes of the object remain that way even after the object has been transformed into its development. When a cylindrical object is unrolled, the circumference becomes a flat line that will determine the length of the development. When a prism is unfolded, its perimeter also becomes a flat line that will determine the length of the development. This flat line is called a ***stretch-out line*** and is defined as a line running the length of an unfolded prism or cylinder. The top and bottom of a prism will appear as a rectangle on the development plane. For cylindrical objects, the top and bottom will appear as circles on the development plane.

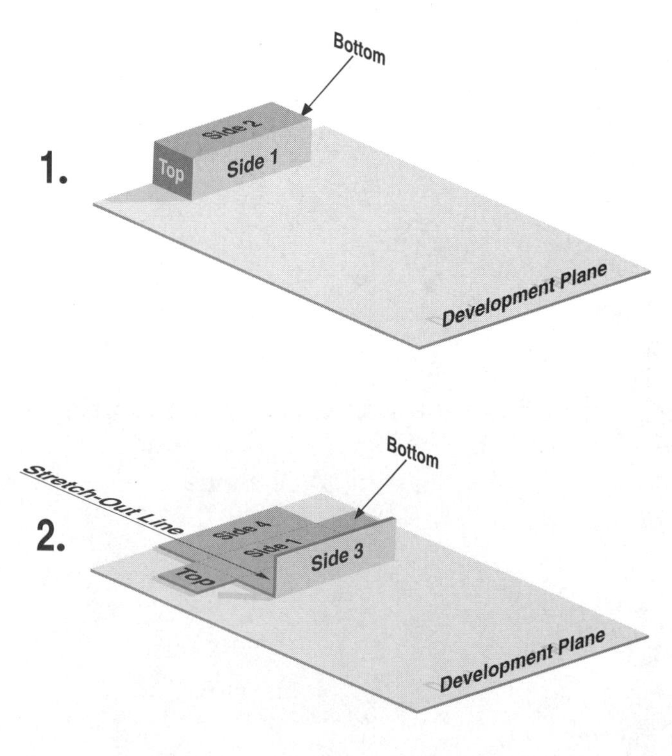


**Figure 6–7a** *Prism.*

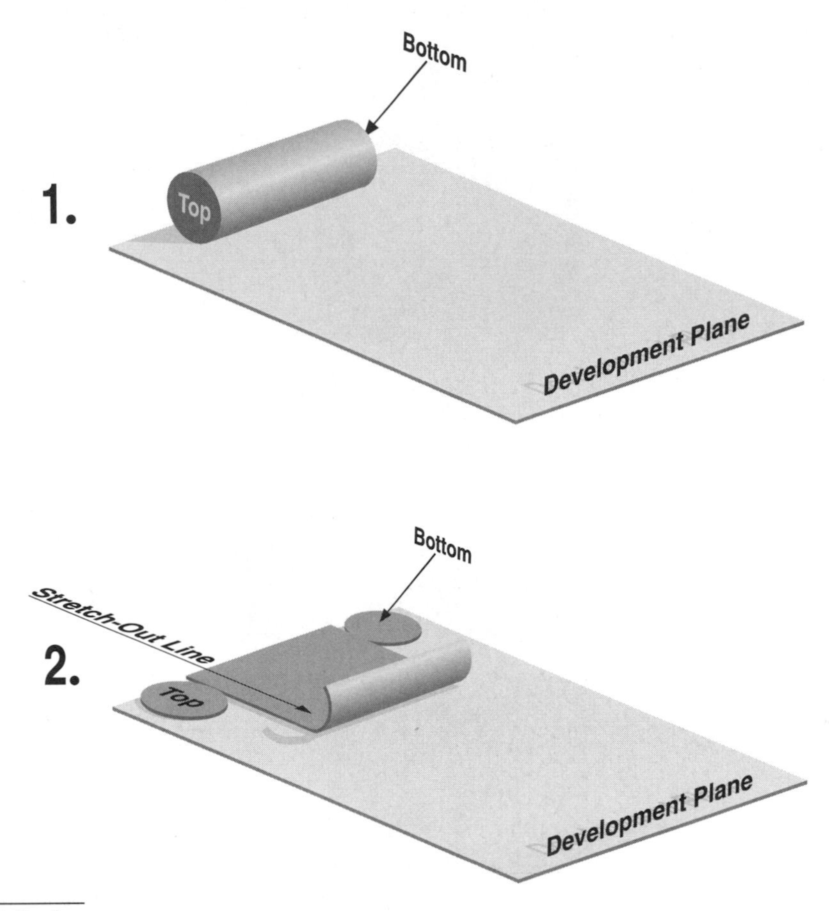


**Figure 6–7b** *Cylinder.*

## Development of a Right and Square Prism

A ***prism*** is a solid object in which the base and top form planes that are the same size and shape, and are parallel to one another, while the sides, or lateral faces, form a parallelogram. A ***right prism*** contains lateral faces that are rectangular (see Figure 6–8), while an ***oblique prism*** contains lateral faces that are not rectangular (see Figure 6–9). A ***right square prism*** contains lateral faces that are rectangular, and the base and top form squares (see Figure 6–10).

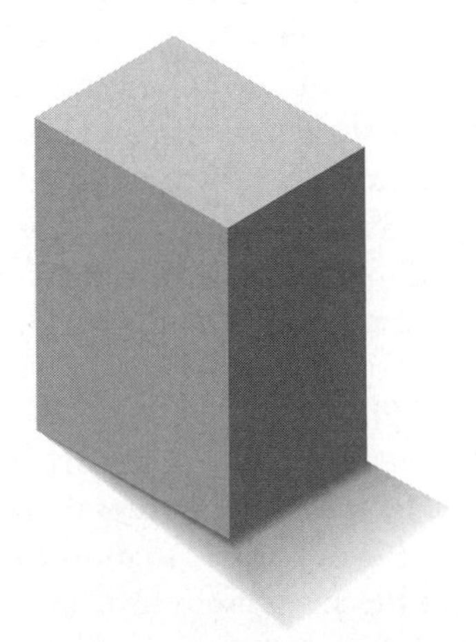

**Figure 6–8** *Right prism.*

**Figure 6–9** *Oblique prism.*

**Figure 6–10** *Right square prism.*

When producing a development of a prism, the lines defining the perimeter and height(s) must be contained in views where each appears in true length. The construction of a development of a right or square prism begins with drawing the stretch-out line equal in length to the perimeter of the prism. The placement of the stretch-out line is not important, but locating it in line with the front view makes measurements easily transferable across to the development. Starting at one end of the stretch-out line, the widths of each lateral face are transferred from the top view to the stretch-out line. The height of the development is transferred from the front view of the prism via projection lines. The bend-lines are added by drawing lines (the widths transferred from the top view) from the positions marked on the stretch-out line, perpendicular to the top of the development. Finally, the appropriate amount of additional material is added to the development for any seaming or hemming operation to be performed, as well as bend allowances.

#### Using AutoCAD to Construct a Development of a Right and Square Prism

To construct a development of a right rectangular or square prism using AutoCAD, start with a drawing containing two views of the prism. One of the views should reveal the true shape and size of the top or base, while the other should show the true height of the prism. In most cases, this would consist of a top view and a front view. Using these two views, create the stretch-out line and transfer all measurements to that line, thereby creating the development. In the following example, a development of the prism shown in Figure 6–11 is constructed using AutoCAD.

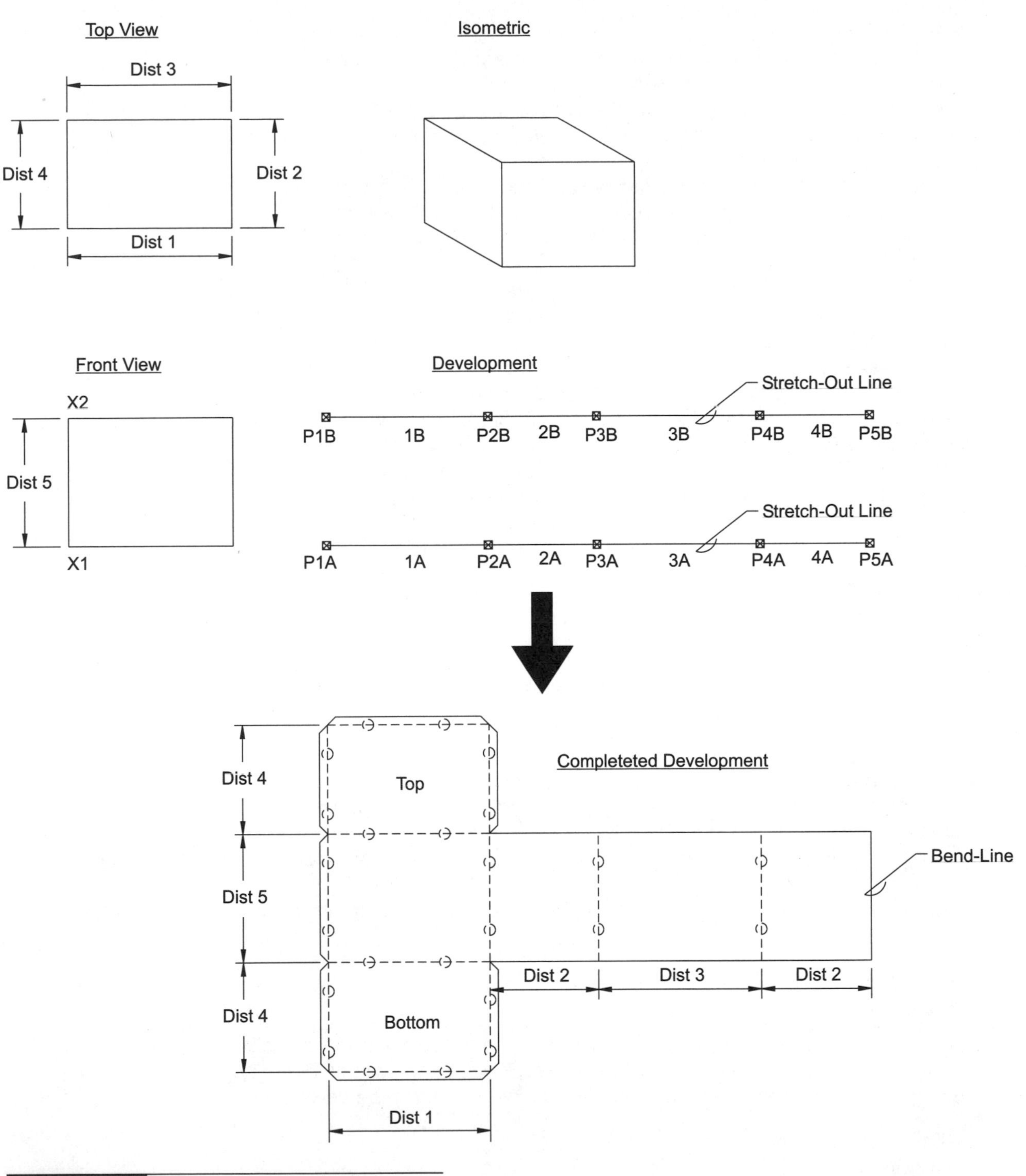


**Figure 6–11** *Development of a right prism.*

## Step #1

Start with drawing DG06-11_Student.dwg on the student CD-ROM. Next, copy the top view to a location where the development is to be generated. Using the ROTATE command, unfold the plan view, forming the stretch-out line of the development.

Command: COPY (ENTER)
Select objects: Other corner: 4 found *(Select the top view using a crossing window.)*
Select objects: *(Press* ENTER *to terminate* COPY *command's selection mode.)*
<Base point or displacement>/Multiple: *(Select an arbitrary point for development's location.)*
Second point of displacement: *(Select an arbitrary point for development's location.)*
Command: ROTATE (ENTER)
Select objects: Other corner: 3 found *(Select lines #2a, #3a, and #4a [see Figure 6–11].)*
Select objects: *(Press* ENTER *to terminate* ROTATE *command's selection mode.)*
Base point: end of *(Select the point labeled P2a as base of rotation.)*
<Rotation angle>/Reference: -90 (ENTER)
(Rotate selection clockwise 90°.)
Command: ROTATE (ENTER)
Select objects: Other corner: 2 found *(Select lines #3a and #4a [see Figure 6–11].)*
Select objects: *(Press* ENTER *to terminate* ROTATE *command's selection mode.)*
Base point: end of *(Select the point labeled P3a as base of rotation.)*
<Rotation angle>/Reference: -90 *(Press* ENTER*. Rotate selection clockwise 90°.)*
Command: ROTATE (ENTER)
Select objects: 1 found *(Select line #4a see Figure 6–11.)*
Select objects: *(Press* ENTER *to terminate* ROTATE *command's selection mode.)*
Base point: end of *(Select the point labeled P4a as base of rotation.)*
<Rotation angle>/Reference: -90 (ENTER) *(Rotate selection clockwise 90°.)*

## Step #2

Use the COPY command to create the top edge of the development by offsetting the stretch-out line by the height of the object. The distance is obtained by selecting one of the lower corners of the prism in the front view as the base point and its corresponding upper corner as the second point.

Command: COPY (ENTER)
Select objects: Other corner: 4 found *(Select lines #1a, #2a, #3a, and #4a [see Figure 6–11].)*
Select objects: *(Press* ENTER *to terminate* COPY *command's selection mode.)*
<Base point or displacement>/Multiple:
end of Second point of *(Using the* END OSNAP *option, select the point labeled X1 as the base [see Figure 6–11].)*
displacement: end of *(Using the* END OSNAP *option, select the point labeled X2 as the second point [see Figure 6–11].)*

## Step #3

The bend-lines are created by drawing lines from the endpoints of the stretch-out line perpendicular to the top edge of the development (points labeled P1A–P5A to points P1B–P5B). Placing dimensions and bend-line nomenclature in the proper locations completes the development.

Command: LINE (ENTER)
From point: end of
To point: end of *(Draw a line from P1a to P1b [see Figure 6–11].)*
To point: *(Press* ENTER *to terminate the* LINE *command.)*
Command: LINE (ENTER)
From point: end of
To point: end of *(Draw a line from P2a to P2b [see Figure 6–11].)*
To point: *(Press* ENTER *to terminate the* LINE *command.)*
Command: LINE (ENTER)
From point: end of

To point: end of *(Draw a line from P3a to P3b [see Figure 6–11].)*
To point: *(Press* ENTER *to terminate the* LINE *command.)*
Command: LINE (ENTER)
From point: end of
To point: end of *(Draw a line from P4a to P4b [see Figure 6–11].)*
To point: *(Press* ENTER *to terminate the* LINE *command)*
Command: LINE (ENTER)
From point: end of
To point: end of *(Draw a line from P5a to P5b [see Figure 6–11].)*
To point: *(Press* ENTER *to terminate the* LINE *command.)*

## Development of a Regular Cylinder

A ***cylinder*** is two circular regions contained on parallel planes with the lateral surface formed by line segments connecting corresponding points on the circumference of the two circular regions (see Figure 6–12). A ***right circular cylinder*** contains lateral lines that are perpendicular to the circular regions (top and base), as shown in Figure 6–12a. An ***oblique cylinder*** contains lateral lines that are not perpendicular to the two circular regions (top and base), as shown in Figure 6–12b.

The procedure for constructing a development for a right circular cylinder is similar to that of a right prism. A stretch-out line is drawn equal in length to the circumference of the base or top of the cylinder. The width of the development is determined by the height of the cylinder, and is retrieved from a view showing the cylinder in true height. When completed, the development will resemble a rectangle (see Figure 6–12c).

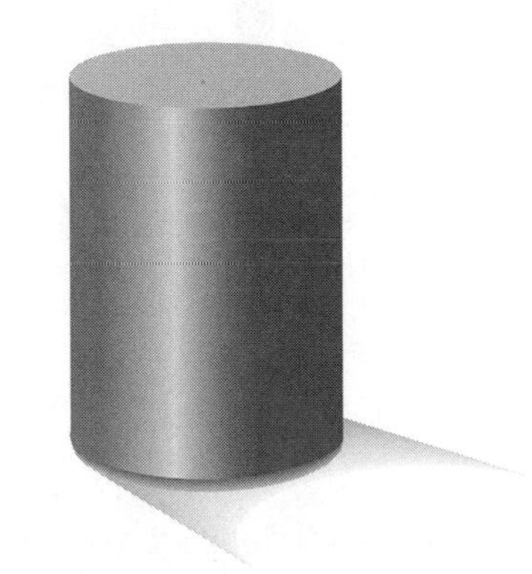

**Figure 6–12a** *Cylinder.*

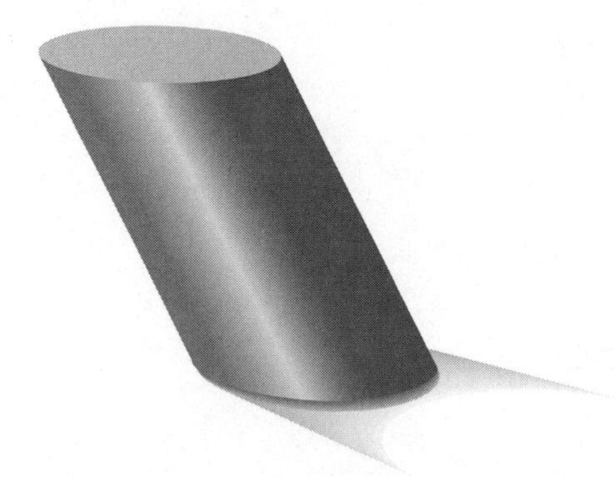

**Figure 6–12b** *Oblique cylinder.*

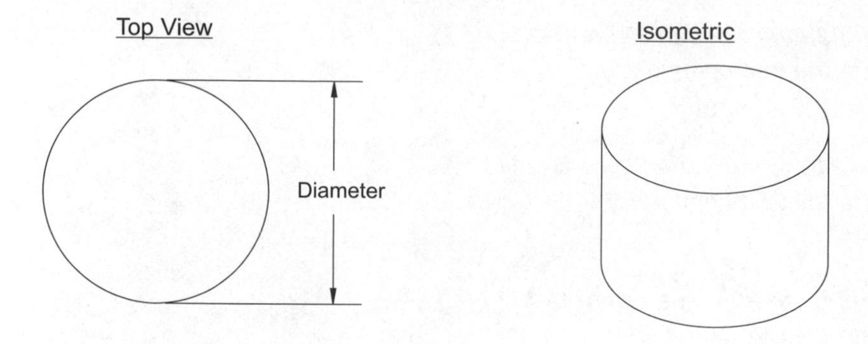


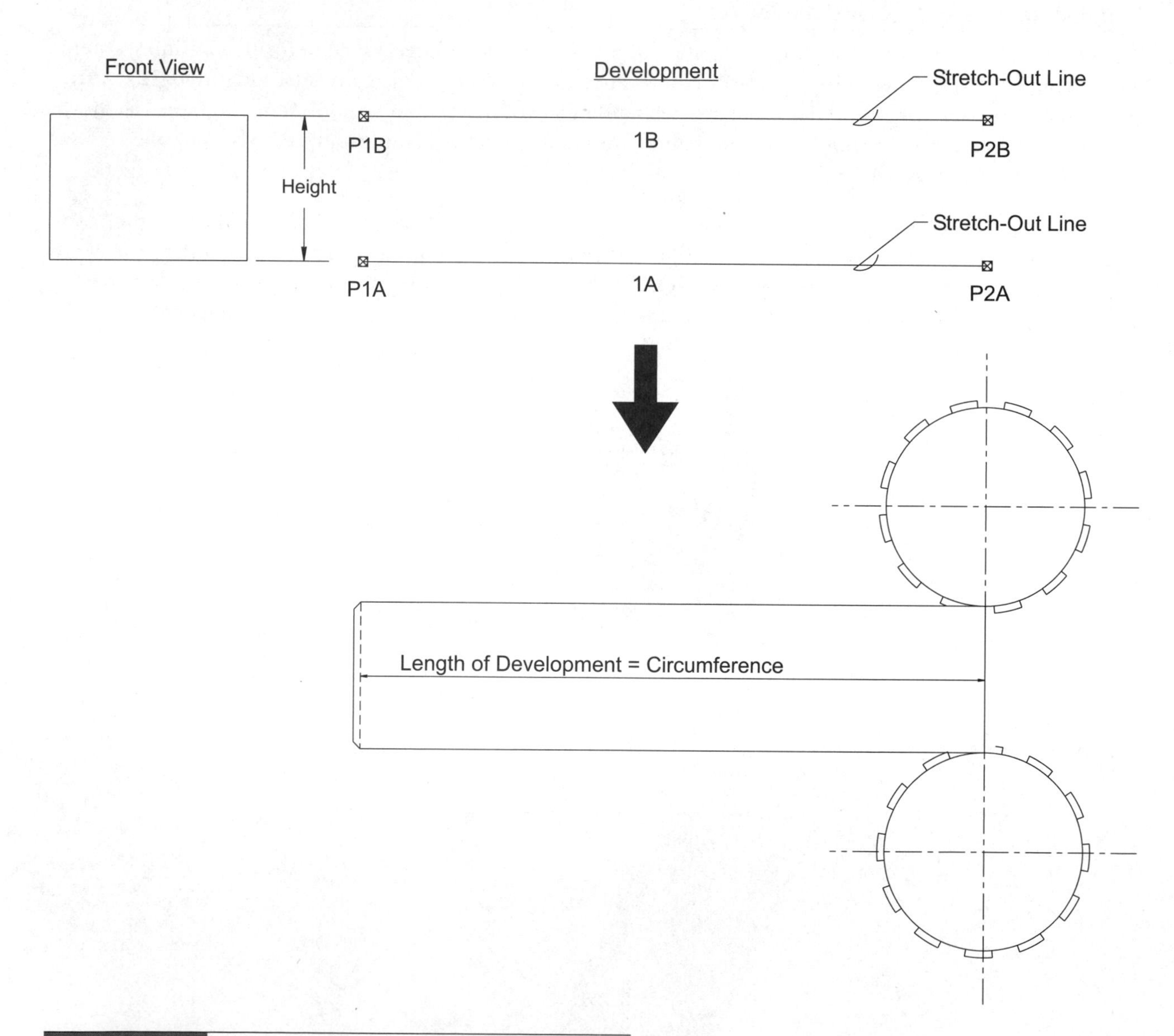


**Figure 6–12c** *Development of a right circular cylinder.*

**Note: Circumference = Pi x Diameter**

## Development of a Truncated Object

Normally, when an object is ***truncated***, a plane slices through the top or bottom at an angle. In the case of a cone, the plane severs the apex from the cone. When a pyramid is truncated, the vertex is removed; for prisms and cylinders the top or bottom surface(s) are eliminated. Figures 6–13, 6–14, 6–15, and 6–16 illustrate different geometric figures that have been truncated. The following section discusses the procedures used for the production of these four common truncated shapes: truncated right and square prisms, truncated cylinders, truncated top and bottom right and square prisms, and truncated top and bottom cylinders.

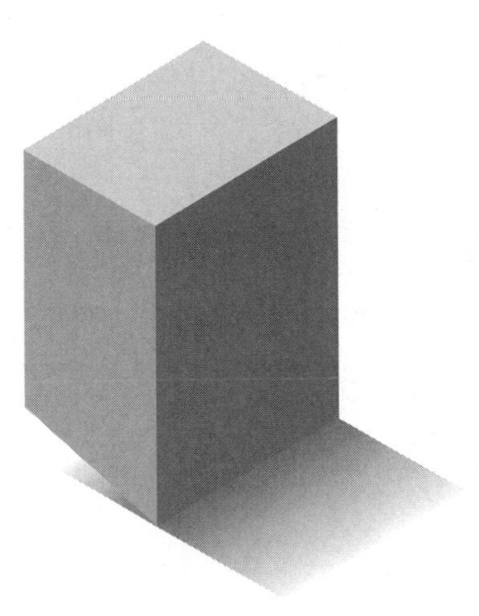

**Figure 6–13** *Truncated prism.*

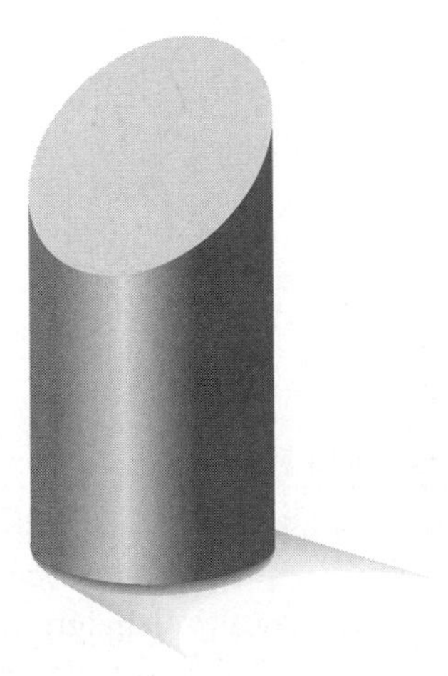

**Figure 6–14** *Truncated cylinder.*

**Figure 6–15** *Top and bottom truncated prism.*

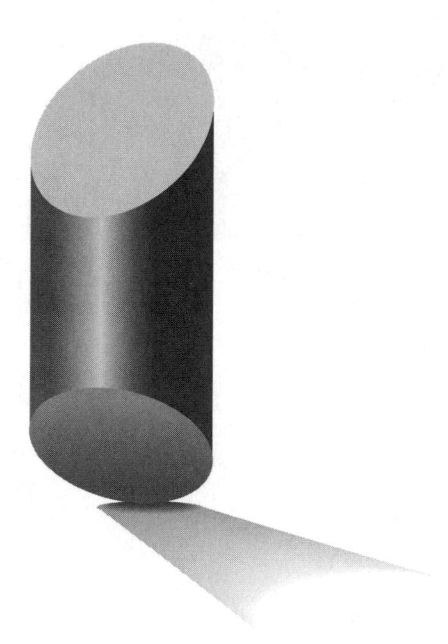

**Figure 6–16** *Top and bottom truncated cylinder.*

## Development of a Truncated Right and Square Prism

The techniques used to produce a development of a truncated right and square prism is comparable to that used for the development of a regular prism. The main difference being that instead of transferring a single height from the front view across to the development, multiple heights must be transferred (see Figure 6–17). After each point of intersection in the bend-line has been transferred to its corresponding point on the height line, the top edge of the development is generated by connecting the points on the height line.

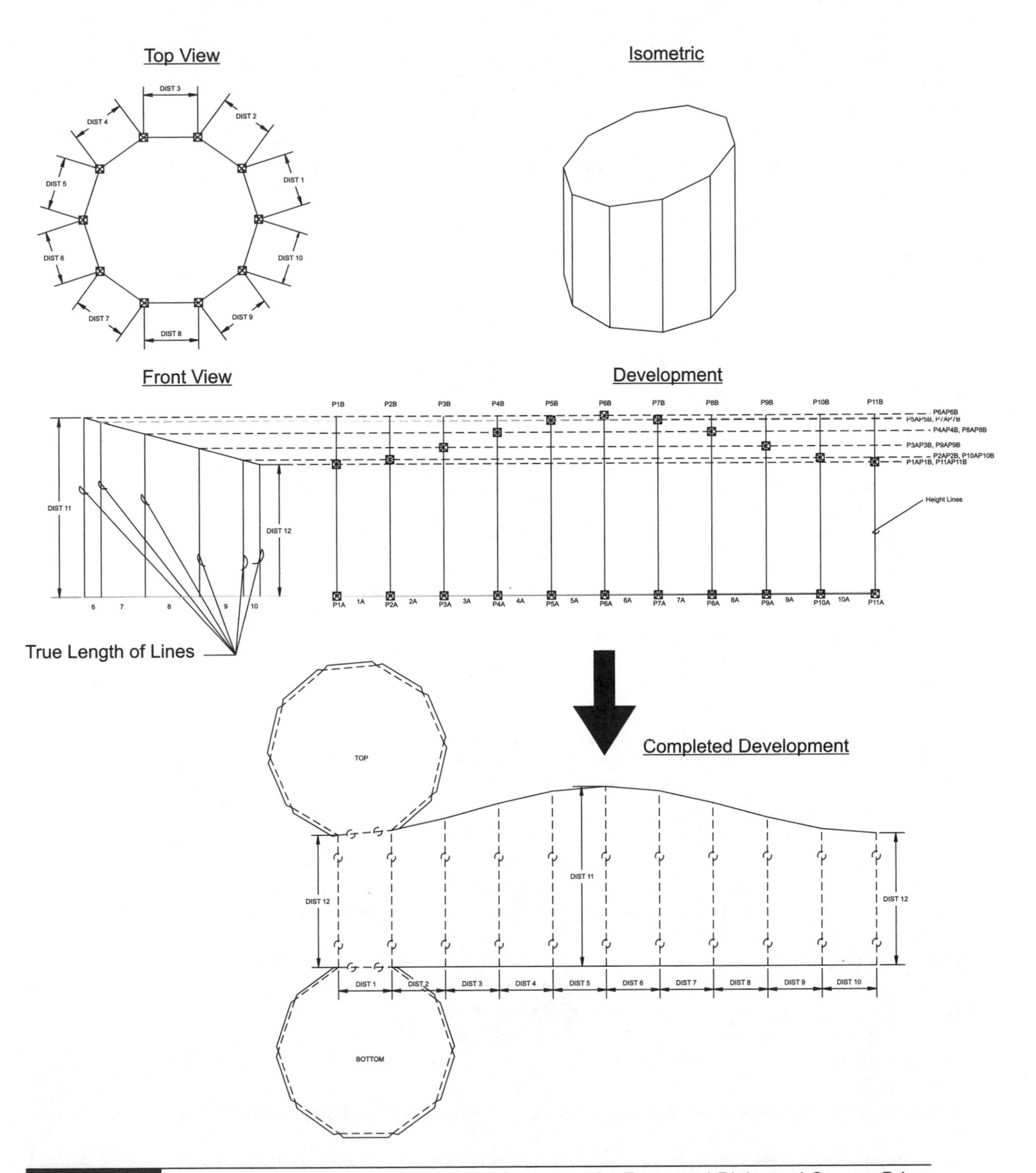


**Figure 6–17** *Using AutoCAD to Construct a Development of a Truncated Right and Square Prism*

The techniques for constructing a development of a truncated prism using AutoCAD are primarily the same as those for constructing a regular prism using AutoCAD. As stated earlier, the main difference is the multiple heights. The distances for the heights are transferred either by projection lines from the front view across to the development, or copied from the front view across to the development. The copy method requires fewer steps, but it also requires more attention to detail. The following example illustrates how a development of a right truncated prism could be constructed using AutoCAD (see Figure 6–17). In this example, the heights are transferred from the front view to the development using projection lines.

## Step #1

Start with the DG06-17_Student.dwg file on the student CD-ROM or create a top and front view of a truncated prism. Next, copy the plan view to a location where the development is to be generated. Using the ROTATE command, unfold the plan view, forming the stretch-out line of the development.

Command: COPY (ENTER)
Select objects: Other corner: 10 found *(Select the top view using a crossing window.)*
Select objects: *(Press ENTER to terminate COPY command's selection mode.)*
<Base point or displacement>/Multiple: *(Select arbitrary point for the development's location.)*
Second point of *(Select arbitrary point for development's location.)*
Command: ROTATE (ENTER)
Select objects: 9 found *(Select lines #9a, #10a, #1a, #2a, #3a, #4a, #5a, #6a, #7a [see Figure 6–17].)*
Select objects: *(Press ENTER to terminate the ROTATE command's selection mode.)*
Base point: end of *(Use END OSNAP option to select left endpoint of line # 9a.)*
<Rotation angle>/Reference: R *(Press ENTER. This command rotates the object in reference to another object.)*
Reference angle <0>: end of *(Using the END OSNAP option, select left endpoint of line #9a.)*
Second point: end of *(Using the END OSNAP option, select right endpoint of line #9a.)*
New angle: *(With ortho on, select a point below line #9a.)*
Command: ROTATE (ENTER)
Select objects: Other corner: 8 found *(Select lines #10a, #1a, #2a, #3a, #4a, #5a, #6a, #7a [see Figure 6–17].)*
Select objects: *(Press ENTER to terminate the ROTATE command's selection mode.)*
Base point: end of *(Use the END OSNAP option to select the left endpoint of line # 10a.)*
<Rotation angle>/Reference: R (ENTER)
Reference angle <0>: end of *(Use the END OSNAP option to select the left endpoint of line #10a.)*
Second point: end of *(Use the END OSNAP option to select the right endpoint of line #10a.)*
New angle: *(With ortho on, select a point below line #10a.)*
Command: ROTATE (ENTER)
Select objects: Other corner: 7 found *(Select lines #1a, #2a, #3a, #4a, #5a, #6a, #7a [see Figure 6–17].)*
Select objects: *(Press ENTER to terminate the ROTATE command's selection mode.)*

Base point: end of *(Use the END OSNAP option to select the left endpoint of line # 1a.)*
<Rotation angle>/Reference: R (ENTER)
Reference angle <0>: end of *(Use END OSNAP option to select the left endpoint of line #1a.)*
Second point: end of *(Use END OSNAP option to select the right endpoint of line #1a.)*
New angle: *(With ortho on, select a point below line #1a.)*
Command: ROTATE (ENTER)
Select objects: Other corner: 6 found *(Select lines #2a, #3a, #4a, #5a, #6a, #7a [see Figure 6–17].)*
Select objects: *(Press ENTER to terminate the ROTATE command's selection mode.)*

Base point: end of *(Use the* END OSNAP *option to select the left endpoint of line # 2a.)*
<Rotation angle>/Reference: R (ENTER)
Reference angle <0>: end of *(Use the* END OSNAP *option to select the left endpoint of line #2a.)*
Second point: end of *(Use the* END OSNAP *option to select the right endpoint of line #2a.)*
New angle: *(With ortho on, select a point below line #2a.)*
Command: ROTATE (ENTER)
Select objects: Other corner: 5 found *(Select lines #3a, #4a, #5a, #6a, #7a [see Figure 6–17].)*
Select objects: *(Press* ENTER *to terminate the* ROTATE *command's selection mode.)*
Base point: end of *(Use the* END OSNAP *option to select the left endpoint of line #3a.)*
<Rotation angle>/Reference: R (ENTER)
Reference angle <0>: end of *(Use the* END OSNAP *option to select the left endpoint of line #3a.)*
Second point: end of *(Use the* END OSNAP *option to select the right endpoint of line #3a.)*
New angle: *(With ortho on, select a point below line #3a.)*
Command: ROTATE (ENTER)
Select objects: Other corner: 4 found *(Select lines #4a, #5a, #6a, #7a [see Figure 6–17].)*
Select objects: *(Press* ENTER *to terminate the* ROTATE *command's selection mode.)*
Base point: end of *(Use the* END OSNAP *option to select the left endpoint of line #4a.)*
<Rotation angle>/Reference: R (ENTER)
Reference angle <0>: end of *(Use the* END OSNAP *option to select the left endpoint of line #4a.)*
Second point: end of *(Use the* END OSNAP *option to select the right endpoint of line #4a.)*
New angle: *(With ortho on, select a point below line #4a.)*
Command: ROTATE (ENTER)
Select objects: Other corner: 3 found *(Select lines #5a, #6a, #7a [see Figure 6–17].)*
Select objects: *(Press* ENTER *to terminate the* ROTATE *command's selection mode.)*
Base point: end of *(Use the* END OSNAP *option to select the left endpoint of line #5a.)*
<Rotation angle>/Reference: R (ENTER)
Reference angle <0>: end of *(Use the* END OSNAP *option to select the left endpoint of line #5a.)*
Second point: end of *(Use the* END OSNAP *option to select the right endpoint of line #5a.)*
New angle: *(With ortho on, select a point below line #5a.)*
Command: ROTATE (ENTER)
Select objects: Other corner: 2 found *(Select lines #6a, #7a [see Figure 6–17].)*
Select objects: *(Press* ENTER *to terminate the* ROTATE *command's selection mode.)*
Base point: *(Use the* END OSNAP *option to select the left endpoint of line #6a.)*
<Rotation angle>/Reference: R (ENTER)
Reference angle <0>: end of *(Use the* END OSNAP *option to select the left endpoint of line #6a.)*
Second point: end of *(Use the* END OSNAP *option to select the right endpoint of line #6a.)*
New angle: *(With ortho on, select a point below line #6a.)*
Command: ROTATE (ENTER)
Select objects: Other corner: 1 found *(Select line #7a, see Figure 6–17.)*
Select objects: *(Press* ENTER *to terminate the* ROTATE *command's selection mode.)*
Base point: end of *(Use the* END OSNAP *option to select the left endpoint of line #7a.)*
<Rotation angle>/Reference: R (ENTER)
Reference angle <0>: end of *(Use the* END OSNAP *option to select the left endpoint of line #7a.)*
Second point: end of *(Use the* END OSNAP *option to select the right endpoint of line #7a.)*
New angle: *(With ortho on, select a point below line #7a.)*

## Step #2

Next, the heights from the lateral surfaces of the prism are transferred from the front view of the development using projection lines. The projection lines are drawn from the endpoints at the corners of the lateral surfaces of the prism (in the front view), straight across and past the far right endpoint of the stretch-out line.

Command: LINE From point: end of
To point: *(Draw a line from P1b in the front view past point P11b.)*
To point: *(Press* ENTER *to terminate the* LINE *command.)*
Command: LINE From point: end of
To point: *(Draw a line from P2b in the front view past point P11b.)*
To point: *(Press* ENTER *to terminate the* LINE *command.)*
Command: LINE From point: end of
To point: *(Draw a line from P3b in the front view past point P11b.)*
To point: *(Press* ENTER *to terminate the* LINE *command.)*
Command: LINE From point: end of
To point: *(Draw a line from P4b in the front view past point P11b.)*
To point: *(Press* ENTER *to terminate the* LINE *command.)*
Command: LINE From point: end of
To point: *(Draw a line from P5b in the front view past point P11b.)*
To point: *(Press* ENTER *to terminate the* LINE *command.)*
Command: LINE From point: end of
To point: *(Draw a line from P6b in the front view past point P11b.)*
To point: *(Press* ENTER *to terminate the* LINE *command.)*

## Step #3

The bend-lines are added to the development by drawing lines from the endpoints of the stretch-out line perpendicular to the projection lines. The spacing of the bend-lines along the development are determined by the true width of the lateral surfaces shown in the top view.

Command: LINE (ENTER)
From point: end of
To point: *(Draw a line from P11a perpendicular to p11B [see Figure 6–17].)*
To point: *(Press* ENTER *to terminate the* LINE *command.)*
Command: LINE (ENTER)
From point: end of
To point: *(Draw a line from P10a perpendicular to P10B [see Figure 6–17].)*
To point: *(Press* ENTER *to terminate the* LINE *command.)*
Command: LINE (ENTER)
From point: end of
To point: *(Draw a line from P09a perpendicular to P09B [see Figure 6–17].)*
To point: *(Press* ENTER *to terminate the* LINE *command.)*
Command: LINE (ENTER)
From point: end of
To point: *(Draw a line from P08a perpendicular to P08B [see Figure 6–17].)*
To point: *(Press* ENTER *to terminate the* LINE *command.)*
Command: LINE (ENTER)
From point: end of
To point: *(Draw a line from P07a perpendicular to P07B [see Figure 6–17].)*
To point: *(Press* ENTER *to terminate the* LINE *command.)*
Command: LINE (ENTER)
From point: end of
To point: *(Draw a line from P06a perpendicular to P06B [see Figure 6–17].)*
To point: *(Press* ENTER *to terminate the* LINE *command.)*
Command: LINE (ENTER)
From point: end of
To point: *(Draw a line from P05a perpendicular to P05B [see Figure 6–17].)*
To point: *(Press* ENTER *to terminate the* LINE *command.)*
Command: LINE (ENTER)

From point: end of
To point: *(Draw a line from P04a perpendicular to P04B [see Figure 6–17].)*
To point: *(Press* ENTER *to terminate the* LINE *command.)*
Command: LINE (ENTER)
From point: end of
To point: *(Draw a line from P03a perpendicular to P03B [see Figure 6–17].)*
To point: *(Press* ENTER *to terminate the* LINE *command.)*
Command: LINE (ENTER)
From point: end of
To point: *(Draw a line from P02a perpendicular to P02B [see Figure 6–17].)*
To point: *(Press* ENTER *to terminate the* LINE *command.)*
Command: LINE (ENTER)
From point: end of
To point: *(Draw a line from P01a perpendicular to P01B [see Figure 6–17].)*
To point: *(Press* ENTER *to terminate the* LINE *command.)*

## Step #4

The top edge of the development is constructed by using the PLINE command to create a polyline from endpoint to endpoint of the bend-lines (see Figure 6–17).

Command: PLINE (ENTER)
From point: END of (From point #P11b)
Current line-width is 0.0000
Arc/Close/Halfwidth/Length/Undo/Width/<Endpoint of line>: END of (To point #P10b)
Arc/Close/Halfwidth/Length/Undo/Width/<Endpoint of line>: END of (To point #P9b)
Arc/Close/Halfwidth/Length/Undo/Width/<Endpoint of line>: END of (To point #P8b)
Arc/Close/Halfwidth/Length/Undo/Width/<Endpoint of line>: END of (To point #P7b)
Arc/Close/Halfwidth/Length/Undo/Width/<Endpoint of line>: END of (To point #P6b)
Arc/Close/Halfwidth/Length/Undo/Width/<Endpoint of line>: END of (To point #P5b)
Arc/Close/Halfwidth/Length/Undo/Width/<Endpoint of line>: END of (To point #P4b)
Arc/Close/Halfwidth/Length/Undo/Width/<Endpoint of line>: END of (To point #P3b)
Arc/Close/Halfwidth/Length/Undo/Width/<Endpoint of line>: END of (To point #P2b)
Arc/Close/Halfwidth/Length/Undo/Width/<Endpoint of line>: END of (To point #P1b)
Arc/Close/Halfwidth/Length/Undo/Width/<Endpoint of line>: *(Press* ENTER *to terminate the* PLINE *command.)*

## Step #5

Place dimensions and bend-line nomenclature in their proper location to complete the development.

### Development of a Truncated Cylinder

To produce a development of a truncated cylinder, start by creating or opening a drawing containing a top and front view of the cylinder. Divide the top view of the cylinder into a minimum of twelve equal segments. Dividing the cylinder into equal parts establishes points along the perimeter of the cylinder. These points can then be used to transfer lateral lines into the front view. The purpose of these lateral lines is to provide a means of transferring the sloping line that represents the truncation to the development. The more segments in which the top view is divided, the more accurate the slope of the development becomes. However, the more divisions that are made to the top view, the more complex the development becomes. Once the points have been projected into the front view, they are then transferred across to the development. The length of the stretch-out line is the same as the circumference of the base of the cylinder. Once the stretch-out line has been established and the heights have been transferred from the front view, the stretch-out line is then divided into the same equal number of segments as the top view. The bend-lines are drawn from the divisions made on the stretch-out line, perpendicular to the height projection lines. Finally, an irregular curve is constructed along the points of intersection of the bend-lines and it's corresponding height projection line. Adding dimensions and bend-line nomenclature in their proper location completes the development (see Figure 6–18).

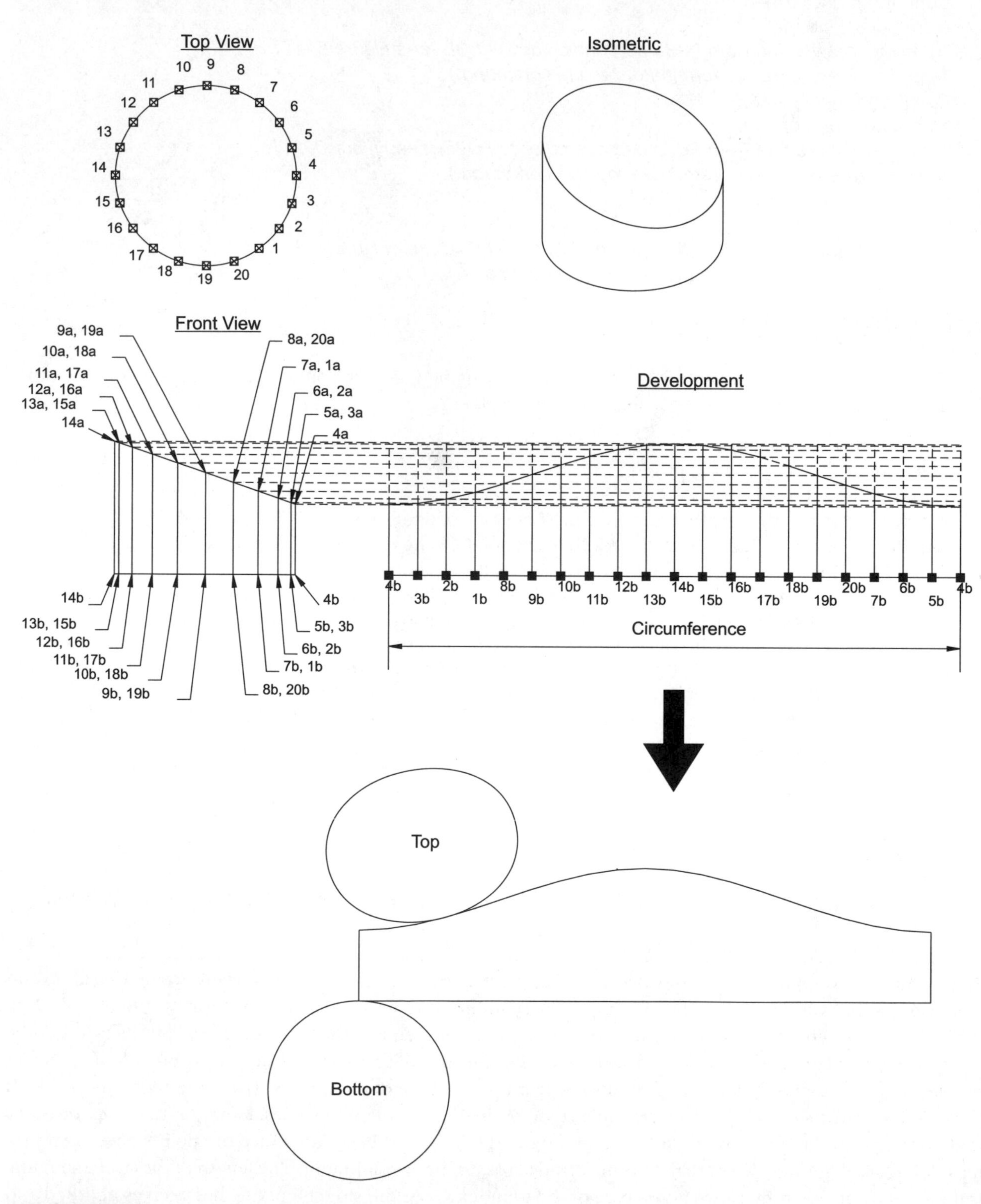


**Figure 6–18** *Development of a truncated cylinder.*

### Using AutoCAD to Construct a Development of a Truncated Cylinder

When using AutoCAD to produce a development of a truncated cylinder, two different methods can be used for dividing the top view of the cylinder into equal segments. The first method uses the DIVIDE command. When using this command, the technician enters the exact number of segments to divide the top view. In the following example, the DIVIDE command is used to divide the top view of a cylinder into twenty equal segments (see Figure 6–18).

Command: DIVIDE (ENTER)
Select object to divide: *(Select top view using a crossing window.)*
<Number of segments>/Block: 20 (ENTER)
The advantages to using the DIVIDE command are:

1. Fewer steps.
2. The NODE OSNAP option can be used when transferring the points from the top view to the front view.

The second method utilizes the ARRAY command. This method first requires that a marker be positioned along the circumference of the circle. The marker is the item that is arrayed around the circle when the command is executed. The marker can be almost anything, but lines work the best. To construct a marker from a line, draw a line from the center of the cylinder to one of its quadrants. Then use the ARRAY command to divide the top of the cylinder into equal segments. In the following example, the ARRAY command is used to divide the top view of a cylinder into twenty equal segments.

Command: LINE (ENTER)
From point: CEN *(Press ENTER. Select circle in top view.)*
of
To point: QUAD *(Press ENTER. Select quadrant 4.)*
of
To point: *(Press ENTER to terminate the LINE command.)*
Command: ARRAY *(Press ENTER. Select line previously drawn.)*
Select objects: 1 found *(Press ENTER to terminate the ARRAY command's selection mode.)*
Select objects: (ENTER)
Rectangular or Polar array (R/P) <R>: P (ENTER)
Center point of array: CEN *(Press ENTER. Select circle in top view.)*
of
Number of items: 20 (ENTER)
Angle to fill (+=ccw, -=cw) <360>: (ENTER)
Rotate objects as they are copied? <Y> (ENTER)

Once the top view of the cylinder has been divided, and the points have been transferred onto the front view, the length of the stretch-out line can easily be obtained by using the LIST command. The LIST command will not only supply the technician with the area and the radius of a circle, it will supply the circumference as well.

Command: LIST (ENTER)
Select objects: *(Select circle in top view.)*
First corner: Other corner: 1 found (ENTER)
Select objects: *(Press ENTER to terminate the LIST command's selection mode.)*
CIRCLE Layer: 0
Space: Model space
Handle = 25D
center point, X=1.9567 Y=8.2978 Z=0.0000
radius 0.6493
***circumference 4.0797***
area 1.3245

After the stretch-out line has been drawn in its proper position, use either the DIVIDE or ARRAY command to divide the stretch-out line into the same number of equal segments as the top view. If the ARRAY command is used, then a line must first be drawn at one end of the stretch-out line. In addition to using a marker, the distance between each bend-line must be known in order to array the marker along the stretch-out line. On the other hand, using the DIVIDE command simply requires selecting the stretch-out line and then inputting the required number of segments.

Now that the stretch-out line has been divided and the corresponding bend-lines and height lines have been transferred, the top of the development can be created by constructing a polyline from intersection to intersection. A polyline is used for the top edge of the development because the PEDIT command can be selected to spline the polyline into an irregular curve.

Command: PEDIT (ENTER)
Select polyline: *(Select the top edge of the development.)*
Close/Join/Width/Edit vertex/Fit/Spline/Decurve/Ltype gen/Undo/eXit <X>: S (ENTER)
Close/Join/Width/Edit vertex/Fit/Spline/Decurve/Ltype gen/Undo/eXit <X>: (ENTER)

## Development of Top and Bottom Truncated Prisms and Cylinders

The development of a top and bottom (base) truncated prism is basically the same as that for the development of a single truncated prism. The main difference is that the stretch-out line is positioned in the center of the development and perpendicular to the front view, instead of the base of the development. This allows the bend-lines to be projected above and below the stretch-out lines (see Figure 6–19).

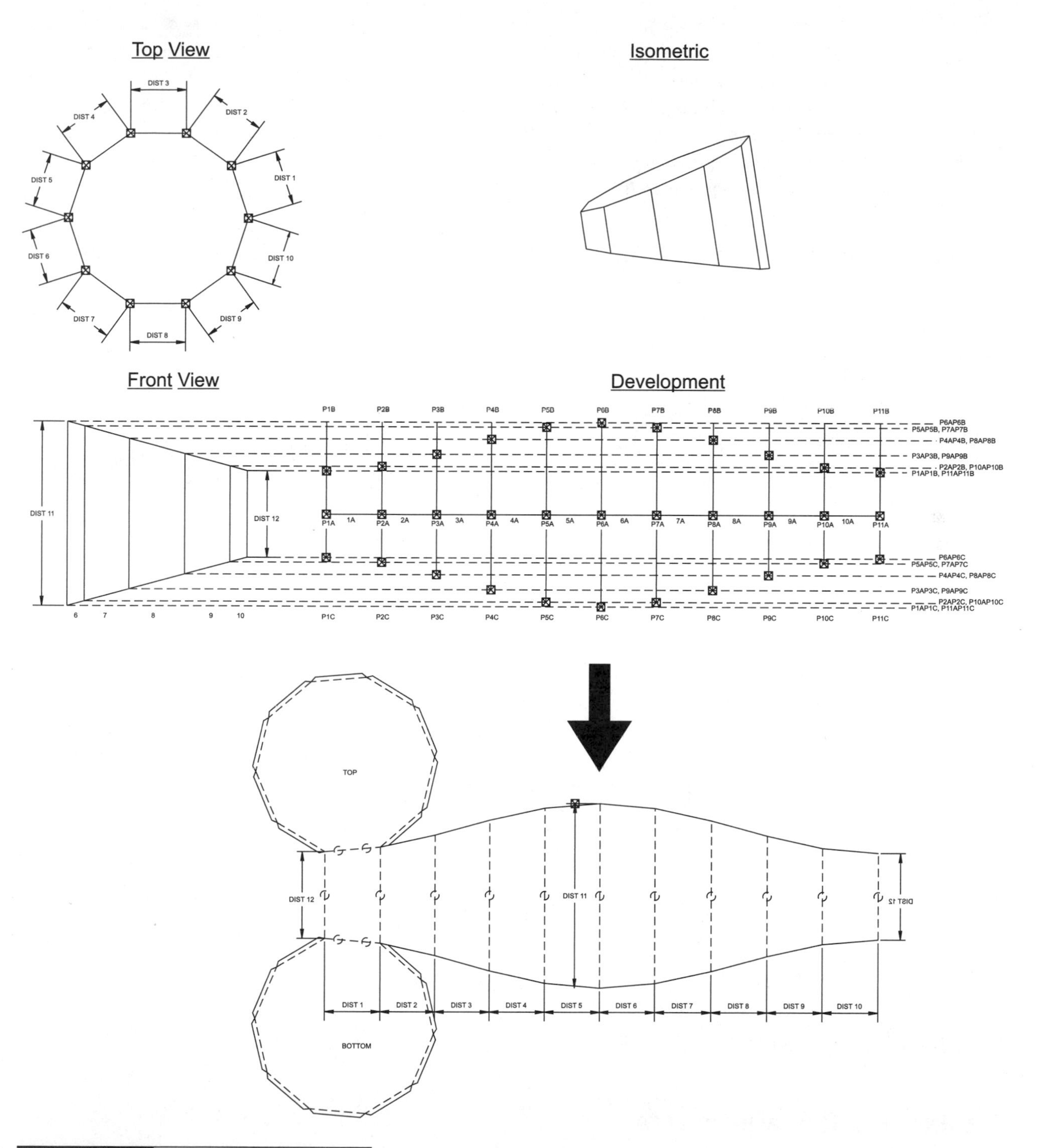


**Figure 6–19** *Truncated right prism.*

## Development of Oblique Prisms and Cylinders

Recall from the previous section that the definition of an oblique object is one where the lateral line or surfaces do not form 90° angles to the base or top of the object. For this reason, when creating a development of an oblique prism or cylinder, the use of an additional view is required. This view is an auxiliary sectional view that contains the true shape and size of the object. Once the sectional auxiliary view has been created, the stretch-out line is drawn in a manner similar to the one discussed earlier for a regular prism or cylinder. The height of the bend-lines is obtained by measuring the distance from the sectional line to the top and base of the object. That measurement is then transferred to the stretch-out line where the bend-lines are created. Drawing a line from the endpoint of one bend-line to the endpoint of another forms the top and bottom edge of the development (see Figure 6–20).

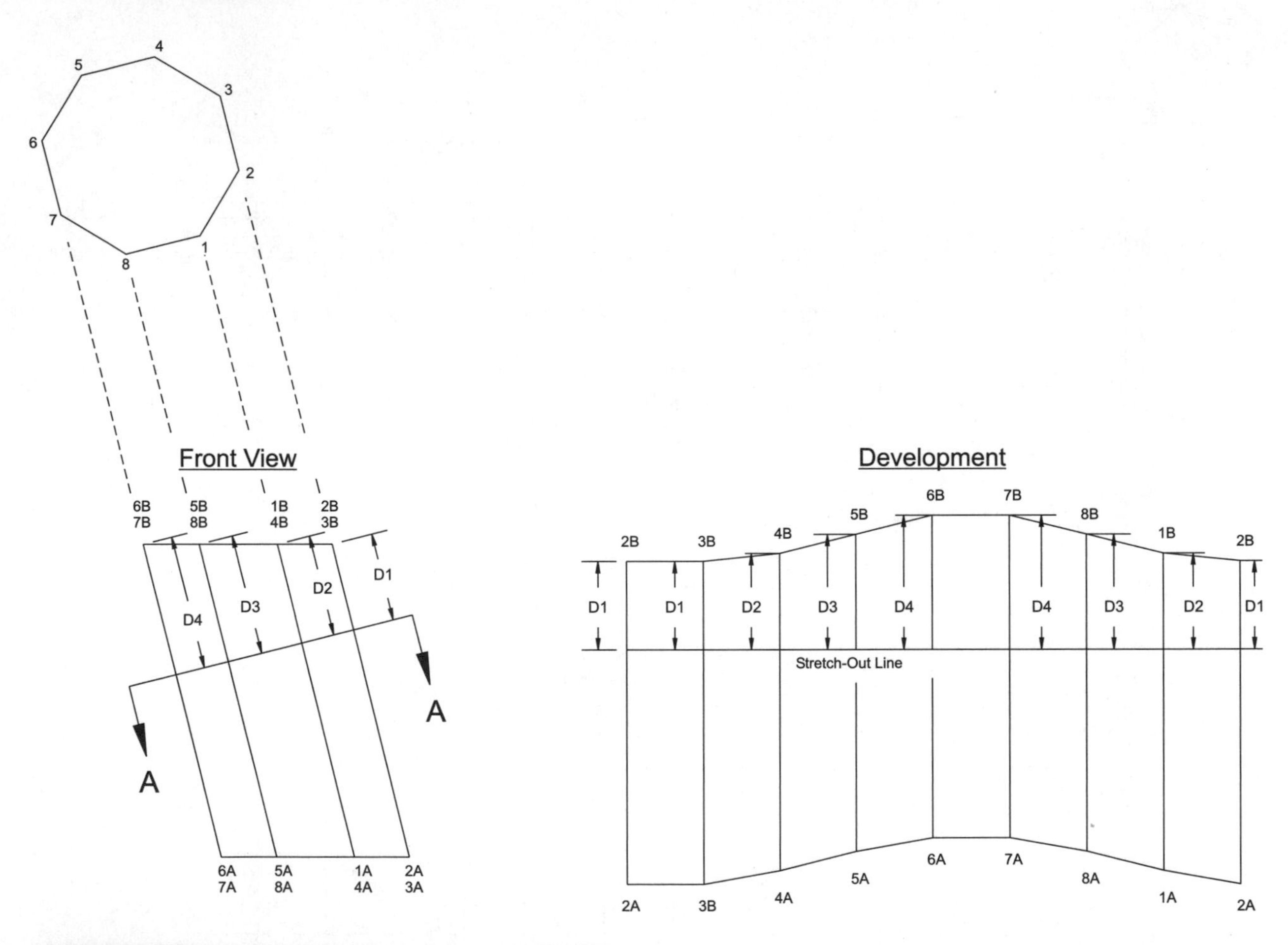


**Figure 6–20** *Development of an oblique prism.*

## Radial-Line Developments

The preceding material has been dedicated to the production of developments which required the use of the parallel line technique. This technique is limited to prisms and cylinders. To produce a development for geometric shapes such as pyramids and cones, a new technique must be employed. The geometric object is positioned with one of its lateral faces on a plane, and then the object is revolved around a point (see Figure 6–21). This technique is called ***radial line technique*** and it involves laying out the true lengths of lines around a radius point.

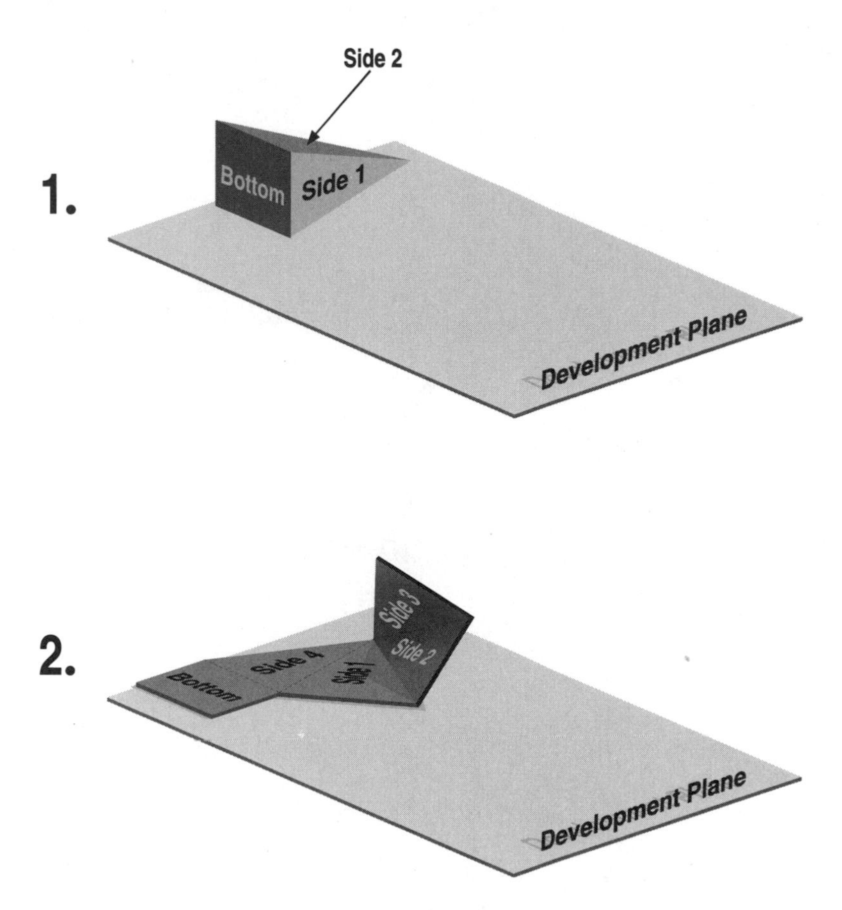


**Figure 6–21** *Pyramid unfolded.*

## Development of a Right Pyramid

A ***pyramid*** is a polyhedron consisting of a polygonal region as the base and triangular regions for its lateral sides. Connecting lines from the vertices of the base to a point called an ***apex*** form the pyramid. The apex of a pyramid doesn't lie in the same plane as the base. The distance from the apex perpendicular to the base is called the ***height***. When the lateral faces of a pyramid form isosceles triangles and the base is a polygon, this shape is called a ***right regular pyramid***. If the lateral regions of a pyramid do not form isosceles triangles, then this shape is known as an ***oblique regular pyramid***. The height along any of the faces of a pyramid is known as the ***slant height***.

To construct a development of a right regular pyramid, start by creating or opening a drawing containing a top and front view. The top view gives the true lengths of the lines forming the base, while the front and top views are used to obtain the true lengths of the corner lines. The true lengths of the corner lines are obtained by revolving the corner lines in the top view into a horizontal position (see Figure 6–22). The endpoints of the corner lines are then transferred into the front view where they intersect projection lines from the base of the pyramid to locate the lower corner (see Figure 6–22). A line is drawn from this corner to the apex, completing the corner line and revealing its true length. Once this has been completed, the technician is ready to construct the stretch-out line by striking an arc, whose radius is equal to the true length of the corner lines. The outer edge of the development is produced by drawing chords (equal in length to the base lines) starting at one end of the stretch-out line and continuing until all base lines have been established. Applying bend-lines and bend allowances as needed, completes the development.

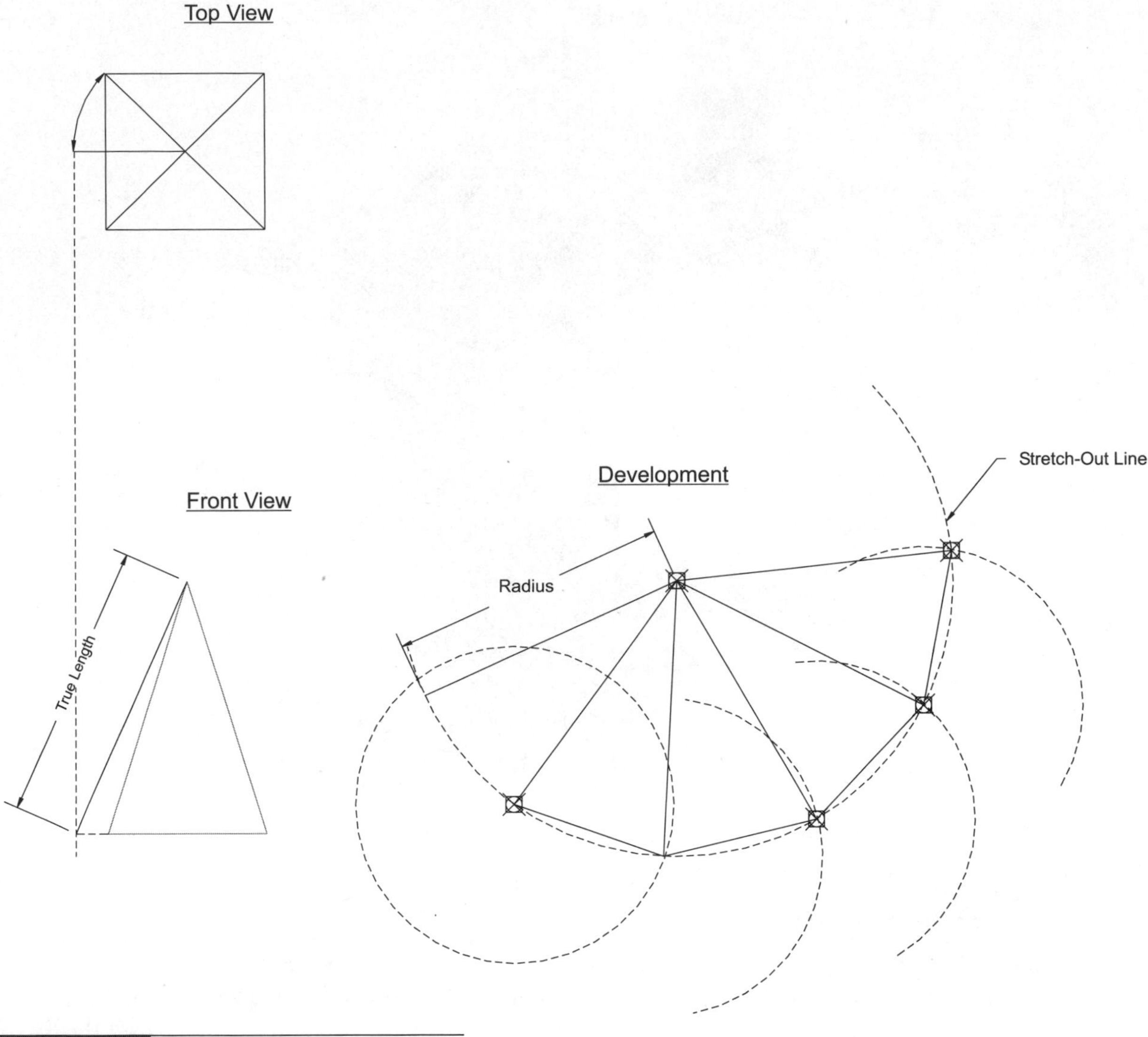


**Figure 6–22** *Development of a pyramid.*

### Using AutoCAD to Construct a Development of a Right Pyramid

AutoCAD makes the production of a development for a right pyramid an easy task to accomplish.

## Step #1

Open the drawing DG06-22_Student.dwg on the student CD-ROM. Starting with the top view, the apex is selected as the base point of rotation, and the corner lines are revolved using the ROTATE command.

Command: ROTATE (ENTER)
Select objects: 1 found *(Select corner line to be revolved.)*
Select objects: *(Press* ENTER *to terminate selection mode.)*
Base point: end of *(Select apex as base point of rotation.)*
<Rotation angle>/Reference: R *(Press* ENTER *to enable the reference option of the* ROTATE *command.)*
Reference angle <0>: end of *(Select apex as first point.)*
Second point: end of *(Select endpoint of corner line to be revolved.)*
New angle: mid of *(Select midpoint of vertical base line.)*

## Step #2

Drawing a line from the endpoint of the revolved corner line past the front view of the pyramid will partially locate the endpoint of the corner in the front view. The exact position of the endpoint of the corner line is determined by using the EXTEND command, and extending the base of the pyramid in the front view to the projection line.

Command: LINE (ENTER)
From point: end of *(Use the END OSNAP option to select the endpoint of the corner previously revolved.)*
To point: *(Select an arbitrary point the in front view below the object.)*
To point: *(Press ENTER to terminate the LINE command.)*
Command: EXTEND (ENTER)
Select boundary edges: (Projmode = UCS, Edgemode = No extend)
Select objects: 1 found *(Select the projection line as the boundary of the extension.)*
Select objects: *(Press ENTER to terminate the EXTEND command selection mode and start the extension process.)*
<Select object to extend> *(Select the base line of the pyramid in front view.)*
<Select object to extend> *(Press ENTER to terminate the EXTEND command.)*

## Step #3

Using the LINE command, the true length of the corner line is represented in the front view by drawing a line from the apex of the pyramid to the intersection of the projection lines. This true corner line is copied to the location where the development is to be created by using the COPY command. As a template for the location of the stretch-out line, a circle can now be created using the CIRCLE command, in which the center is located at one endpoint of the copied corner line, and the outside edge is located at the other endpoint.

Command: LINE (ENTER)
From point: INT of *(Select the intersection point of the projection line with the base line as first point.)*
To point: END of *(Select the apex of pyramid as the second point.)*
To point: *(Press ENTER to terminate the LINE command.)*
Command: COPY (ENTER)
Select objects: 1 found *(Select true corner line created.)*
Select objects: *(Press ENTER to terminate the selection mode.)*
<Base point or displacement>/Multiple: *(Select an arbitrary point as base point.)*
Second point of *(Select a point where development is to be created.)*
displacement:
Command: CIRCLE (ENTER)
3P/2P/TTR/<Center point>: end of *(Select endpoint of line copied.)*
Diameter/<Radius>: end of *(Select other endpoint of line copied.)*

## Step #4

The lengths of the chords (circles that will act as templates to create the baselines) are transferred to the development by constructing circles in the top view with the center positioned at one endpoint of a base line and the outer edge positioned at the other endpoint. Using the MOVE command, one of the circles is moved from the top view onto the stretch-out line (the center of the circle is positioned anywhere along the stretch-out line). The next circle is then moved onto the stretch-out line by using its center as the base point, and the intersection of the first circle with the stretch-out line as the second point. This procedure is repeated until all circles have been located, each time using the intersection of the previous circle and stretch-out line as the second point.

Command: CIRCLE (ENTER)
3P/2P/TTR/<Center point>: end of *(Select the endpoint of a base line in the top view.)*
Diameter/<Radius> <1.0000>: end of *(Select other endpoint of base line.)*
Command: CIRCLE (ENTER)
3P/2P/TTR/<Center point>: @ *(Press* ENTER. *Enter last point selected.)*
Diameter/<Radius> <1.0000>: end of *(Select other endpoint of base line.)*
Command: CIRCLE (ENTER)
3P/2P/TTR/<Center point>: @ *(Press* ENTER. *Enter last point selected.)*
Diameter/<Radius> <1.0000>: end of *(Select other endpoint of base line.)*
Command: CIRCLE (ENTER)
3P/2P/TTR/<Center point>: @ *(Press* ENTER. *Enter last point selected.)*
Diameter/<Radius> <1.0000>: end of *(Select other endpoint of base line.)*
Command: MOVE (ENTER)
Select objects: 1 found *(Select first circle drawn.)*
Select objects: *(Press* ENTER *to terminate selection mode.)*
Base point or displacement: cen of *(Uses center of circle as base point.)*
Second point of displacement: nea to *(Select any point along perimeter of "stretch-out line" circle.)*

Command: MOVE (ENTER)
Select objects: 1 found *(Select second circle drawn.)*
Select objects: *(Press* ENTER *to terminate selection mode.)*
Base point or displacement: cen of *(Uses center of circle as base point.)*
Second point of displacement: int of *(Select intersection of first circle's edge with stretch-out line as second point of displacement.)*
Command: MOVE (ENTER)
Select objects: 1 found *(Select third circle drawn.)*
Select objects: *(Press* ENTER *to terminate selection mode.)*
Base point or displacement: cen of *(Uses center of circle as base point.)*
Second point of displacement: int of *(Select intersection of previous circle's edge with stretch-out line as second point of displacement.)*
Command: MOVE (ENTER)
Select objects: 1 found *(Select fourth circle drawn.)*
Select objects: *(Press* ENTER *to terminate selection mode.)*
Base point or displacement: cen of *(Uses center of circle as base point.)*
Second point of displacement: int of *(Select intersection of previous circle's edge with stretch-out line as second point of displacement.)*

## Step #5

The bend-lines are completed by drawing lines from the intersection of each circle to the center of the circle that forms the stretch-out line. The bottom edge is created by drawing a line from the center of each circle to the center of its adjacent circle. The necessary fold-lines and annotation, as well as any bend allowances that may be needed, are added to complete the development.

Command: PLINE (ENTER)
From point: cen of *(Creates a polyline connecting each point selected.)*
Current line-width is 0.0000
Arc/Close/Halfwidth/Length/Undo/Width/<Endpoint of line>: INT of
Arc/Close/Halfwidth/Length/Undo/Width/<Endpoint of line>: INT of
Arc/Close/Halfwidth/Length/Undo/Width/<Endpoint of line>: INT of
Arc/Close/Halfwidth/Length/Undo/Width/<Endpoint of line>: INT of
Arc/Close/Halfwidth/Length/Undo/Width/<Endpoint of line>: INT of
Arc/Close/Halfwidth/Length/Undo/Width/<Endpoint of line>: C (ENTER)

## Development of a Truncated Right Pyramid

A truncated pyramid is a pyramid that has had the apex removed by a plane. To produce a development of a truncated pyramid, the procedure used in the previous section is also employed here. This time, however, when the corner line is revolved, both ends of the corner are projected into the front view. The top edge of the development is constructed by transferring the distances (in the front view) between the apex and the top edge of the pyramid, to the bend-lines in the development (see Figure 6–23).

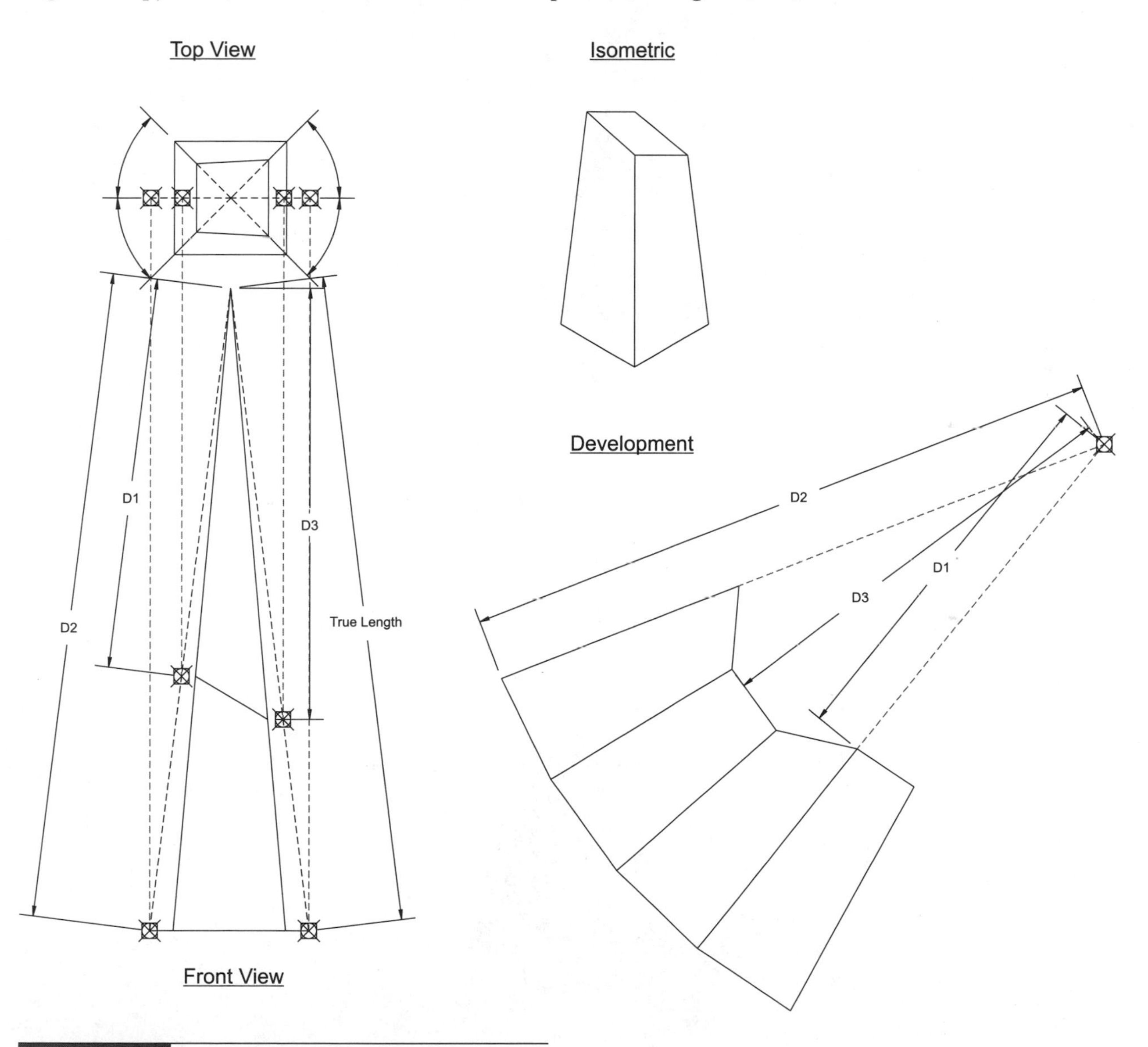


**Figure 6–23** *Development of truncated pyramid.*

## Development of a Right Circular Cone

A ***cone*** is a geometric shape consisting of a circular base and a point (located on a parallel plane) called an ***apex*** (see Figure 6–24a). The two are connected by line segments extending from the apex and tangent to the outside edge of the base. A ***right circular cone*** is a cone in which a line extended from the apex will be perpendicular to the base (see Figure 6–24b). If the line is not perpendicular to the base, then this type of cone is called an ***oblique circular cone*** (see Figure 24c). Like a right pyramid, the ***height of a right cone*** is the distance from the center of the base to the apex. The distance from the apex to a point on the outside edge of the circular base is known as the ***slant height***.

The development for a right circular cone consists of a stretch-out line equal in length to the circumference of the base of the cone revolved around a single point. The radius of the stretch-out line is equal to the slant height of the cone. Seam lines extend from the ends of the stretch-out line to the apex (see Figure 6–24d).

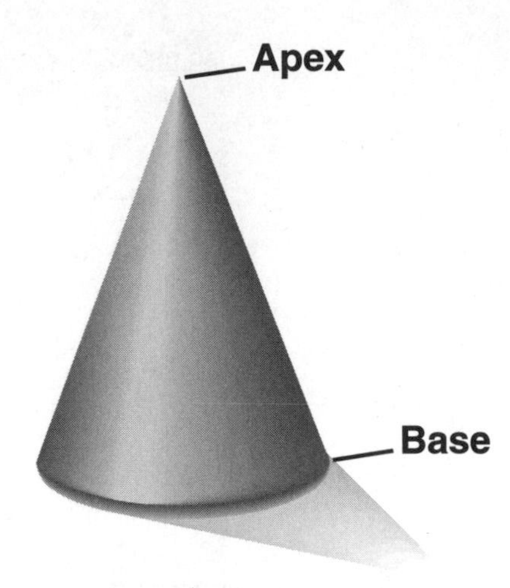


**Figure 6–24a** *Cone.*

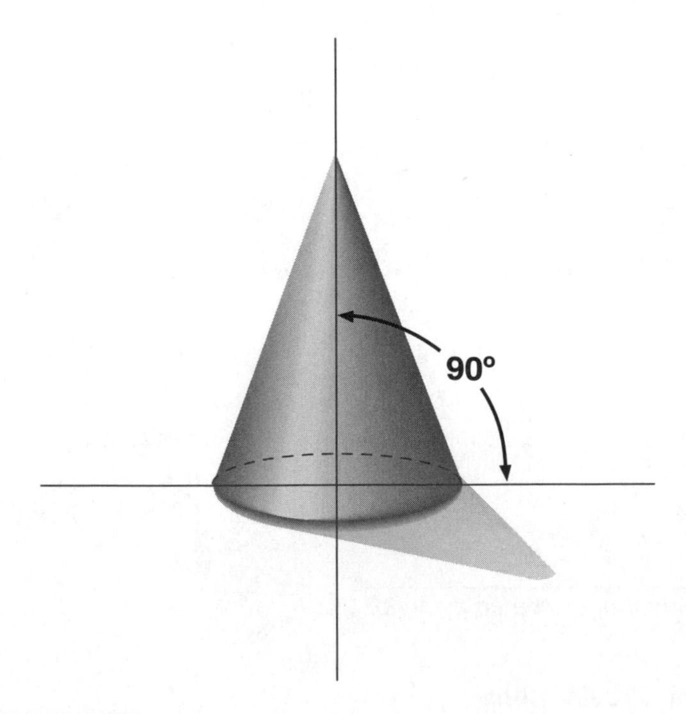


**Figure 6–24b** *Right circular cone.*

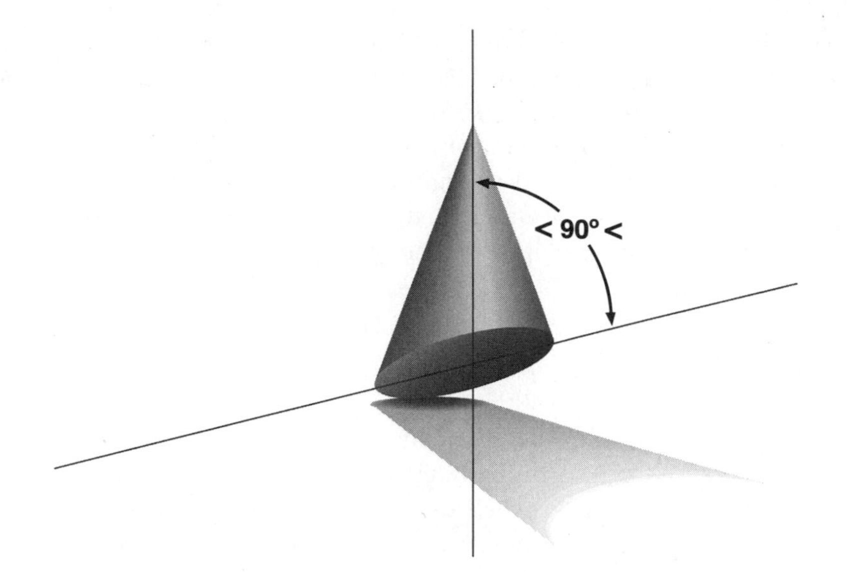


**Figure 6–24c** *Oblique circular cone.*

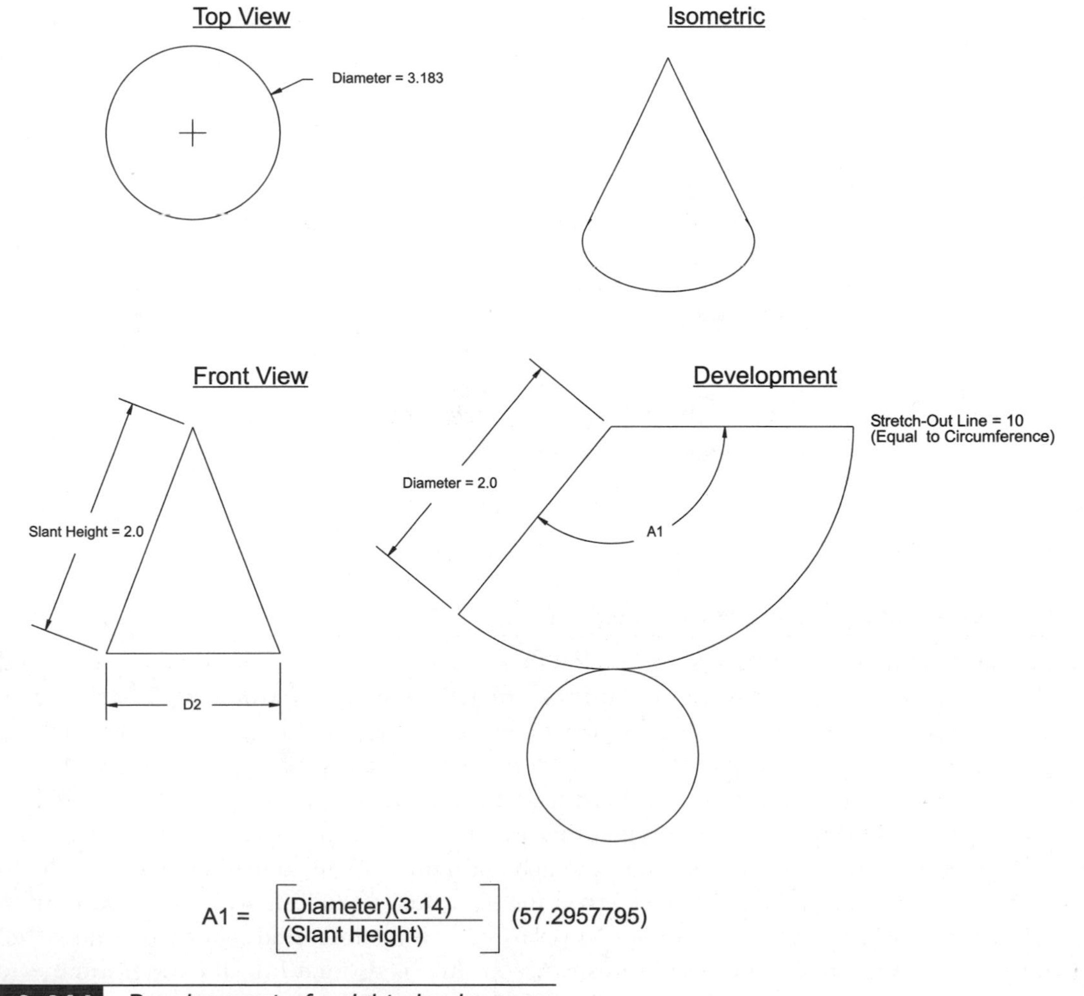


$$A1 = \left[\frac{(\text{Diameter})(3.14)}{(\text{Slant Height})}\right](57.2957795)$$

**Figure 6–24d** *Development of a right circular cone.*

To construct a development of a right circular cone, begin with a drawing containing a top and front view. The top view is used to obtain the true length of the diameter of the base. The diameter of the base is used to calculate the circumference of the base by using the formula D ( (Diameter × Pi). The front view is used to obtain the true length of the slant height of the cone. In the location of the development, an arc is struck that contains a radius equal to the slant height and a length equal to the circumference. The development is completed by adding extra material for the seaming method to be used and drawing lines from the ends of the stretch-out line to the center of the development (the apex).

#### Using AutoCAD to Construct a Development of a Right Circular Cone

The method used to construct a development of a right circular cone in AutoCAD is similar to the one used for pyramids. Using the COPY command, one of the inclined lines forming the outer edge of the cone in the front view is copied to the location where the development will be positioned. This line serves a dual purpose; first, it will be used to determine the radius of the stretch-out line. This is accomplished by using the CIRCLE command and selecting one end of the line as the center of the circle and the other as the circumference. Second, this line will be used to indicate one side of the development. Selecting and rotating a copy of this line into its desired position creates the other side of the development. The amount that this line is rotated is determined by the formula (S/r) × 57.29577951. In this formula, the S is equal to the circumference of the base of the cone; it can be obtained by using the AutoCAD LIST command and selecting the base circle in the top view. The r is equal to the true length of the cone's slant height. The 57.29577951 is a conversion factor to convert radians to degrees (see Chapter seven for an explanation of radians). For example, given a cone with a circumference of 10 and a slant height of 2, the angle between the two ends of the stretch-out line is calculated as 10/2.0 * 57.29577951. This yields an answer of 286.4788976° (see Figure 6–24d).

```
Command: LIST (ENTER)
Select objects: 1 found (Select base circle in top view.)
Select objects: (Press ENTER to terminate the LIST command's selection mode.)
            CIRCLE    Layer: 0
                  Space: Model space
            Color: 3 (green)    Linetype: BYLAYER
            Handle = 407
       center point, X=   3.2319  Y=   8.1051  Z=   0.0000
       radius    1.5915 (Equal to the radius of the base circle.)
   circumference   10.0000 (Equal to the circumference of the base circle.)
        area    7.9578
```

This information is now used to construct the development.

### Development of a Truncated Right Circular Cone

The method used for the production of a truncated right circular cone is an extension of the one used for the right circular cone. First, a top and front view are constructed to reveal the true length of the object's base circle and slant height. Next, an arc is constructed that is equal in length to the circumference of the base and with a radius equal to the slant height of the cone. The development is not completed by simply connecting the endpoints of the stretch-out line to the apex. Instead, the top edge is produced by borrowing a technique used for truncated right cylinders. To use this technique, the base of the cone (in the top view) must be divided into equal segments. Lines are drawn from each endpoint (of the cone's base circle in the top view) to the apex. The points at which these lines intersect the truncated top, along with their counterparts along the base circle, are transferred to the front view. For clarity it is recommended that each point in both views be labeled. Once this has been completed, the stretch-out line is divided into the same number of equal segments as the top view. Using the distance from the truncation of the cone to the apex in the front view, the top edge of the development is constructed by transferring these distances to the development in the same manner as that used for the construction of a truncated right cylinder. After all the distances have been transferred, the top edge of the development is constructed by drawing an irregular curve that connects each transferred point.

### Development of Oblique Pyramids and Cones

The production of an oblique pyramid or cone is slightly different from that of a right circular cone or right pyramid. In the oblique versions of each shape, the radial lines forming the sides of the object are not all equal in length. Therefore, the true length of each radial line or element must be determined before the development can be produced. The easiest way to accomplish this task is by revolving the radial lines in the top view into a horizontal position and transferring their endpoints into the front view. For a pyramid this is not a difficult task, since the lateral lines of each corner are simply revolved into a horizontal position. A cone is slightly more complex because the base circle must be divided into equal segments, followed by connecting the endpoints of the segments to the apex (in the top view). These segments are then revolved into a horizontal position (in the top view), and their endpoints are transferred into the front view using projection lines. The remaining portion of the development for both objects is identical to the production of truncated objects.

## Triangulation Developments

When a development of an object cannot be produced using the parallel-line or radial line techniques, then a different method, called ***triangulation method***, must be employed. This method produces a development by using a series of triangles to approximate the layout. The triangulation method is used extensively for the production of developments of ***transitional fittings***. A ***transitional fitting*** is a connector used to connect two dissimilar shapes or sizes of sheet metal ducts. For example, when connecting a round pipe to a square pipe, a transitional fitting would be required to form a smooth conversion from one pipe to another. These fittings are most often used with HVAC Ducts.

### Development of Sheet-Metal Transitions

Since an infinite number of sheet metal configurations can be produced, this section is limited to the production of a development of one type of transitional fitting. The basic technique remains the same for other configurations. To produce a development for the transitional fitting in Figure 6–26a, a top and front view of the object are constructed first. The top view is used to determine the true lengths of the lines forming the base of the object. Although the front view does not contain the true length of the bend-lines, it will be used in conjunction with the top view to derive the true length (see Figure 6–26b). The circular top is divided into sixteen equal segments (see Figure 6–26c). Sixteen equal segments is a common midrange used to produce an accurate and smooth surface without adding unnecessary steps to the drawing, and sixteen is divisible by four—an important factor for this method of development. Once the top has been divided, then the positions of the points are transferred from the top view to the front view via projection lines (see Figures 6–26d, 6–24e, 6–24f, and 6–24g). Next, starting in one corner, the division points in the top view are connected to their corresponding corners (see Figure 6–26h) to generate the bend-lines. After all points have been connected to their corresponding corners, the true lengths of these bend-lines are determined by revolving the lines around their connecting corner into a horizontal position (see Figure 6–26i). The endpoints are then transferred from the top view to the front view by projecting a line from the revolved lines (in the top view) perpendicular to the extended horizontal line that forms the top edge of the transitional fitting. A line drawn from the corresponding corner point (in the front view) to the point at which the projection lines intersect the extended top line represents the true length of the bend-lines (see Figure 6–26j). Starting at the location where the development is to be created, the seam line is constructed (the seam line can be drawn at any length), as shown in Figure 6–26j. Next, the base line that connects the seam to the first corner (in the top view) is drawn perpendicular to the existing seam line in the development. The point on this base line opposite the intersection of the seam line becomes the first corner (see Figure 6–26k). All of the bend-lines for the first corner will emanate from this point. The true length of the first bend-line is transferred by constructing an arc with a radius equal to the length of the bend line. The center of this arc is placed on the first corner point in the development. The point at which the arc intersects the seam line determines the opposite endpoint of the first bend-line in the development (see Figure 6–26L). The successive bend-lines are transferred in a similar manner, but instead of using the intersection of their arcs with the seam line, a second arc is constructed from the endpoint of the previous bend-line. The radius of the arc will equal the chordal length between points in the top view. This procedure is repeated until all points are transferred and the development is complete (see Figure 6–26m).

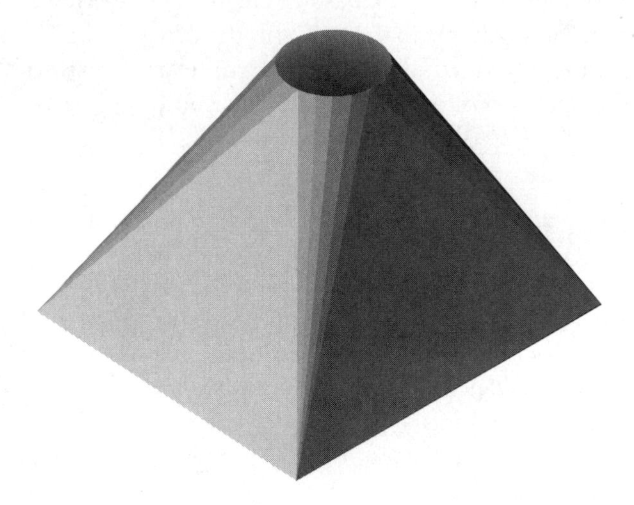

**Figure 6–26a** *Transitional fitting.*

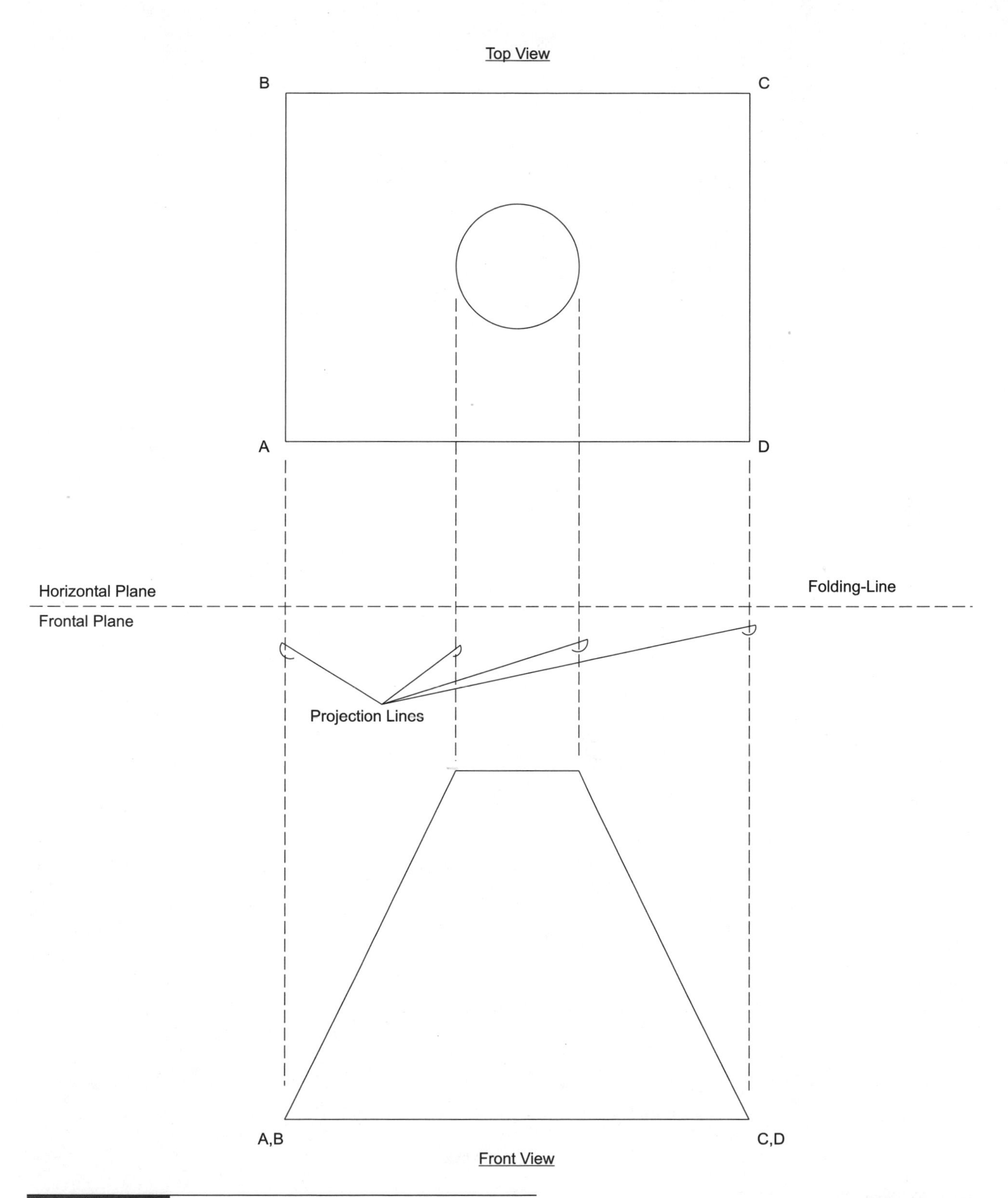


**Figure 6–26b** *Top and front view of transitional fitting.*

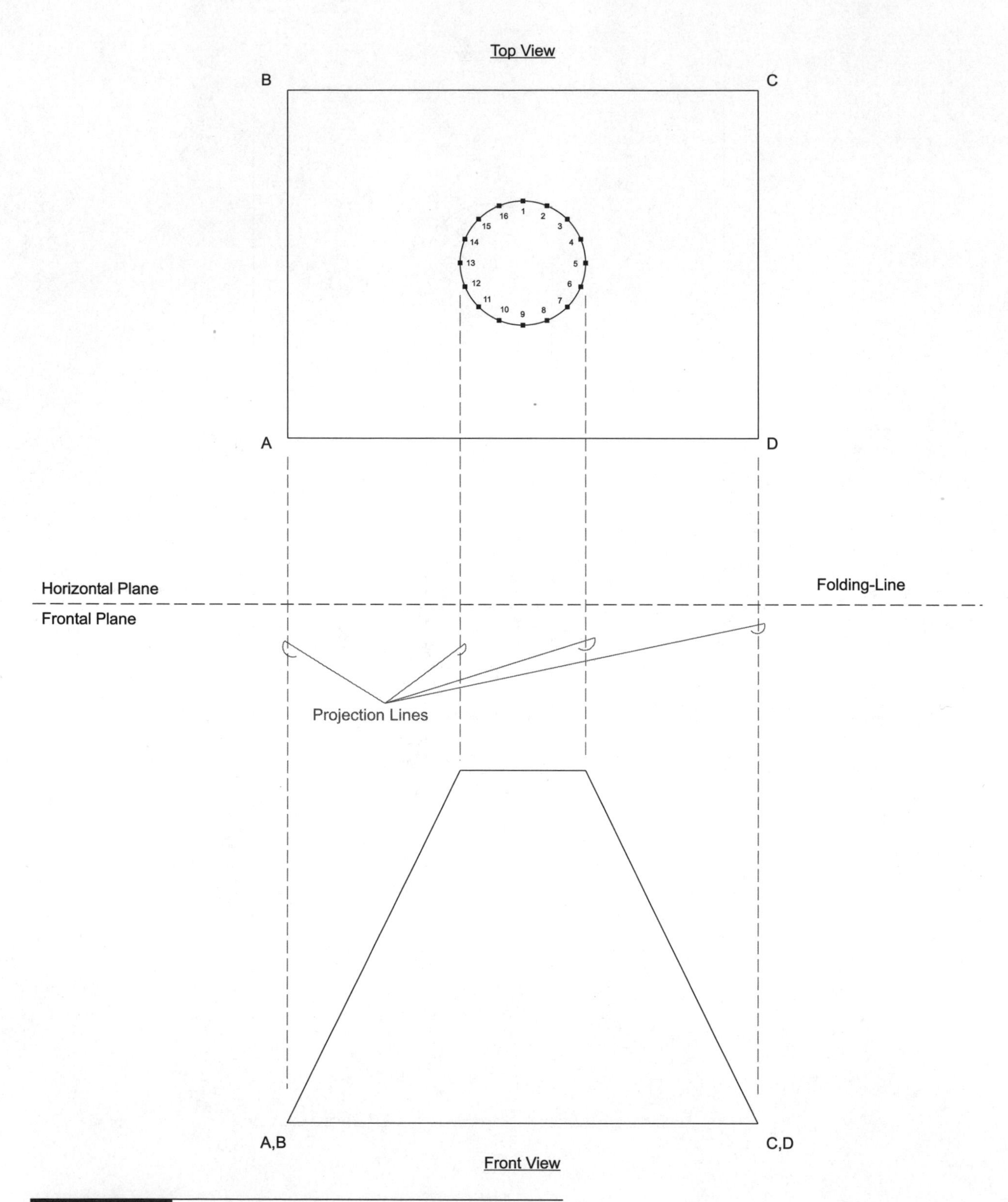


**Figure 6–26c** *Top is divided into sixteen equal sections.*

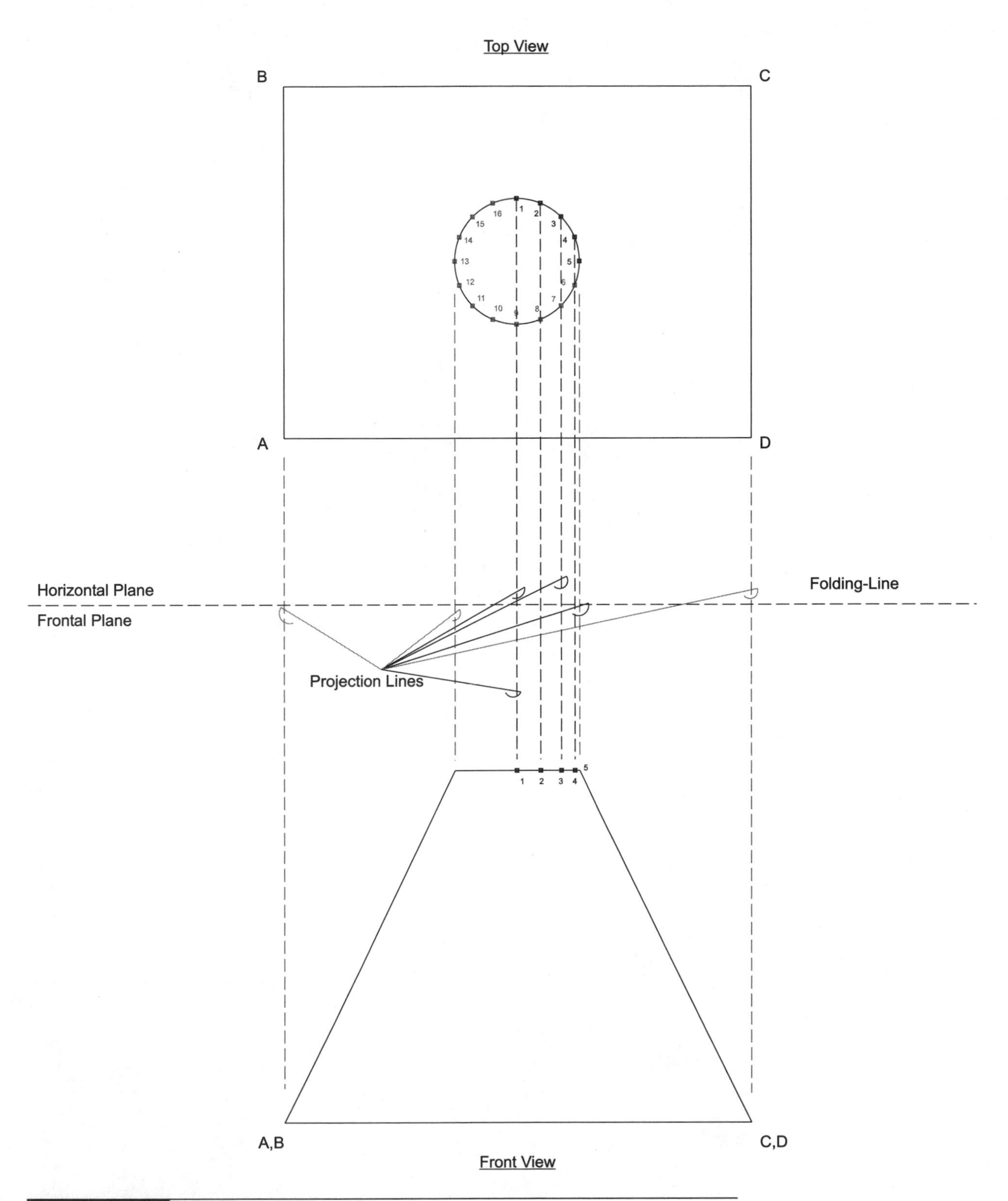


**Figure 6–26d** *Division points 1, 2, 3, and 4 are transferred to the front view.*

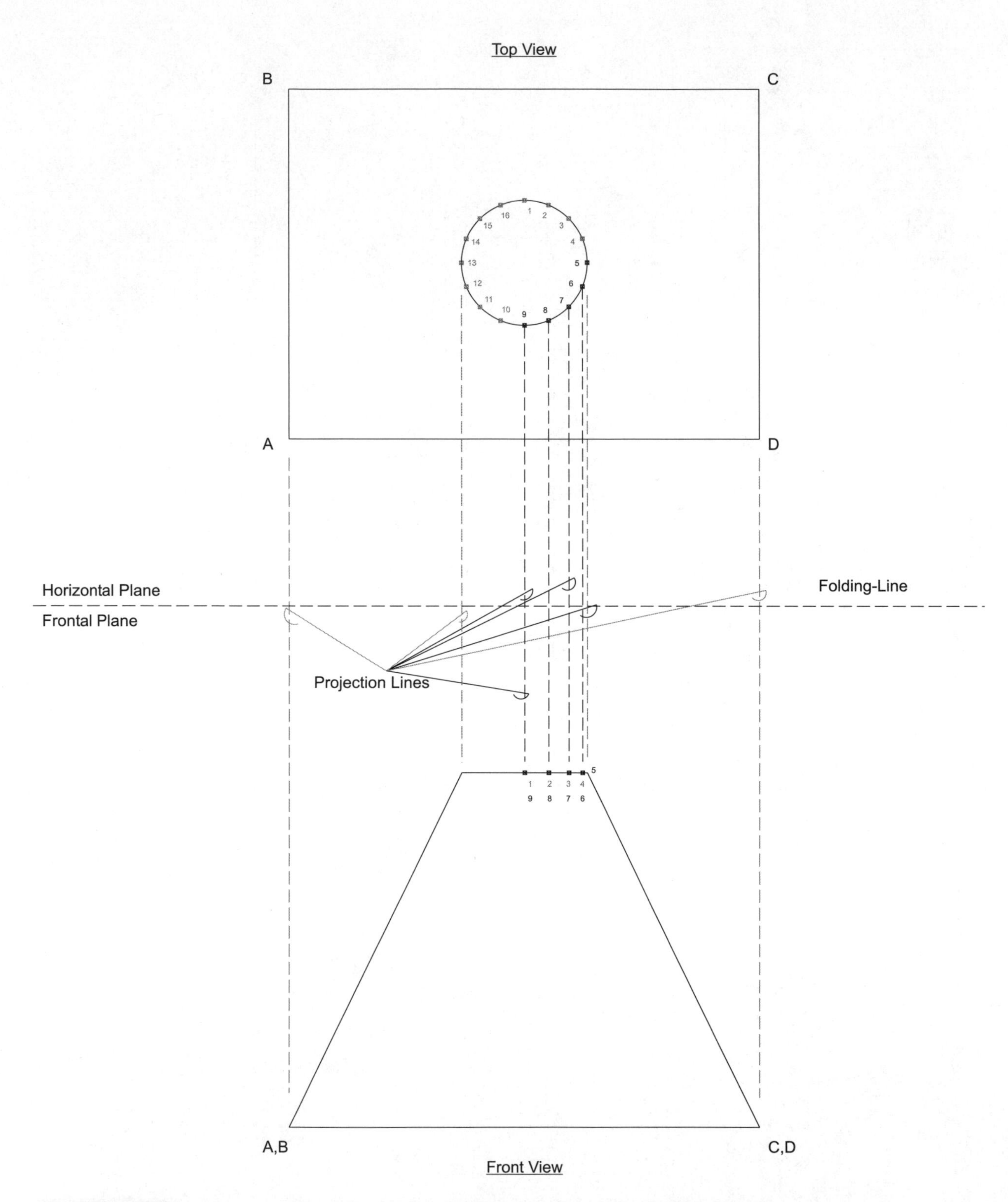


**Figure 6–26e** *Division points 5, 6, 7, 8, and 9 are transferred to the front view.*

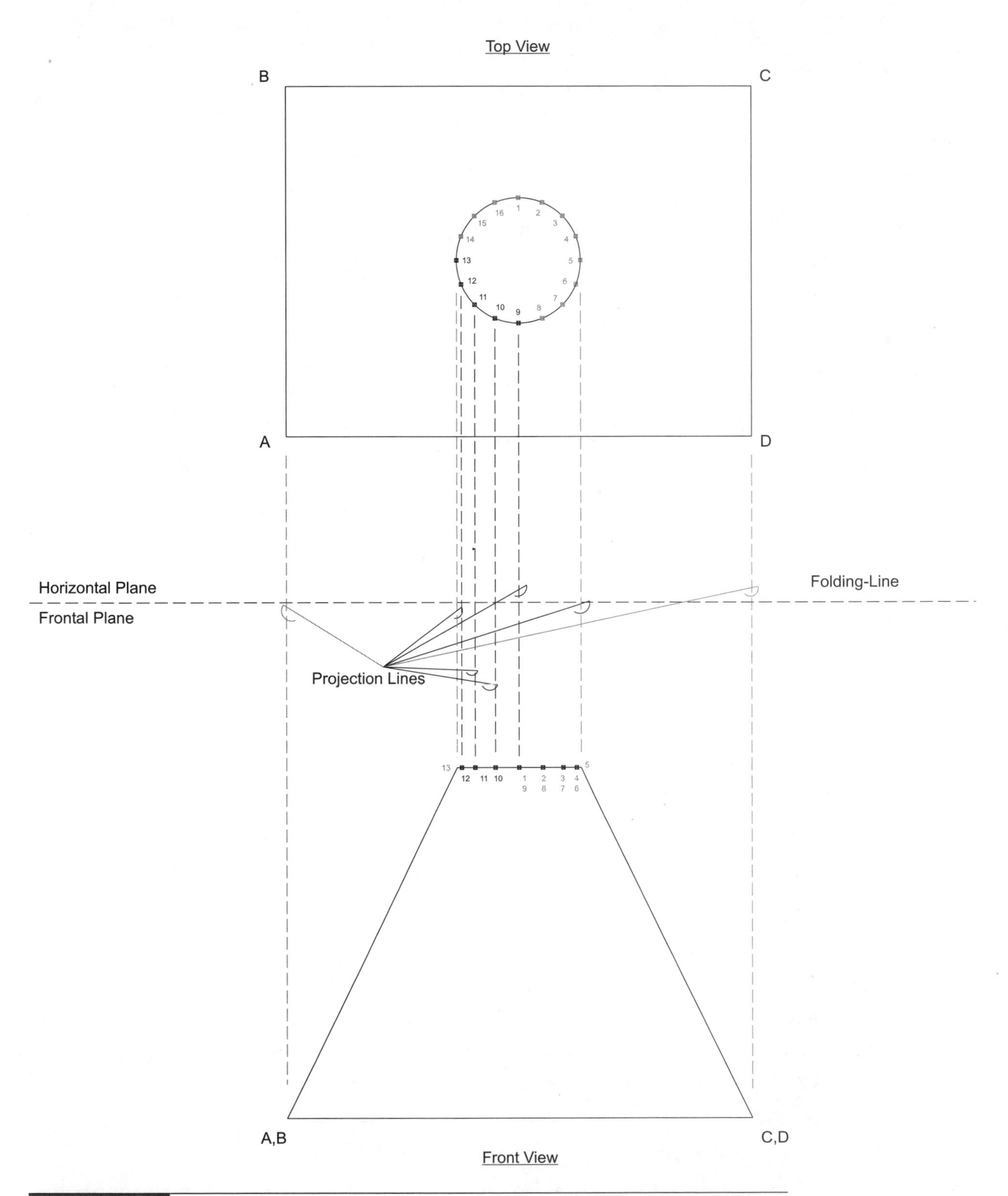


**Figure 6–26f** *Division points 10, 11, 12, and 13 are transferred to the front view.*

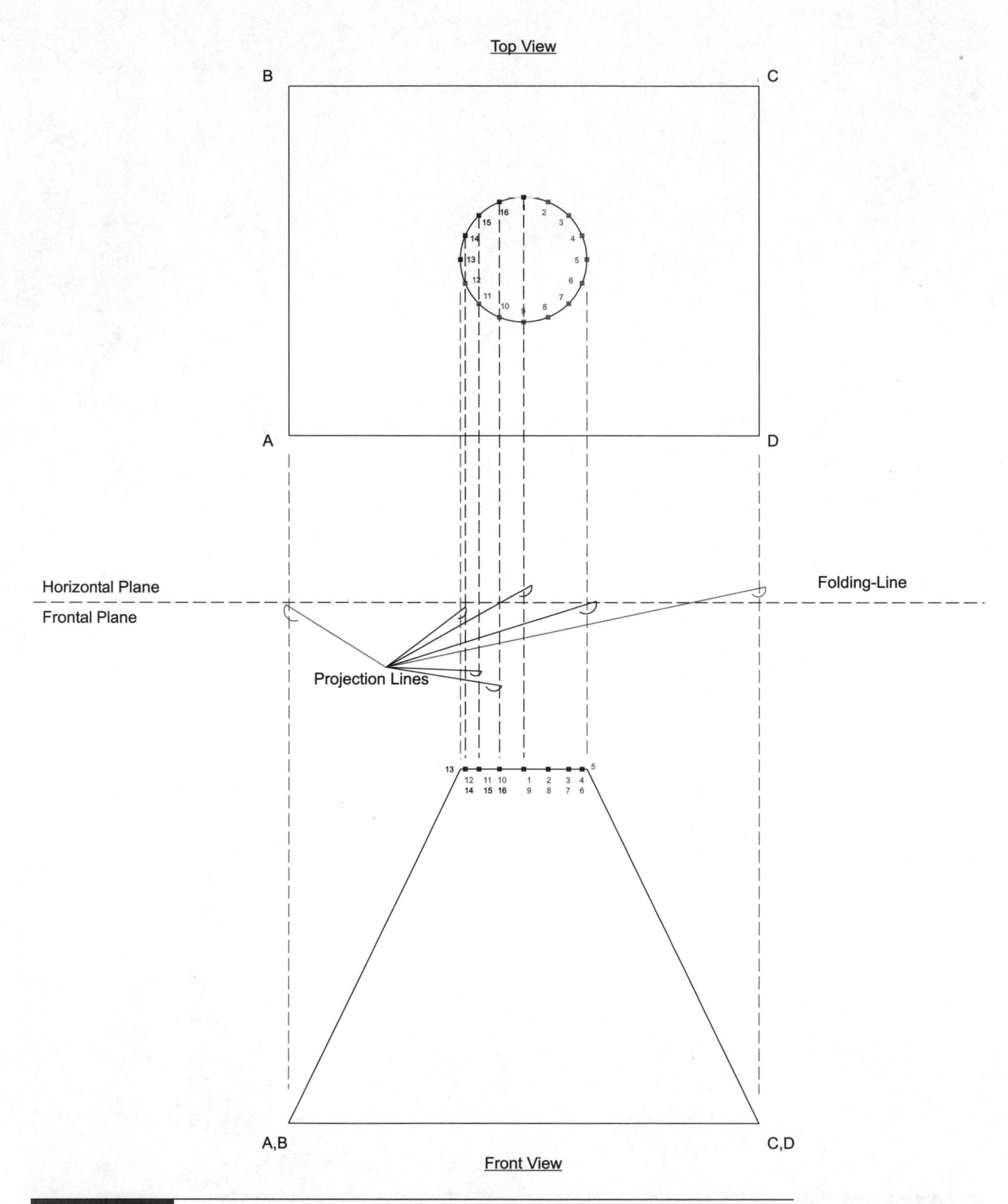


**Figure 6–26g** *Division points 13, 14, 15, and 16 are transferred to the front view.*

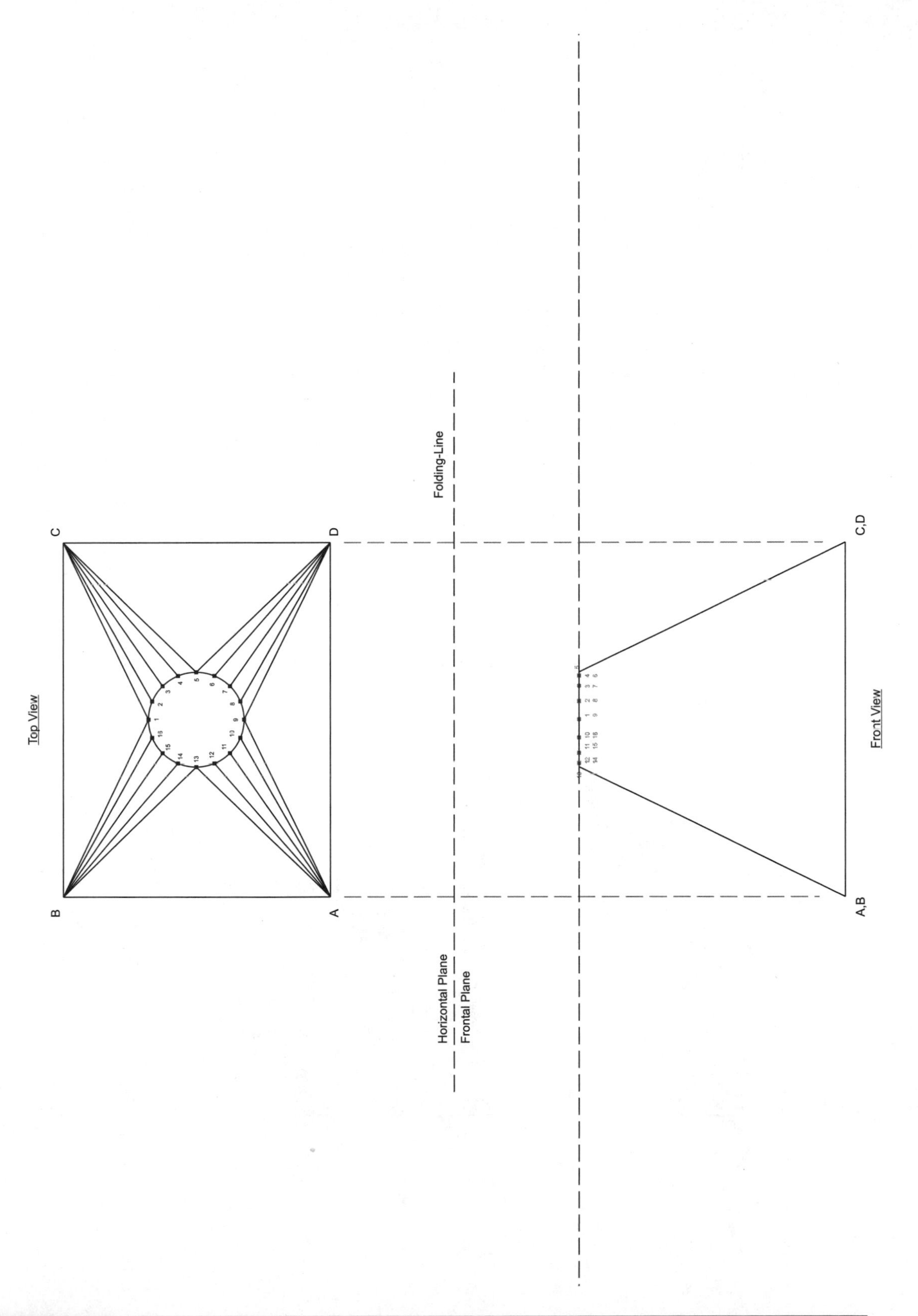


**Figure 6–26h** *The division points are connected to their corresponding corners in the top view.*

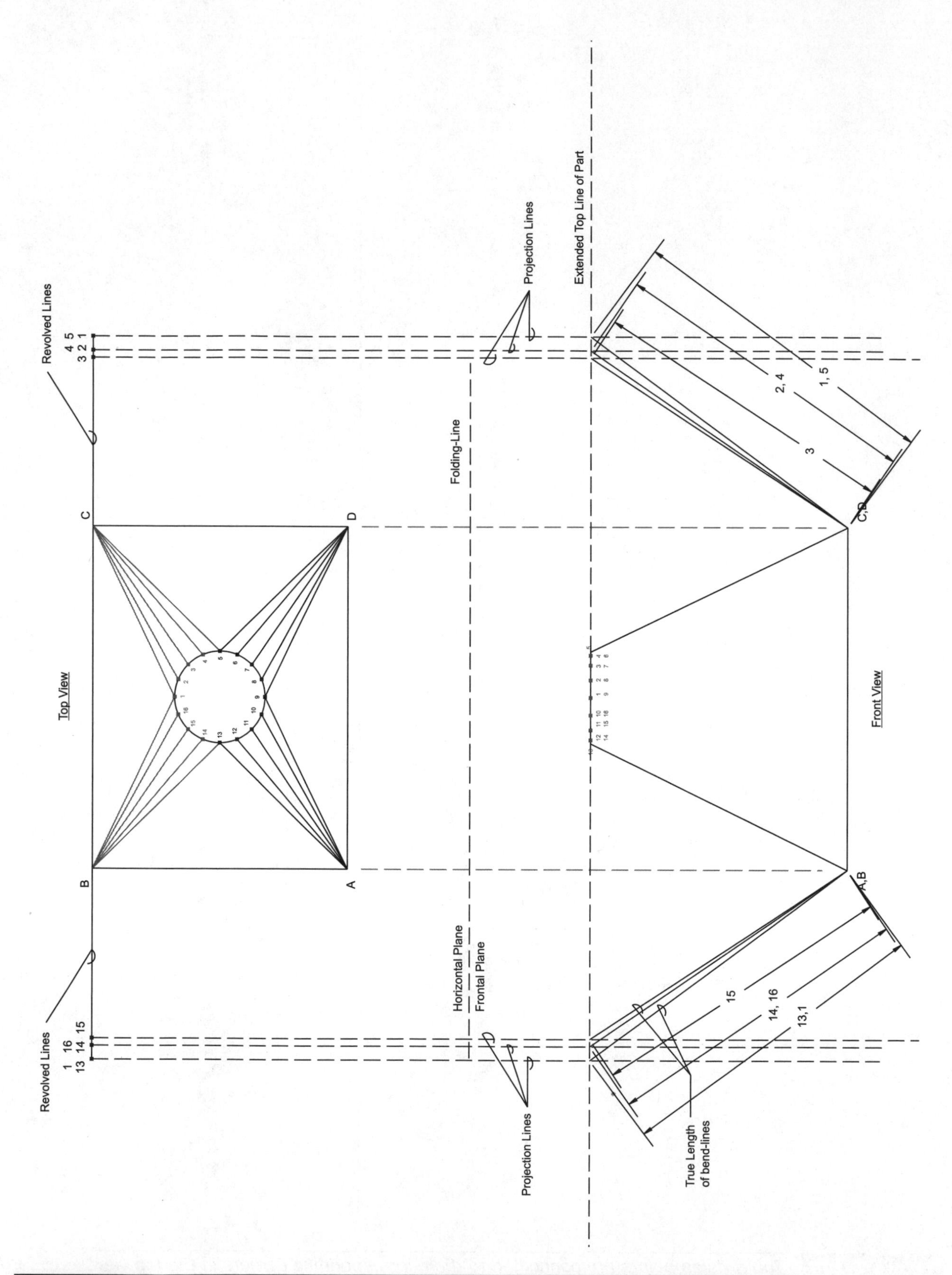


**Figure 6–26i** *True lengths of the bend-lines are found by using the revolution method.*

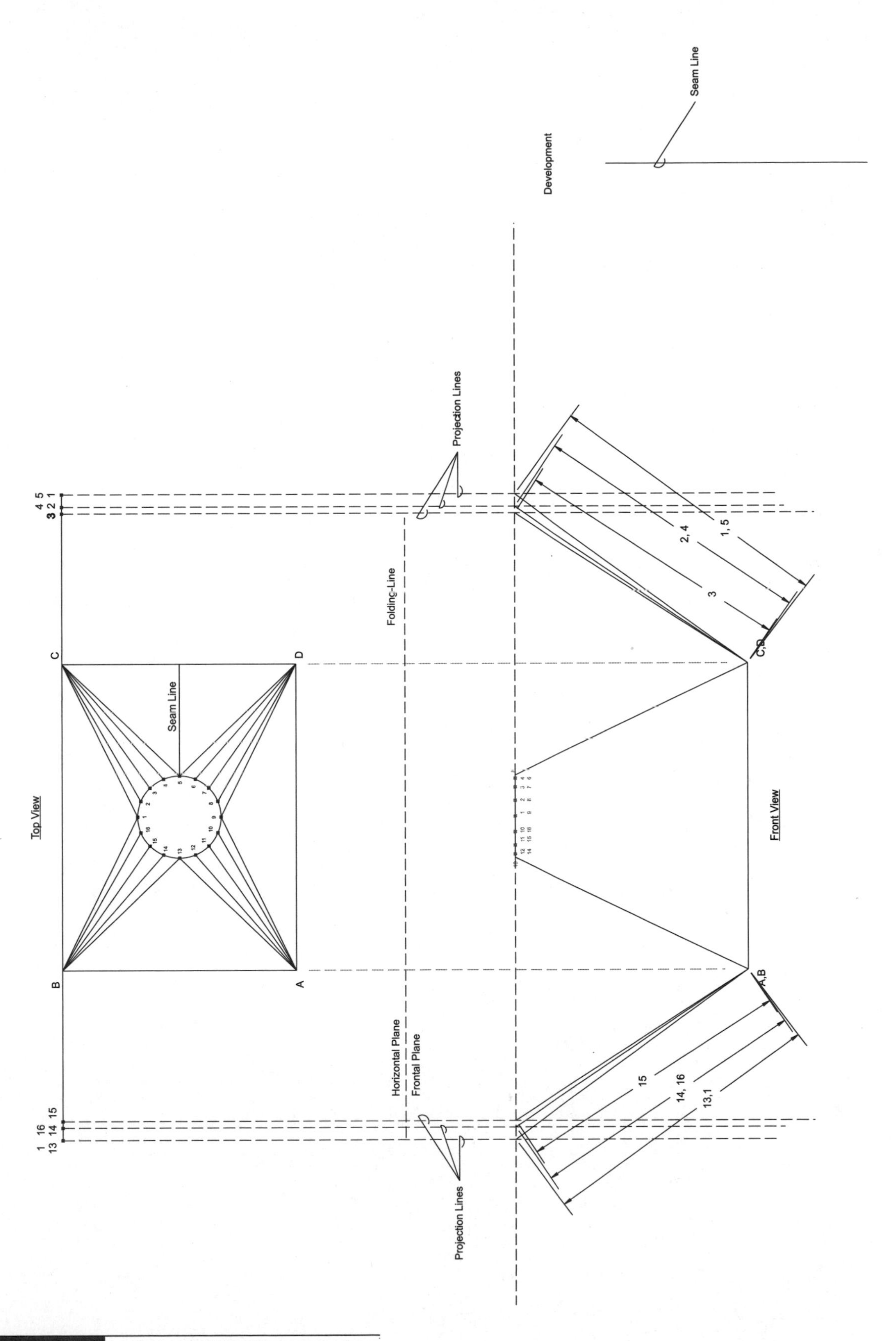


**Figure 6–26j** *Constructing the seam line.*

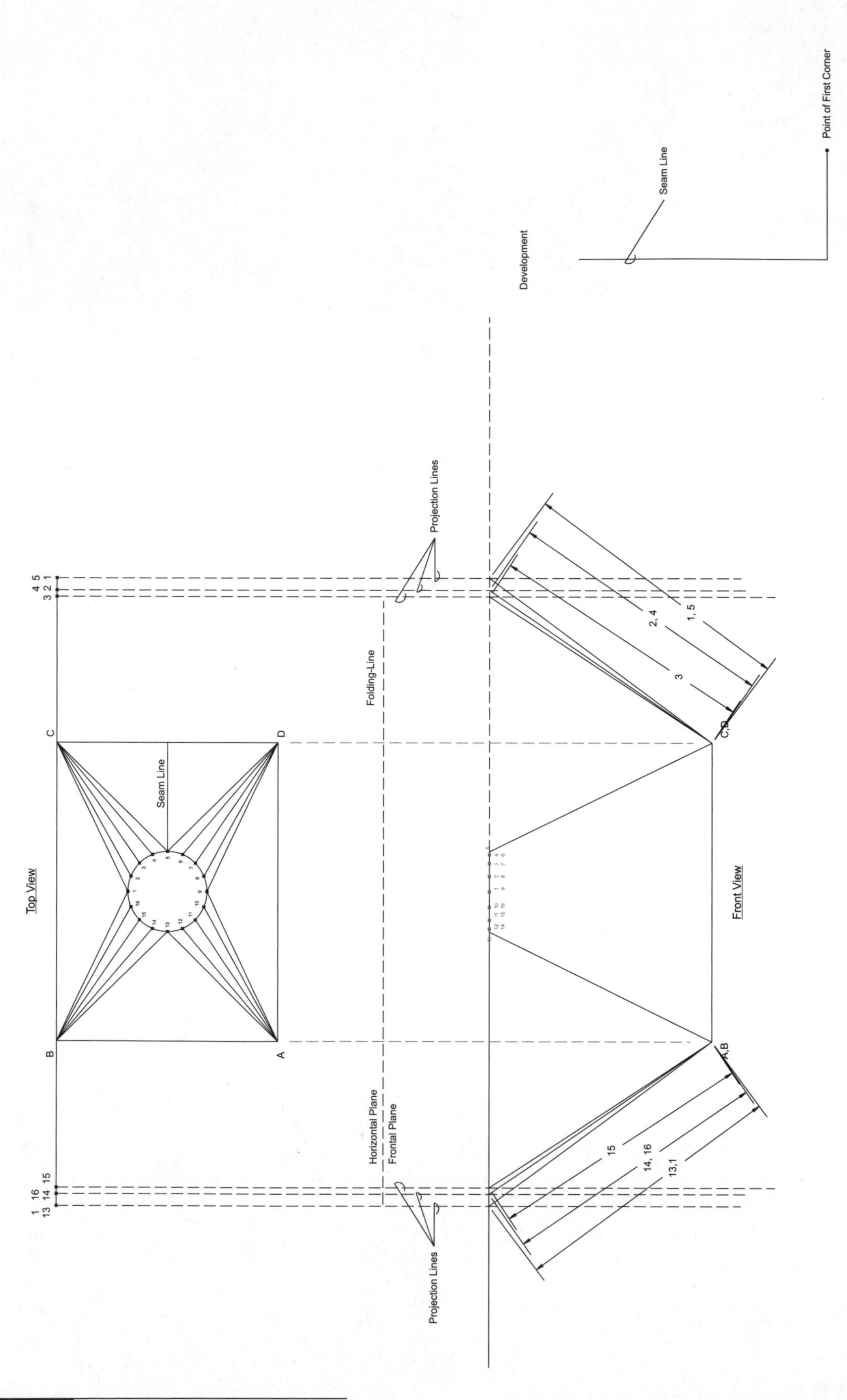


**Figure 6–26k** *Constructing the base line.*

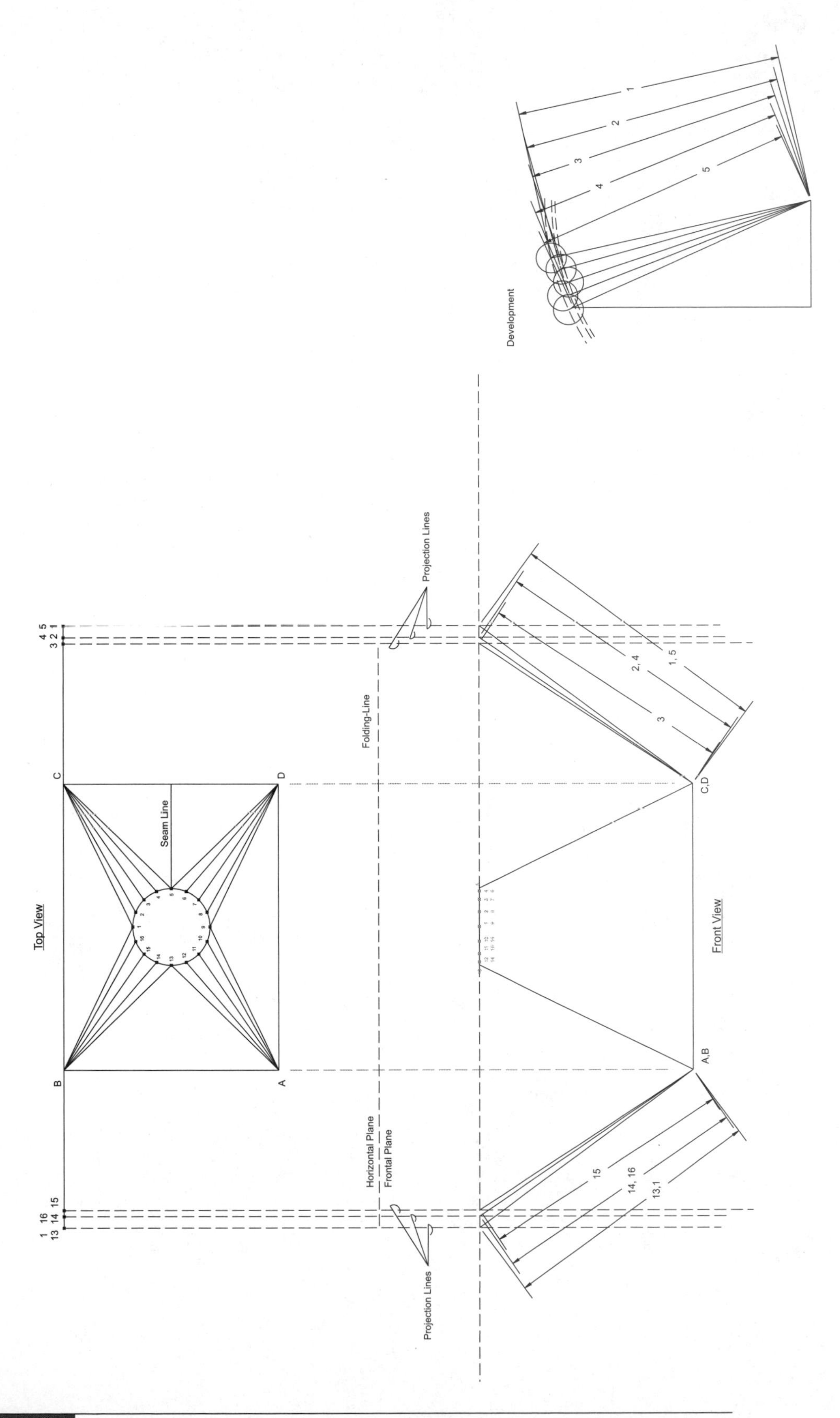


**Figure 6–26L** *Transferring the chordal lengths and bend-lines to the development.*

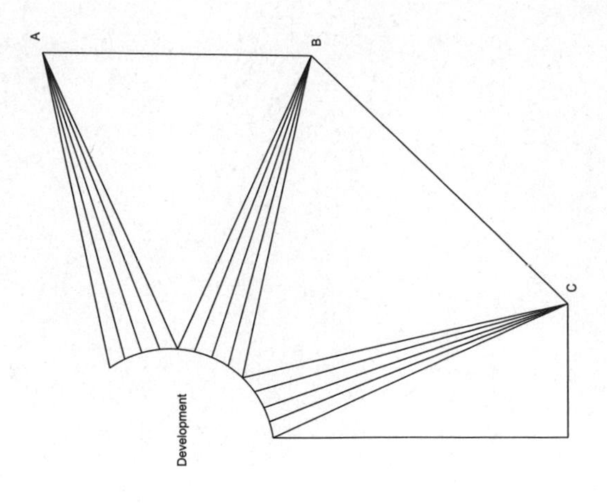


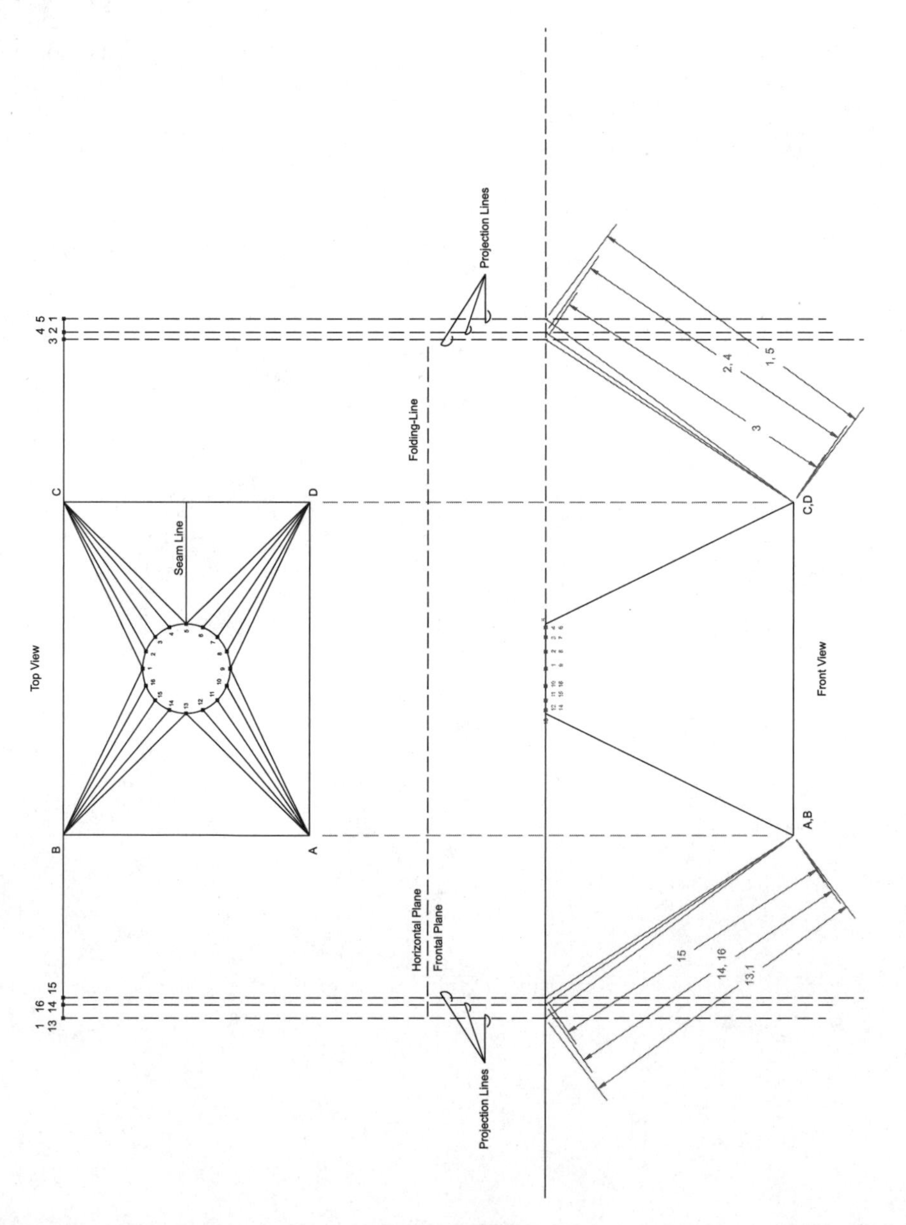


**Figure 6–26m** *Three-quarter complete development.*

## Linking a Database to a Development Using AutoCAD

The ability to track information is essential in the world of engineering. Often, this information includes part numbers, dimensions, cost, material strengths, or any other attribute that maybe associated with an object. The most commonly used tool for this task is the ***database.*** A ***database*** is a collection of text and numerical data stored in a list created and managed by an application called a ***database manager system*** or ***DBMS.*** In reality, AutoCAD is a database manager that stores information about an object (entity type, point coordinates, layer names, styles, fonts, text size, etc.) into a file called a drawing file (.dwg). As a sophisticated database management application, AutoCAD can also manage database files created by other DBMS and/or even link these databases to entities contained within an AutoCAD drawing file. AutoCAD accomplishes this by using a programming format called ***Structured Query Language***, or ***SQL***, that is based on the international standard ISO/IEC 9075-SQL (SQL2).

AutoCAD's ability to link objects in a drawing to a database gives the software unlimited possibilities. For example, a manufacturer of spur gears could link a database containing the following information: number of teeth, pitch diameter, pressure angle, hole depth, chordal addendum, chordal thickness, and working depth to an AutoCAD drawing representing a group of gears. This would allow AutoCAD to display the specifications contained in the database as text within that drawing. Another example is in the production of sheet metal parts, a manufacturer of a group of prism-shaped objects could produce a single development linked to a database, which contains information regarding the dimensioning, thickness, and bend allowances of the objects produced. This offers several advantages: it reduces the amount of drawings required (for similar parts), minimizes time spent searching for information, maximizes productivity, and decreases the margins of error. The following sections are designed to introduce to the inner working relationship between databases, SQL and AutoCAD.

### Introduction to Databases

As stated earlier, a database is a collection of text and numerical data arranged in a list. This list is presented as a ***table*** that can be broken down into ***columns*** and ***rows.*** A ***column*** is the vertical grouping of entities, while the horizontal grouping is known as a ***row.*** When one or more columns are used in a table to identify a specific row, then that group of columns is called a ***key.*** All the information entered into a database is called a ***record.*** This information composes a variable called a ***field.***

### Math Functions

A database has the ability to organize electronic data according to requirements set forth by a technician. The manufacturer of spur gears could have the computer sort the information in his database and arrange it according to any field or combination of fields specified. Not only are databases useful for the organization of data, they can also perform mathematical calculations. For example, given the formula: Allowance = (0.64 × Thickness) + (0.5 × 3.14 × Radius of Bend), a database can be set up to calculate bend allowances. Most databases use the same syntax for the basic math operations, ***addition*** +, ***subtraction*** –, ***multiplication*** *, and ***division*** /. Formulas written for a database typically follow the same format despite the brand used (field name + field name; field name – field name; field name * field name; field name / field name). The differences lie in the syntax that designates a formula. To create a formula in Microsoft Access, the equals sign is placed before the formula. This signifies to the software that a formula is to follow. Using the fields "Thickness" and "Bend_Radius," the formula in the above example would appear as = (Thickness* 0.64) + (0.5 * 3.14* Bend_Radius) when written in database format. This formula would take the record of the field thickness and multiply it by 0.64. Next, it would multiply 0.5 by 3.14 followed by multiplying their product by the record in the field Bend_Radius. Finally, it would add the two products and generate the result.

## Introduction to DBCONNECT

Before AutoCAD can access a database it must first be configured using Microsoft's ODBC and OLEDB programs, which includes creating a data source and providing the necessary driver information. Once the database has been configured, AutoCAD can then access the database's information regardless of its format or the platform that created it. Currently, AutoCAD 2000 supports seven different database systems (Microsoft Access 97, dBASE V and III, Microsoft Excel 97, Oracle 8.0 and 7.3, Paradox 7.0, Microsoft Visual FoxPro 6.0, and SQL Server 7.0 and 6.5), with each varying in the way in which they are configured. Therefore a single procedure for configuring a database cannot be provided in this textbook. However, information regarding the procedures for configuring the different databases can be obtained in the AutoCAD acad_asi.hlp file and ODBC and OLE DB help files. Once a database has been successfully configured, a configuration file is produced containing the extension .udl. By default, AutoCAD searches for configuration files in the Data Links folder, which resides in the folder where AutoCAD was installed, but a different location may be specified by using the OPTIONS command. After this has been completed, AutoCAD can access the database by using the database connectivity manager. The data connectivity manager is a dockable, resizable window that allows the user to connect, edit, view, and even link AutoCAD entities to database table records. The manager also allows the user to execute SQL queries (see Figure 6–27).

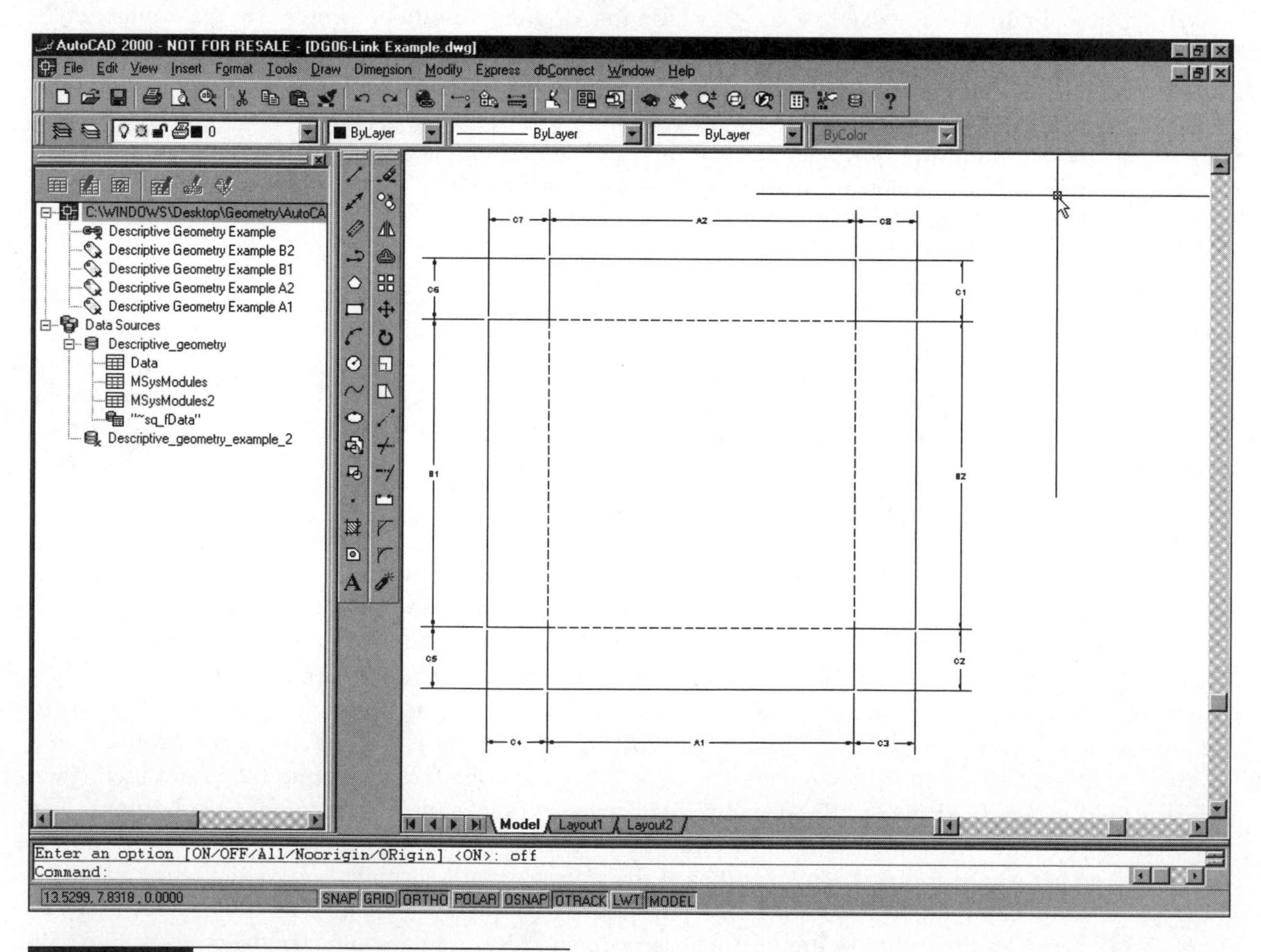


**Figure 6–27** *Database connectivity manager.*

### Launching the Database Connectivity Manager

Once the configuration file has been created, the first step in utilizing the information contained within a database is to launch the database connectivity manager. To initialize the connectivity manager the DBCONNECT command must be used. In addition to launching the dbconnect manager, this command also reconfigures the pull-down menu options to include a new category labeled dbconnect. It is under this category that the user is given the options: configure data sources, templates, queries, links, labels, view data, and synchronize and convert links (see Figure 6–28).

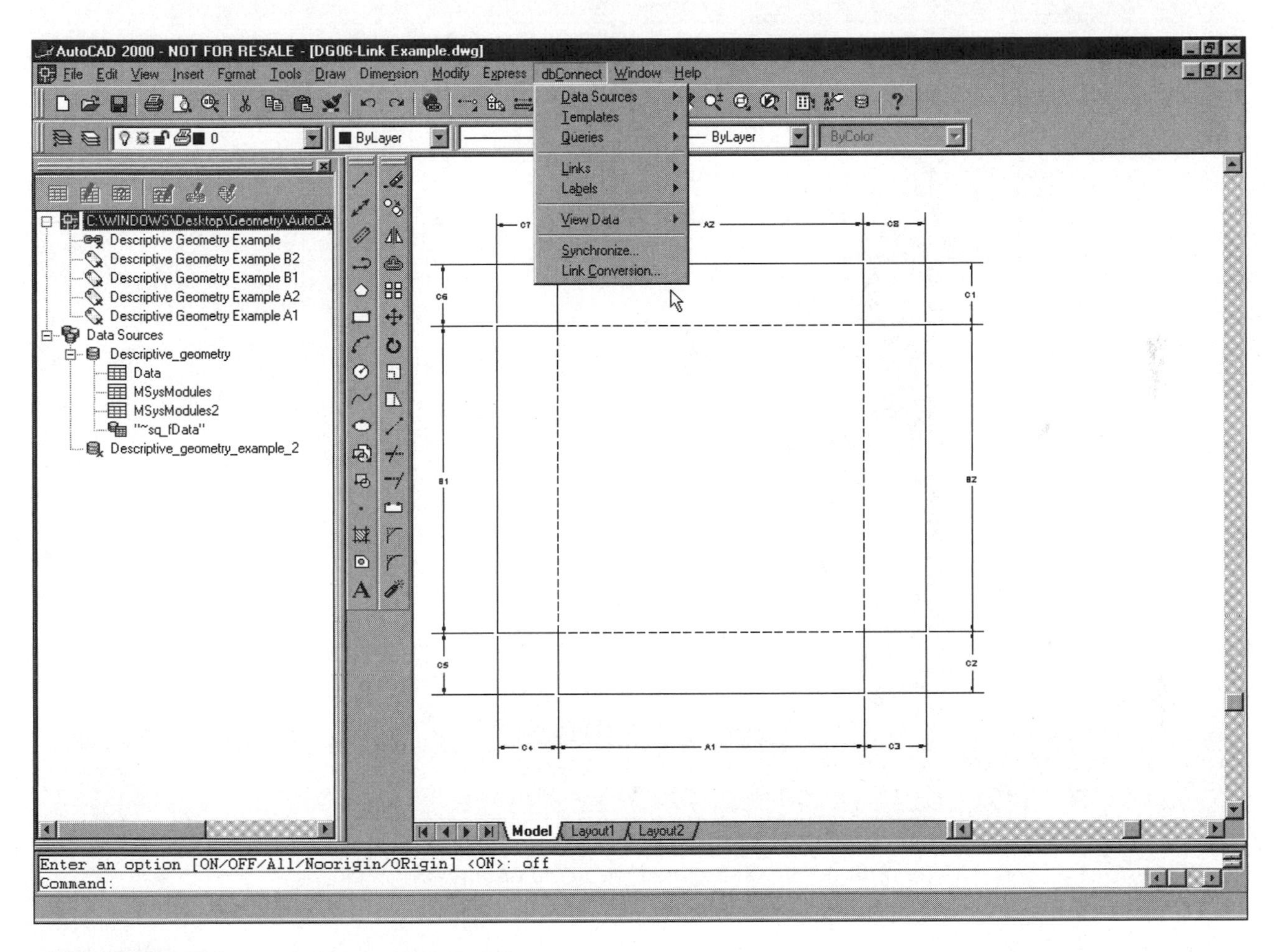


**Figure 6–28** DBCONNECT *pull-down menu*

### Working with the Data Source

After the configuration file has been created and the dbconnect manager is initialized, a connection must be made to the source so that AutoCAD can use its data. There are three ways a data source may be connected. First, by moving the cursor to the left-hand side of the dbmanager, the user can double-click on the data source that is to be connected. Second, by performing a single right-mouse-click on the data source that is to be connected, a menu will appear with "connect" listed as one of its options (see Figure 6–29). Third, by selecting the pull-down menu option "dbconnect," selecting "data sources" and then "connect," a dialog box appears where all data sources are displayed, and the user double-clicks on the name of the desired data source (see Figure 6–30).

Once a data source has been configured, it is sometimes necessary to change one or more of its parameters. Autodesk has provided a means of changing the configuration file (data source) without having to exit AutoCAD. Changes can be made to the data source by either performing a single right-hand mouse-click on the data source name or selecting "configure" from the dbconnect pull-down menu (see Figure 6–31).

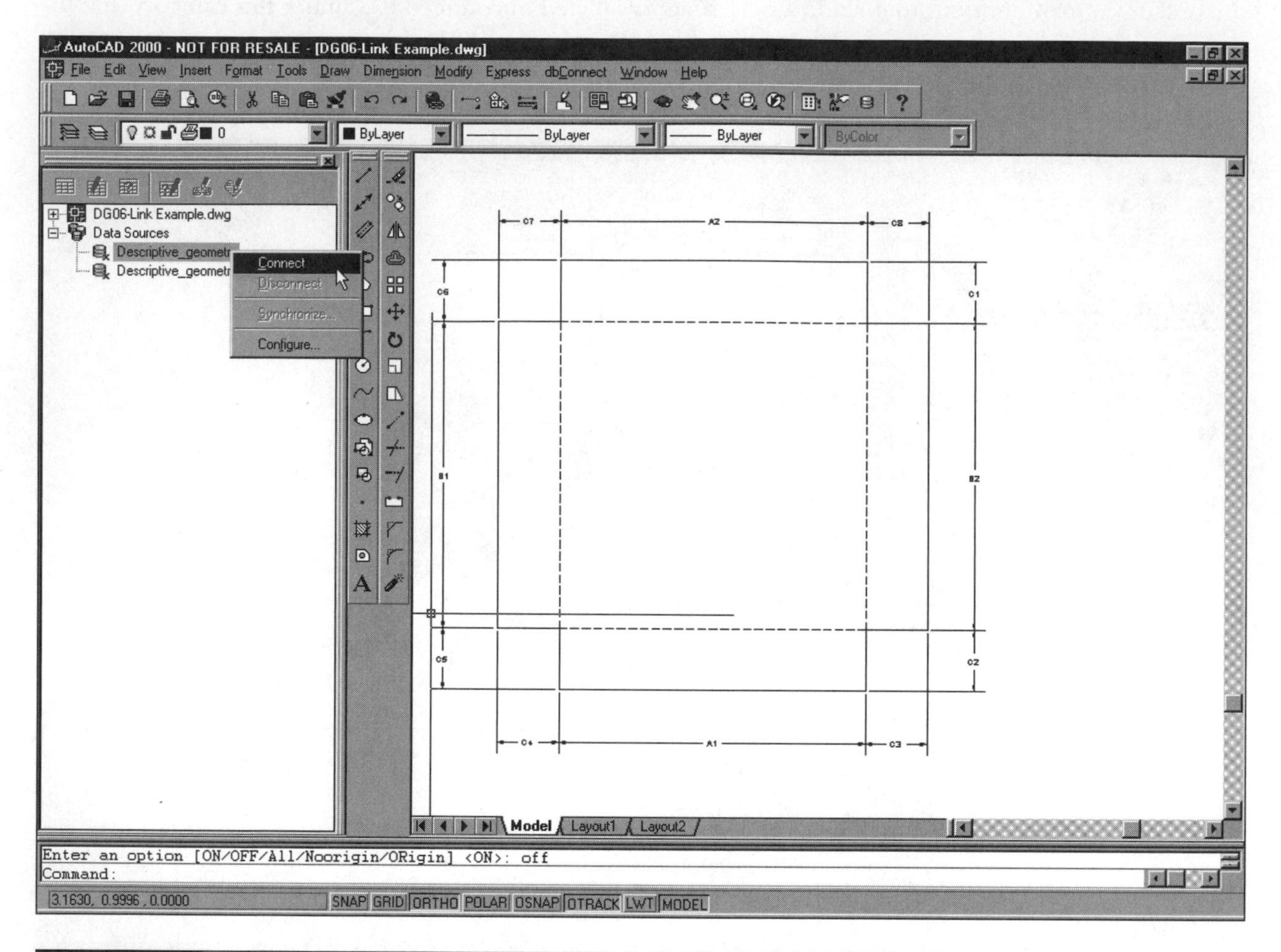


**Figure 6–29** *Connecting a data source using a single right-hand mouse click.*

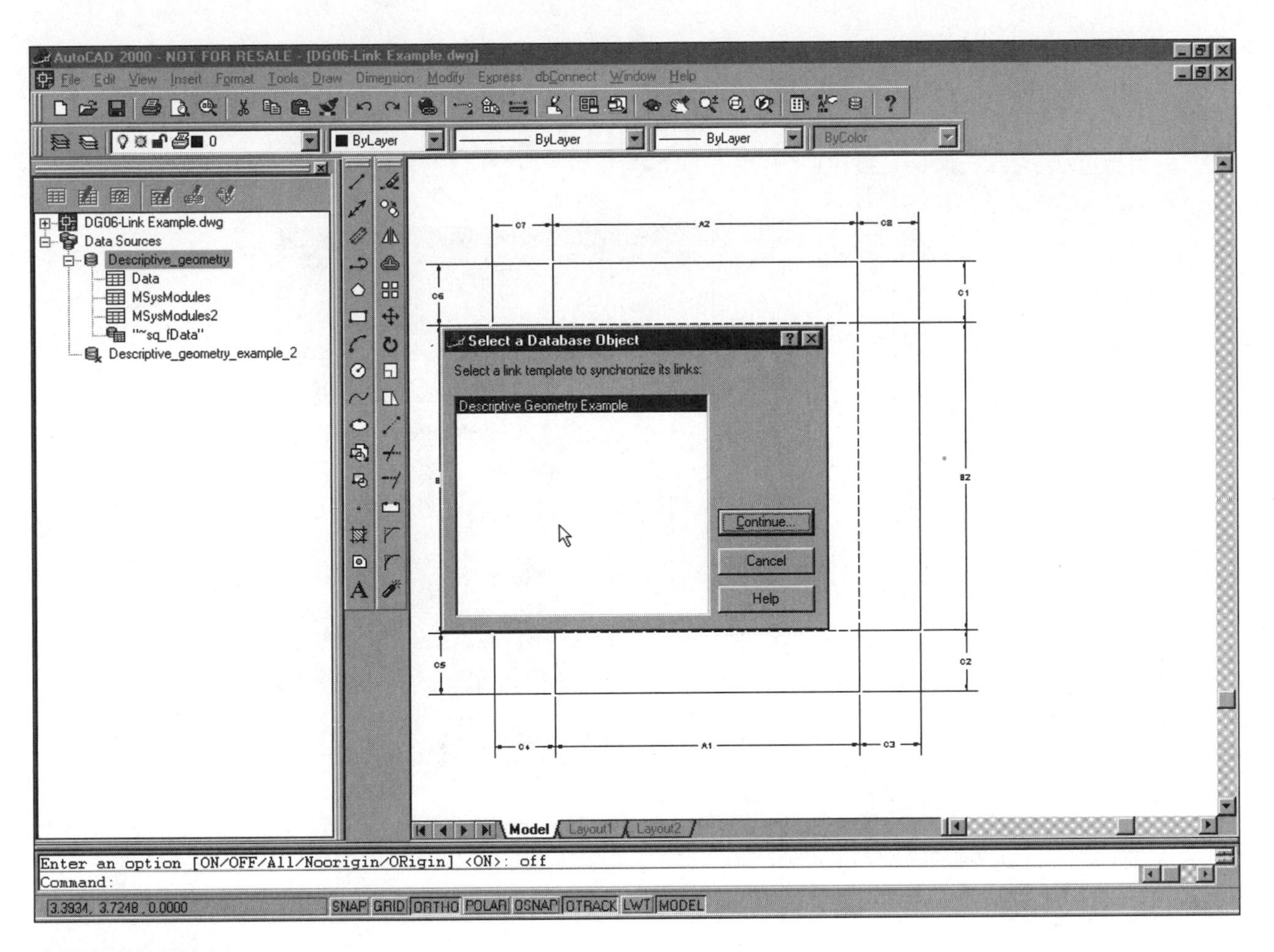


**Figure 6–30** *"Select" dialog box.*

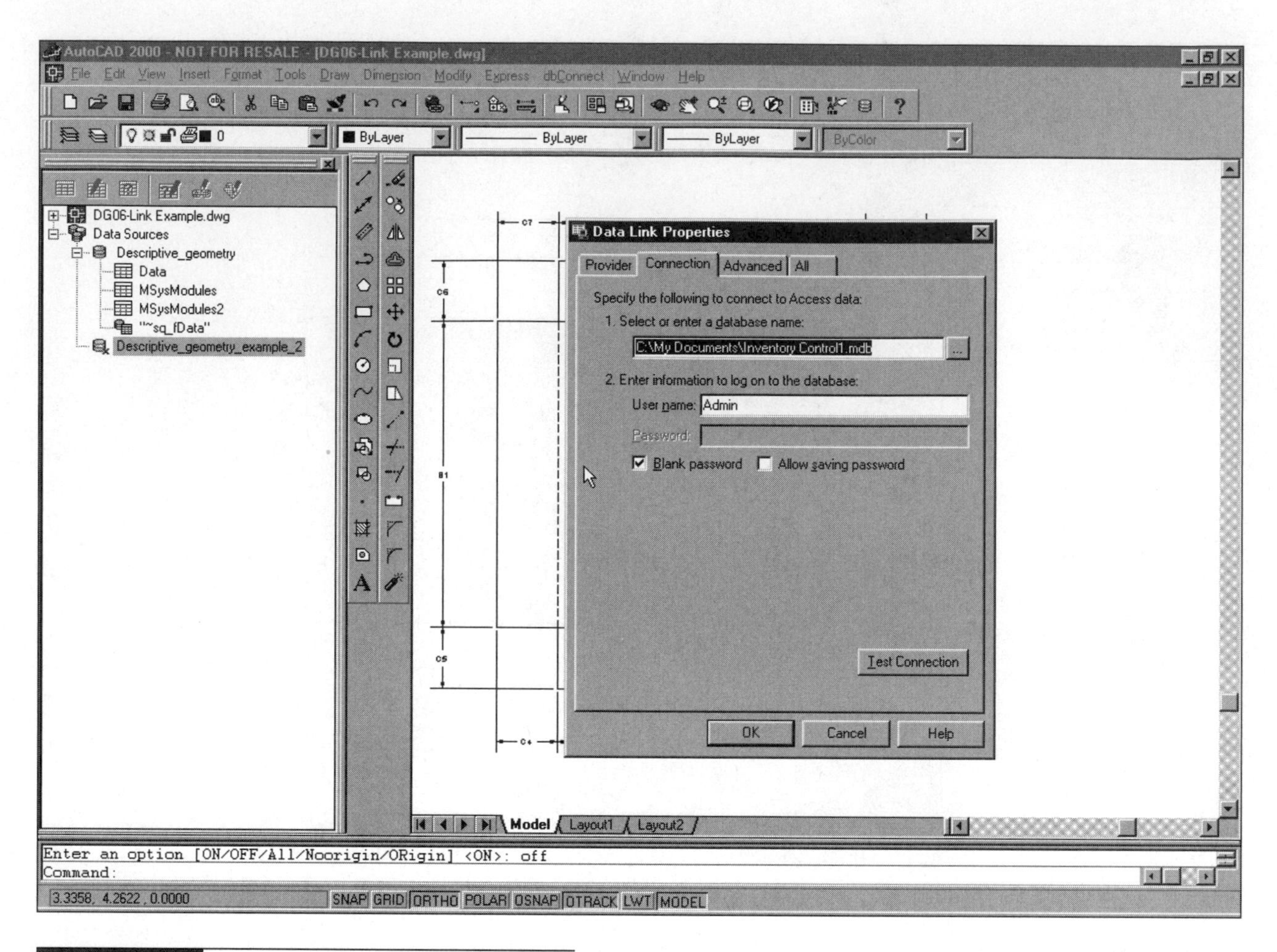


**Figure 6–31** *Data link properties dialog box.*

#### Editing and Viewing Information Contained within a Database

It is only after a data source has been connected that the user is able to view or edit information contained within a database. To view the information contained within a database table, the user would perform a single right-mouse-click on the name of the table to view activating a menu that offers "view table" as an option. When a database table is to be edited, use the same procedure to view a table or double-click on the table name to produce the same results (see Figure 6–32).

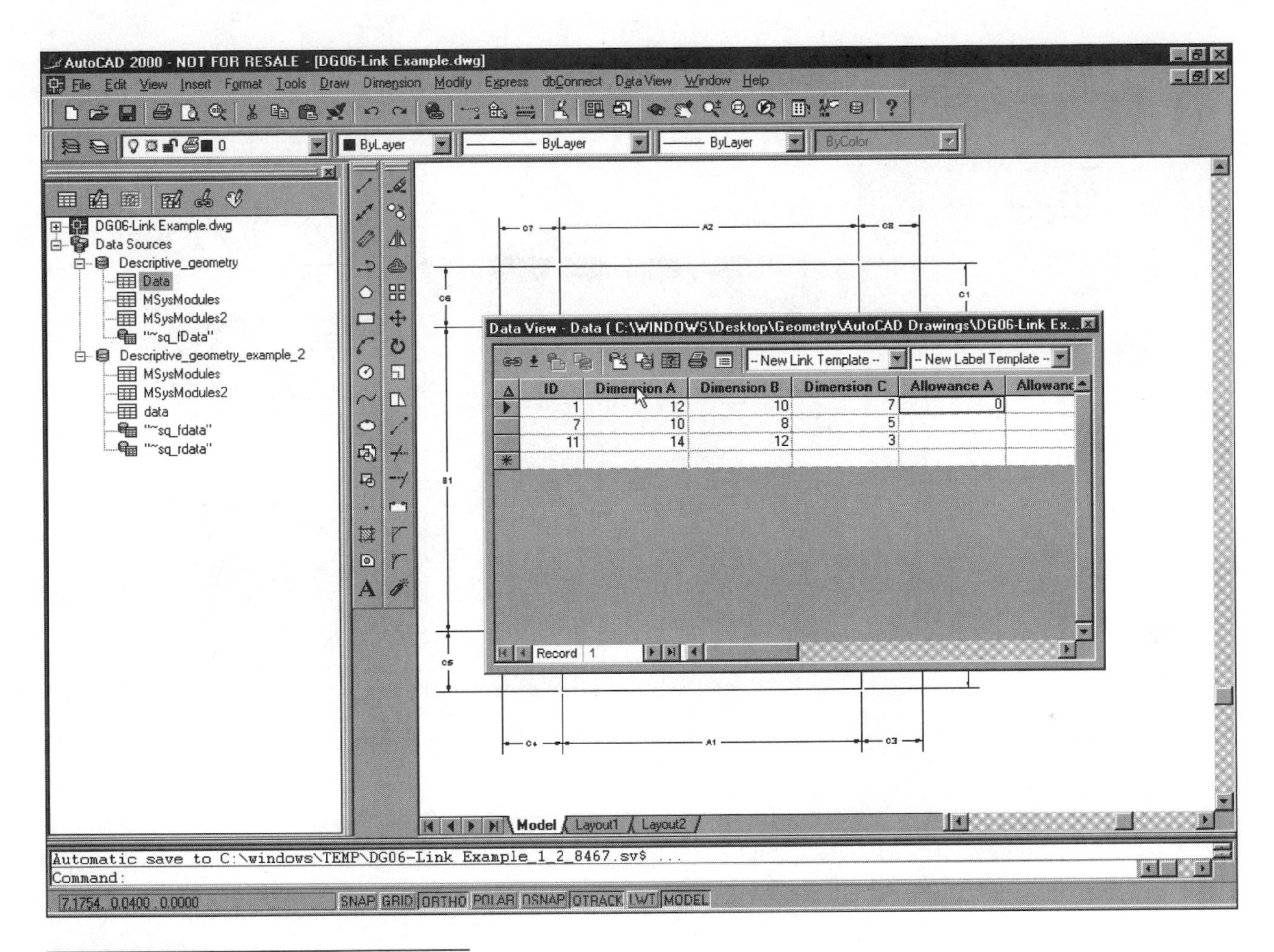


**Figure 6–32** *Data view dialog box.*

### Creating a Link to a Graphical Object

The main objective of the connectivity feature is to associate database information with AutoCAD entities, to provide a means of managing both AutoCAD and database information in one package. To establish an association between AutoCAD entities and database information, a link must be created. A link can only be created between graphical objects (lines, polylines, etc.), and once a link is created it is tightly associated with the object to which it is connected. When a link is created between an AutoCAD graphical object and a database table, that link is a dynamic link. In other words, if information is changed in a database table that is linked to an AutoCAD entity, then the information stored within the drawing concerning that object is updated to match the database table. Before a link can be established between an AutoCAD entity and a database table record, a link template must be created. A link template is used to identify which fields from a database table are associated with the links that are connected to the AutoCAD graphical entity. To create a link template, the user can either right-click on the table name displayed in the dbconnect and then choose "New Link Template" or select the dbconnect pull-down menu followed by "templates" and then "new link template" (see Figure 6–33). Once the link template has been created, a link is constructed by opening the table in which the template was created and selecting the link icon located in the upper left-hand corner of the data view dialog box (see Figure 6–34).

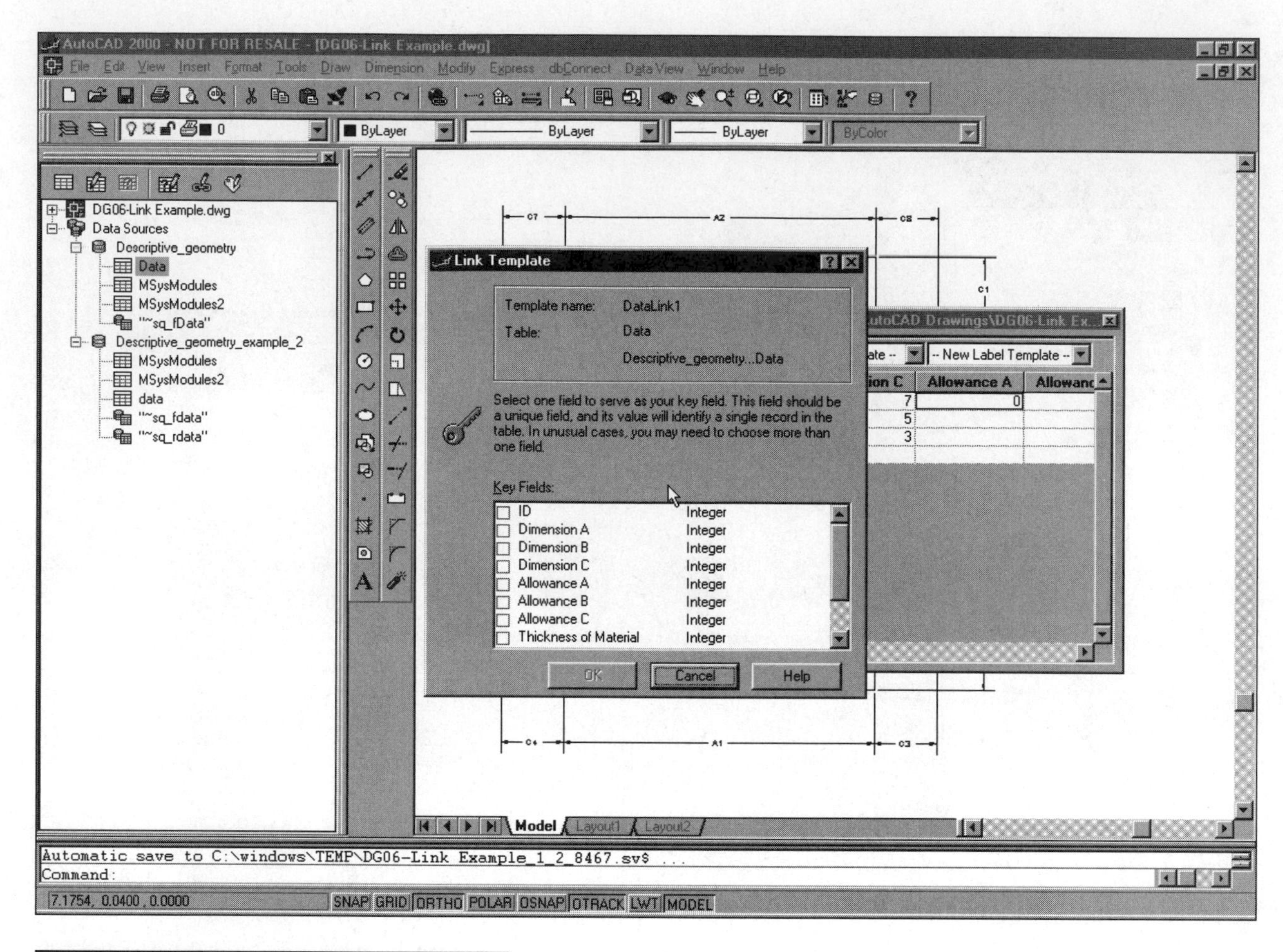


**Figure 6–33** *Link template dialog box.*

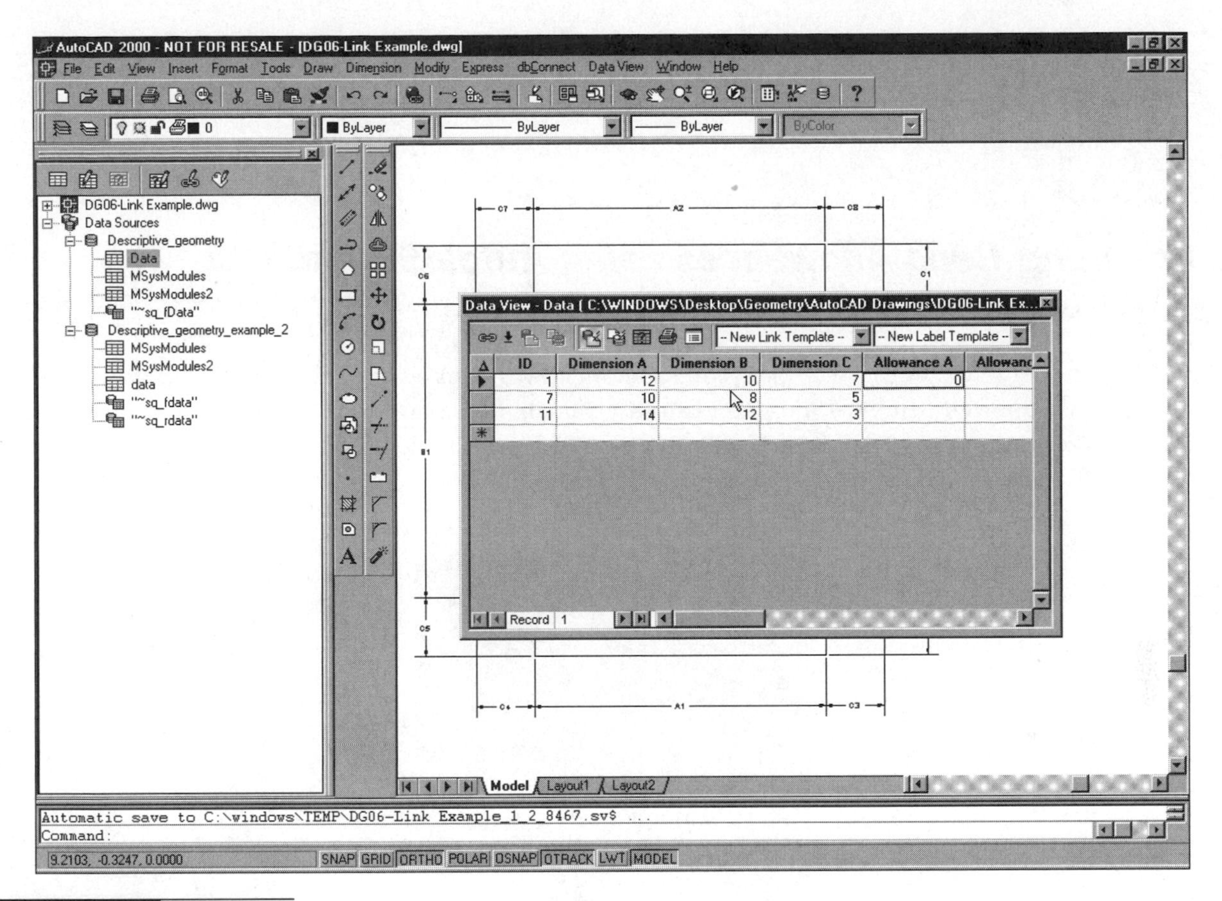


**Figure 6–34** *Link icon.*

## Sheet Metal Drawings

By combining the information presented in the previous sections concerning bend allowances, developments, and databases with elements from Chapter one, sheet metal drawings can now be constructed. Sheet metal drawings are used by the fabrication department to build the object presented. The complexity and accuracy of the drawing depend upon the part to be constructed. Sheet metal parts can be divided into one of two categories, precision and non-precision. A precision part is manufactured to very tight tolerances; therefore bend allowances must be calculated. These parts are commonly associated with the electronics industry, where space is a limiting factor. Non-precision parts are those that are not produced to close tolerance, and bend allowances are not usually calculated. These products are commonly associated with the HVAC industry.

The production of sheet metal drawings is basically the same for both precision and non-precision parts. The difference lies in the calculation of bend-allowances, but both drawings contain a development of the part, and at least one view showing the part in its final form (usually an isometric). The development shows all necessary bend-lines, dimensions, and any special instructions regarding the steps or processes involved in completing the part. When these drawings are produced manually, a scale is selected before the drawing is started (depending upon the size of the part, a scale other that 1:1 may be used). When these drawings are produced with AutoCAD, they are constructed full scale in model space with one or more layouts constructed in paper space containing the required views and at a scale dictated by the paper size on which the layout is to be plotted. Today, the concept of a paperless manufacturing environment requires more than ever that these drawings be created full scale. Often, another computer application uses these drawings to generate source code for a CNC (Computer Numeric Control) operation. Databases are incorporated into these drawings that not only contain dimensioning information, but also contain special instructions regarding the manufacturing steps and/or processes involved.

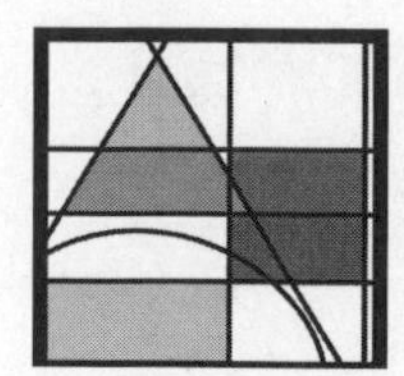

# advanced applications

## Calculating Bend Allowances using AutoLISP and DCL Programming Languages

Calculating the bend allowances of a precision sheet metal part can be a time-consuming task. This is one area where a customized AutoLISP program can save valuable time and even eliminate possible mistakes. Provided below is an example of an AutoLISP program designed to calculate bend–allowances (see Figures 6–35 and 6–36).

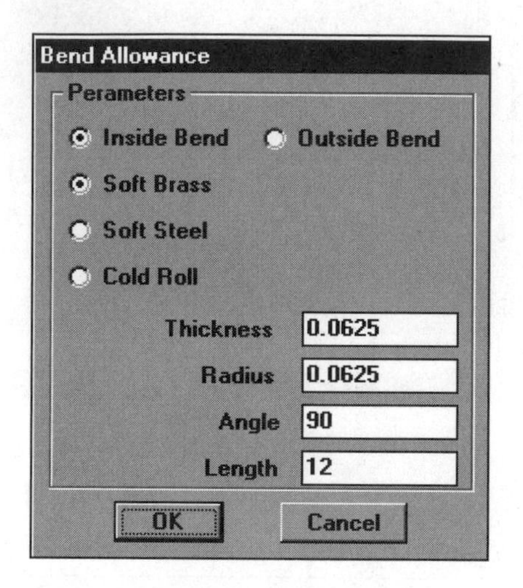


**Figure 6–35** *Bend allowance dialog box*

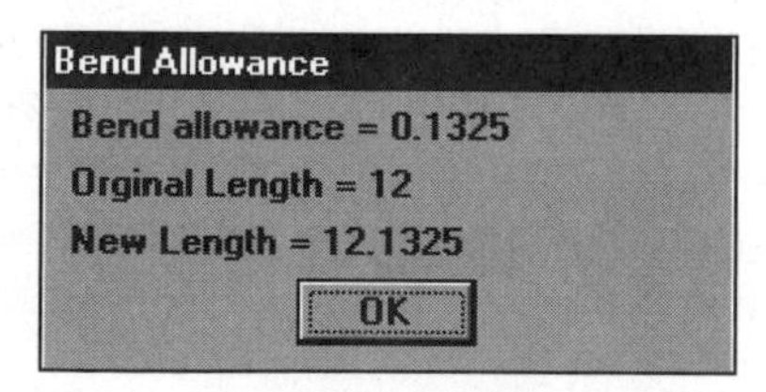


**Figure 6–36** *Bend allowance results*

## AutoLISP Program

```
;;;*********************************************************************
;;;
;;;     Program Name: DG06.lsp
;;;
;;;     Program Purpose: This program allows the user to calculate the
;;;                      necessary bend allowance for the three classifications
```

```
;;;                         of metal described in chapter 6.  The program uses
;;;                         formulas based on the formulas found in the Machinery's
;;;                         Handbook 24th edition.
;;;
;;;     Program Date: 10/25/98
;;;
;;;     Written By: James Kevin Standiford
;;;
;;;*************************************************************************
;;;*************************************************************************
;;;
;;;                     Main Program
;;;
;;;*************************************************************************
(defun c:bend (/ all mat rad fir sec thi ang cold soft softs tog len in
                out)
  (setq
    dcl_id4
     (load_dialog
       "c:/windows/desktop/geometry/student/autolisp programs/DG06.dcl"
     )
  )
  (if (not (new_dialog "bend" dcl_id4))
    (exit)
  )
  (set_tile "Soft" "1")
  (set_tile "in" "1")
  (set_tile "mat" "0.0625")
  (set_tile "rad" "0.0625")
  (set_tile "len" "12")
  (set_tile "ang" "90")
  (action_tile "accept" "(retreve) (done_dialog)")
  (start_dialog)
  (if (= tog 1)
    (progn
      (if (not (new_dialog "lay" dcl_id4))
       (exit)
      )
      (setq fir1 "Bend allowance = "
            sec1 "Orginal Length = "
            thi1 "New Length = "
      )
      (set_tile "fir" (strcat fir1 (rtos all)))
      (set_tile "sec" (strcat sec1 len))
      (set_tile "thi" (strcat thi1 (rtos (+ leng all))))
      (action_tile "accept" "(done_dialog)")
      (start_dialog)
    )
  )
  (princ)
)
(defun retreve ()
  (setqSoft  (get_tile "Soft")
```

```
        SoftS (get_tile "SoftS")
        Cold  (get_tile "Cold")
        out   (get_tile "out")
        in    (get_tile "in")
        mat   (get_tile "mat")
        rad   (get_tile "rad")
        ang   (get_tile "ang")
        len   (get_tile "len")
   )
   (if (= out "1")
     (setq leng (- (atof len) (+ (atof rad) (atof mat))))
   )
   (if (= in "1")
     (setq leng (atof len))
   )
   (if (= soft "1")
     (setq all (* (+ (* 0.55 (atof mat)) (* 0.5 pi (atof rad)))
                 (/ (atof ang) 90.0)
              )
     )
   )
   (if (= Softs "1")
     (setq all (* (+ (* 0.64 (atof mat)) (* 0.5 pi (atof rad)))
                 (/ (atof ang) 90.0)
              )
     )
   )
   (if (= cold "1")
     (setq all (* (+ (* 0.71 (atof mat)) (* 0.5 pi (atof rad)))
                 (/ (atof ang) 90.0)
              )
     )
   )
   (setq tog 1)
)
(princ "\nTo excute enter BEND at the command prompt ")
(princ)
```

## Dialog Control Language Program

```
//%%%%%%%%%%%%%%%%%%%%%%%%%%%%%%%%%%%%%%%%%%%%%%%%%%%%%%%%%%%%%%%%%%%%%%%%
//
//       Activates dialog box
//
//       Descriptive Geometry Chapter 6 DCL File Bend Allowances
//
//        Calculates Bend Allowances
//
//%%%%%%%%%%%%%%%%%%%%%%%%%%%%%%%%%%%%%%%%%%%%%%%%%%%%%%%%%%%%%%%%%%%%%%%%
bend : dialog {
label = "Bend Allowance";
    : boxed_column {
      label = "Perameters";
```

```
      children_fixed_width = true;
      children_alignment = left;
    : radio_row {
    : radio_button {
      key = "in";
      label = "Inside Bend";
    }
    : radio_button {
      key = "out";
      label = "Outside Bend";
    }
  }
    : radio_column {
    : radio_button {
      key = "Soft";
      label = "Soft Brass";
    }
    : radio_button {
      key = "SoftS";
      label = "Soft Steel";
    }
    : radio_button {
      key = "Cold";
      label = "Cold Roll";
    }
    }
    : edit_box {
      label = "Thickness";
      alignment = right;
      key = "mat";
      edit_width = 10;
    }
    : edit_box {
      label = "Radius";
      alignment = right;
      key = "rad";
      edit_width = 10;
    }
    : edit_box {
      label = "Angle";
      alignment = right;
      key = "ang";
      edit_width = 10;
    }
    : edit_box {
      label = "Length";
      alignment = right;
      key = "len";
      edit_width = 10;
    }
  }
    is_default = true;
    ok_cancel;
```

```
}
lay : dialog {
      label = "Bend Allowance";
    : text {
      key = "fir";
      width = 30;
    }
    : text {
      key = "sec";
      width = 30;
    }
    : text {
      key = "thi";
      width = 30;
    }
    is_default = true;
    ok_only;
}
```

# Review Questions

Answer the following questions on a separate sheet of paper. Your answers should be as complete as possible.

1. What is a bend allowance and why is it important?

2. List the steps involved in constructing the following developments: prism, truncated right cone, transitional fitting.

3. Explain the difference between precision and non-precision sheet metal parts.

4. Define the following terms: table, column, row, key, field, and record.

5. Name the categories of materials from which a part may be manufactured.

6. What are the advantages of linking a database to an AutoCAD entity?

7. Given the following information, calculate the bend allowances:
   Material = Soft Brass, Radius = 0.125, Angle = 45°, Thickness = 0.0625
   Material = Cold Roll Steel, Radius = 0.25, Angle = 34°16', Thickness = 0.125
   Material = Soft Steel, Radius = 0.375, Angle = 90°, Thickness = 0.0313

8. List the steps involved in constructing a development of the following: truncated prism, truncated pyramid, oblique prism.

9. What is a radial line development and when is it used?

10. What are developments and why are they used?

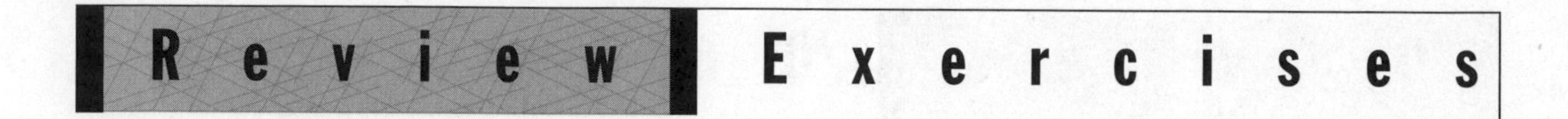


## Parallel Line Developments

Using a separate sheet of paper, complete the following developments. All drawings with an * can be found on the student CD-ROM. Create a development for each of the following parts.

***#1**

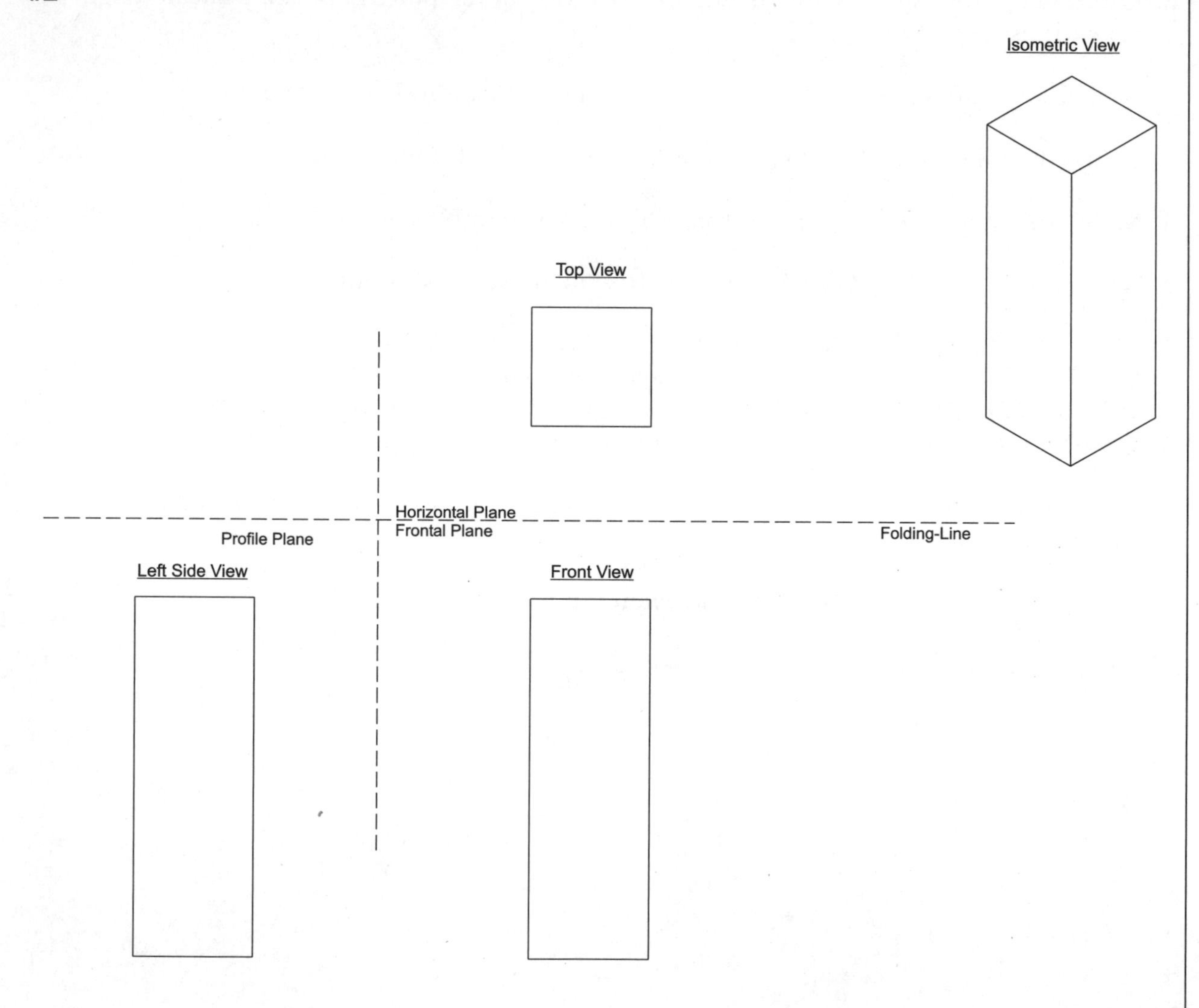

*#2

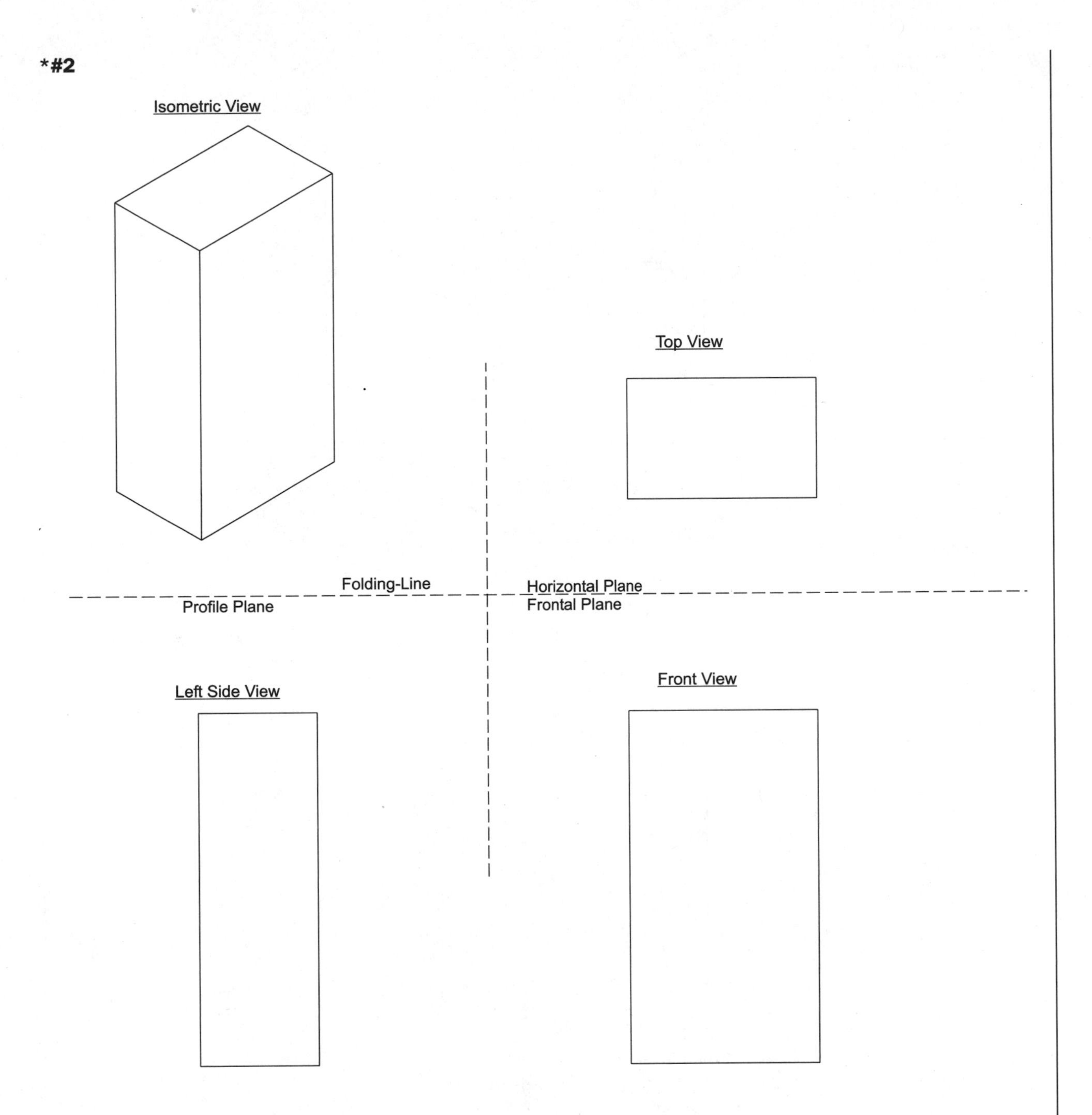
Isometric View
Top View
Folding-Line
Horizontal Plane
Profile Plane
Frontal Plane
Left Side View
Front View

***#3**

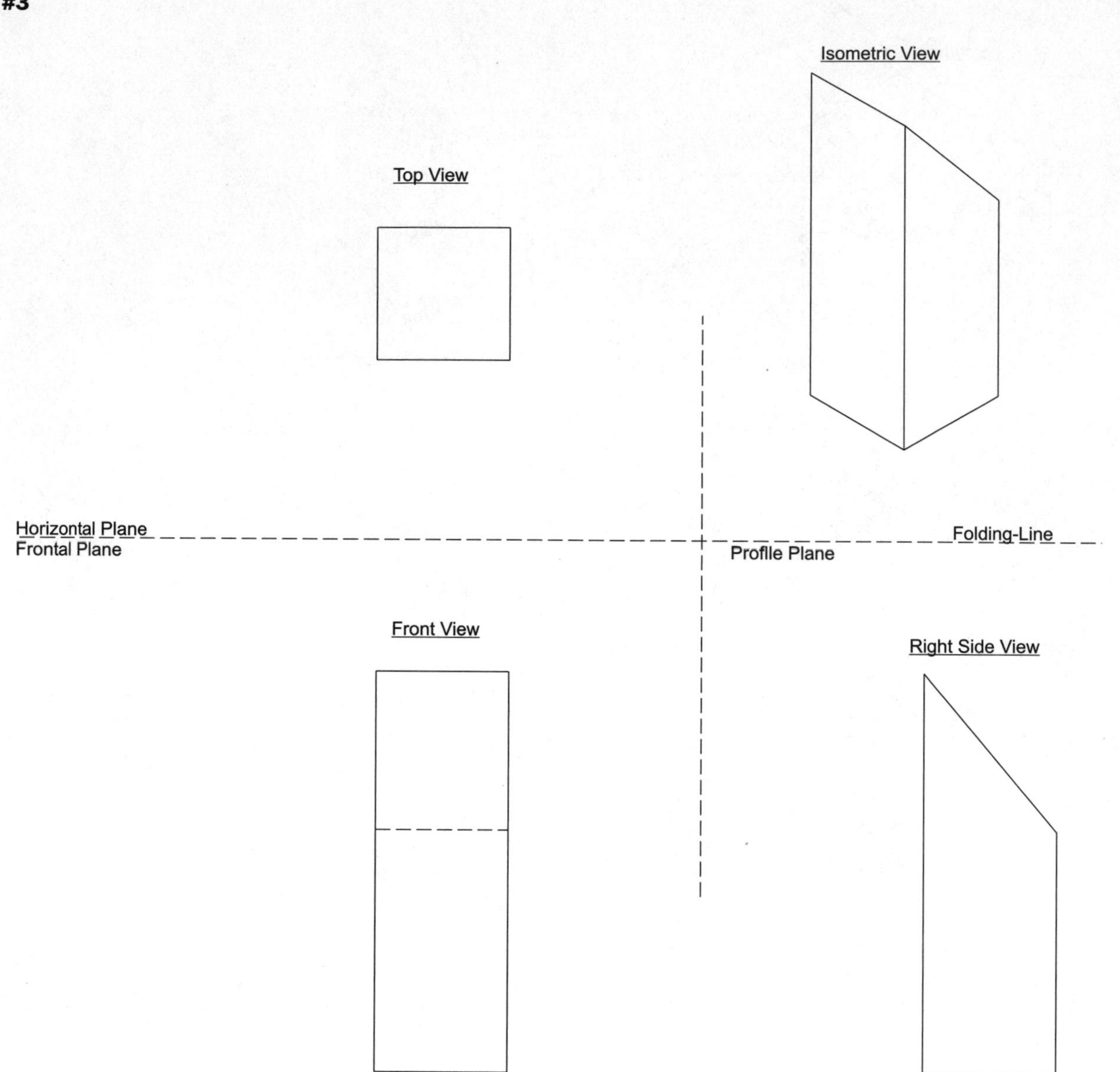

Isometric View
Top View
Horizontal Plane
Frontal Plane
Folding-Line
Profile Plane
Front View
Right Side View

*#4

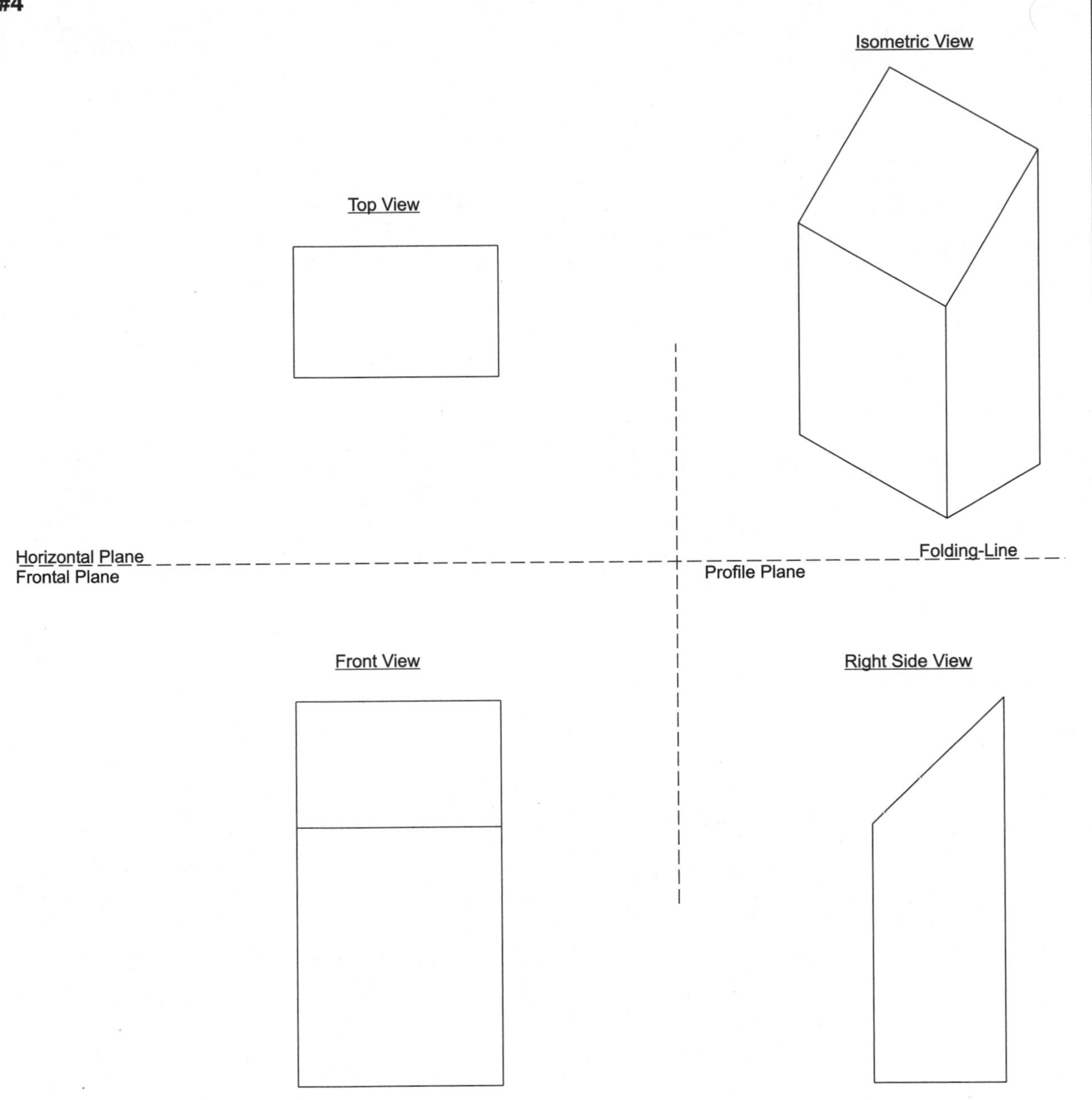
Isometric View
Top View
Horizontal Plane
Frontal Plane
Folding-Line
Profile Plane
Front View
Right Side View

*#5

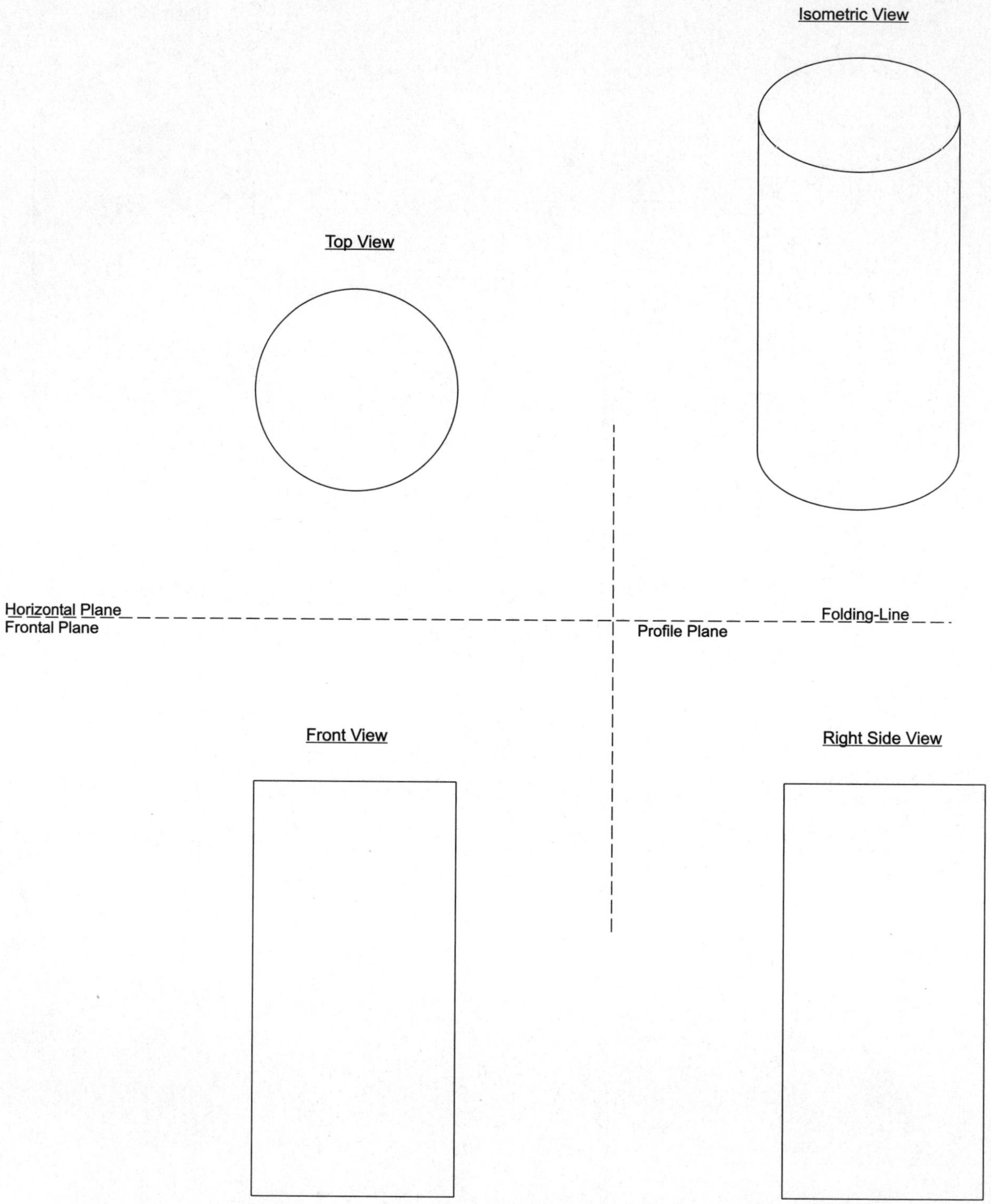

Isometric View

Top View

Horizontal Plane

Frontal Plane

Folding-Line

Profile Plane

Front View

Right Side View

*#6

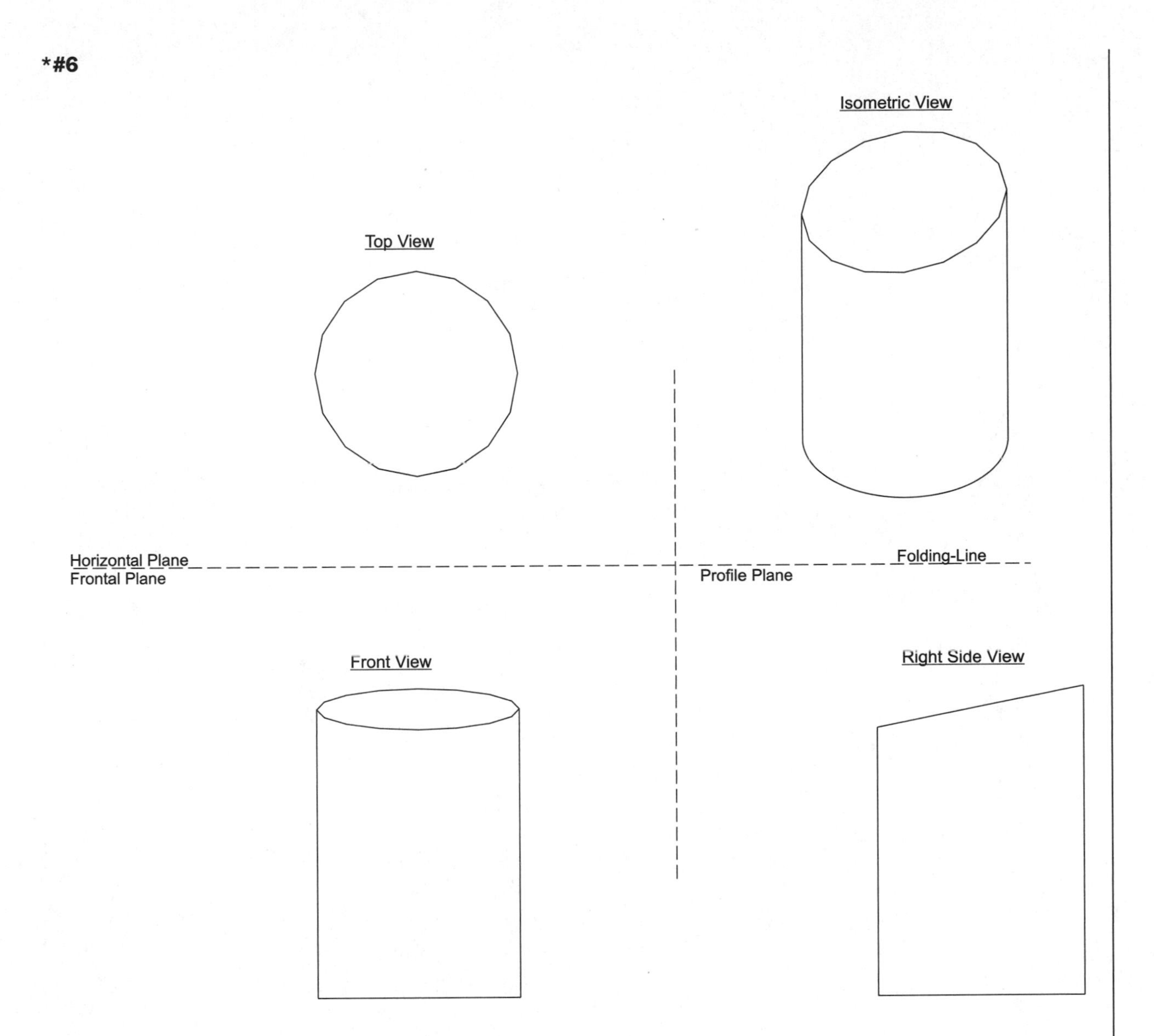
Isometric View
Top View
Horizontal Plane
Frontal Plane
Folding-Line
Profile Plane
Front View
Right Side View

***#7**

Folding-Line

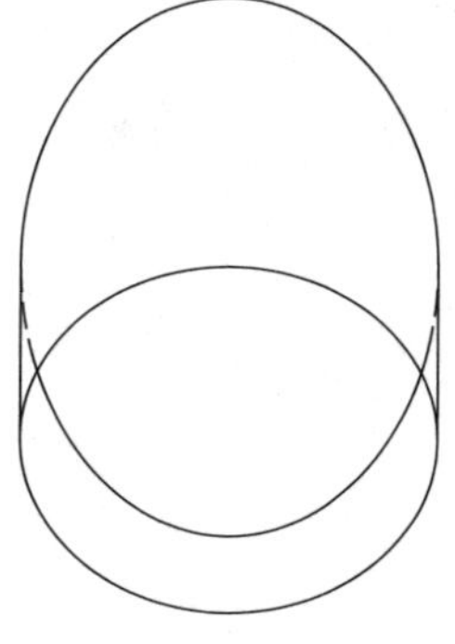

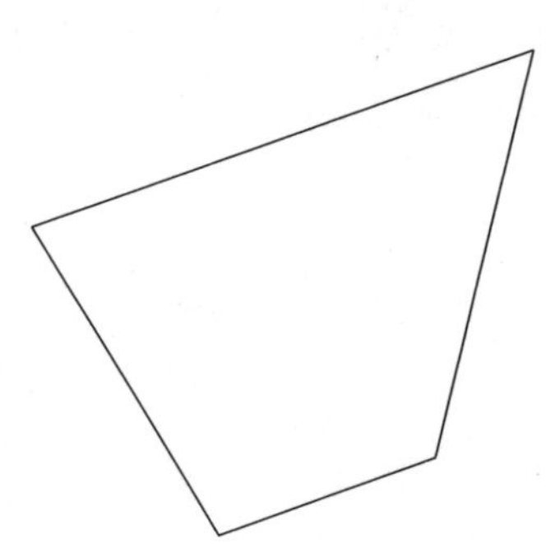

Horizontal Plane

Frontal Plane

## Radial Line Developments

Create a development for each of the following parts.

***#8**

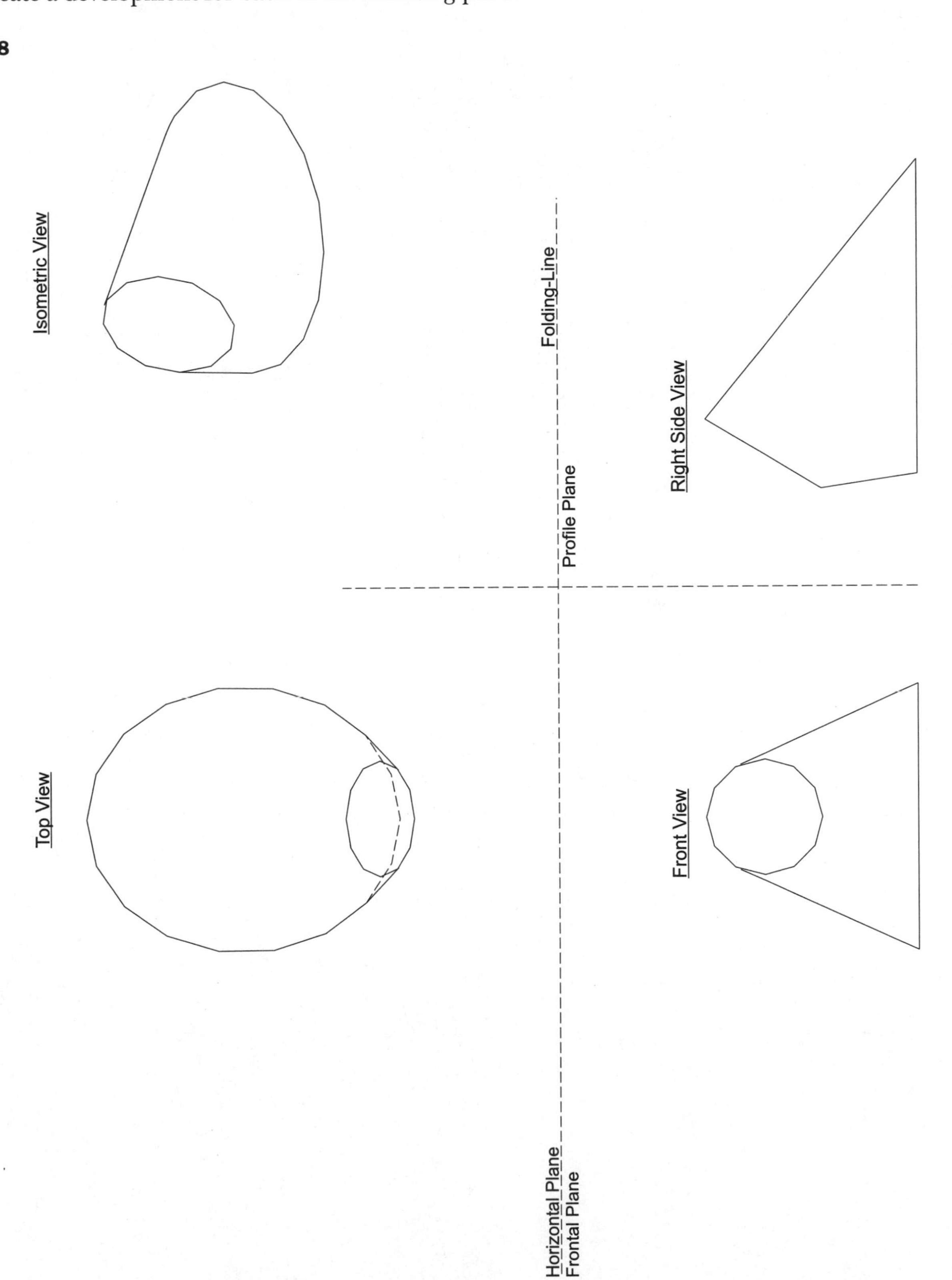

*#9

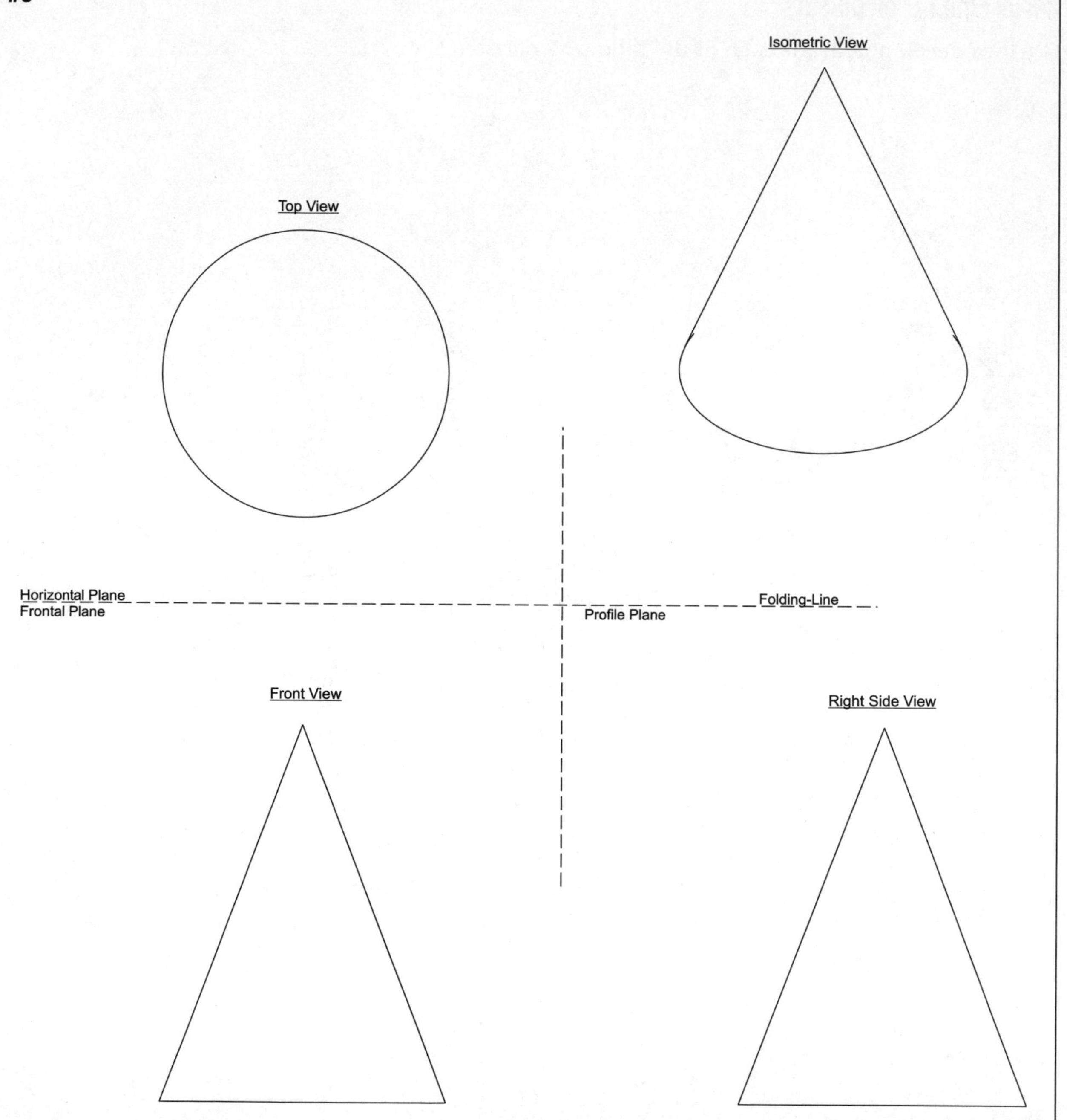

Isometric View
Top View
Horizontal Plane
Frontal Plane
Profile Plane
Folding-Line
Front View
Right Side View

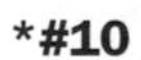

*#10

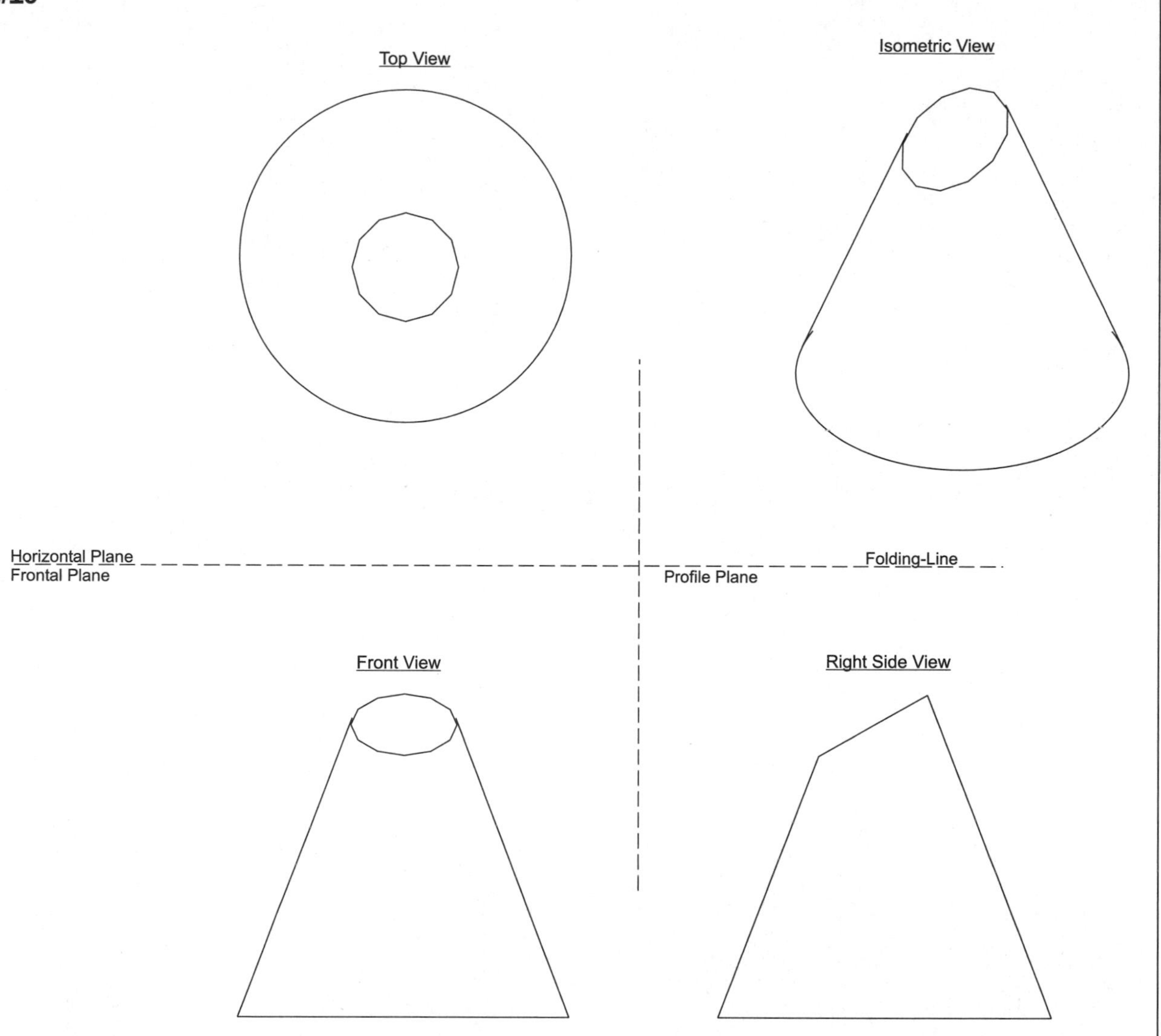
Top View
Isometric View
Horizontal Plane
Frontal Plane
Folding-Line
Profile Plane
Front View
Right Side View

***#11**

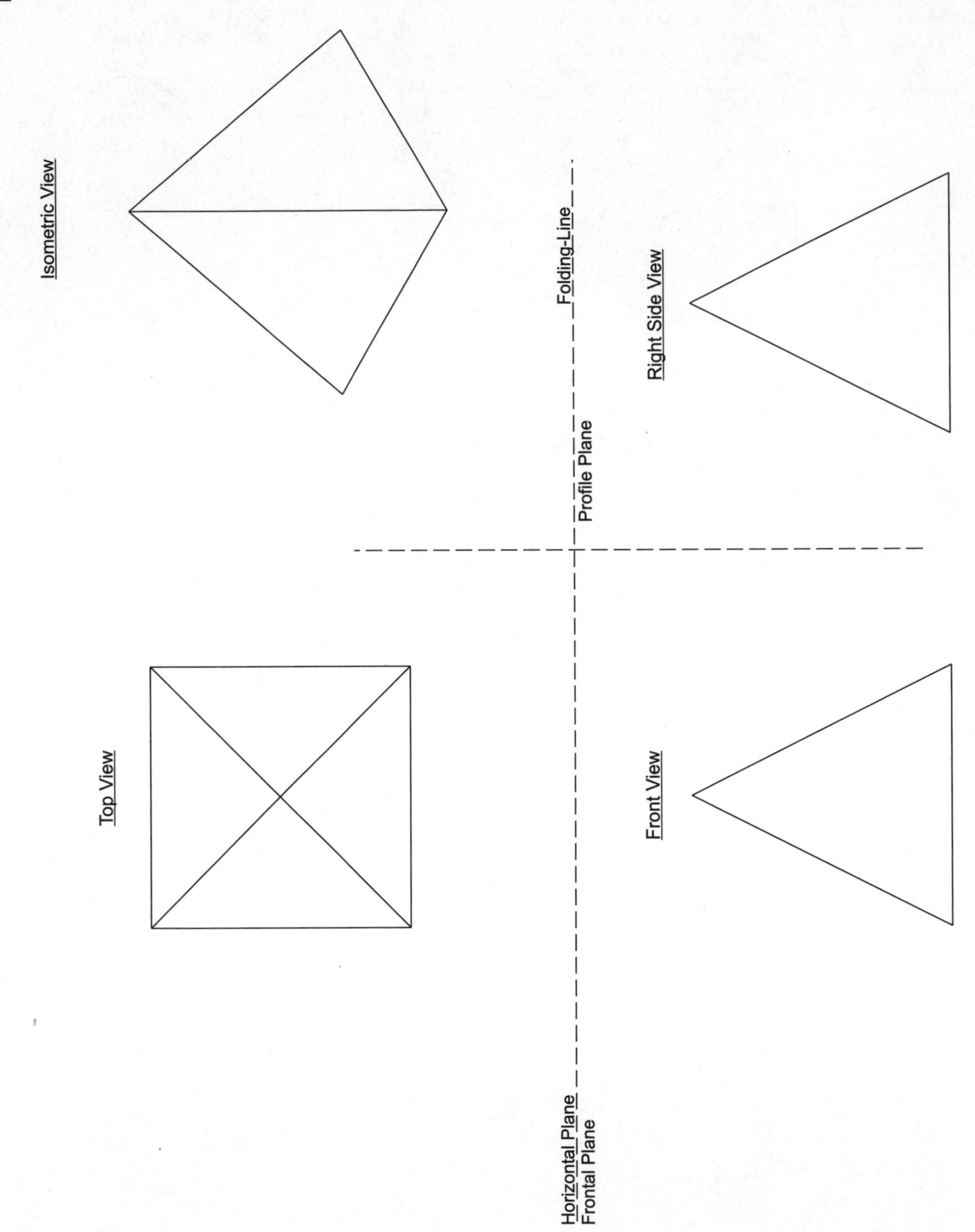

Isometric View
Folding-Line
Right Side View
Profile Plane
Top View
Front View
Horizontal Plane
Frontal Plane

*#12

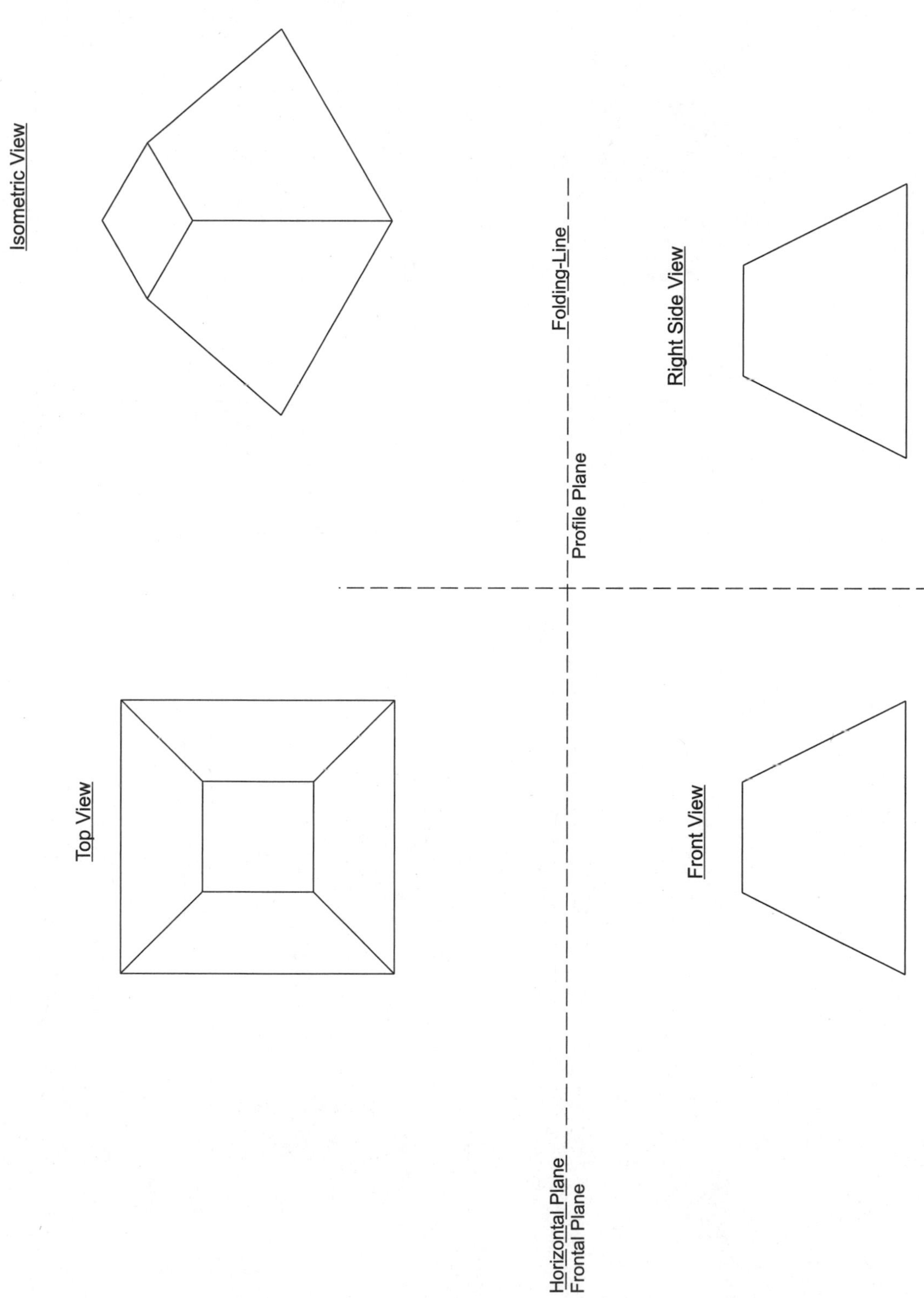
Isometric View
Folding-Line
Profile Plane
Right Side View
Top View
Front View
Horizontal Plane
Frontal Plane

*#13

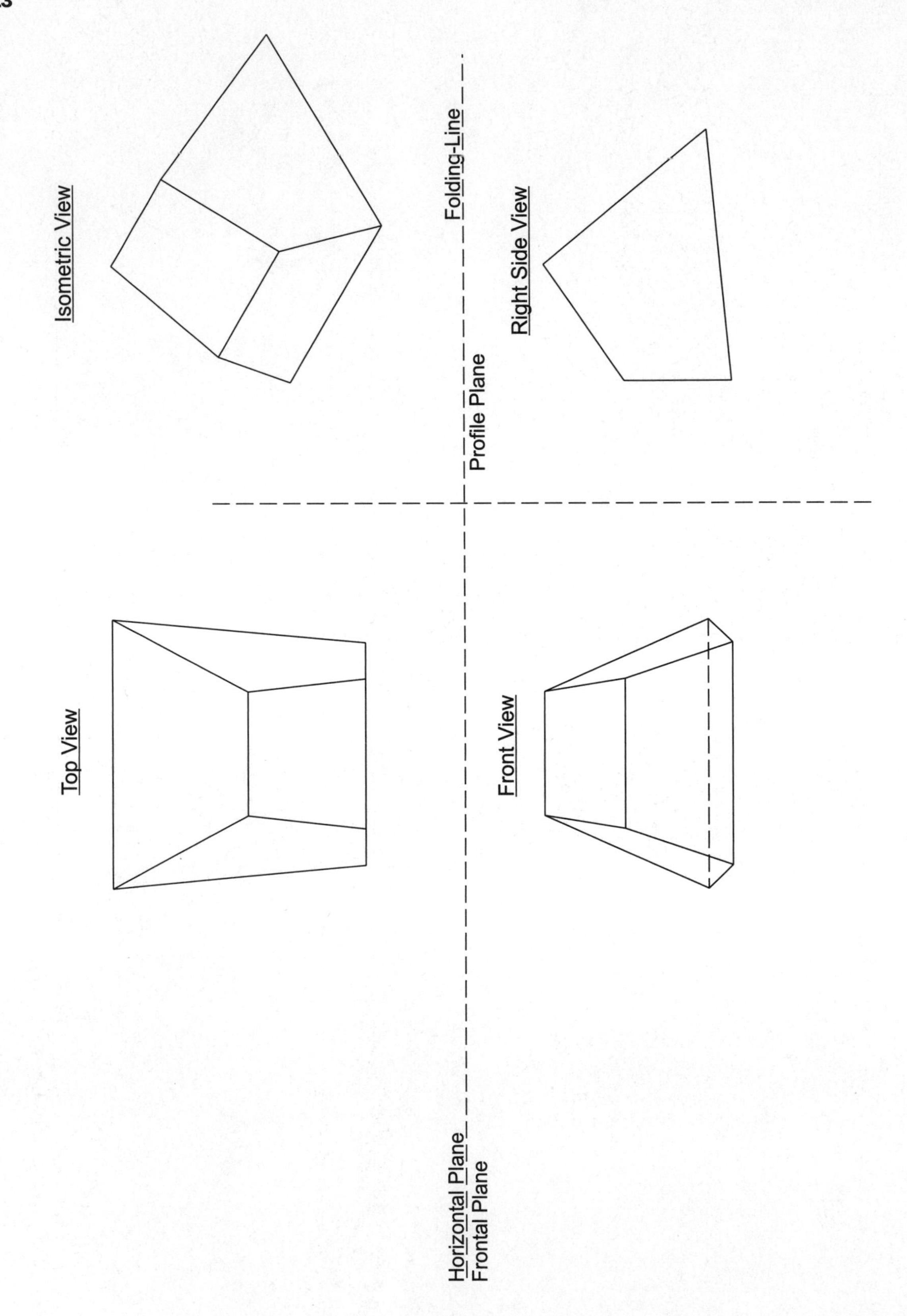

Isometric View
Folding-Line
Right Side View
Profile Plane
Top View
Front View
Horizontal Plane
Frontal Plane

## Triangulation Developments

Create a development for each of the following parts.

***#14**

Folding-Line

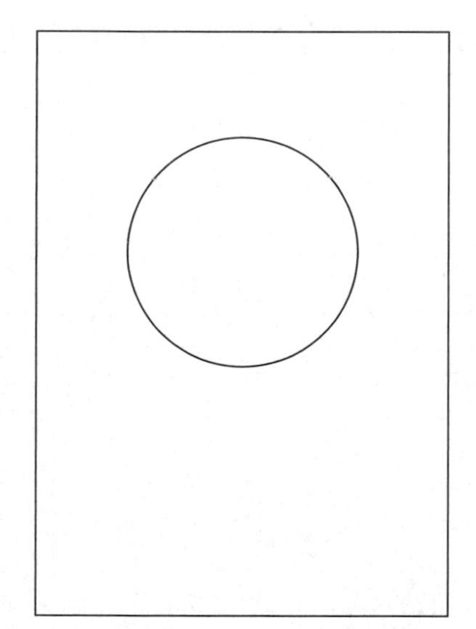

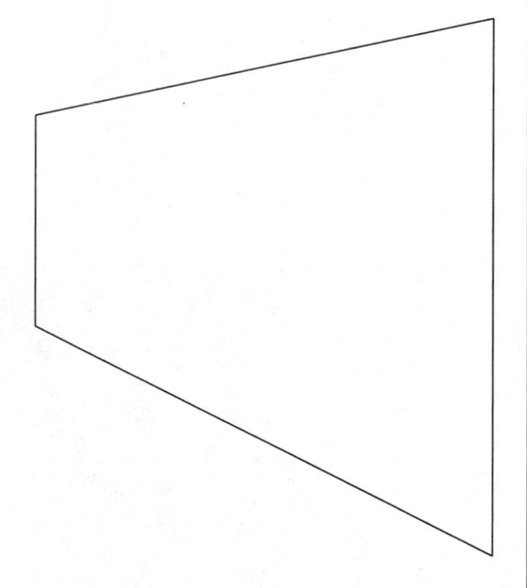

Horizontal Plane

Frontal Plane

***#15**

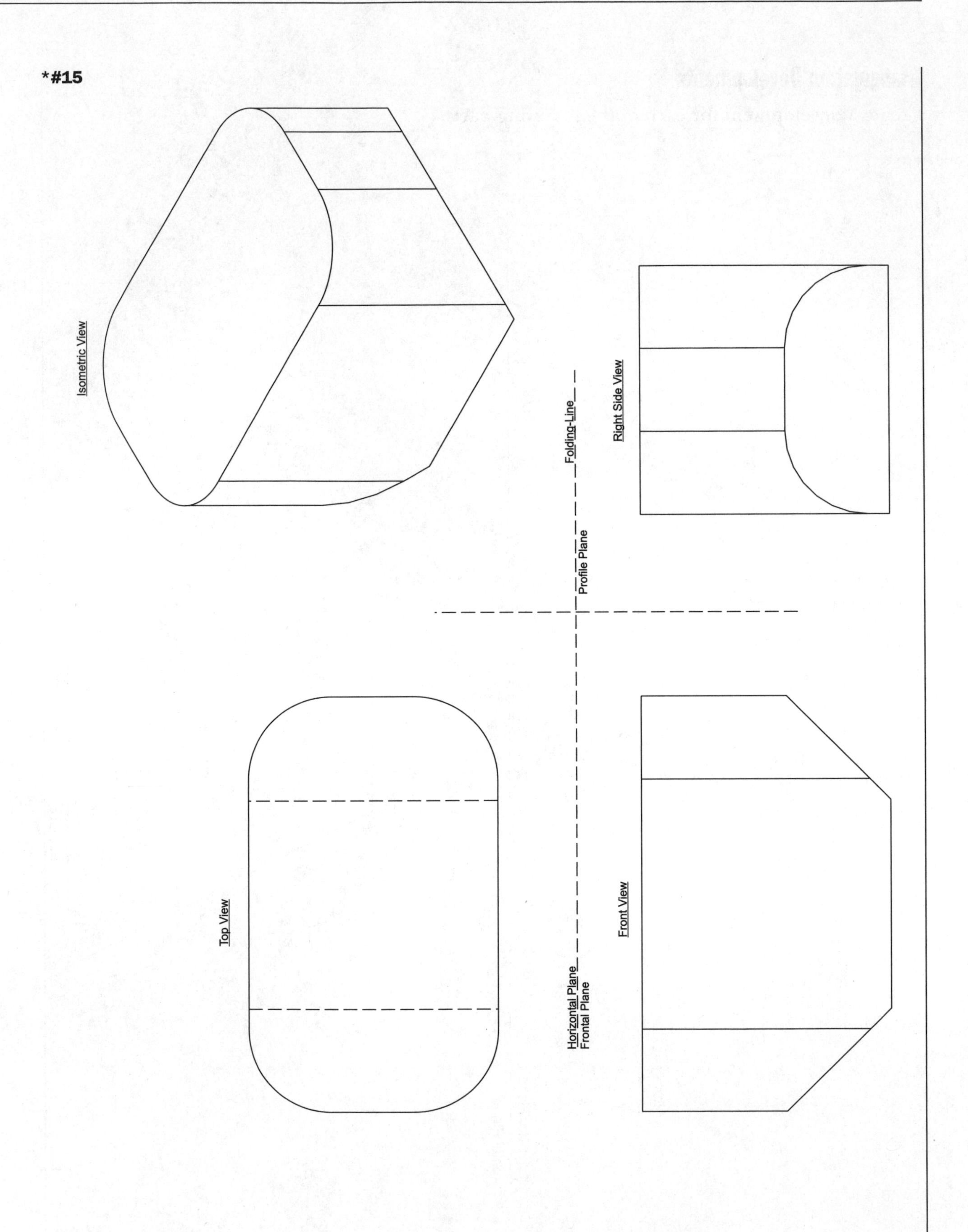

Isometric View
Folding-Line
Right Side View
Profile Plane
Top View
Horizontal Plane
Frontal Plane
Front View

*#16

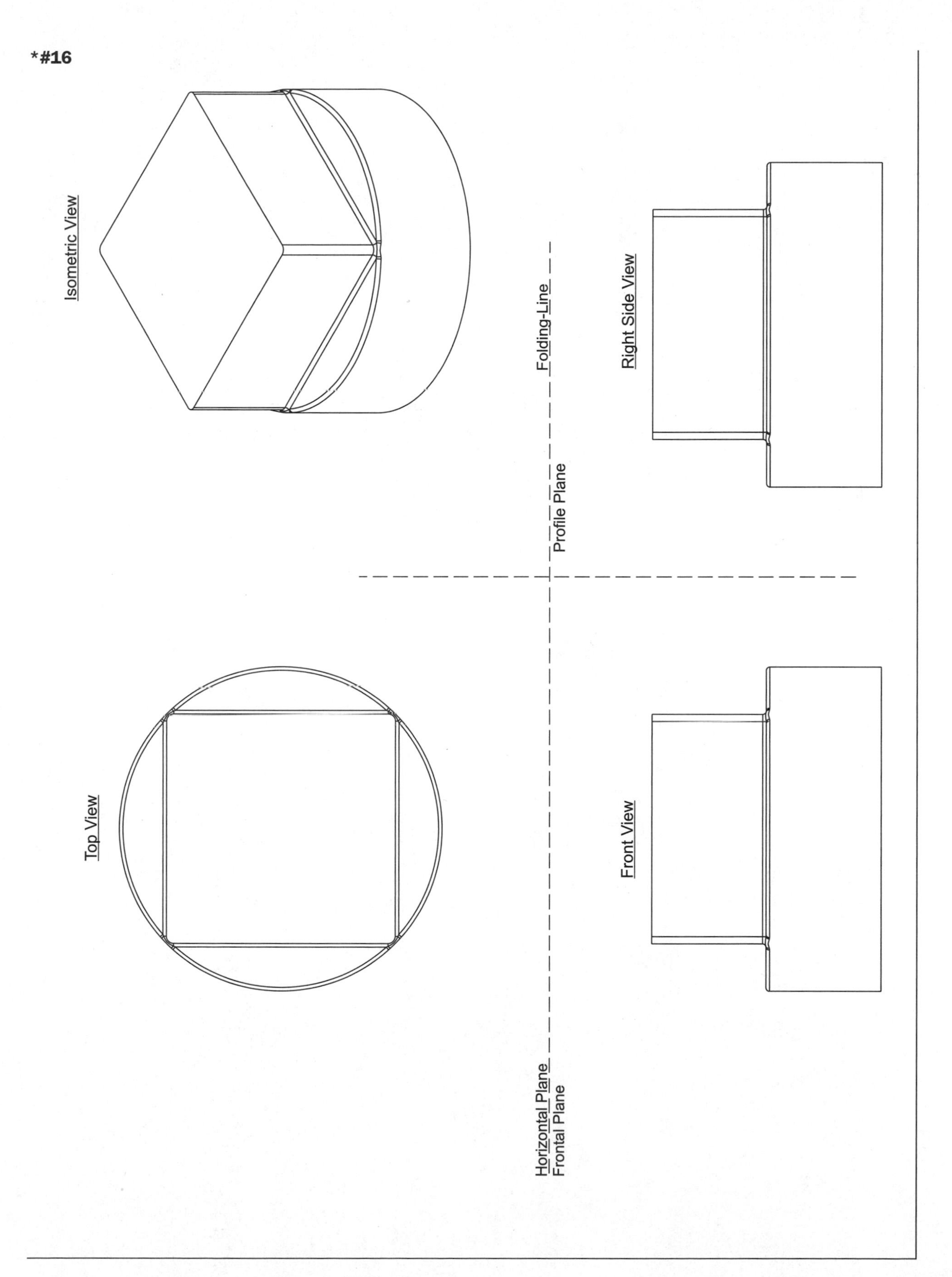
Isometric View
Top View
Folding-Line
Profile Plane
Right Side View
Front View
Horizontal Plane
Frontal Plane

# Vector Geometry

## OBJECTIVES

**After completing this chapter the student will be able to do the following:**

- *Describe the difference between vector and scalar quantities.*
- *Properly create vectors graphically.*
- *Define concurrent, coplanar, noncoplanar, components, and resultant.*
- *Identify the three types of triangles.*
- *Find the missing angle of a triangle when the other two angles are given.*
- *Use Pythagorean's theorem to find the missing side of a triangle when the other two are given.*
- *Identify the hypotenuse, adjacent, and opposite sides of a triangle.*
- *Find the resultant of a concurrent coplanar vector system using the parallelogram method.*
- *Find the resultant of more than two concurrent coplanar vectors using the polygon method.*
- *Use the trigonometric functions sine, cosine, and tangent to resolve vectors into their X- and Y-components.*
- *Use Pythagorean's theorem to find the magnitude of a vector.*
- *Use the arcsine, arccosine, and arctangent functions to find the direction of a vector.*
- *Define the basic math functions standard to all spreadsheets.*
- *Set up equations in an Excel spreadsheet.*
- *Combine trigonometry with the capabilities of spreadsheets to determine the resultant of concurrent coplanar vectors.*
- *Describe the characteristics of a force and the units used to measure them.*
- *State and apply Newton's First, Second, and Third Laws of motion.*
- *Define equilibrium, space diagram, and free-body diagram.*

## KEY WORDS AND TERMS

- Angular degrees
- Arctangent
- Aristotle
- Cell
- Columns
- Components
- Concurrent noncoplanar vectors
- Concurrent vectors
- Converting radians to degrees
- Coplanar vectors
- Cosine
- Cosine function
- Dynamics
- Equilibrium
- Force
- Free-body diagrams
- Galileo
- Gradients
- Inertia
- Kilonewtons
- Kilopounds
- Moment
- Newton's First Law of Motion
- Newton's Second Law of Motion
- Newton's Third Law of Motion
- Newton
- Noncoplanar vectors
- Parallelogram method
- Polygon method
- Pounds
- Pythagorean theorem
- Radians
- Resultant
- Rows
- Scalar
- Sine
- Sine function
- Space diagram
- Statics
- Tangent
- Tangent function
- Tons
- Trigonometry
- Vector
- Vector-polygon method

## Introduction to Vectors

Until now, lines have been used as the main tool for determining the precise location of an object's attributes by projecting them into other views. Lines have been used as visual rays, projection lines, object lines, folding lines, bend lines, and stretch-out lines, as well as the boundaries of planes. However, a line can be used to represent much more than this; it can also be used to describe motion. A fundamental concern of engineering is the study of motion, whether it is the movement of a machine or the flow of electrons through a wire. The study of motion and its effects is a major focus in the technical world. This study has led to the development of the electric motor, the automobile, and the space shuttle, just to name a few critical inventions. In short, everything has come into existence because of the direct or indirect study of motion. This study of motion has in turn lead to the development of a type of line known as a ***vector***.

By definition a ***vector*** is a line that graphically represents magnitude and direction. Both magnitude and direction are necessary to describe a vector. For example, a plane traveling one hundred miles could not be shown as a vector, because its direction was not mentioned, even though the magnitude (the distance or length of the line) has been established as one hundred miles. If the magnitude is defined without a direction, then this is known as a ***scalar*** quantity. Examples of scalar quantities are elapsed time, volume, distance, area, speed, and temperature. If it had been stated that the plane was traveling one hundred miles northeast, then this could be represented as a vector, because both magnitude and direction have been established. Motion isn't the only quantity that can be represented by a vector. Vectors can be used to represent many different quantities, one of which is force. For example, if a one-hundred-pound object is placed on a table, the downward force produced from its weight can be represented graphically with a vector. In this example, magnitude and direction are a result of gravity rather than object motion, as in the example of the airplane. Other examples of quantities that can be represented with vectors are cash flow, inductance, capacitance, velocity, acceleration, and air force.

A vector consists of two main parts, a line and an arrowhead. The arrowhead is used only to show the direction of the vector, its size is not important. Vectors of different magnitudes will have the same size arrowheads used throughout a drawing. This uniformity enhances the clarity and overall esthetics of a layout. The line represents the magnitude of the vector, its length being proportional to the numeric value of the quantity it represents. For example, a vector having a magnitude of one-hundred foot-pounds would be twice as long as a vector having a magnitude of fifty foot-pounds. This often leads to questions concerning the scale to use when drawing a vector. Any scale can be used as long as it is consistent throughout the solution of a single problem. If a scale of one inch is equal to one-hundred pounds is chosen to construct a system of forces, the vector used to represent a force of two-hundred pounds would have a line two-inches long. This length would be determined by dividing the force by the scale chosen (the number of total pounds of force divided by the number of pounds required to equal one inch) or 200 pounds / 100 pounds. If a second force in the system has a magnitude of fifty pounds, then its vector would have a line that is one-half-inch long. Again, this length is determined by dividing the magnitude by the scale, 50 pounds / 100 pounds.

Like the atom, all vectors can be split into three separate entities or ***components***, as they are called. They are the X-component, Y-component, and Z-component. If a vector can be split into its components, then it stands to reason that several components can be combined to create a single entity. This combination yields a product referred to as a ***resultant***. In other words, a resultant is the sum created by combining two or more components into a single vector. Often a system of vectors will act through a common point; these vectors are called ***concurrent vectors***. When a system of vectors share the same plane they are known as ***coplanar vectors***. Likewise, a system of vectors that do not share the same plane are known as ***noncoplanar vectors***.

## Finding the Resultant of Concurrent and Coplanar Vectors

### Introduction to Sine, Cosine, and Tangent

This chapter introduces the student to techniques for solving vector problems using trigonometry and descriptive geometry. Later in this chapter, these techniques will be adapted for the use of computer applications such as AutoCAD, AutoLISP programming, and Excel spreadsheets. Before these tools can be used, a few concepts must be addressed, namely a basic understanding of the trigonometric principles regarding the use of ***sine***, ***cosine***, ***tangent***, the ***Pythagorean theorem***, and ***converting radians to degrees***.

There are three types of units that are used to express an angle: ***angular degrees***, ***radians***, and ***gradients***, but typically only angular degrees and radians are used. Since the circumference of any circle contains 360 degrees, an ***angular degree*** is equal to 1/360th of the circumference of a circle. This means that by drawing a circle (of any size) and dividing its circumference into three hundred and sixty equal segments (called ***arc lengths***), the angle formed by constructing a line from the center of the circle to the endpoints of one arc length would produce a wedge equal to one degree (see Figure 7–1). The other common unit of measure for an angle is the ***radian***. The circumference of a circle (of any size) when divided by the length of its radius will always yield a quotient equal to 6.283 or 6.283 segments (arc lengths). It is the angle that is suspended between the segments and the center of the circle that is of concern. A ***radian*** is the angle opposite to an arc length when that arc length is equal to the length of the radius (see Figure 7–2). A radian is equal to 1/6.283 of that circumference or 57.2958°. Half of a circle would contain exactly half of these segments, 6.283/2 or 3.1415 (pi). Likewise, half of the angular degree circle would contain exactly half of its segments, 360°/2 or 180°. The number of segments in each circle is used as a basis for conversions between the two units of measure. To convert angular degrees to radians, divide 3.14 radians by 180 angular degrees, then multiply the quotient by the angle to be converted. If an angle of 45° is to be converted to radians, the equation will be $[(3.14) \div 180°) \times 45°]$ or $(0.0175 \times 45°)$. The degrees symbol is then removed and the suffix Rad is added. To convert from radians to angular degrees, divide 180° angular degrees by 3.14 radians; multiply the quotient by the angle to be converted. If the radian angle is 4.65 Rads, then the equation would look like this: [(180 ÷ 3.14) × 4.65 rad] or (57.295 × 4.65). The suffix Rad is then removed and the degrees symbol is added.

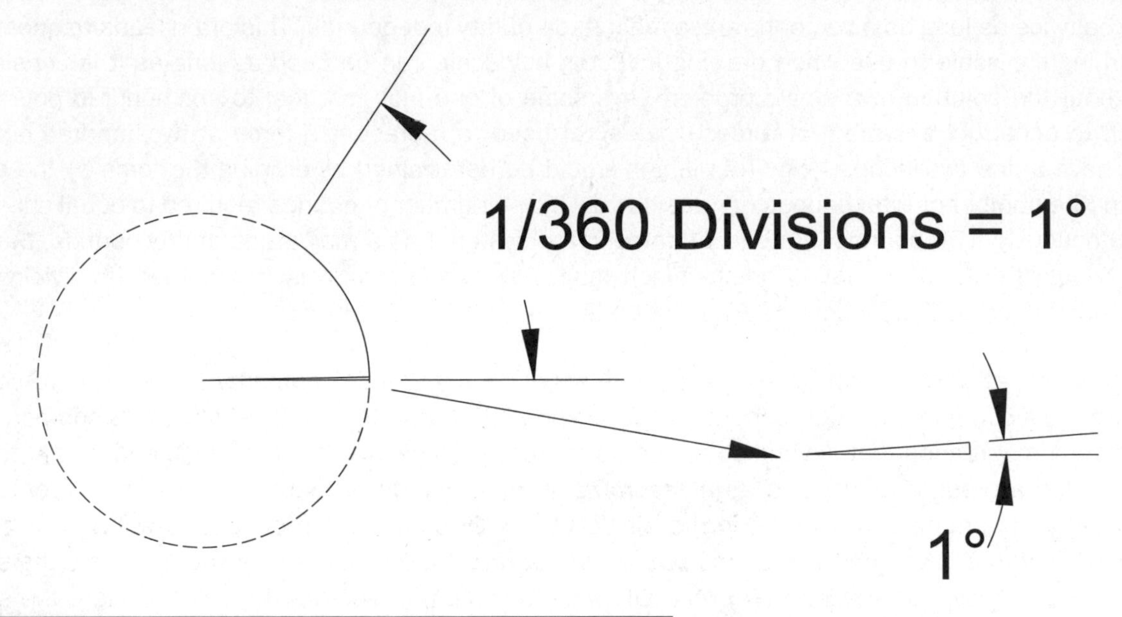


**Figure 7-1** *Wedge produced by dividing a circle 360 times.*

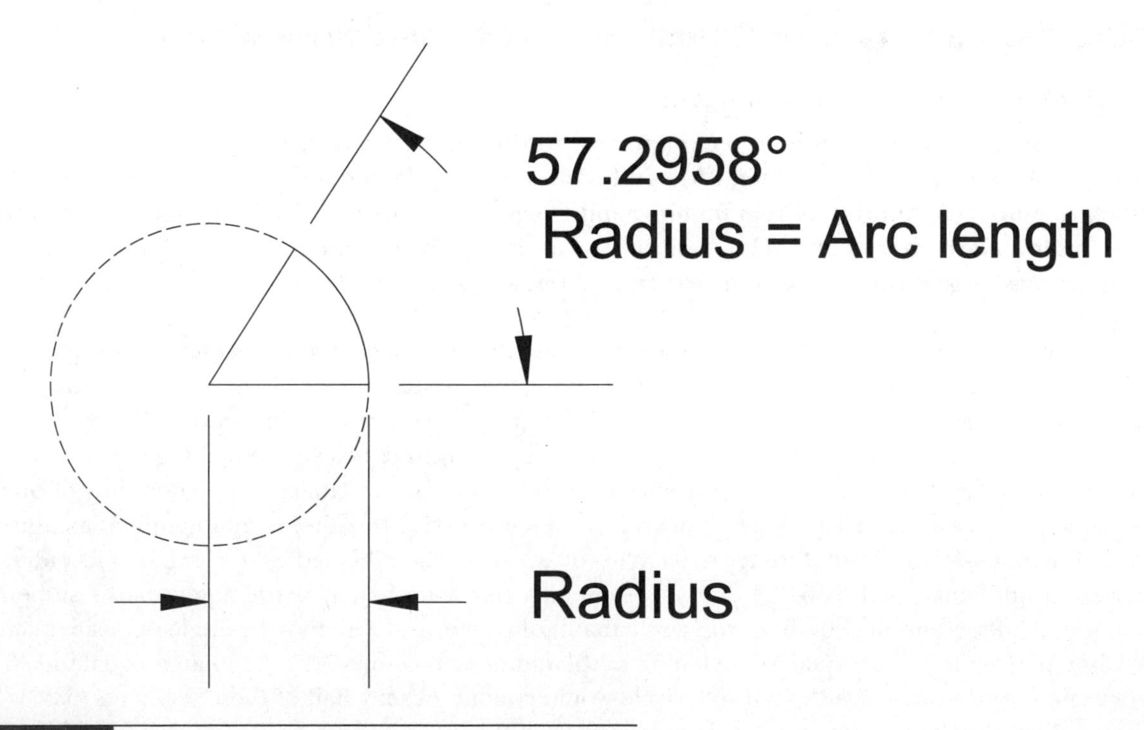


**Figure 7-2** *Arc length is equal to the radius of a circle.*

***Trigonometry*** is the branch of mathematics concerned with the relationship between the sides and angles of triangles. As indicated by its name, a triangle contains three angles. When these three angles are added, their sum is equal to 180 degrees, or angle A + angle B + angle C = 180°. By using this fact, a missing angle can be calculated if the other two angles are given. For example, the missing angle of a triangle containing a 70° and a 55° angle is found by adding the two known angles, and then subtracting their product from 180° (180° – [70°+55°]). After performing the calculation, the missing angle is found to be 55°.

Each side of a triangle has a name: the hypotenuse, opposite, and adjacent sides. In a right triangle the hypotenuse is always the longest side. The other two sides, opposite and adjacent, are labeled relative to the acute angle (any angle less than 90°) being focused on in a given calculation (see Figures 7–3 and 7–4). The adjacent side is the side adjacent to the angle in question. The opposite is the side opposite to that angle.

A major principle built upon in trigonometry is that in a right triangle, there is a direct relationship between the angles and the lengths of the sides. This relationship can be summed up in three fundamental trigonometric functions: ***sine***, ***cosine***, and ***tangent***. The ***sine function*** is defined as the ratio of the side opposite to an acute angle divided by the hypotenuse or (Sine A = a/c), as shown in Figures 7–3 and 7–4. The ***cosine function*** is defined as the ratio of the side adjacent to an acute angle divided by the hypotenuse or (Cosine A = b/c), and the ***tangent function*** is defined as the ratio of the side opposite to an acute angle divided by the side adjacent (Tan A = a/b). Another important concept in trigonometry is the ***Pythagorean theorem***. It states that the square of the hypotenuse of a right triangle is equal to the sum of the square of the other two sides, or $R^2 = X^2 + Y^2$. In this equation R is equal to the hypotenuse, and X and Y are equal to the adjacent and opposite sides of the triangle. With this equation, the missing side of a right triangle can be determined if the other two sides are given by applying the ***Pythagorean theorem***. In order to calculate the hypotenuse, algebra would have to be used to isolate the variable R. In other words, the hypotenuse is found by solving for R, not $R^2$. To isolate R, the square root of both sides of the equation must be taken, and doing so would yield the equation $R = \sqrt{(X^2 + Y^2)}$. For example, given a right triangle that contains an adjacent side (b) equal to three inches and an opposite side (a) equal to four inches, the hypotenuse can be found by using the formula: ($R = \sqrt{(3^2 + 4^2)}$, $R = \sqrt{25}$, R=5), as shown in Figure 7–5.

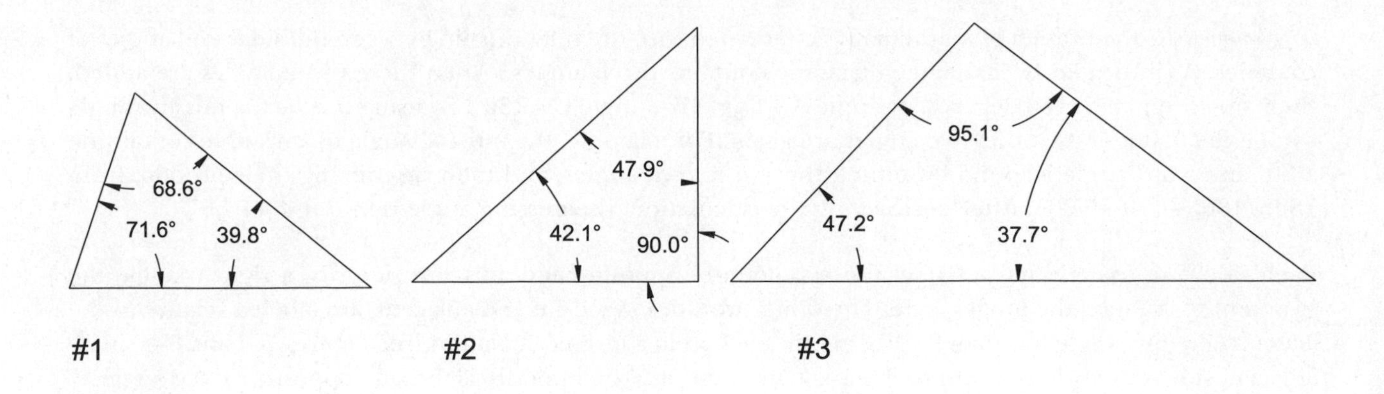


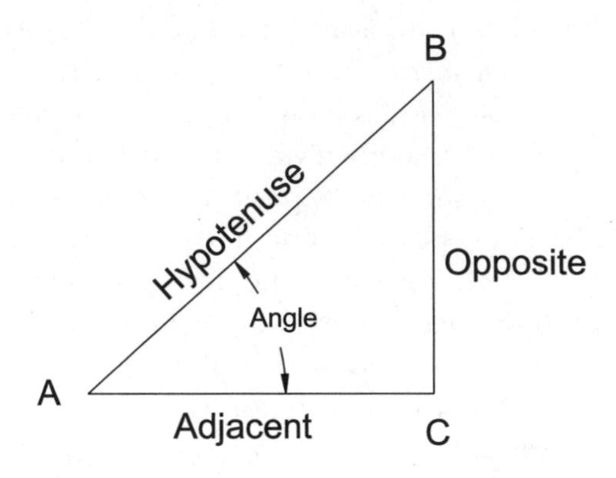


$$\text{Sine} = \frac{\text{Opposite}}{\text{Hypotenuse}}$$

$$\text{Cosine} = \frac{\text{Adjacent}}{\text{Hypotenuse}}$$

$$\text{Tan} = \frac{\text{Opposite}}{\text{Adjacent}}$$

**Figure 7–3** *#1 Acute triangle—all angles are less that 90°.*
*#2 Right triangle—one angle is 90°.*
*#3 Obtuse triangle—one angle is greater than 90°.*

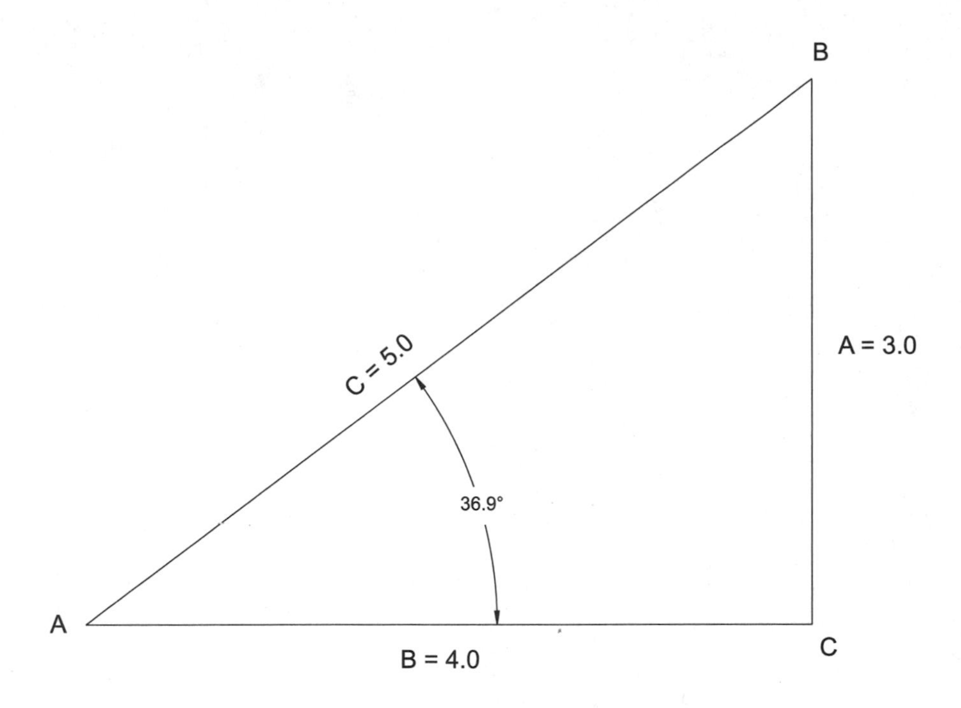


$$\text{Sine } 36.9^\circ = \frac{3.0}{5.0} \qquad \text{Cosine } 36.9^\circ = \frac{4.0}{5.0} \qquad \text{Tan } 36.9^\circ = \frac{3.0}{4.0}$$

$$\text{Sine } 36.9^\circ = 0.6 \qquad \text{Cosine } 36.9^\circ = 0.80 \qquad \text{Tan } 36.9^\circ = 0.75$$

**Figure 7–4** *Examples of sine, cosine, and tangent.*

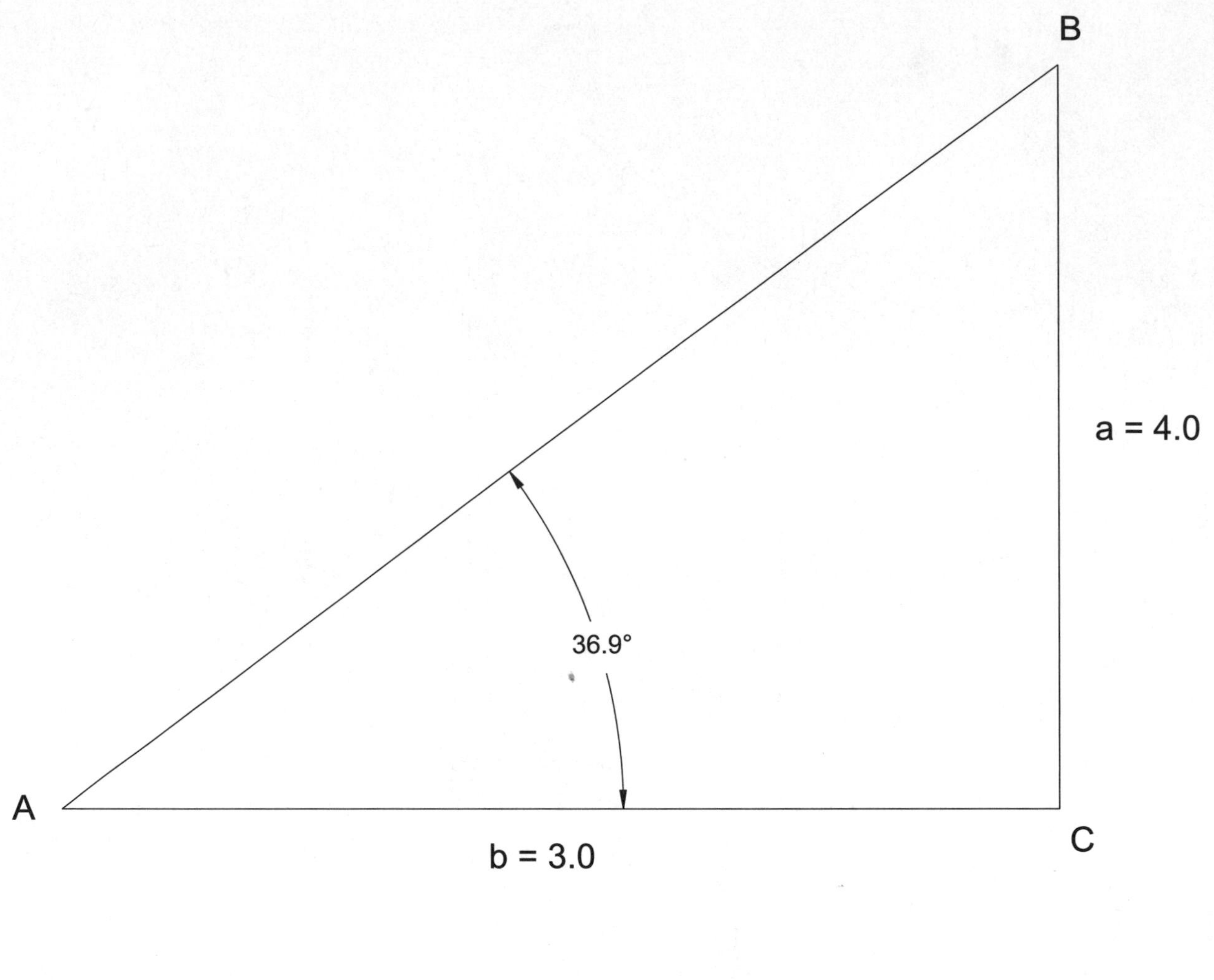


$$R = \sqrt{B^2 + A^2}$$

$$R = \sqrt{9 + 16}$$

$$R = \sqrt{25}$$

$$R = 5$$

**Figure 7–5** *Example of the Pythagorean theorem.*

## Using the Parallelogram Method to Add Vectors

Now that the groundwork has been established, attention can be given to dealing with vectors through the use of addition. The first type of vector is the concurrent vector. Recall from the previous definition that concurrent vectors are vectors that act through a common point. The resultant can be found by executing one of two methods. The first is the ***parallelogram method***. This technique adds two concurrent vectors by arranging them into a parallelogram (see Figure 7–6). Each vector is "mirrored" onto the end of the opposite vector. This method can be employed if the vectors are both concurrent and their line of action is either in the same or opposite directions. Once the parallelogram has been constructed, the resultant is determined by drawing a diagonal line from the point that the two vectors intersect to the corner created by the intersection of the "mirrored" lines.

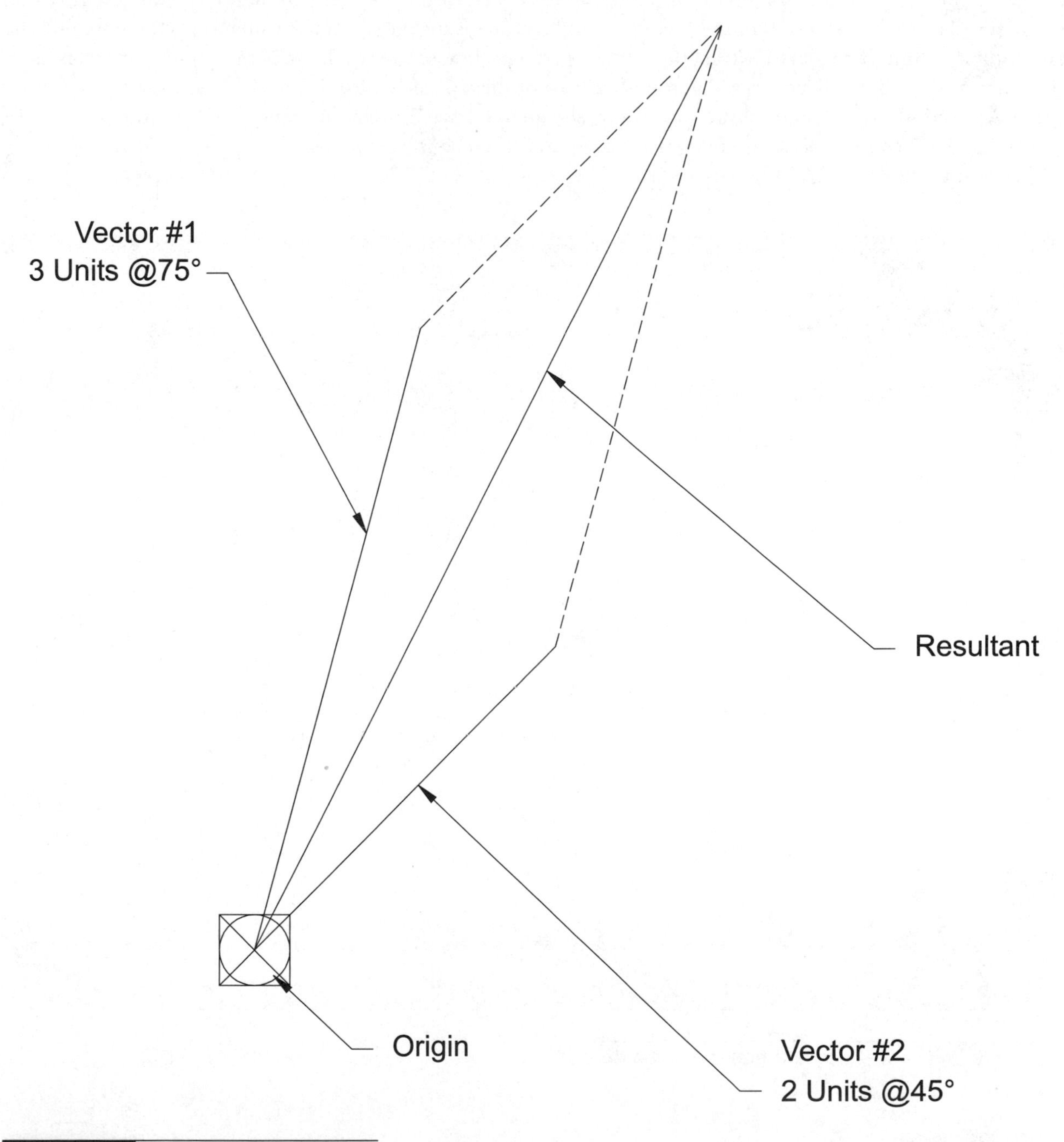


**Figure 7-6** *Parallelogram method.*

## Using the Parallelogram Method in AutoCAD

The parallelogram method can be easily adapted to AutoCAD and provides a useful tool for solving concurrent vector problems quickly. First, a line is drawn with its length representing the magnitude and its angle representing the direction of the first vector. Next, using the beginning point of the first vector drawn, a line representing the second vector is created in a matter similar to the first. Using the COPY command, a parallelogram is constructed by copying the first vector from its origin to the endpoint of the second vector. This procedure is repeated for the second vector to complete the parallelogram. Finally, either draw a diagonal line from the origin of the first vector to the opposite corner, or use the DIST command to select the origin of the first vector to the endpoint of the second vector. If a line is drawn instead of using the DIST command, the LIST command must be used to extract the angle and magnitude of the resultant. If the DIST command is used, the magnitude and the direction will automatically be displayed (see Figure 7–7). The following example uses AutoCAD to find the resultant of the two vectors shown in Figure 7–6. The first vector is two units long with a direction of 45° counterclockwise from the X-axis, and the second vector is three units long with a direction of 75° counterclockwise from the X-axis.

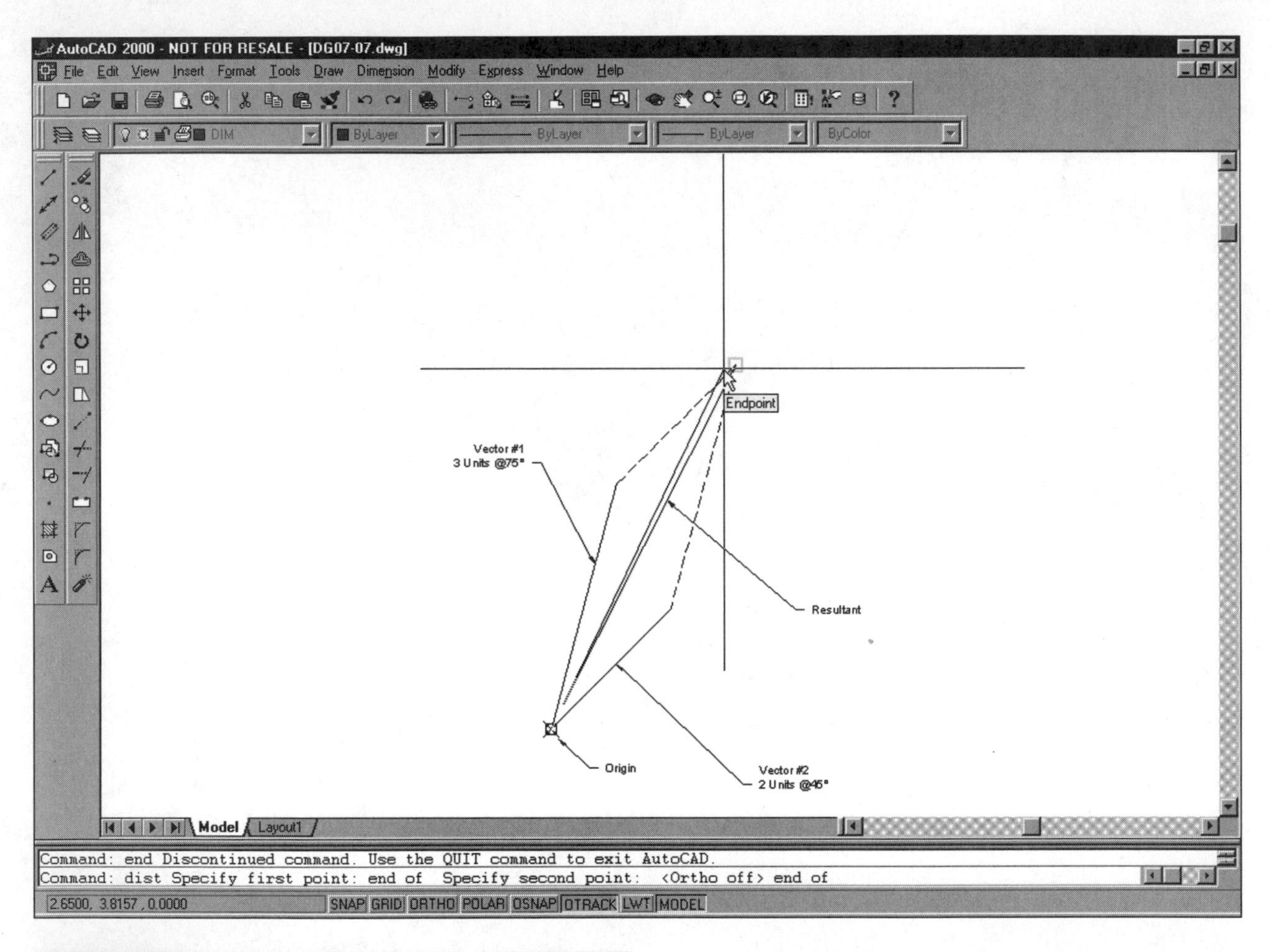


**Figure 7–7** *Parallelogram method using AutoCAD.*

## Step #1

Start by drawing the two concurrent coplanar vectors using the LINE command and polar coordinates.

Command: PLINE (ENTER)
From point: *(Select an arbitrary point.)*
Current line-width is 0.0000

Arc/Close/Halfwidth/Length/Undo/Width/<Endpoint of line>: @2<<45 *(Press* ENTER. *Use the polar coordinate system to draw a line 2 units long at an angle of 45° counterclockwise from the X-axis.)*
Arc/Close/Halfwidth/Length/Undo/Width/<Endpoint of line>: (ENTER)
Command: PLINE (ENTER)
From point: end of *(Select the origin of the previous line drawn using the* END OSNAP.*)*
Current line-width is 0.0000
Arc/Close/Halfwidth/Length/Undo/Width/<Endpoint of line>: @3<<75 *(Press* ENTER. *Use the polar coordinate system to draw a line 3 units long at an angle of 75° counterclockwise from the X-axis.)*
Arc/Close/Halfwidth/Length/Undo/Width/<Endpoint of line>:

## Step #2

Use the COPY command to copy the two previously drawn vectors thereby forming a parallelogram.

Command: COPY (ENTER)
Select objects: 1 found *(Select the first vector drawn.)*
Select objects: *(Press* ENTER *to terminate the selection mode.)*
<Base point or displacement>/Multiple: end of *(Select the starting point of the first vector drawn using the* END OSNAP.*)*
Second point of displacement: end of *(Select the end point of the second vector drawn using the* END OSNAP.*)*
Command: COPY (ENTER)
Select objects: 1 found *(Select the second vector drawn.)*
Select objects: *(Press* ENTER *to terminate selection mode.)*
<Base point or displacement>/Multiple: end of *(Select the starting point of the second vector drawn using the* END OSNAP.*)*
Second point of displacement: end of *(Select the end point of the second vector drawn using the* END OSNAP.*)*

## Step #3

Use the LINE command to draw a line from the origin of the first vector to the opposite corner, forming a diagonal. Finally, use the LIST command to obtain the length and direction of the resultant.

Command: LINE (ENTER)
From point: end of
To point: end of *(Select the origin of the first two vectors drawn using the* END OSNAP.*)*
To point: *(Use* END OSNAP *to connect the line to the opposite corner.)*
Command: LIST (ENTER)
Select objects: 1 found *(Select diagonal line.)*

Select objects: (ENTER)
LINE Layer: 0
Space: Model space
Handle = 253
from point, X= 13.4524 Y= 1.6245 Z= 0.0000
to point, X= 15.6430 Y= 5.9365 Z= 0.0000
*Length = 4.8366, Angle in XY Plane = 63*
Delta X = 2.1907, Delta Y = 4.3120, Delta Z = 0.0000

**Note: In this example, AutoCAD computed the resultant to be 4.8366 units long with a direction of 63° counterclockwise from the X-axis.**

## Adding More Than Two Concurrent Vectors

The above method is used only when combining a two-vector system. To find the resultant of three or more vectors, a slightly different approach must be taken. Instead of arranging the vectors to form a parallelogram, three or more vectors are placed end to end to form an open polygon. The resultant is the line that closes this figure. This procedure is called the ***polygon method*** (see Figure 7–8).

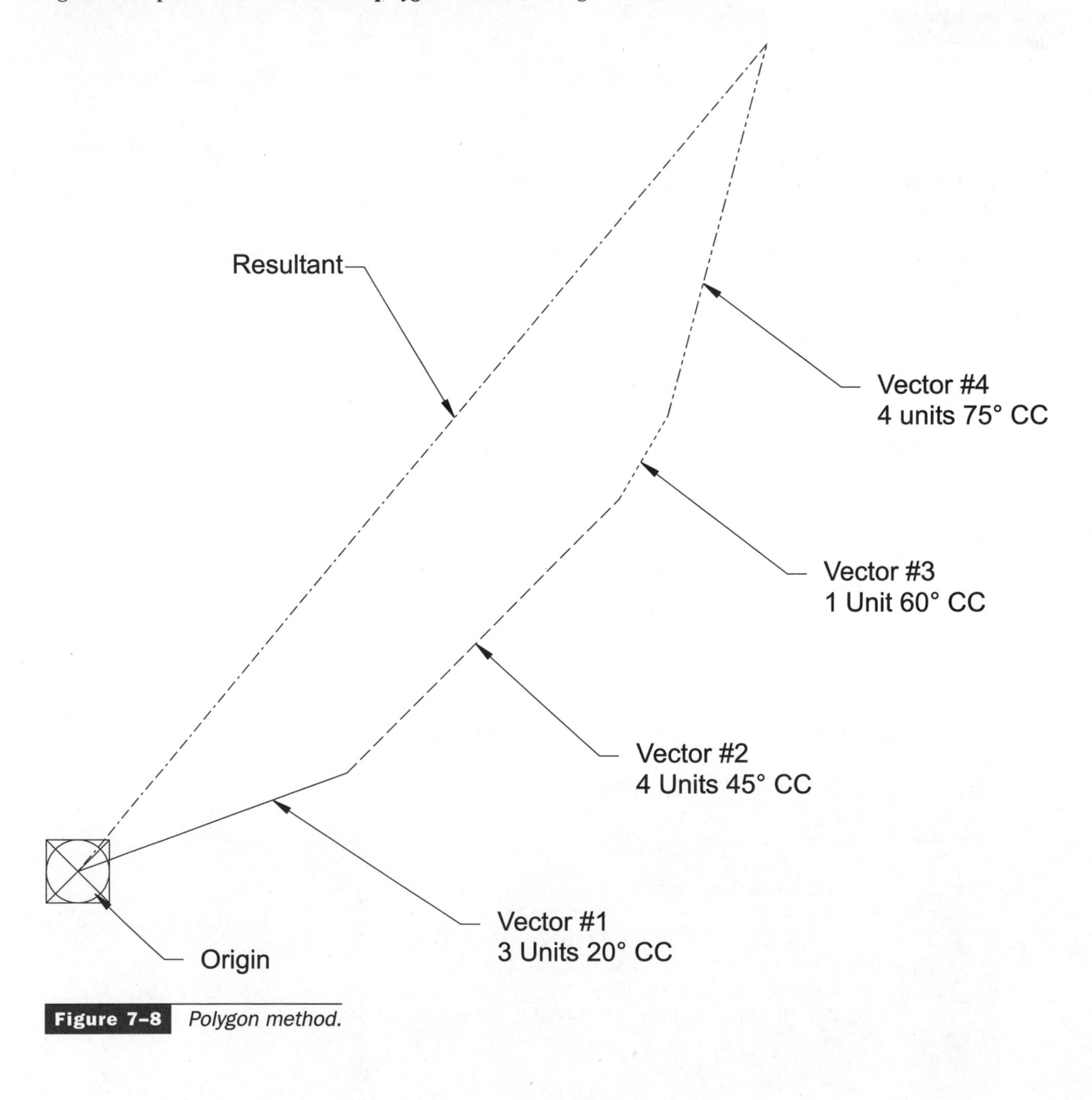


**Figure 7–8** *Polygon method.*

### Using AutoCAD to Solve Vectors Using the Polygon Method

The polygon method can be easily adapted to AutoCAD as well. This time, when the first vector is completed, instead of terminating the LINE command, use polar coordinates to continue drawing the remaining vectors. To close the polygon, use the C option of the LINE command. Again, either the LIST command or DIST command can be used to determine the magnitude and direction of the resultant. If the LIST command is used, select the last line drawn (this is the line closing the polygon). If the DIST command is used, select the origin of the first vector drawn and the endpoint of the last vector drawn (see Figure 7–9). The following example uses AutoCAD to find the resultant of the four-vector system shown in Figure 7–8. The first vector is three units long with a direction of 20° counterclockwise from the X-axis. The second vector is four units long with a direction of 45° counterclockwise from the X-axis. The third vector is one unit long with a direction of 60° counterclockwise from the X-axis, and the fourth vector is four units long with a direction of 75° counterclockwise from the X-axis.

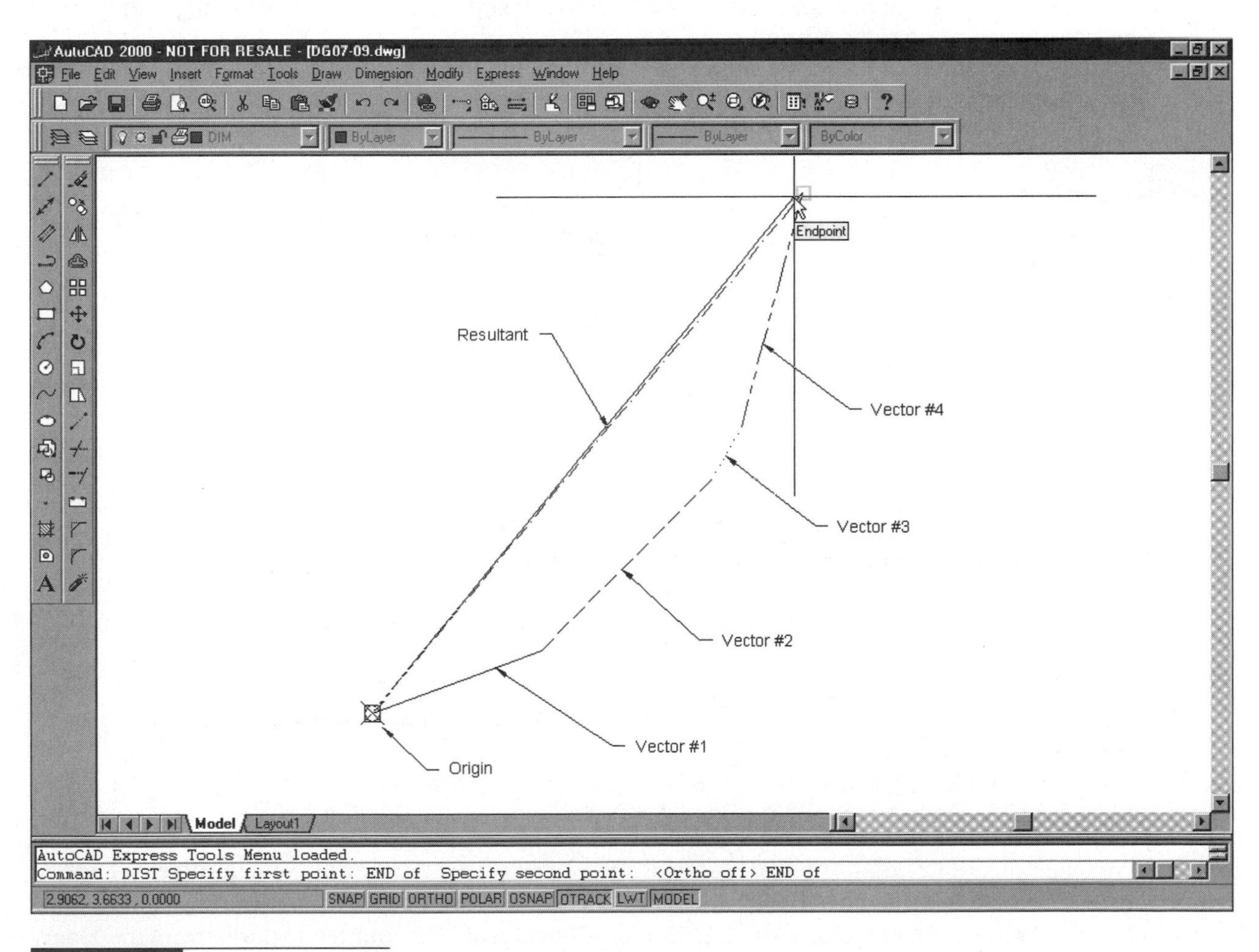


**Figure 7–9** *Polygon method.*

## Step #1

Use the LINE command to begin drawing the vectors. After the last vector has been drawn, terminate the LINE command and complete the polygon by using the Close option.

Command: LINE (ENTER)
From point: *(Select an arbitrary point.)*
To point: @3<<20 *(Press* ENTER. *Use the polar coordinate system to draw a line 3 units long that is 20° counterclockwise from the X-axis.)*
To point: @4<<45 *(Press* ENTER. *Use the polar coordinate system to draw line 4 units long that is 45° counterclockwise from the X-axis.)*
To point: @1<<60 *(Press* ENTER. *Use the polar coordinate system to draw a line 1 unit long that is 60° counterclockwise from the X-axis.)*
To point: @4<<75 *(Press* ENTER. *Use the polar coordinate system to draw line 4 units long that is 75° counterclockwise from the X-axis.)*
To point: C *(Press* ENTER. *This option closes the polygon and terminates the* LINE *command. If* ENTER *is pressed without first inputting the C, then the command is terminated without closing the polygon.)*

## Step #2

Use the LIST command or the DIST command to reveal the information regarding the resultant.

***Using the* LIST *Command***

Command: LIST (ENTER)
Select objects: 1 found *(Select the last line drawn.)*
Select objects: (ENTER)

```
            LINE    Layer: 0
                  Space: Model space
           Handle = 274
       from point, X=  30.5733  Y=  11.2762  Z=   0.0000
         to point, X=  23.3905  Y=   2.6920  Z=   0.0000
    Length =  11.1929,  Angle in XY Plane =    230
          Delta X =  -7.1828, Delta Y =  -8.5842, Delta Z =   0.0000
```

***Using the* DIST *Command***

Command: DIST First point: end of Second point: end of *(Select the origin of the first vector and the endpoint of the last vector.)*
Distance = *11.1929*, Angle in XY Plane = *50*, Angle from XY Plane = 0
Delta X = 7.1828, Delta Y = 8.5842, Delta Z = 0.0000

In the above example, notice how AutoCAD produced two different answers for the same problem. The DIST command reported a magnitude of 11.1929 and a direction of 50° counterclockwise from the X-axis, while the LIST command reported a magnitude of 11.1929 and a direction of 230° counterclockwise from the X-axis. So which answer is correct and why did AutoCAD generate two different answers for the same problem? The 50° angle is the correct answer. The reason that the software derived two different answers is because the LIST command calculates the angle from the endpoint of the last vector to the origin of the first vector. This placed the vector in quadrant three, resulting in an angle of 230°. Because all vectors in this example are positioned in the first quadrant, this should alert the technician that the correct answer will be between 0° and 90°. It is at this point that the following analog can be made: The direction of the resultant will be greater than the smallest angle and less than the greatest angle. Therefore, 50° should be correct because it is between 20° and 75°. The answer obtained from the LIST command can be salvaged by subtracting 180° from 230°; this would reposition the vector into the first quadrant and produce the correct answer of 50°.

#### Using Trigonometry to Find the Resultant of Concurrent Coplanar Vectors

The resultant to a system of vectors can also be found by using a more traditional method: trigonometric functions. When using trigonometry, two rules must be followed. First, vectors can be added if their line of action is either the same or opposite (180° apart). To do this, the magnitudes are added together and the direction remains the same as the components. For example, a vector having a direction of 35° and a magnitude of 10 units could be combined with another vector whose direction is 35° with a magnitude of 20 units. The resultant would be a vector with a direction of 35° and a magnitude of 30 units. If two vectors that are opposite in direction are combined, then the magnitude is obtained by subtraction, and the direction will be the same as the larger of the two components. For example, a vector with a magnitude of four units and a direction of 210° counterclockwise from the X-axis, combined with a vector five units in magnitude and a direction of 30° counterclockwise from the X-axis, would yield a resultant that has a magnitude of one unit and a direction of 30° counterclockwise from the X-axis.

The second rule is used when combining two vectors whose lines of action are not the same or opposite. In this case, they must be resolved into their X- and Y-components using trigonometric functions. The vector can be thought of as a right triangle. The magnitude is the hypotenuse and the direction of the vector is set equal to angle "A". For example, to resolve a vector with a magnitude of five units and a direction of 25° counterclockwise from the X-axis, the hypotenuse is set equal to five and angle "A" is set equal to 25° (see Figure 7–10). When the opposite and adjacent sides are added to this hypotenuse, as demonstrated in Figure 7–11, it becomes apparent that the Y-component of the vector can be found by rearranging the sine function; by rearranging the cosine function, the X-component of the vector can be obtained. Once the vector has been resolved into its X- and Y-components, it can be combined with other vectors that have also been resolved. This produces one vector whose line of action is along the X-axis and one whose line of action is along the Y-axis. By using the Pythagorean theorem, the components are combined to produce the resultant.

Finally, the direction of the vector can be found by applying the ***arctangent*** function. The ***arctangent*** function is the inverse of the tangent function. It needs only the adjacent and opposite sides of a triangle to calculate an angle. To use the arctangent function, first divide the Y-value by the X-value. This yields a quotient that is used by the arctangent function to calculate the angle. This quotient can be entered into most calculators, and then the second function (2ndF) button can be pressed, followed by the tangent button, to produce the arctangent. A similar procedure would be followed for using the arcsine and arccosine functions (see Figure 7–12).

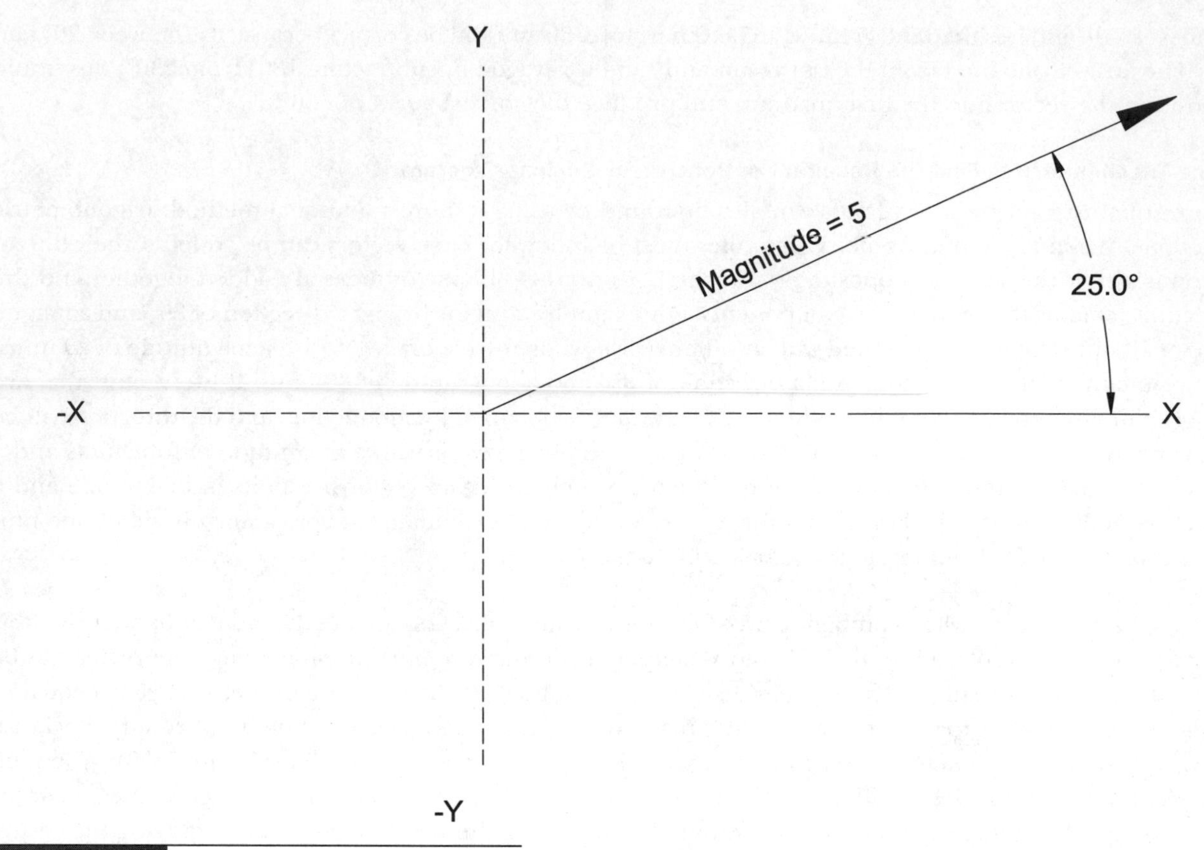


**Figure 7–10** *Vector with a magnitude of 5*

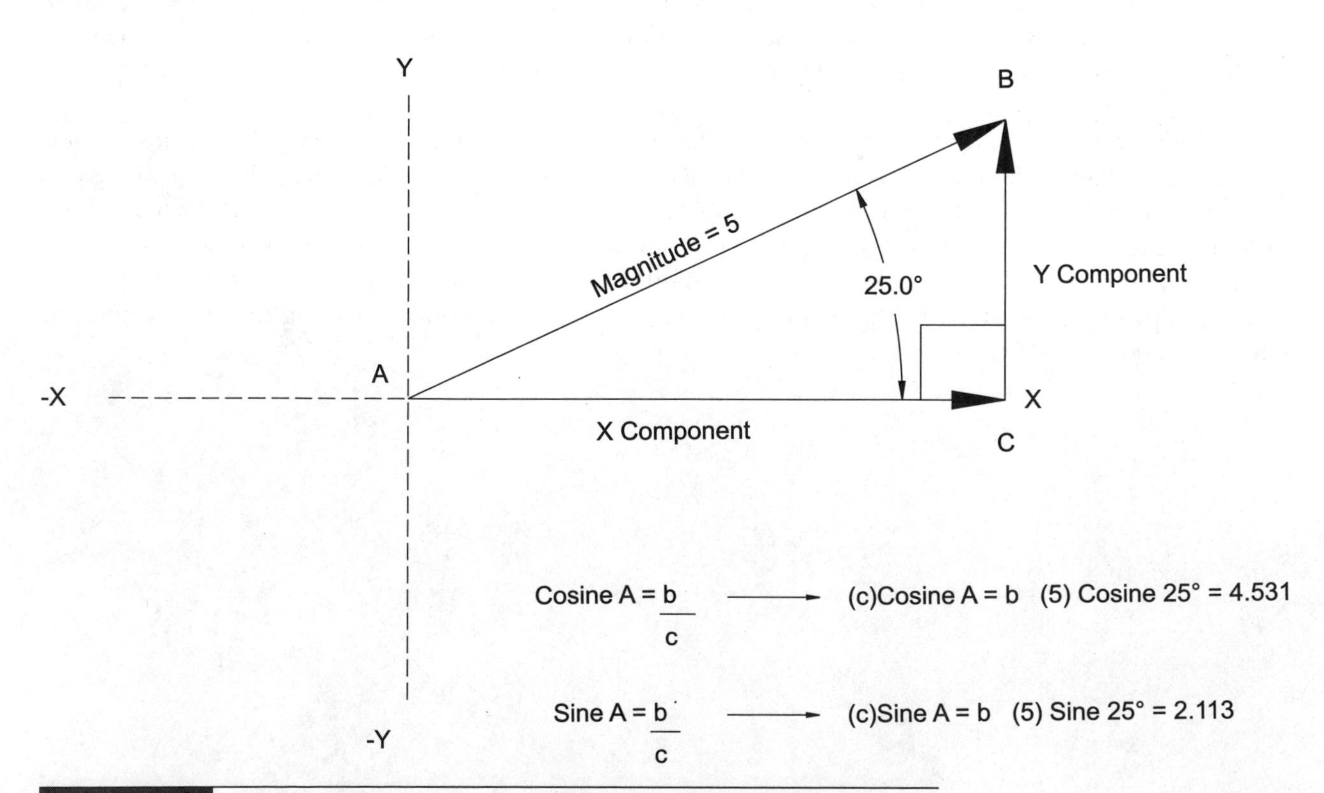


**Figure 7–11** *Using trigonometry to determine the components of a vector*

**Figure 7–12** *Finding the resultant using a calculator*

### Using Excel to Find the Resultants of Concurrent Coplanar Vectors

A spreadsheet is a collection of data and formulas arranged into ***rows*** and ***columns***. This information can be in the form of numbers or text. In a spreadsheet, ***columns*** are identified with letters and are arranged vertically, while ***rows*** are arranged horizontally and are identified by numbers. A ***cell*** is the intersection of a row and a column. Cells are identified by the column and row on which they reside. The cell A1 identifies data at the intersection of the "A" column and row number "one" (see Figure 7–13).

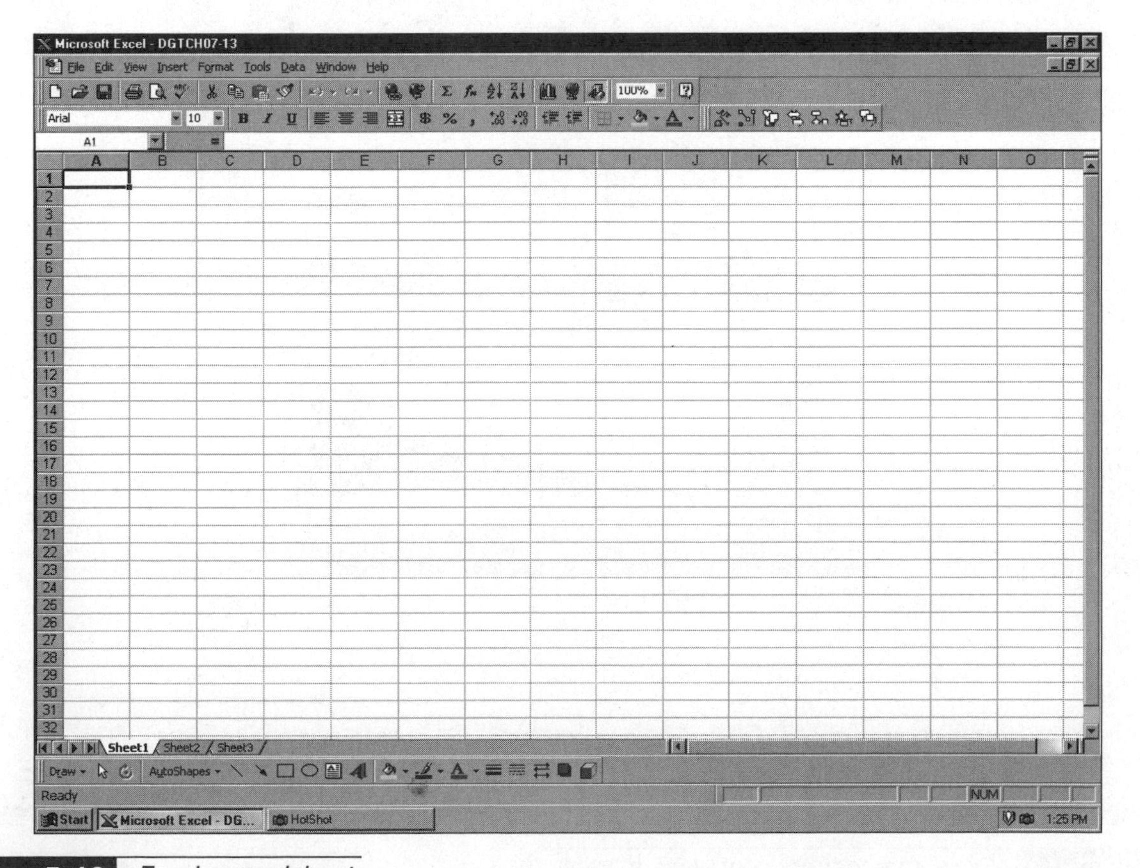


**Figure 7–13** *Excel spreadsheet.*

Most spreadsheets use the same syntax for the basic math and trigonometry functions. In a spreadsheet, basic math operations such as *addition, subtraction, multiplication,* and *division* are accomplished by using the syntax: cell + cell (A1+A2); cell – cell (A1 – A2); cell * cell (A1 * A2); cell / cell (A1 / A2). The syntax for the trigonometry operations such as sine, cosine, and tangent are usually entered as: sin (angle), cos (angle), and tan (angle). When using trigonometry functions, it should be noted that spreadsheets do not return the angle measurement in degrees, instead they are returned in radians. To convert from radians to degrees, the number passed to the trigonometry function must be divided by 57.32.

Advanced math functions that are common to most spreadsheets are power and square roots. The power function will raise a number to a power. The following syntax is used for this operation: cell^(number). The square root function is performed by using the following syntax sqrt (number).

Formulas set up in a spreadsheet typically follow the same format regardless of the brand of spreadsheet used. The differences are in the syntax that designate a formula. To create a formula in Excel, the syntax is =+. In other words, the technician must enter the =+ at the beginning of the formula so that Excel will recognize that the data entered is a formula. Microsoft Works uses the equal sign to designate a formula, while Lotus and Quattro Pro use the plus sign to indicate a formula. When constructing a formula in a spreadsheet, normally one or more columns are set aside for the input and one or more columns are set aside for the formulas performing the calculations. For example, to calculate the Y-value of the equation $X^2 + X - 4 = Y$, the A column can be used to represent the X-value in the equation while the B column can be used for the actual equation. This would yield the formula +A2^2 + A2 – 4 (see Figure 7–14). By applying what has been covered, a spreadsheet can be developed that will resolve and combine concurrent coplanar vectors.

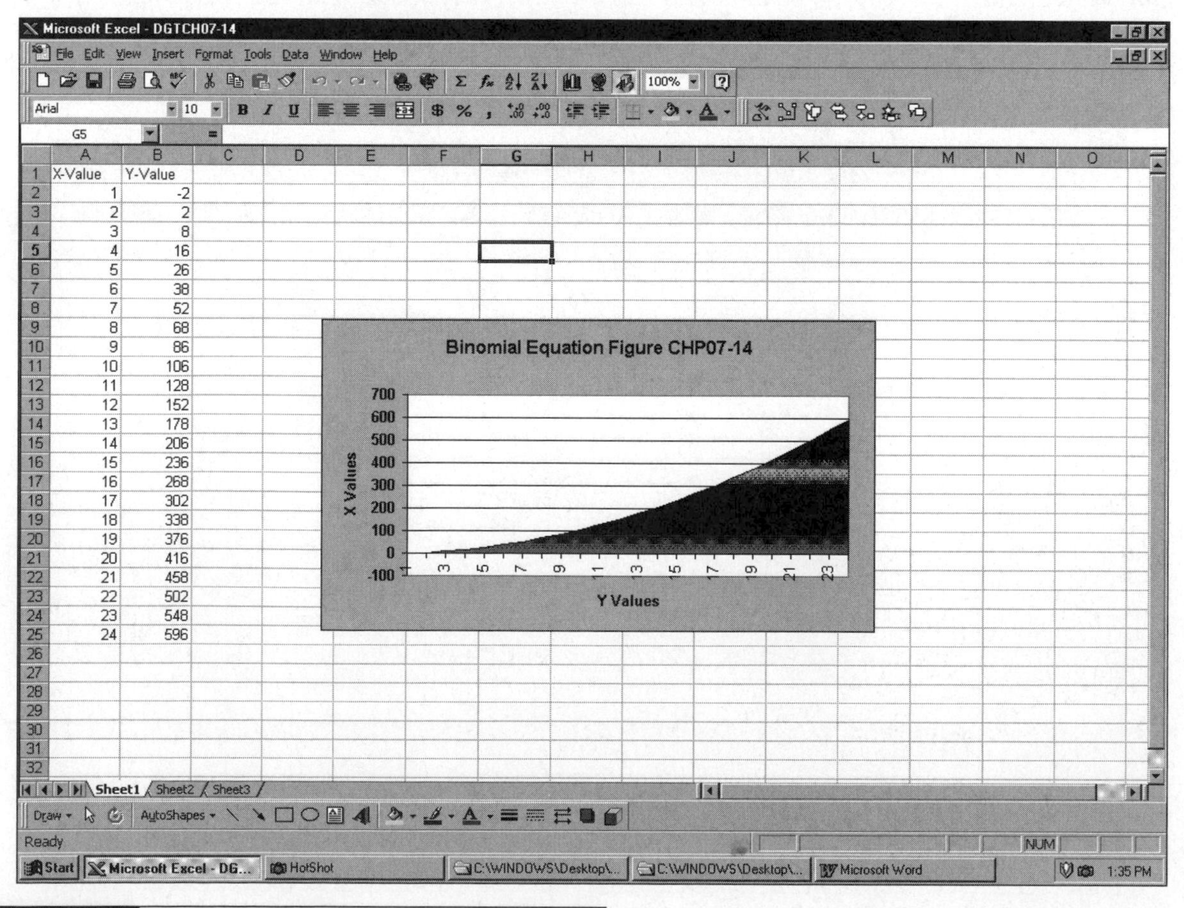

| X-Value | Y-Value |
|---|---|
| 1 | -2 |
| 2 | 2 |
| 3 | 8 |
| 4 | 16 |
| 5 | 26 |
| 6 | 38 |
| 7 | 52 |
| 8 | 68 |
| 9 | 86 |
| 10 | 106 |
| 11 | 128 |
| 12 | 152 |
| 13 | 178 |
| 14 | 206 |
| 15 | 236 |
| 16 | 268 |
| 17 | 302 |
| 18 | 338 |
| 19 | 376 |
| 20 | 416 |
| 21 | 458 |
| 22 | 502 |
| 23 | 548 |
| 24 | 596 |



**Figure 7–14** *Excel spreadsheet containing a graph.*

**Building a Spreadsheet in Microsoft Excel and Verifying the Answer with AutoCAD**

This section combines the concepts covered for both spreadsheets and trigonometry. The following example will illustrate how a system of concurrent coplanar vectors can be solved using an Excel spreadsheet. The first vector has a magnitude of 54 psi and a direction of 35° counterclockwise from the X-axis. The second vector has a magnitude of 23 psi and a direction of 23° counterclockwise from the X-axis. In this example, the first column of the spreadsheet will be used to hold the magnitude of the vector (A2 = 54 and A3 = 23), while the second column is dedicated to the angle of the vector from the X-axis (B2=35 and B3=23). The third column will be used to determine the X-component of the vector by using the equation =+(cos (B2 / 57.32) * A2). This equation calculates the value of the X-component by first dividing the value of the cell B2 by 57.32 (converting the angle into radians). Next, this value is passed to the cosine function, where the value of cosine for that angle is determined. Finally, the value for cosine for that particular angle is multiplied by the value of cell C2, thus yielding the X-component of the vector. The fourth column will be used to determine the Y-component of the vector, in a manner similar to the way the X-component was calculated. In this equation, the value of B2 is divided by 57.32, passed to the sine function, and then multiplied by A2. Cells A1 and B1 were not used in this equation because those cells are used as ***headers***. A ***header*** is a title used to describe the content of a column. It is intended to help keep track of the information residing within that column.

By adding four more columns to this spreadsheet, the total X-value, the total Y-value, the magnitude of the resultant, and its direction from the X-axis, can also be calculated. To find the total X-value, the formula =+C2+C3 is inserted into cell E5. To find the total Y-value, the formula =+D2+D3 is placed in cell F5. To calculate the magnitude of the resultant, the Pythagorean theorem ($R^2 = X^2 + Y^2$) must be employed. Adapting this equation to a spreadsheet format yields the following formula: =+SQRT (E5^2+F5^2). This formula first squares the values in the cells F5 and E5, then combines their product, and takes the square root of the sum. This formula is placed in cell G5. To calculate the direction of the resultant, an equation must be derived that incorporates the arctangent function, =+(ATAN (F5/E5))*57.32. This equation divides the value of cell F5 by cell E5. The quotient is then passed to the arctangent function. Notice how the value returned by the arctangent function is multiplied by 57.32. This is because the results are in radians and must be converted to degrees. This formula is assigned to cell H5. Once completed, the spreadsheet should look like the one shown in Figure 7–15. If entered correctly, the spreadsheet will determine that the resultant for this example has a magnitude of 76.64701127 psi and a direction of 31.42300151° from the X-axis.

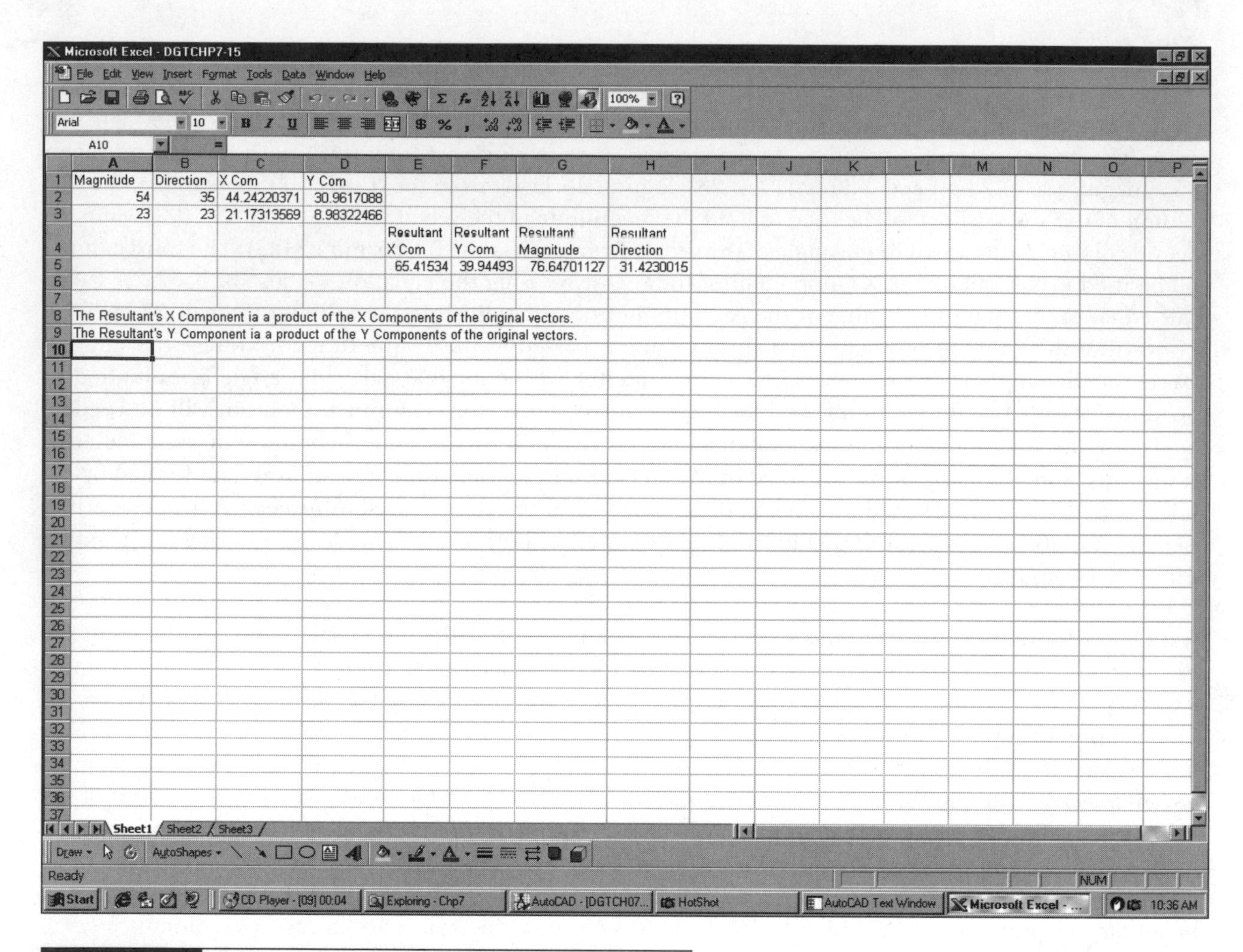


**Figure 7–15** *Excel spreadsheet used to resolve vectors.*

This result can be verified by using AutoCAD. First, a line 54 units long at an angle of 35° counterclockwise to the X-axis is created using the LINE command. Without terminating the LINE command, another line segment 23 units long at an angle of 23° counterclockwise to the X-axis is drawn (see Figure 7–16). This is followed by the C option of the LINE command, which completes the polygon and terminates the command. Finally, using the LIST command, the line segment that connects the last vector to the first vector is selected to reveal the results (see Figure 7–17).

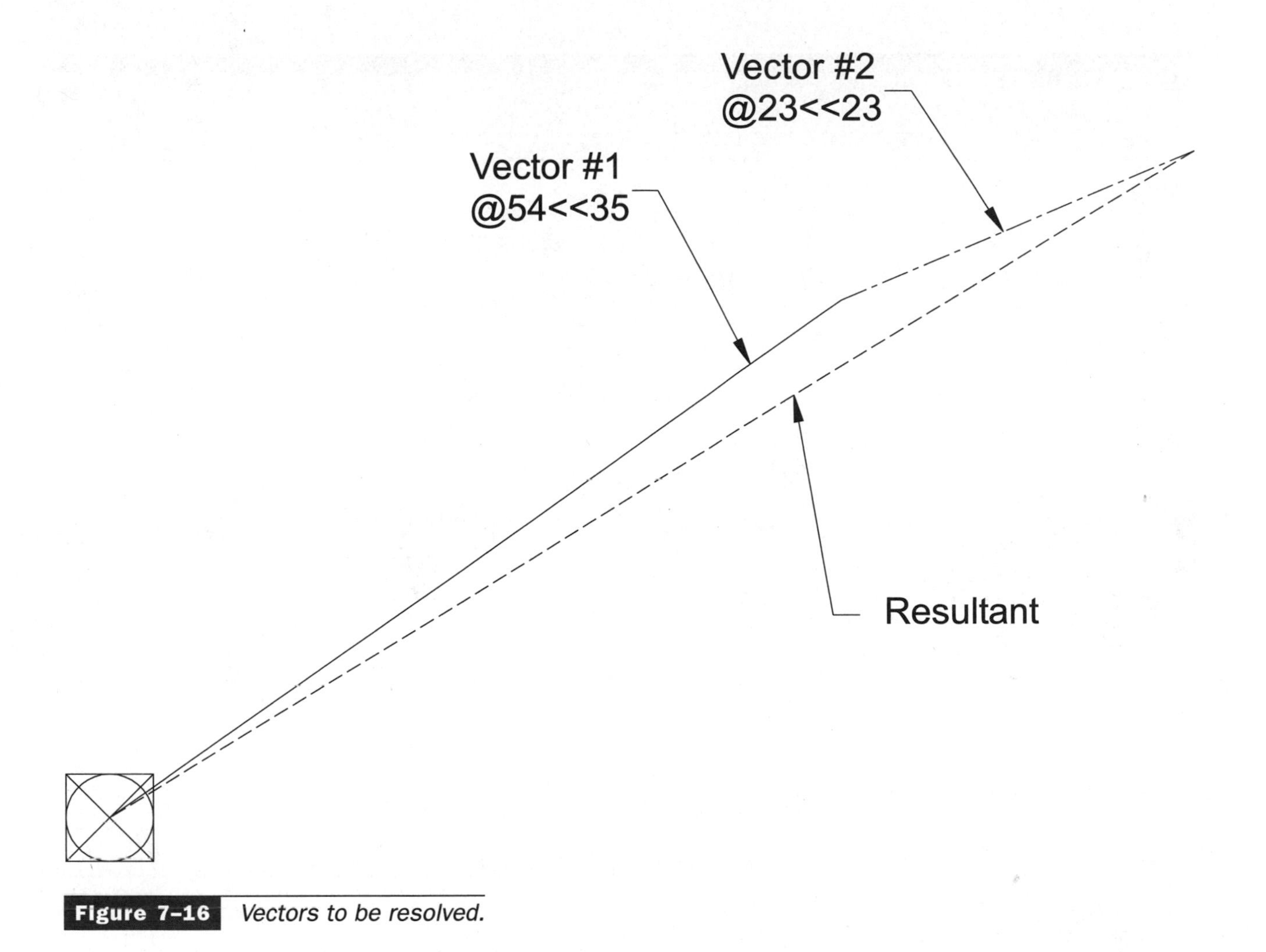


**Figure 7-16** *Vectors to be resolved.*

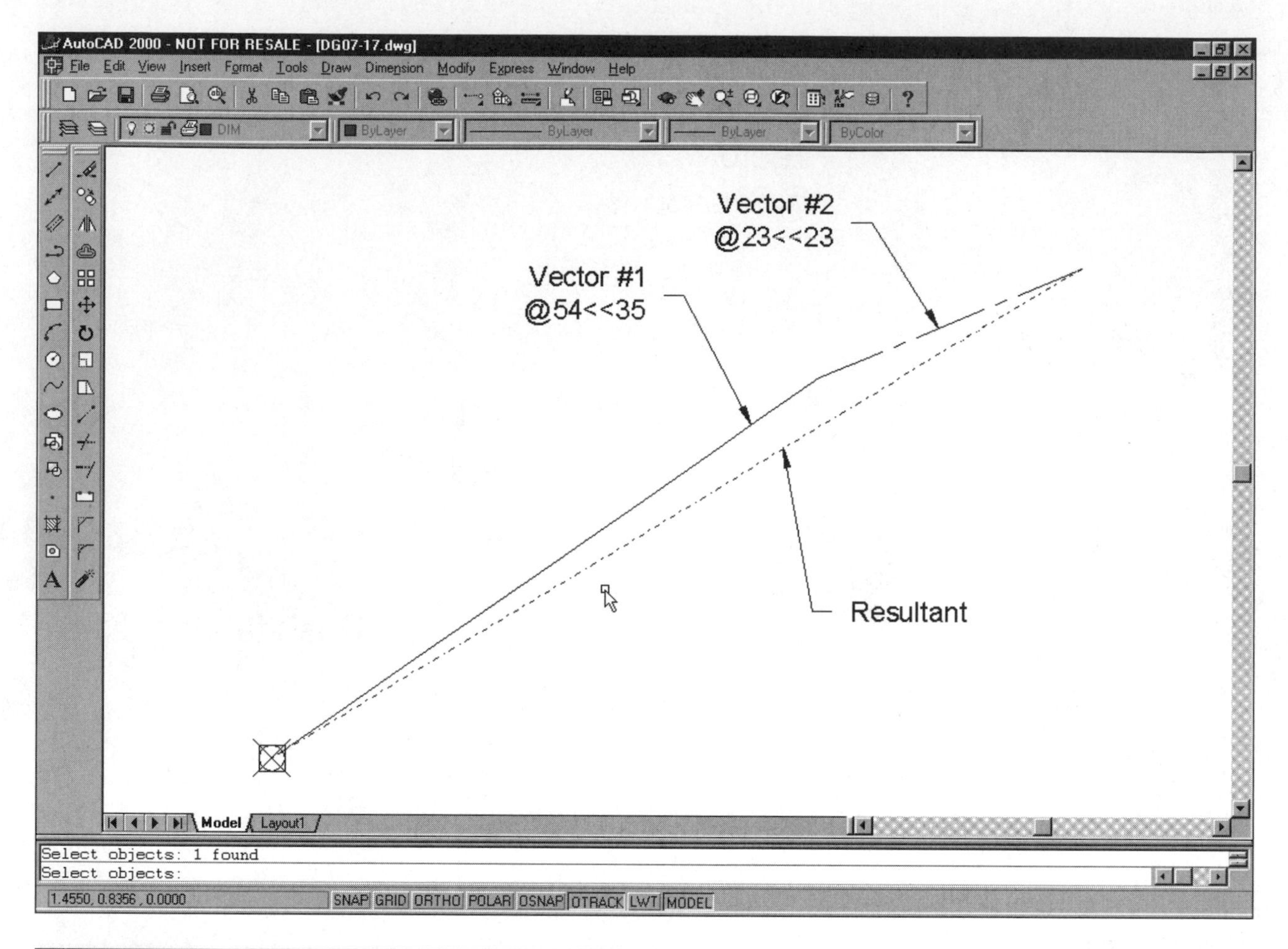


**Figure 7-17** *Resolving vector problem using AutoCAD.*

Command: LINE (ENTER)
From point: *(Select an arbitrary point.)*
To point:@54<<35 *(Press* ENTER. *Use the polar coordinate system to draw a line 54 units long at an angle of 35° counterclockwise from the X-axis.)*
To point: @23<<23 *(Press* ENTER. *Use the polar coordinate system to draw a line 23 units long at an angle of 23° counterclockwise from the X-axis.)*
To point: C *(Press* ENTER. *Closes the polygon and terminates the* LINE *command.)*
Command: LIST (ENTER)
Select objects: 1 found *(Select the last line drawn.)*
Select objects: (ENTER)

```
          LINE      Layer: 0
                    Space: Model space
           Handle = 2A
     from point, X=  72.2521  Y=  43.0010  Z=   0.0000
       to point, X=   6.8463  Y=   3.0410  Z=   0.0000
  Length =  76.6467,  Angle in XY Plane =    211
         Delta X = -65.4058, Delta Y = -39.9599, Delta Z =  0.0000
```

**Note: Recall from the previous section where AutoCAD was used that to solve a vector using the polygon method, the answer must be subtracted from 180° in order to reposition the resultant into the proper quadrant and produce the correct answer.**

## Finding the Resultant of Noncoplanar Vectors

To complicate matters, not all vectors are concurrently coplanar. Another class of vectors exist which share a common point but reside on different planes. This type of vector system is known as ***concurrent noncoplanar vectors*** (see Figure 7–18). These vectors can be combined using a method called the ***vector-polygon method.*** This method constructs a three-dimensional polygon similar to the one used earlier in the polygon method. The vector-polygon method can also be adapted to AutoCAD to provide a very efficient vehicle for combining noncoplanar vectors. Using AutoCAD, draw a line representing the first vector, then rotate the UCS and draw the second vector, rotate the UCS again, and draw the next vector, repeating this process until all the vectors have been drawn. To find the resultant, either use the DIST command and select the origin of the first vector followed by the endpoint of the last vector, or create a line between these two points and then use the LIST command (see Figure 7–19).

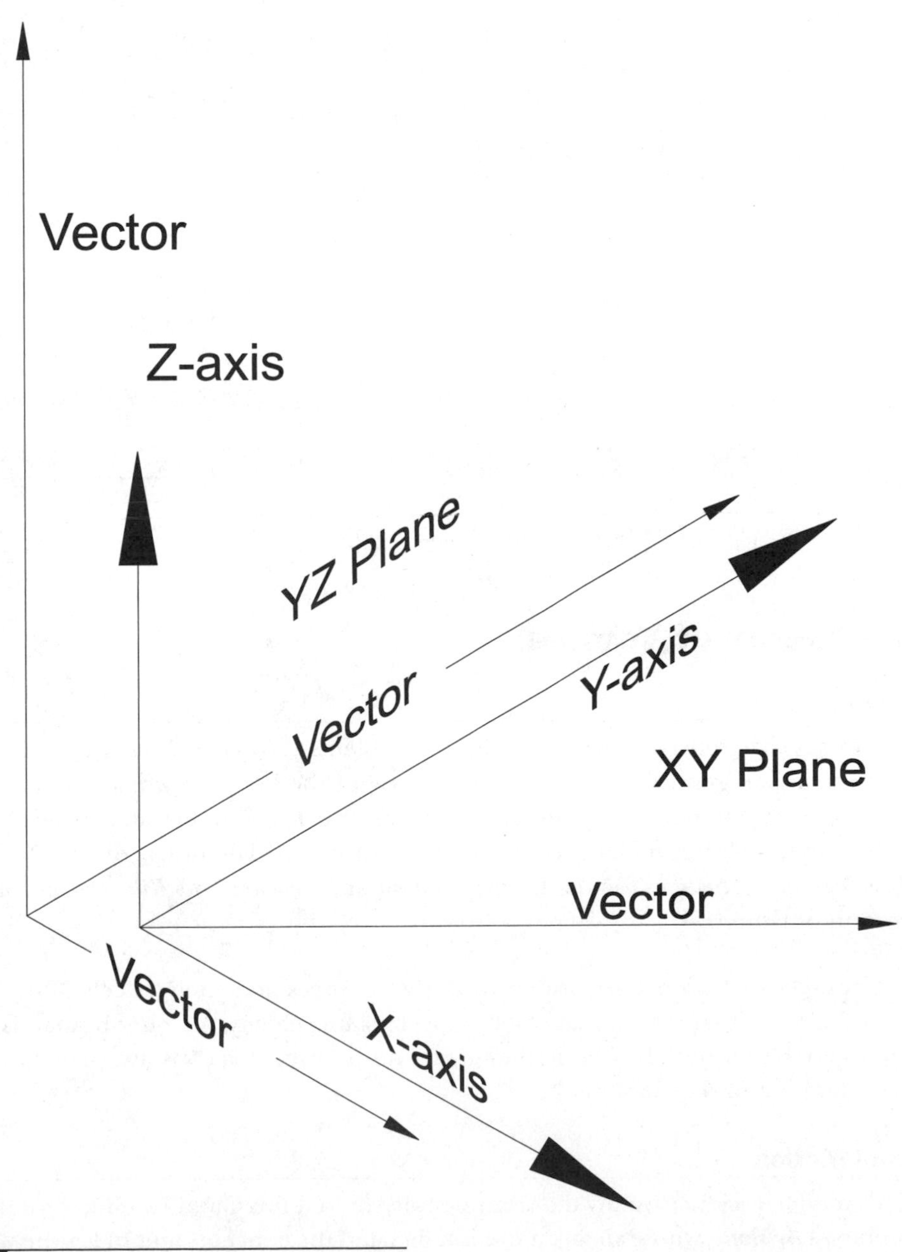


**Figure 7–18** *Noncoplanar system of vectors.*

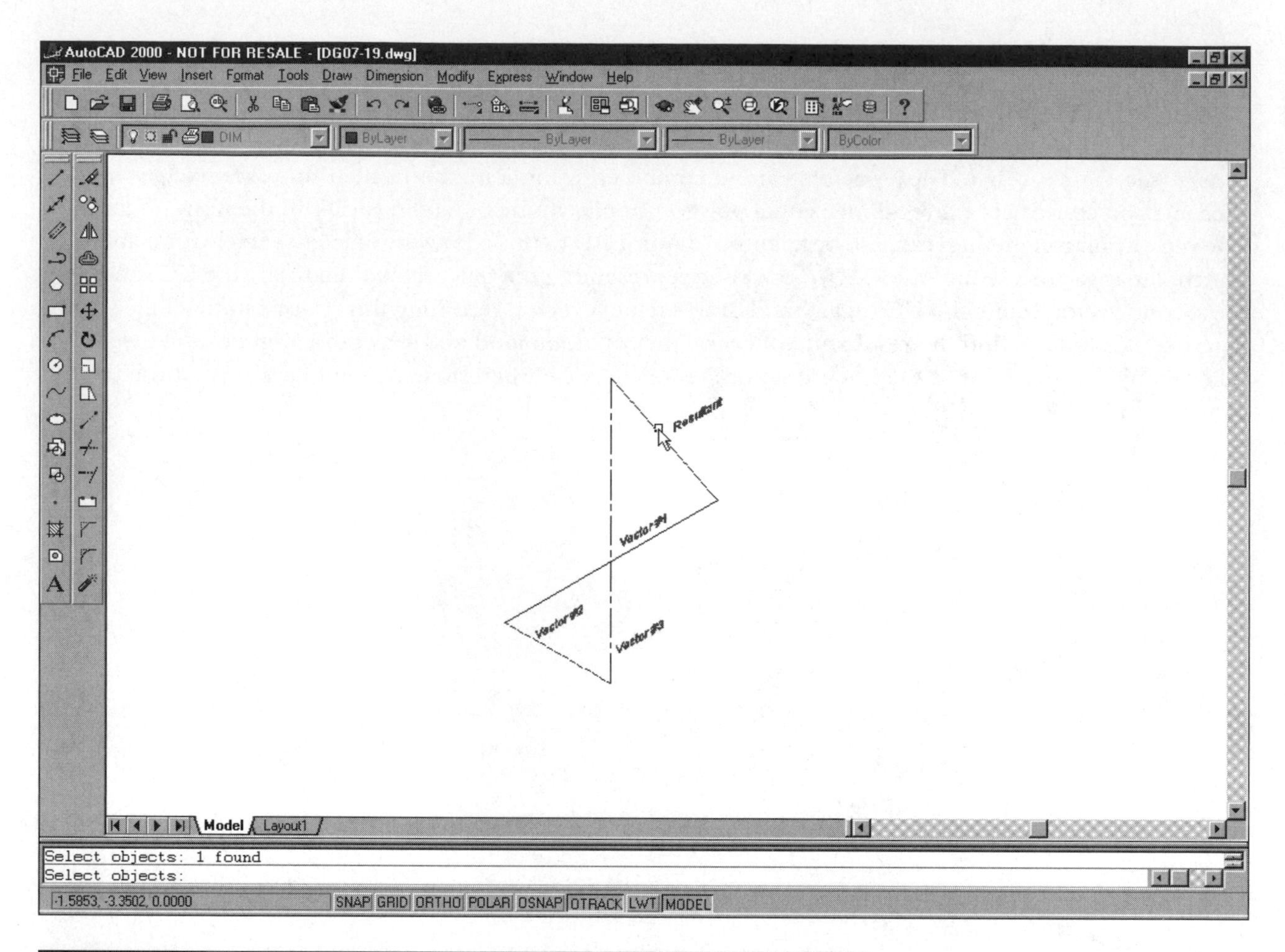


**Figure 7-19** *Solving a noncoplanar system of vectors using AutoCAD.*

# Applying the Theory of Vectors

## Characteristics of a Force

The concept of vectors is best illustrated when applied to problems of force. A ***force*** is a vector quantity which is inclined to produce a change in the motion or the shape of the body on which it acts. There are four classifications of forces: gravitational, electromagnetic, strong (nuclear), and weak (nuclear). Forces can be applied, non-applied, external, internal, concentrated, or distributed. The units commonly used to express force are ***pounds***, ***kilopounds***, and ***tons*** for the English system and ***newtons*** and ***Kilonewtons*** for the SI system (International System of Units).

There are several branches of science dedicated to studying forces and their effects. One branch, ***statics***, studies force and its effects on rigid bodies at rest or at a constant velocity. Another branch is ***dynamics***—the study of force and its effects on rigid bodies neither at rest nor moving at a constant velocity. In other words, objects that are accelerating or decelerating.

## Newton's Laws of Motion

In the early days of modern science, many different people studied force and its effects on objects. Among these early pioneers was ***Aristotle***, whose views on motion dictated the concepts taught for the next 1500 years.

Aristotle reasoned that the natural state of an object was at rest, and would remain that way unless acted upon by another force or object. Aristotle's conclusions were based on logic and his observations of nature. It was not until ***Galileo*** that Aristotle's views on motion were challenged. Galileo concluded that bodies in motion had a tendency to stay in motion, and bodies at rest tended to remain at rest unless acted on by another force. Galileo referred to this tendency as ***inertia***. However, it was ***Newton*** that made the final contribution to the study of motion. It is this contribution that formed what is known today as ***Newton's Laws of Motion.***

### First Law of Motion

Newton's First Law of Motion states that *if the algebraic sum of all forces in a system is equal to zero, then the system will retain its present state.* In other words, a body will remain either in motion or at rest until it is acted upon by an outside force. This is an important concept when doing a force analysis. It basically states that the net force of a static system is equal to zero. A static system is one that is either at rest or travels at a constant velocity; it is not accelerating or decelerating. This concept is summarized in the equation: the sum of the forces is equal to zero or $\Sigma F=0$.

### Second Law of Motion

Newton's Second Law of Motion states that *if the algebraic sum of the forces in a system is not equal to zero, then the net result will be that the system will accelerate or decelerate. The acceleration or deceleration of the system will be directly proportional to the force that is applied and inversely proportional to the mass of the system. The system will accelerate in a straight line to the force (in the same direction).* In this law, Newton states that when an unbalanced force is applied to an object, the result is acceleration. The amount of acceleration is determined by dividing the force applied by the object's mass (how much matter an object contains). ***Acceleration*** is defined as a change in velocity. This law is summarized with the equation: force is equal to mass times acceleration, or $F=Ma$. This concept explains why the astronauts aboard the space shuttle are able to float in space while traveling at a speed of 17,500 miles per hour. The shuttle is moving at a constant rate (velocity), so the force due to acceleration is equal to zero. In this example, the mass of the object does not change, it stays constant. The lack of acceleration causes the force to equal zero, and permits the astronauts to float.

The equation for Newton's second law can be refined to include acceleration due to gravity. The equation becomes $F=Mg$, or force is equal to the mass of an object times the acceleration due to gravity. This equation is used with free-falling objects. The force that is calculated is also known as weight, so the equation now becomes $W=Mg$, or weight is equal to the mass of an object times the acceleration due to gravity. This explains why people in space are weightless, while people on earth have weight; people in space are not being acted upon by gravity while people on earth are.

### Third Law of Motion

Newton's Third Law of Motion is perhaps the most famous all of the laws of motion. It states that *when an object transmits a force to another object, the net result is the receiving object will transmit a force of equal magnitude in the opposite direction.* More specifically, one might say that *for every action there is an equal and opposite reaction.* This concept can be illustrated by a common experience—standing. In order to stand, an individual must exert force through their feet against the ground. This force is counteracted by the ground applying a force of equal magnitude in the opposite direction; this permits a person to stand. Imagine what would happen if the ground did not exert this force back (see Figure 7–20)?

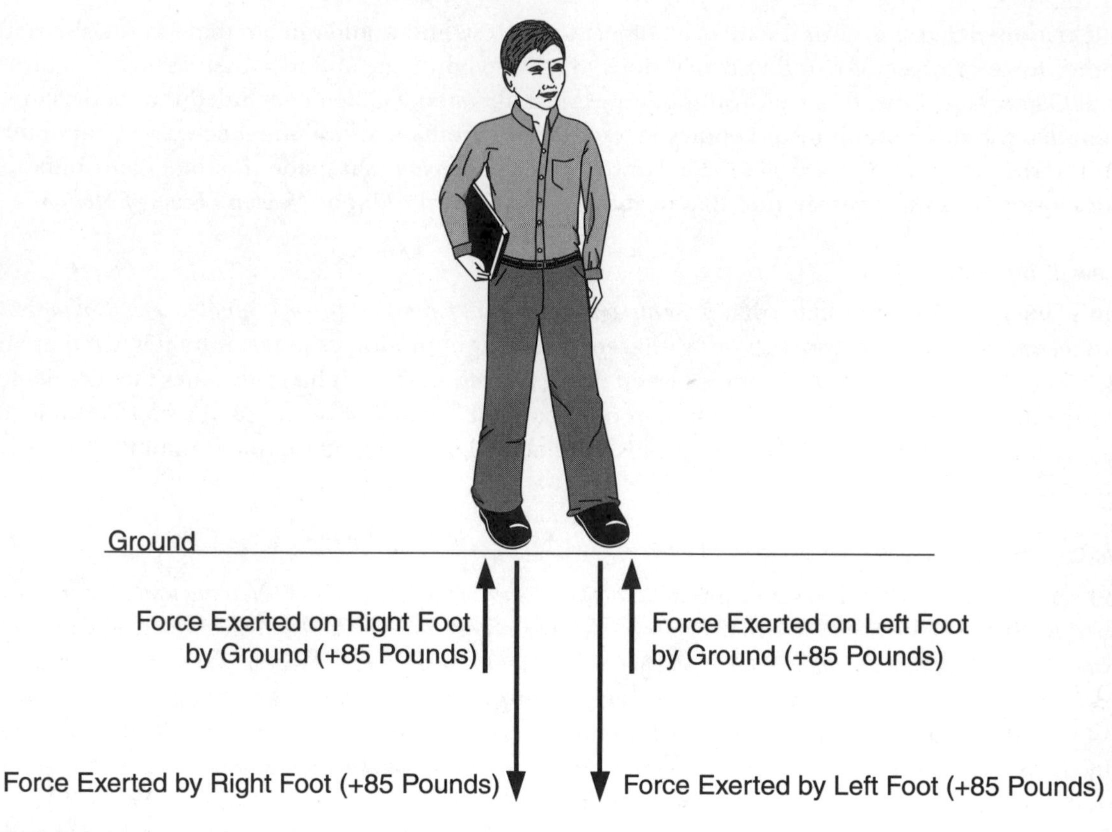


**Figure 7-20** *Equal and opposite reaction*

## Determining the Direction of a Force

When solving force problems, it is necessary to determine the direction a force is applied. The most common method in the engineering field is to assign a positive or negative sign to the magnitude of the force, but any method may be used. The only requirement is consistency; it is critical to use the same method throughout the calculation. In the *Positive/Negative system*, a force acting along the positive X-axis is considered positive, while a force acting along the negative X-axis is considered negative. The same conclusions hold true for forces in the Y- and Z-directions. The system used for moments (which will be covered in the next section) is clockwise positive and counterclockwise negative.

## Equilibrium of a System of Forces

If a system of forces is not accelerating, then it is said to be ***static*** or in ***equilibrium***. To create a system of forces that is in equilibrium, Newton's first and third laws of motion must be applied. All unbalanced forces must be removed by applying a force of equal magnitude and opposite in direction to each force in the system. The algebraic sum of all forces must be equal to zero ($\Sigma F=0$) for the system to be considered in equilibrium. Since a force is a vector and vectors can be resolved into their X- and Y-components, then the equation for equilibrium must be rewritten to allow for forces acting in different directions. Once rewritten, the equation can be stated as: the sum of the forces along the X-axis must equal zero, the sum of the forces along the Y-axis must equal zero, and the sum of the forces along the Z-axis must equal zero, or $\Sigma F_X=0$, $\Sigma F_Y=0$, and $\Sigma F_Z=0$. These formulas can be used when solving force problems where one force is missing. A basic rule of both science and mathematics states that when solving a problem that contains multiple unknown values, a separate equation must be used for each unknown value. For example, a triangle with only one angle and one side given could not be solved by using the equation $R = \sqrt{(X^2 + Y^2)}$. This is because both the hypotenuse and one of the sides are missing. Instead, one of the trigonometric functions along with the Pythagorean theorem must be applied to find the missing side.

It is common when solving a force problem that more than one value for a force will be missing along a given axis. When this occurs, the ***moment*** about one of the missing force must be considered. A ***moment*** is the tendency of a force to cause rotation about a point. The rotation can be either clockwise or counterclockwise. When dealing with forces, a moment is found by multiplying the force times the distance from the point of rotation, or (force) (distance). Therefore, to keep an object in equilibrium a new statement and equation must be added to the list: the sum of all moments about a point must be equal to zero or $\Sigma M_P=0$.

The concepts of equilibrium and moments can be combined with the information about vectors to calculate many common problems associated with structural, mechanical, civil, industrial, and electrical engineering. A considerable amount of time should be spent studying these types of problems to become proficient with their calculation.

## Free-Body Diagrams

Constructing a diagram is extremely critical when solving problems involving force. The diagram will contain information about the object to which the force is being applied, along with information concerning the force itself. These diagrams can be divided into two categories, ***space diagrams*** and ***free-body diagrams***. In a ***space diagram*** the physical relationship of the object and the physical location of the force is illustrated. These diagrams are used to show the actual setup of the system. However, forces can sometimes emanate from a single point, which would require a ***free-body diagram*** (see Figure 7–21). This common point in a free body diagram is also used as the origin for a Cartesian coordinate system.

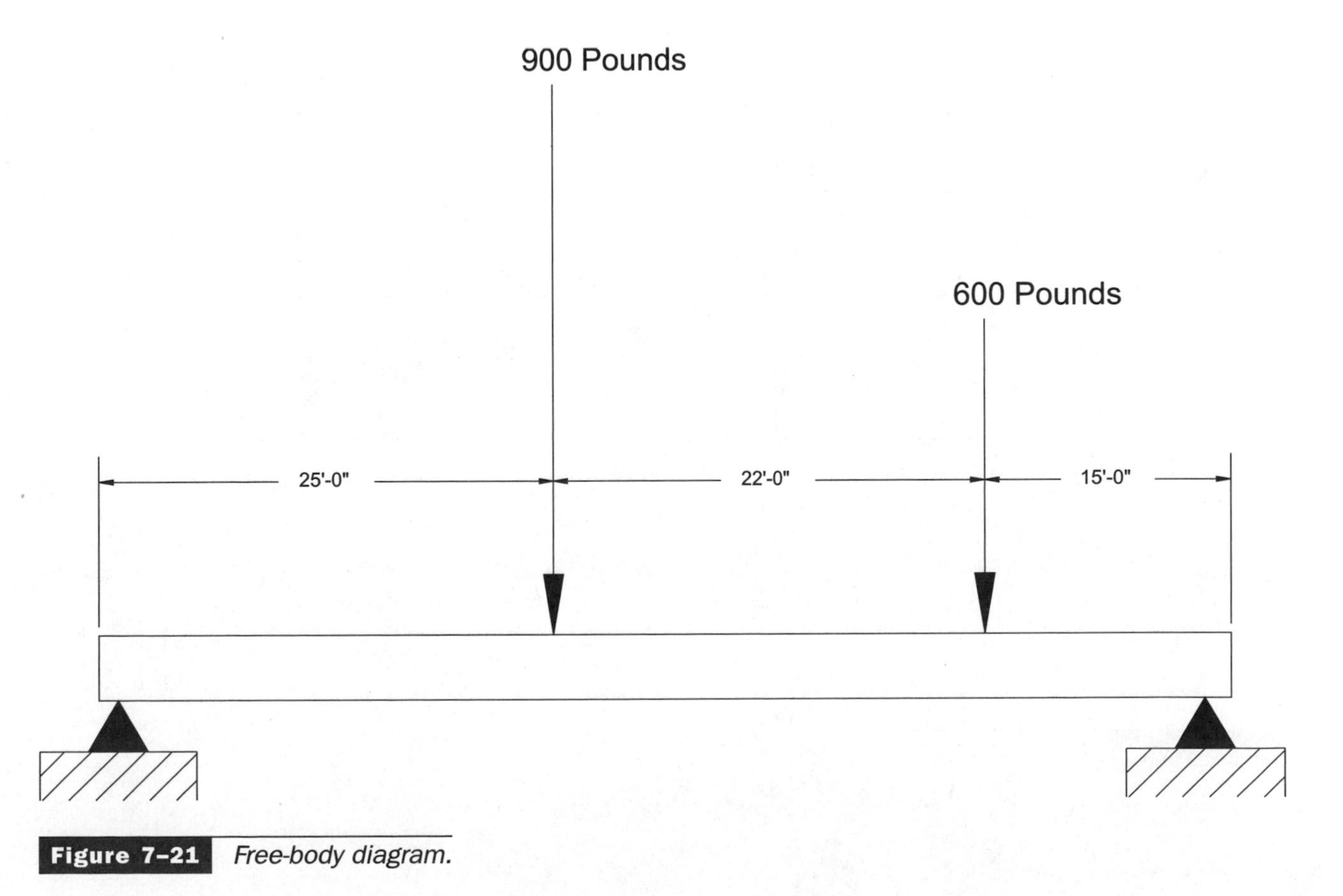


**Figure 7–21** *Free-body diagram.*

#### Using Excel to Solve Force Problems

Spreadsheets can be extremely helpful in obtaining solutions to force problems. By adapting and slightly modifying the concepts previously discussed, a spreadsheet can be developed that will answer "What if" questions, unlike a database, which is primarily used to store persistent data. For example, given a beam supported at both ends with several forces acting upon it, a technician using spreadsheets could easily move the locations of the forces and/or change their magnitudes to study the effects it will have upon the beam (see Figure 7–22). Although a single spreadsheet cannot be developed to solve every force problem encountered, several can be developed to solve the most common problems performed by the technician.

With slight modifications, the spreadsheet developed earlier for solving vector problems can be used to calculate the forces required at $R_1$ and $R_2$ to keep the object shown in Figure 7–23 in equilibrium. In this example, all forces must be resolved into their X- and Y-components. The force labeled $F_1$ will have an X- and Y-component, while the force label $F_2$ will have only a Y-component because it reacts perpendicular to the object. The force required at $R_1$ is found by adding the Y-components of the force labeled $F_1$ and $F_2$. The force required at $R_2$ would simply be the X-component of the force labeled $F_1$. This problem will require the use of both equations, $\Sigma F_X=0$ and $\Sigma F_Y=0$, because it has two unknown values.

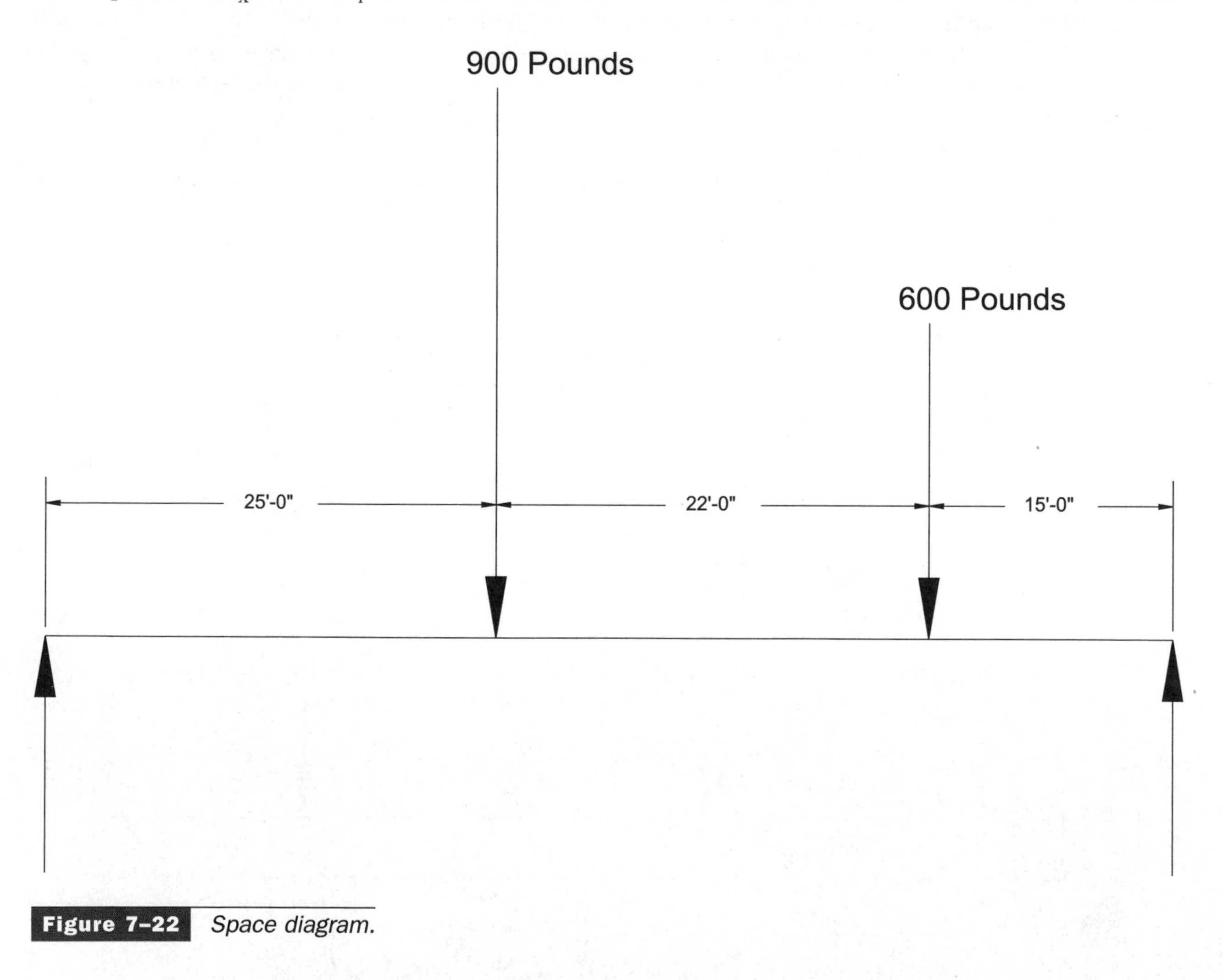


**Figure 7–22** *Space diagram.*

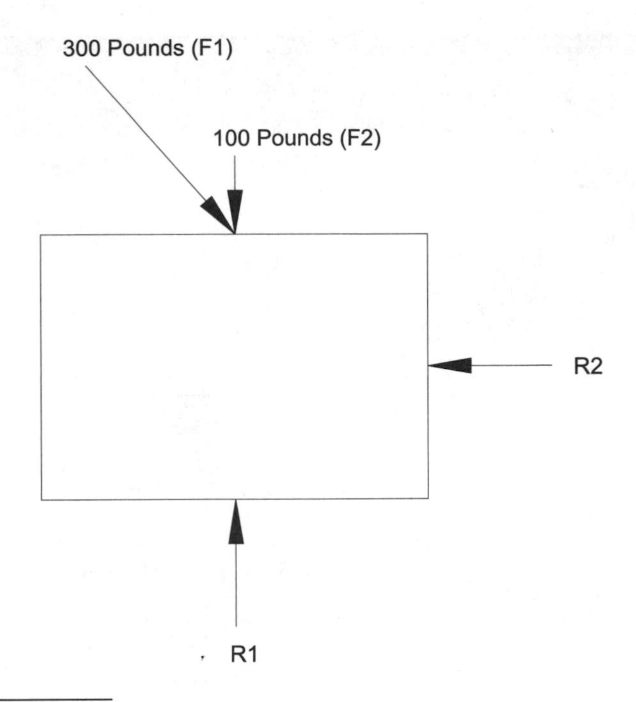


**Figure 7–23** *Free-body diagram.*

Taking this spreadsheet one step further, it can be developed to solve a system of forces that involves moments by adding the equation $\Sigma M_p=0$. Figure 7–24 will be used to illustrate this type of problem. First it must be decided what information is necessary to do the calculations and then how to physically layout the spreadsheet. For this spreadsheet, both known forces and distance from the end of the beam to the force are needed. For the physical layout of this spreadsheet, the format shown in Figure 7–25 is used for simplicity.

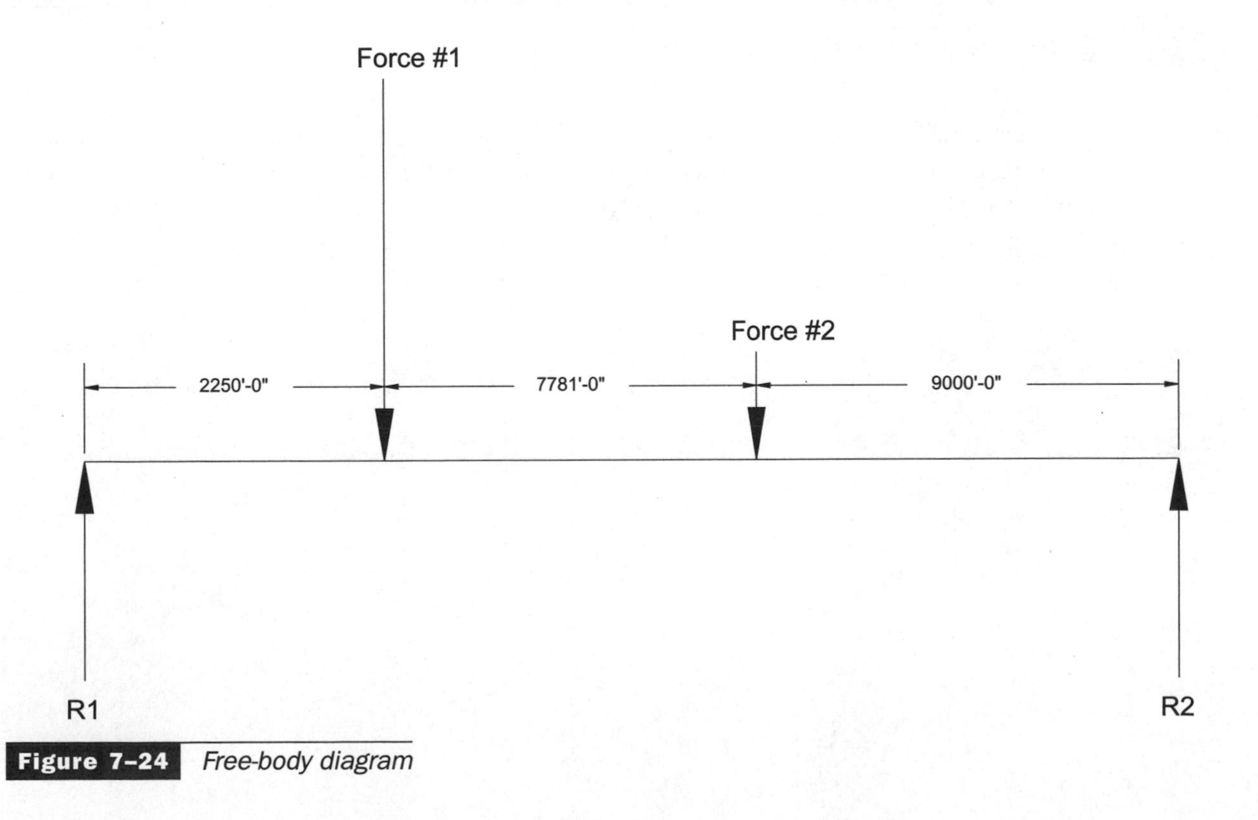


**Figure 7–24** *Free-body diagram*

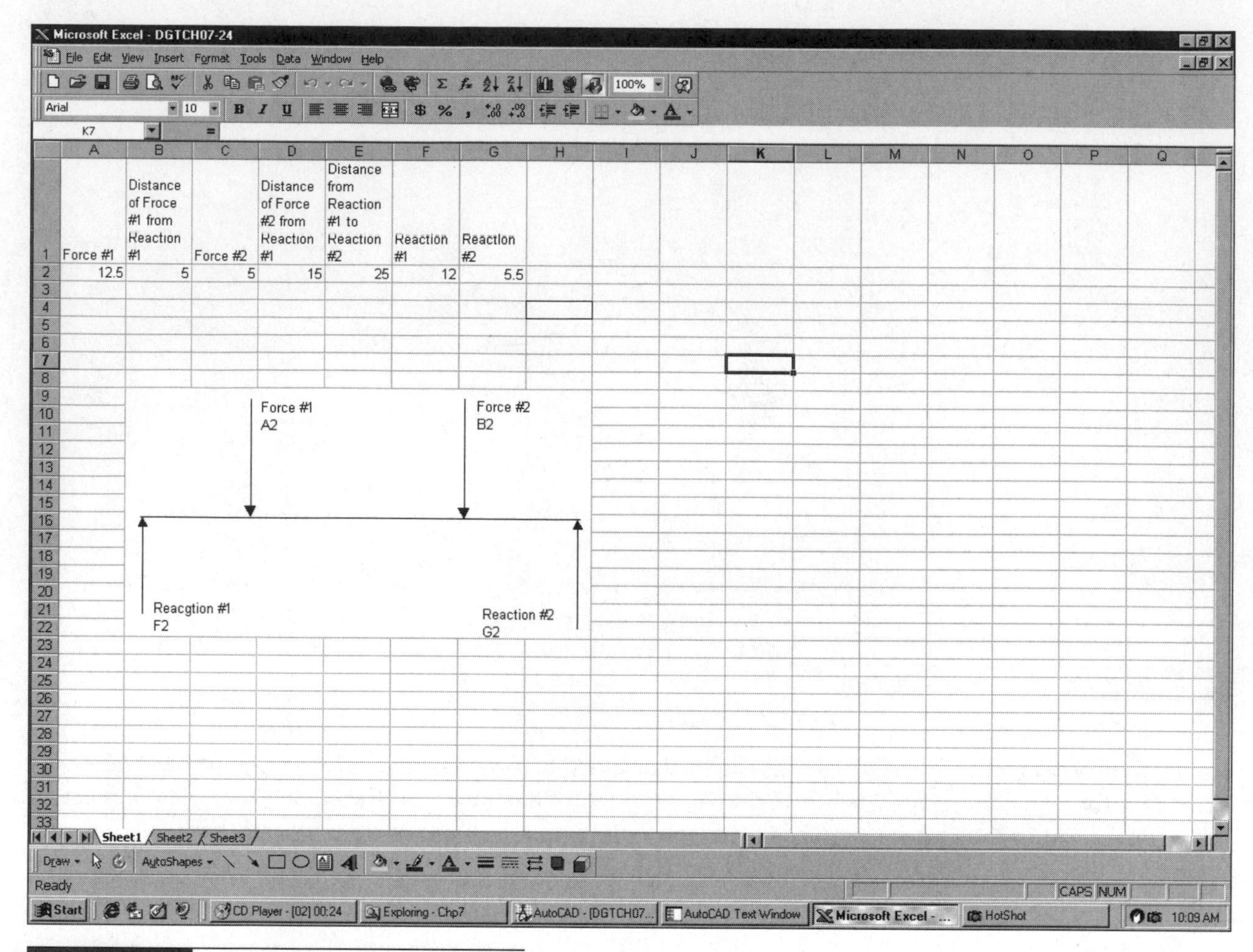


**Figure 7-25** *Load calculated using Excel*

Before the required reaction at point $R_1$ can be calculated, the formula $\Sigma M_p=0$ must first be used to calculate the reaction at $R_2$. After this information has been obtained, the formula $\Sigma F_Y=0$ can be used to calculate the reaction at $R_1$. The equations $\Sigma M_p=0$ and $\Sigma F_Y=0$, when expressed in spreadsheet format, will resemble the following: =(((A2)*(B2))+((C2)*(D2)))/E2, and =((A2)+(C2))–(G2).

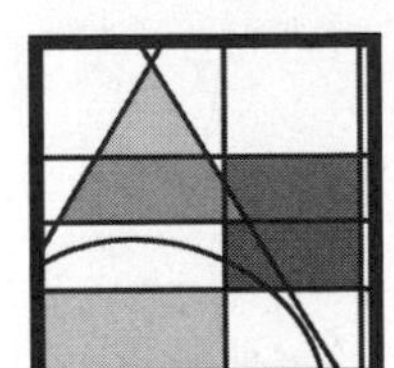

# advanced applications

## Solving Concurrent Coplanar and Coplanar Vector Problems Using AutoLISP and DCL Programming Languages

There are many programs available on the market today that will do force analysis. These programs range in price from a hundred dollars to several thousand dollars. They also range in complexity from the very simple to very complex finite element modeling. However, with proficiency in vector analysis and a good understanding of programming, AutoLISP, ADS, Visual Basic, or Visual AutoLISP can be used to solve these problems. When properly planned, AutoCAD, along with the use of third-party software or in-house software, can save many hours of calculation and even help prevent costly mistakes. Provided below is an example of an AutoLISP program that could be further developed and enhanced into a good vector analysis program. When executed in AutoCAD, the program will produce the dialog boxes shown in Figures 7–26 and 7–27.

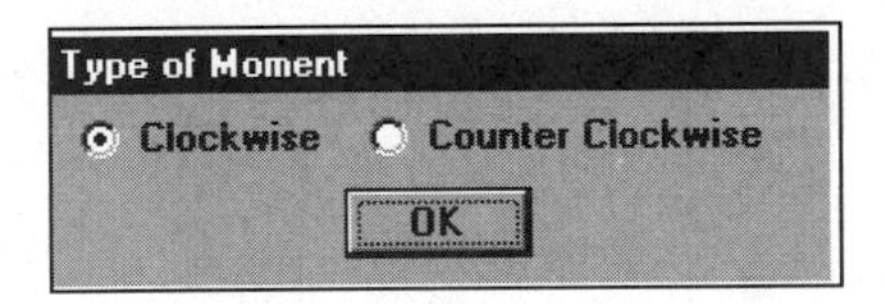


**Figure 7–26** *Direction dialog box.*

Force Analysis Results
The force needed at point A is 35.2102
The force needed at point B is -19.4984
OK

**Figure 7–27** *Result dialog box.*

## AutoLISP Program

```
;;;*************************************************************************
;;;
;;;     Program Name: DG07.lsp
;;;
;;;     Program Purpose: This program will analysis a system of forces and
;;;                      the necessary reaction in order for the system to
;;;                      be in equilibrium.
;;;
;;;     Program Date: 10/25/98
;;;
;;;     Written By: James Kevin Standiford
;;;
;;;*************************************************************************
```

```
;;;*************************************************************************
;;;
;;;                    Main Program
;;;
;;;*************************************************************************
(defun c:force ()
  (setqnum3 0
      num2 0
      ent  (nentsel "\nSelect one-line reaction beam : ")
  )
  (if (/= ent nil)
    (progn
      (setq pt (assoc 10 (entget (car ent)))
           pt2 (assoc 11 (entget (car ent)))
           enty (nentsel "\nSelect first vector : ")
      )
      (if (= enty nil)
       (princ "\nNo Vector was selected : ")
      )
      (while (/= enty nil)
       (if (/= enty nil)
         (progn
           (setq enty  (entget (car enty))
                dcl_id4
                       (load_dialog
                         "c:/windows/desktop/geometry/student/autolisp programs/DG07.dcl"
                       )
           )
           (if(not (new_dialog "force" dcl_id4))
             (exit)
           )
           (set_tile "clo" "1")
           (action_tile "accept" "(for) (done_dialog)")
           (start_dialog)
           (if(= clo "1")
             (progn
              (setq num  (- (setq num
                                 (abs (- (car (cdr (cdr (assoc 10 enty))))
                                         (car (cdr (cdr (assoc 11 enty))))
                                      )
                                 )
                            )
                            (* num 2)
                         )
                    num3 (+ (abs num) num3)
                    dis  (car (cdr (assoc 10 enty)))
                    num1 (* num dis)
              )
             )
           )
           (if(= ccl "1")
             (progn
              (setq num  (abs      (- (car (cdr (cdr (assoc 10 enty))))
```

```
                                (car (cdr (cdr (assoc 11 enty))))
                            )
                        )
                num3 (+ (abs num) num3)
                dis  (car (cdr (assoc 10 enty)))
                num1 (* num dis)
            )
          )
        )
        (setq num2 (+ num2 num1)
             enty (nentsel "\nSelect next Vector : ")
        )
        (set_tile "clo" "0")
        (set_tile "ccl" "0")
      )
    )
   )
   (setq posx (car pt2)
        ans2 (/ num2 posx)
   )
   (if (not (new_dialog "info" dcl_id4))
    (exit)
   )
   (setq firl "The force needed at point A is "
        secl "The force needed at point B is "
   )
   (set_tile "fir" (strcat firl (rtos (- num3 ans2))))
   (set_tile "sec" (strcat secl (rtos ans2)))
   (action_tile "accept" "(done_dialog)")
   (start_dialog)
   (setq tog 0)
  )
 )
 (princ)
)
(defun for ()
 (setqclo (get_tile "clo")
      ccl (get_tile "ccl")
 )
)
```

## Dialog Control Language Program

```
//%%%%%%%%%%%%%%%%%%%%%%%%%%%%%%%%%%%%%%%%%%%%%%%%%%%%%%%%%%%%%%%%%%%
//
//      Activates dialog box
//
//      Descriptive Geometry Chapter 7 DCL File Vector Force
//
//
//
//%%%%%%%%%%%%%%%%%%%%%%%%%%%%%%%%%%%%%%%%%%%%%%%%%%%%%%%%%%%%%%%%%%%
force : dialog {
```

```
label = "Type of Moment";
    : radio_row {
    : radio_button {
      key = "clo";
      label = "Clockwise";
    }
    : radio_button {
      key = "ccl";
      label = "Counter Clockwise";
    }
 }
    is_default = true;
    ok_only;
}
info : dialog {
      label = "Force Analysis Results";
    : text {
      key = "fir";
      width = 45;
    }
    : text {
      key = "sec";
      width = 45;
    }
    is_default = true;
    ok_only;
}
```

# Review Questions

Answer the following questions on a separate sheet of paper. Your answers should be as complete as possible.

1. Explain the difference between scalar quantities and vectors.
2. Define the following terms: concurrent, coplanar, noncoplanar, concurrent coplanar, and concurrent noncoplanar.
3. State Newton's three laws of motion and give an example of how each is applied.
4. What is the parallelogram method and when can it be used?
5. What is the polygon method and when can it be used?
6. Define these terms: columns, rows, and spreadsheet.
7. What are the three types of triangles?
8. When is a free-body diagram used in place of a space diagram?
9. Write the formula $X^3 + X^2 - X + 3/X$ in spreadsheet form.
10. When are the arc functions used?

# Review Exercises

## Vectors and Force

**#1 through #10**

Find the missing components of the right triangles shown below using the sine, cosine, and tangent functions. All problems marked with an * are supplied on the student CD-ROM.

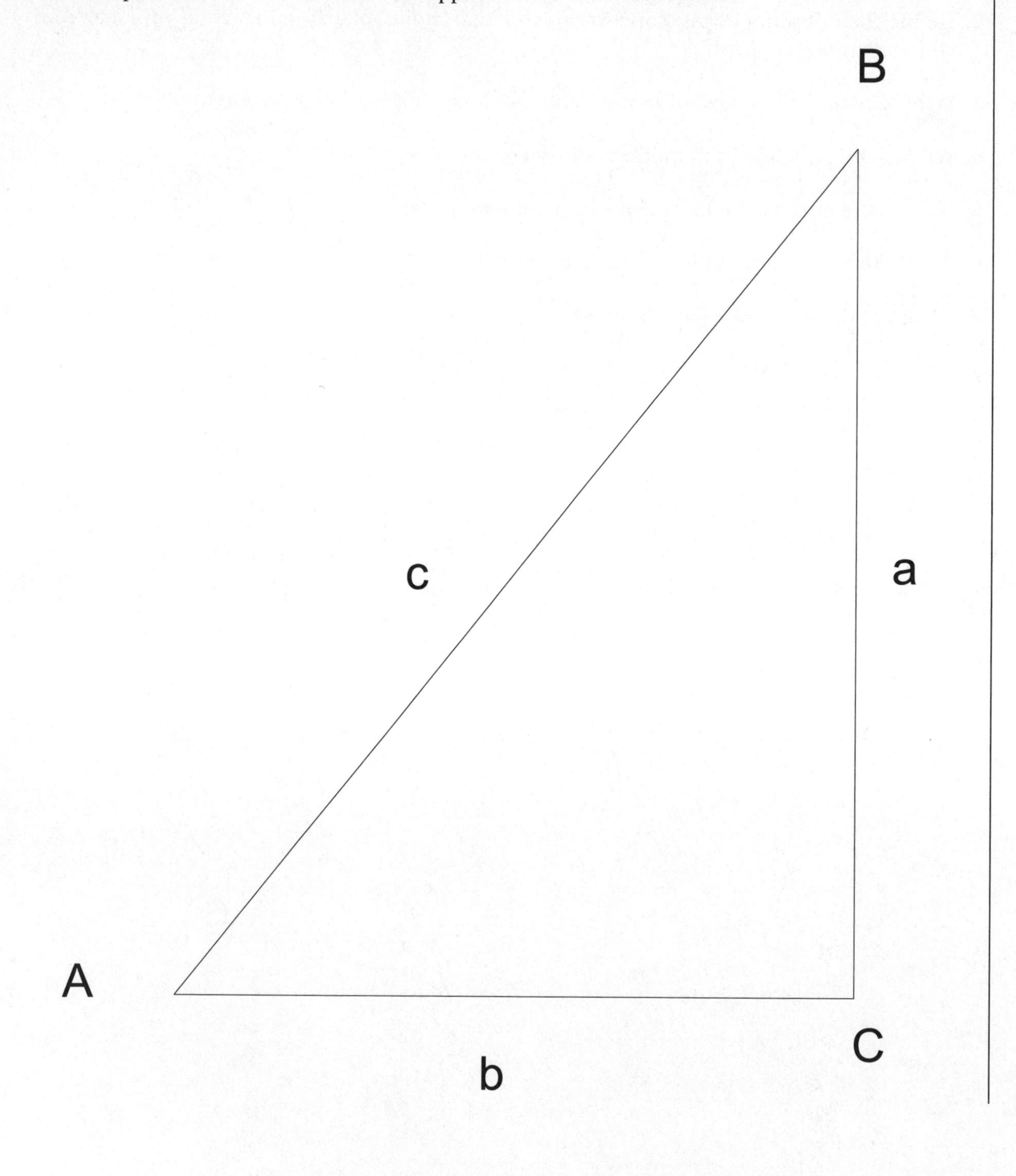

## Right Triangle Chart

| Problem # | Angle A | Angle B | Angle C | Side a | Side b | Side c |
|---|---|---|---|---|---|---|
| 1 | 53.1301° | | | 4 | 3 | 5 |
| 2 | | | | 2 | 10 | |
| 3 | 45° | | | 3 | 3 | |
| 4 | | 35.7535° | | 4.1667 | | 5.1344 |
| 5 | | | | 10.783 | 3 | |
| 6 | 29° | | | | | 2.22451 |
| 7 | | 30° | | | 1.5675 | |
| 8 | 76° | | | | | 2.7517 |
| 9 | | 77° | | 0.1574 | 0.6752 | |
| 10 | 82° | | | 3.5142 | | |

**#11 through #20**

Using AutoCAD, verify the results yielded from the above calculations.

***#21 through #25**

Using AutoCAD, solve the following vectors using either the parallelogram or polygon method.

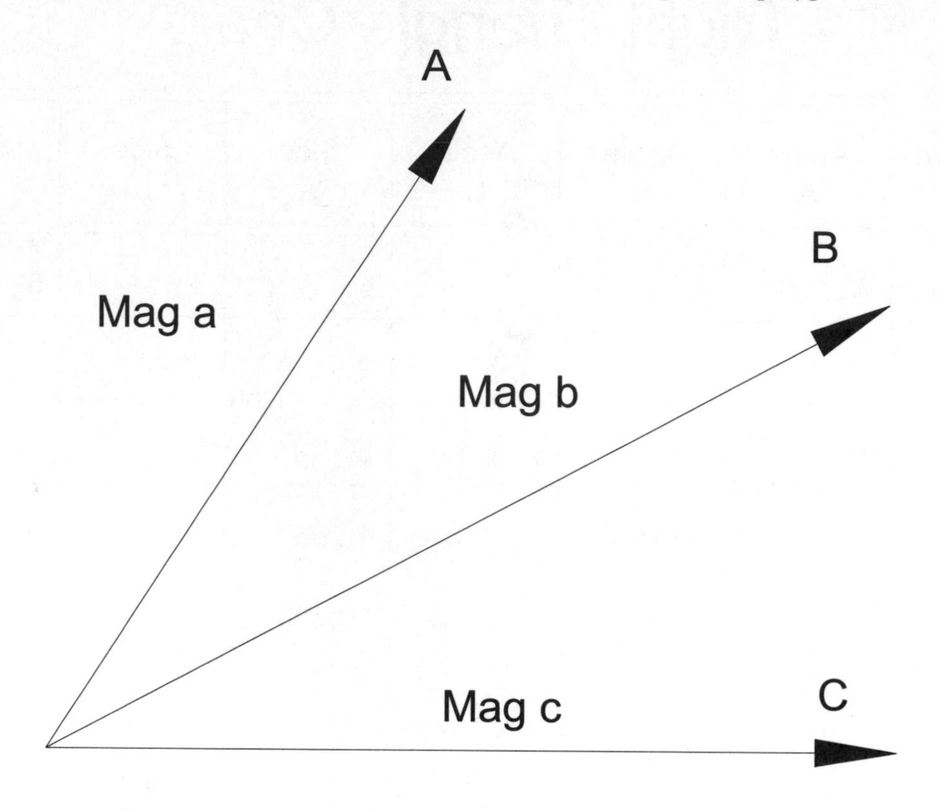


| Vector Chart | | | | | | | | |
|---|---|---|---|---|---|---|---|---|
| Problem # | Vector A Direction | Vector B Direction | Vector C Direction | Mag. a | Mag. b | Mag. c | Resultant | |
| | | | | | | | Mag. | Direction |
| 1 | 30° | 5° | not used | 10 | 1 | not used | | |
| 2 | 45° | 78° | 55° | 22.5 | 41 | 2.68 | | |
| 3 | 57° | 0° | 17° | 35 | 25.8 | 3.61 | | |
| 4 | 12° | 57° | not used | 15 | 14.9 | not used | | |
| 5 | 24° | 38° | 5° | 8 | 12.7 | 3.33 | | |

**#26 through #30**

Using Excel, verify the answers obtained by AutoCAD in problems 21 through 25.

***#31 through #35**

Calculate the missing force in the following problems.

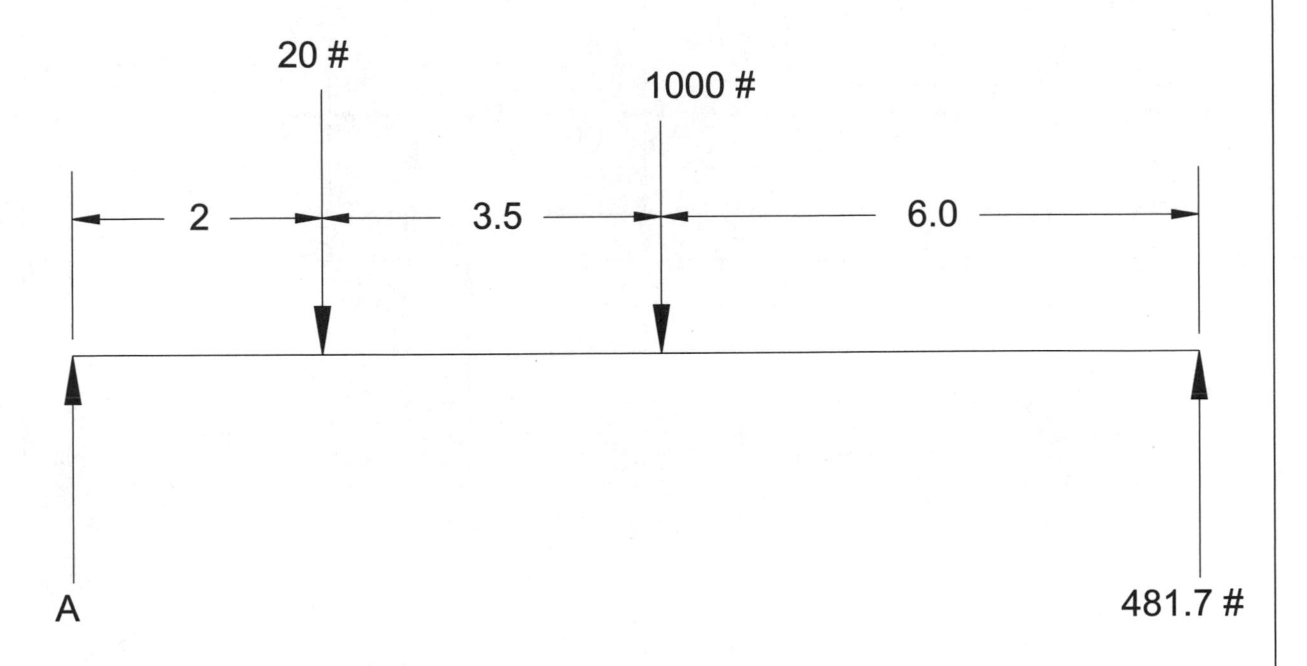


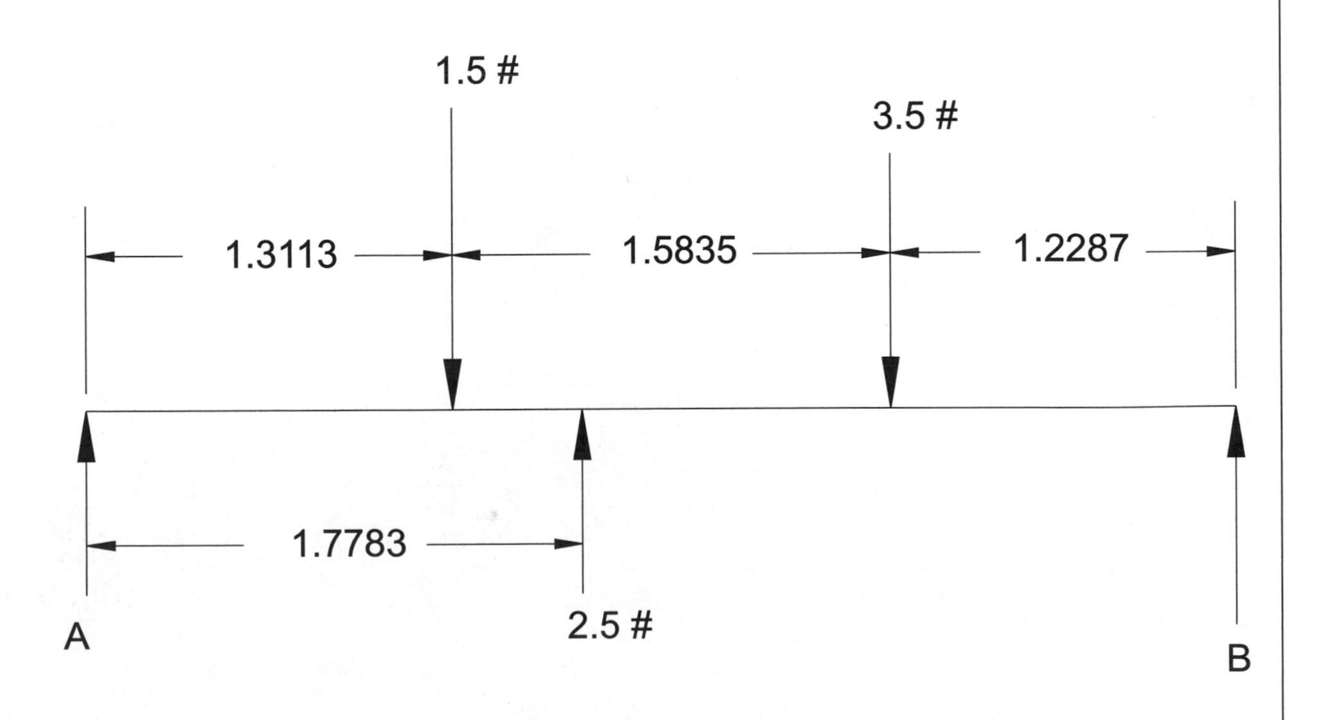

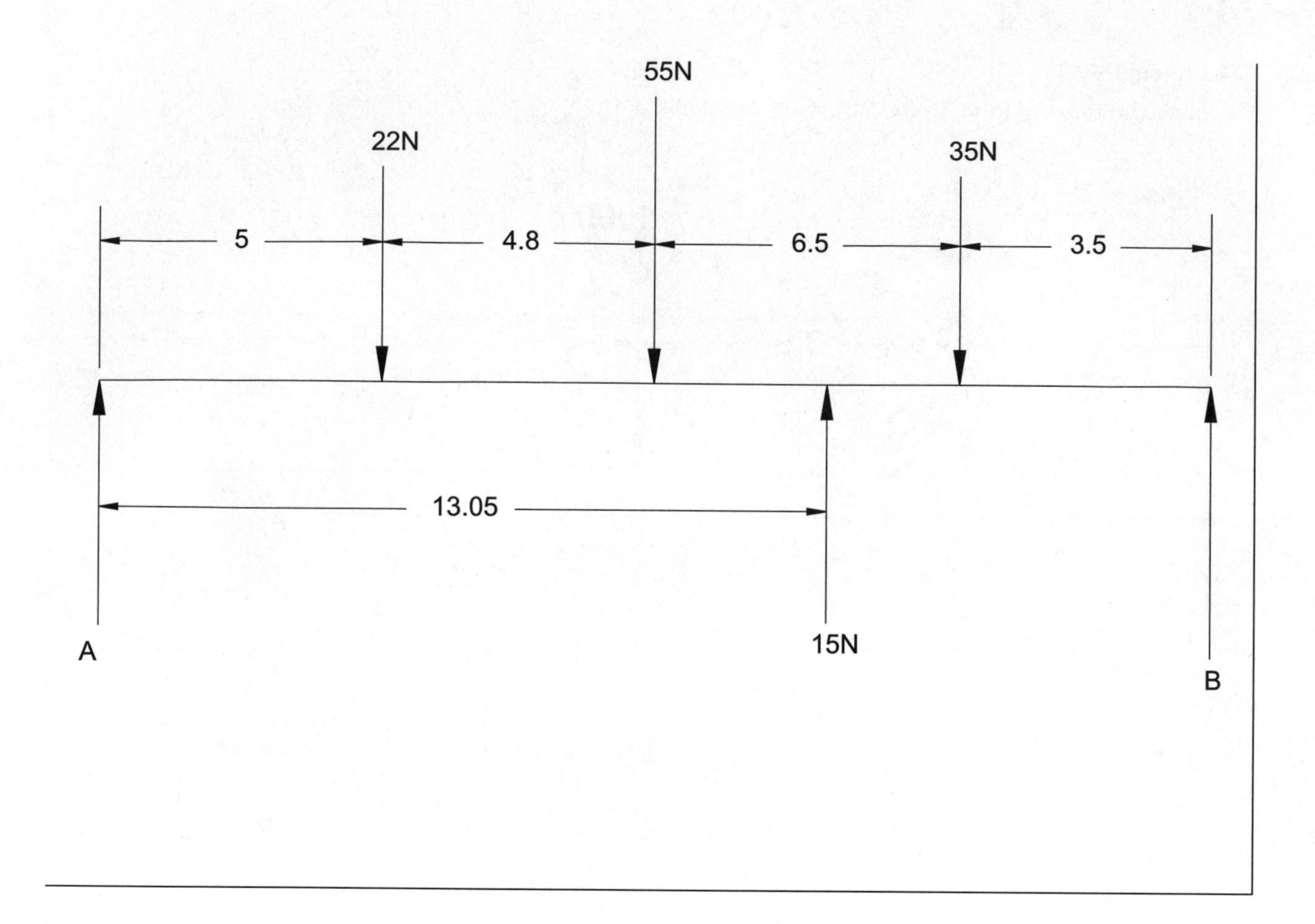
55N
22N
35N
5
4.8
6.5
3.5
13.05
A
15N
B

# Geometric Construction: Selected Topics

## Section 1: Introduction to Geometric Construction

Several methods have been developed to construct geometric shapes using manual drafting equipment and techniques. These methods can be grouped together by one category known as ***geometric construction.*** Although many of these techniques have been automated into AutoCAD commands (MIDpoint option of the OSNAP command for bisecting lines, ELLIPSE command for constructing ellipses, COPY command for transferring angles and triangles, DIVIDE command for dividing lines into equal segments, and POLYGON command for constructing polygons), there are still occasions when geometric construction is needed to produce certain types of geometric shapes. The following section describes how these techniques can be applied to an AutoCAD drawing.

### Bisecting an Angle using AutoCAD

**STEP #1**

Draw a circle with the center located at the vertex of the angle "O". The radius of the circle should be about half the distance of one of the sides of the angle (see Figure A–1).

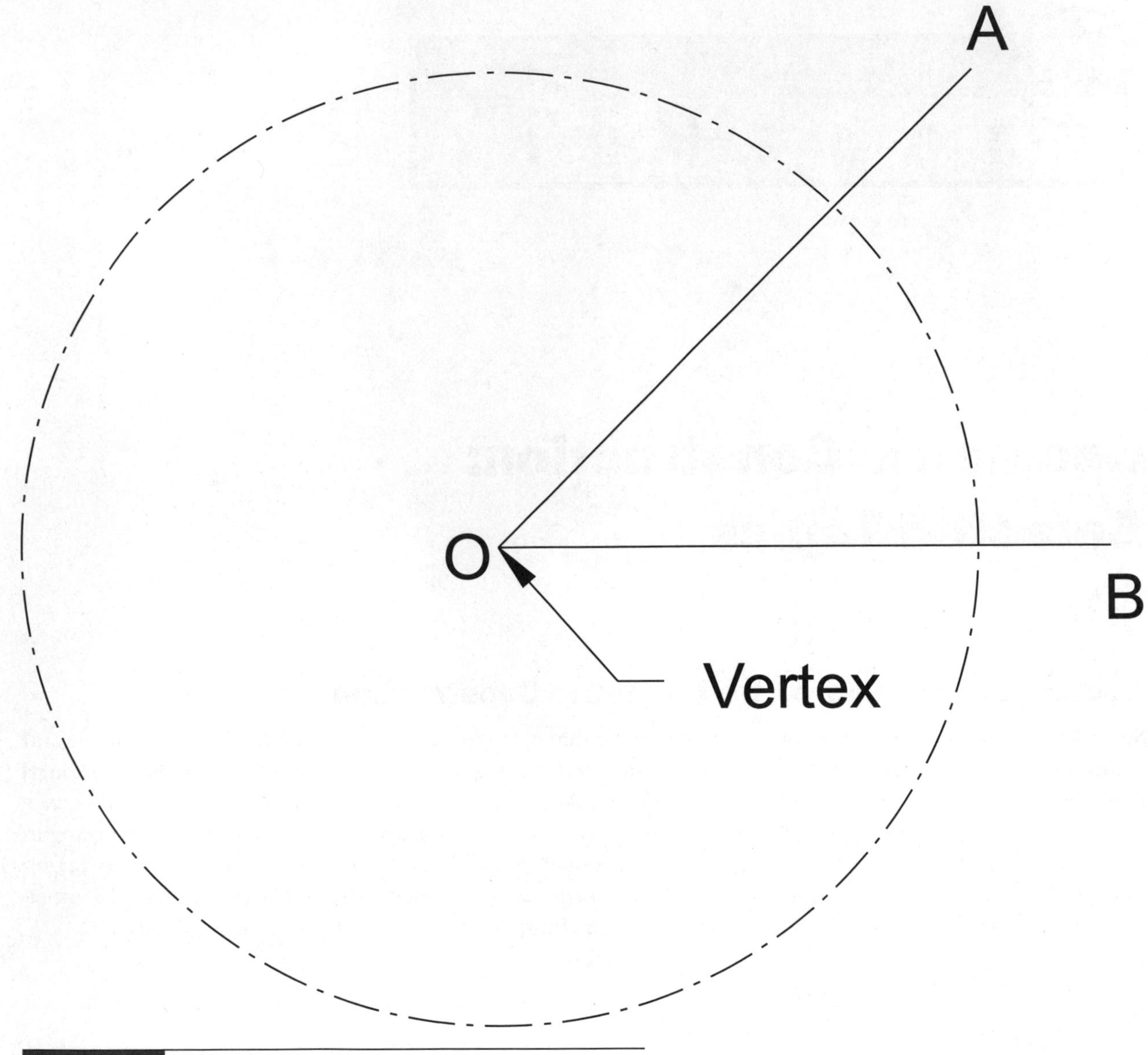


**Figure A–1** *Step #1: Bisecting an angle using AutoCAD.*

## STEP #2

Draw a circle (any size radius may be used) with the center located at the intersection of the first circle and one of the sides of the angle (line "AO", see Figure A–2).

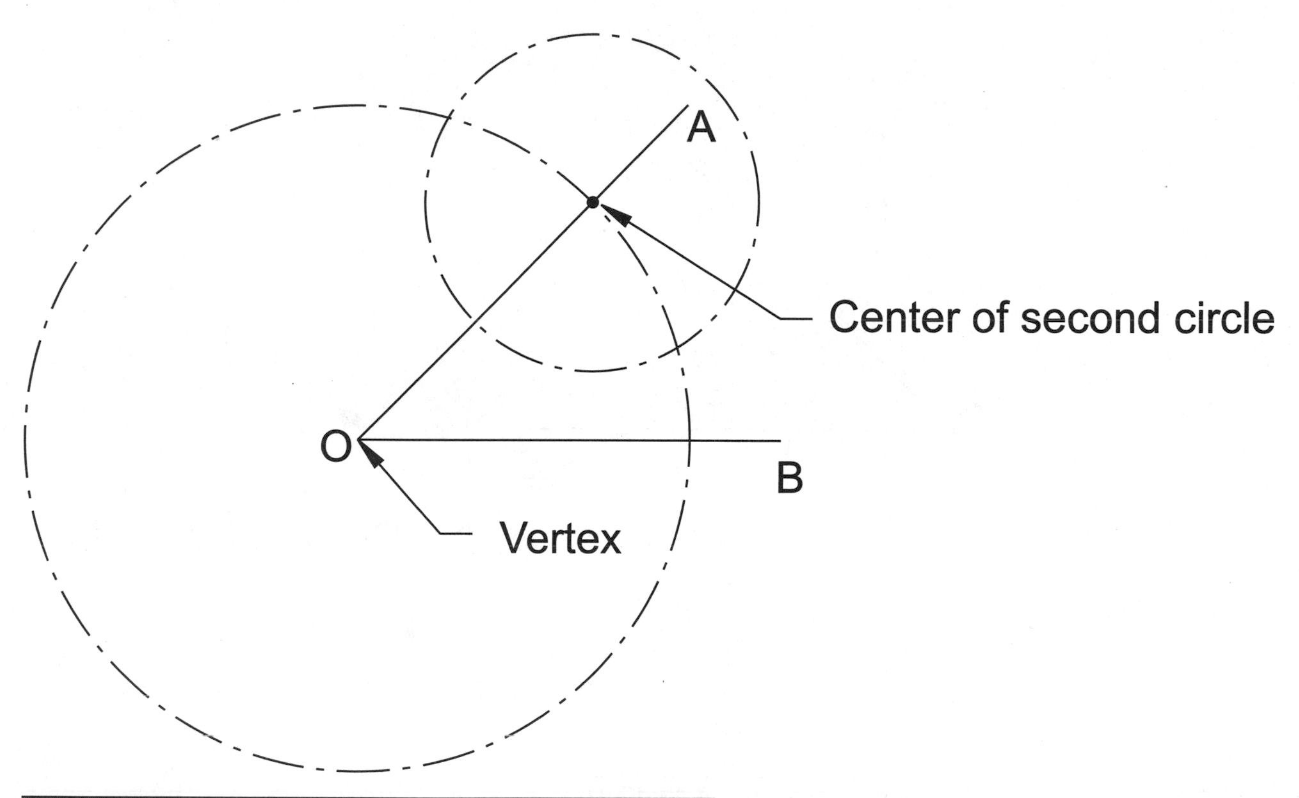


**Figure A–2** *Step #2: Bisecting an angle using AutoCAD.*

## STEP #3

Draw a circle (use the same size radius as step #2) with the center located at the intersection of the first circle and the other side of the angle (line "BO", see Figure A–3).

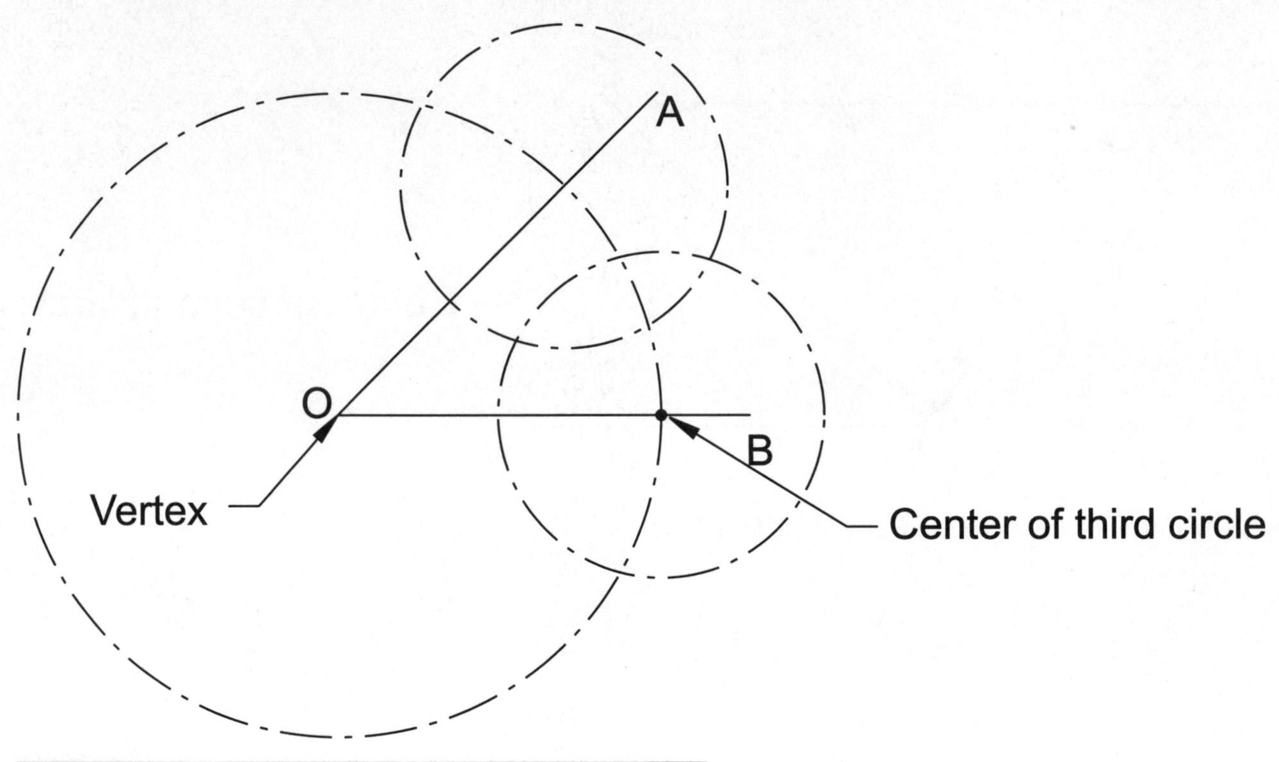


**Figure A–3** *Step #3: Bisecting an angle using AutoCAD.*

**Note: The circles from step #2 and step #3 should intersect; if they don't, readjust the radius of the circles (in step #2 and #3) so that they do intersect.**

## STEP #4

Draw a line from the intersection of the circles in step #2 and step #3 to the vertex bisecting the angle (see Figure A–4).

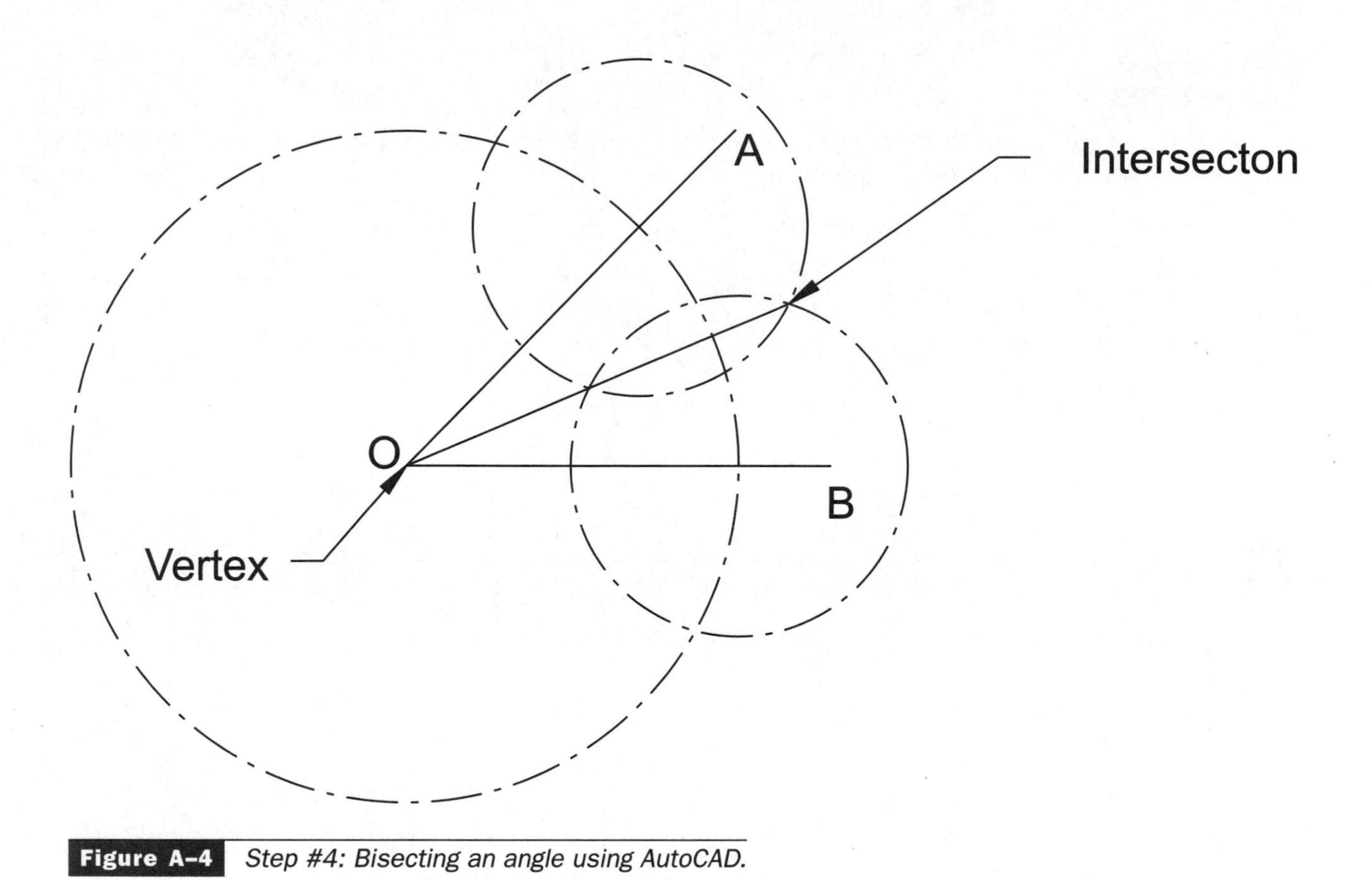


**Figure A–4** *Step #4: Bisecting an angle using AutoCAD.*

## Constructing a Triangle Given Three Sides using Manual Drafting Techniques in AutoCAD

### STEP #1

From the location where the triangle is to be constructed, use the LINE command to draw a line equal in length to one of the sides of the triangle, (see Figure A–5).

Used in step #1

Used in step #2

Used in step #3

A B

**Figure A–5** *Step #1: Line drawn equal in length to one of the sides of the triangle.*

## STEP #2

Using the CIRCLE command, draw a circle at the end of the first line (B) that has a radius equal to the length of the second line (see Figure A–6).

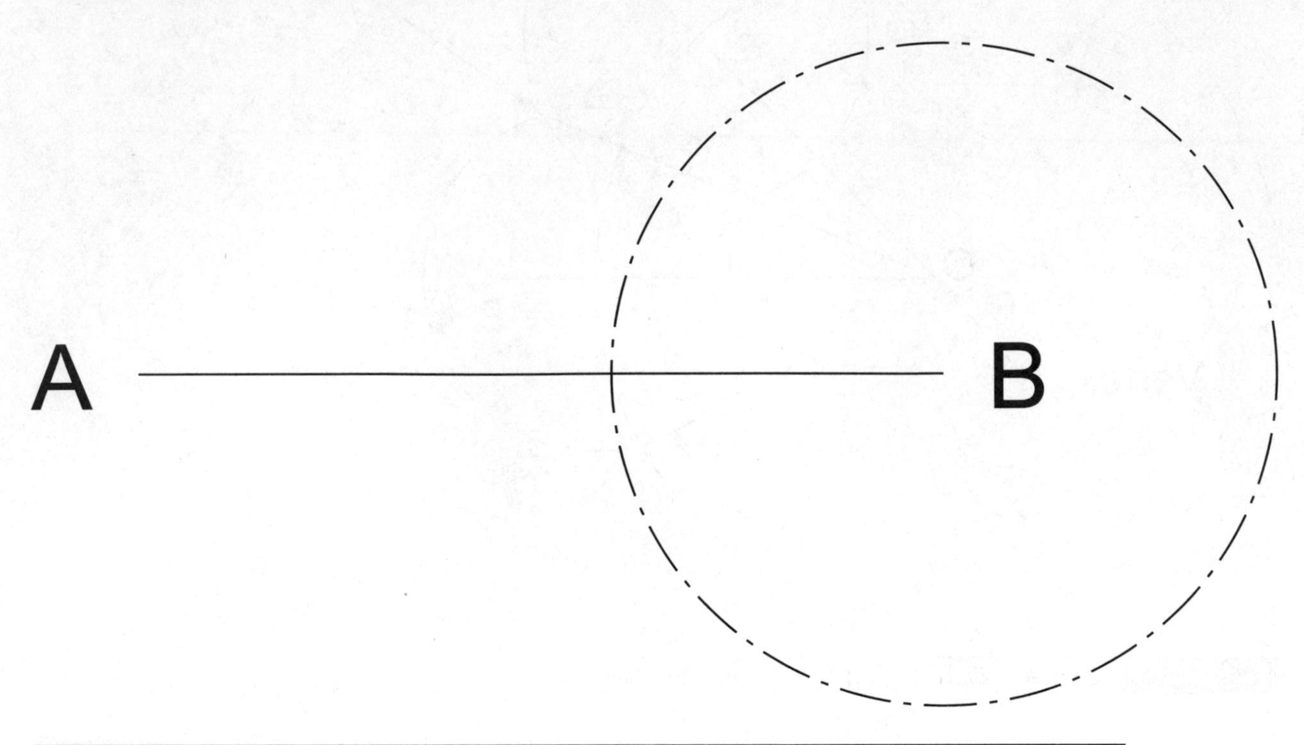


**Figure A–6** *Step #2: Constructing a circle with a radius equal in length to the second line.*

## STEP #3

Using the CIRCLE command, draw a circle that has a radius equal to the radius of the third line at the opposite end of the first line drawn (A, see Figure A–7).

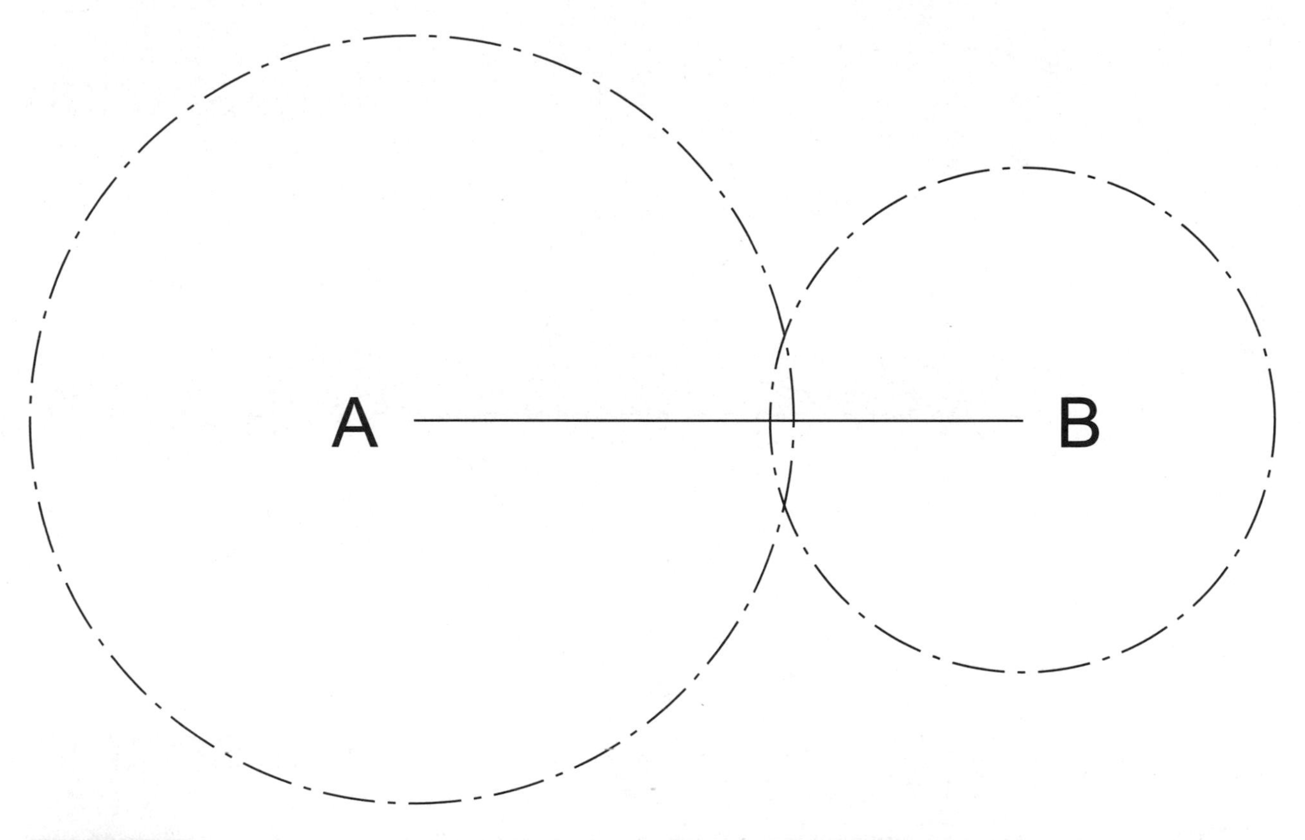


**Figure A–7** *Step #3: Constructing a circle with a radius equal in length to the third line.*

STEP #4

Using the LINE command, draw a line from one of the endpoints of the first line drawn to the intersection of the two circles (from step #2 and #3) to the opposite endpoint (see Figure A–8).

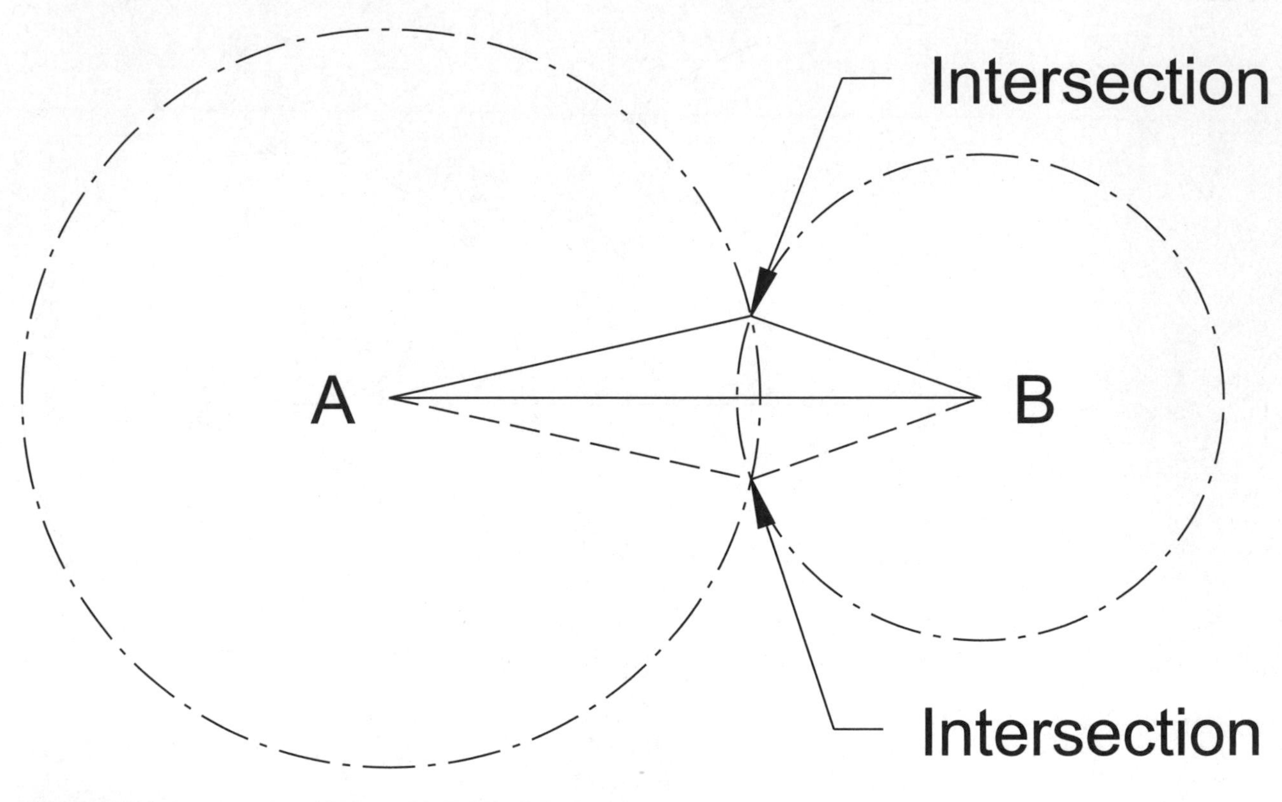


**Figure A–8** *Step #4: The completed intersection.*

**Note: Either points of intersection may be used.**

**Constructing a Triangle Given Three Sides—Advanced Application**

Many of the techniques featured in this section can be automated using AutoLISP programming to save valuable time and money. The following program illustrates this point by automating the previous procedure. The program prompts the technician to enter the three lengths. Next, it determines if the triangle will close, and if so, displays the triangles' three angles. If the triangles will not close, it displays a warning that states that the triangle will not close. Finally, it gives the option to create the triangle. This program is located on the student CD-ROM.

## AutoLISP Program

```
;;;************************************************************************
;;;
;;;    Program Name : DGA2.lsp
;;;
;;;    Program Purpose : Calculates the angles formed by the intersection
;;;                      of three given lines
;;;
;;;    Date: 06/02/98
;;;
;;;
;;;************************************************************************
(defun c:apx1 (/ value1 value2 value3 value4 value5 value6 vi)
  (setvar "cmdecho" 0)
  (setq
    dcl_id4
     (load_dialog
       "c:/windows/desktop/geometry/student/autolisp programs/DGA2.dcl"
     )
  )
  (if (not (new_dialog "AP1" dcl_id4))
    (exit)
  )
  (set_tile "sidea" "1")
  (set_tile "sideb" "1")
  (set_tile "sidec" "1")
  (action_tile "accept" "(apx) (done_dialog)")
  (start_dialog)
  (if (= vi 1)
    (progn
      (if (/= sidea 0)
       (progn
         (if (/= sideb 0)
           (progn
             (if (/= sidec 0)
              (progn
                (setqvalue1 (/ (+ (expt sidea 2)
                                  (* -1 (expt sideb 2))
                                  (* -1 (expt sidec 2))
                               )
                               (* (* sideb sidec) -2)
                            )
                )
                (if (> 1 value1)
                  (progn
                    (setq value2 (sqrt (abs (- 1 (expt value1 2)))))
                    (if (/= value1 0)
                     (progn
                       (setq
                         angle_cal (cvunit (atan (/ value2 value1))
                                    "radians"
                                    "degrees"
```

```
                  )
       )
     )
    )
    (if (= value1 0)
     (setq angle_cal 90.0)
    )
    (setq value3 (/ (+ (expt sideb 2)
                       (* -1 (expt sidea 2))
                       (* -1 (expt sidec 2))
                    )
                    (* (* sidea sidec) -2)
                 )
    )
    (if (> 1 value3)
     (progn
       (setq value4 (sqrt (- 1 (expt value3 2))))
       (if (/= value3 0)
         (progn
           (setq angle1
                 (cvunit (atan (/ value4 value3))
                         "radians"
                         "degrees"
                 )
           )
         )
       )
       (if (= value3 0)
         (setq angle1 90.0)
       )
       (setqvalue5 (/ (+ (expt sidec 2)
                         (* -1 (expt sidea 2))
                         (* -1 (expt sideb 2))
                      )
                      (* (* sidea sideb) -2)
                   )
       )
       (if (> 1 value5)
         (progn
           (setq value6 (sqrt (- 1 (expt value5 2))))
           (if (/= value5 0)
            (progn
              (setq
                angle2 (cvunit
                         (atan (/ value6 value5))
                         "radians"
                         "degrees"
                       )
              )
            )
           )
           (if (= value5 0)
            (setq angle2 90.0)
```

```
        )
        (if (not (new_dialog "info1" dcl_id4))
         (exit)
        )
        (setq a "Angle A = ")
        (setq b "Angle B = ")
        (setq c "Angle C = ")
        (set_tile      "angle"
                (strcat a (rtos angle_cal))
        )
        (set_tile      "angle1"
                (strcat b (rtos angle1))
        )
        (set_tile      "angle2"
                (strcat c (rtos angle2))
        )
        (action_tile
         "accept"
         "(creat) (done_dialog)"
        )
        (start_dialog)
        (if (= create "1")
         (progn
           (setqpt
                 (getpoint "\nSelect base point : "
                 )
           )
           (command "line"
                   pt
                   (strcat "@"
                           (rtos sidea)
                           "<<"
                           (rtos angle_cal)
                   )
                   ""
           )
           (command "line"
                   "@"
                   (strcat "@"
                           (rtos sideb)
                           "<<-"
                           (rtos angle1)
                   )
                   pt
                   ""
           )
         )
        )
        (princ)
      )
    )
  )
)
```

```
            )
          )
          (if (< 1 value1)
            (progn
              (if (not (new_dialog "info2" dcl_id4))
               (exit)
              )
              (action_tile "accept" "(done_dialog)")
              (start_dialog)
            )
          )
          (if (< 1 value3)
            (progn
              (if (not (new_dialog "info2" dcl_id4))
               (exit)
              )
              (action_tile "accept" "(done_dialog)")
              (start_dialog)
            )
          )
          (if (< 1 value5)
            (progn
              (if (not (new_dialog "info2" dcl_id4))
               (exit)
              )
              (action_tile "accept" "(done_dialog)")
              (start_dialog)
            )
          )
          (princ)
        )
       )
      )
     )
    )
   )
   (if (= sidea 0)
    (progn

      (if (not (new_dialog "info3" dcl_id4))
        (exit)
      )
      (setq err1 "ERROR ! Side a Must Be Specified ")
      (set_tile "err1msg" err1)
      (action_tile "accept" "(done_dialog)")
      (start_dialog)
    )
   )
   (if (= sideb 0)
    (progn

      (if (not (new_dialog "info3" dcl_id4))
        (exit)
```

```
          )
          (setq err2 "ERROR ! Side b Must Be Specified ")
          (set_tile "err1msg" err2)
          (action_tile "accept" "(done_dialog)")
          (start_dialog)
        )
      )
      (if (= sidec 0)
       (progn

          (if (not (new_dialog "info3" dcl_id4))
            (exit)
          )
          (setq err3 "ERROR ! Side c Must Be Specified ")
          (set_tile "err1msg" err3)
          (action_tile "accept" "(done_dialog)")
          (start_dialog)
        )
      )
    )
  )
  (princ)
)
(defun apx ()
  (setqsidea (atof (get_tile "sidea"))
       sideb (atof (get_tile "sideb"))
       sidec (atof (get_tile "sidec"))
       vi    1
  )
)
(princ "\nTo excute enter apx1 at the command prompt ")
(princ)
(defun creat ()
  (setq create (get_tile "create"))
)
Dialog Control Language Program
//%%%%%%%%%%%%%%%%%%%%%%%%%%%%%%%%%%%%%%%%%%%%%%%%%%%%%%%%%%%%%%%%%%%%%%
//
//     Activates dialog box
//
//     Descriptive Geometry Appendix Section 2
//
//      Calculates Angles
//
//%%%%%%%%%%%%%%%%%%%%%%%%%%%%%%%%%%%%%%%%%%%%%%%%%%%%%%%%%%%%%%%%%%%%%%
AP1 : dialog {
label = "Constructing a Triangle";
    : boxed_column {
      label = "Lengths";
      children_fixed_width = true;
      children_alignment = left;
    : radio_column {
    : edit_box {
```

```
        key = "sidea";
        label = "Side A";
      }
      : edit_box {
        key = "sideb";
        label = "Side B";
      }
      : edit_box {
        key = "sidec";
        label = "Side C";
      }
    }
  }
      is_default = true;
      ok_cancel;
}
info1 : dialog {
        label = "Triangle Analysis Results";
      : text {
        key = "angle";
        width = 10;
      }
      : text {
        key = "angle1";
        width = 10;
      }
      : text {
        key = "angle2";
        width = 10;
      }
      : toggle {
        key = "create";
        label = "Create Triangle ";
      }
      is_default = true;
      ok_only;
}
info2 : dialog {
        label = "Triangle Analysis Results";
      : text {
        label = "ERROR!  Triangle Does Not Close";
      }
      is_default = true;
      ok_only;
}
info3 : dialog {
        label = "Triangle Analysis Results";
      : text {
        key = "err1msg";
      }
      is_default = true;
      ok_only;
}
```

## Constructing an Equilateral Triangle without using the Polygon Command (Helpful Hints for Constructing Isometric Polygons)

A polygon is a multi-sided geometric figure containing three or more sides and angles. Examples: triangles, squares, pentagons, hexagons, and so forth. An equilateral triangle is a three-sided polygon in which all three sides are equal in length. Equilateral triangles can be constructed by drawing a (base) circle with a radius equal to 0.577350269 or *$(2/3 \sqrt{length^2 - \{length \div 2\}^2})$ times the length of one of its sides. Using the divide command, the circle is then divided into three equal segments. The segments are then connected using the pline command to form the equilateral triangle. The following is the sequence of steps outlining the procedure described above.

**Note: The coefficient was calculated using the formula $2/3 \sqrt{1^2 - (1 \div 2)^2}$.**

### Circle—Divide Method (Non-Isometric)

**STEP #1**

Using the length of one side of the equilateral triangle, calculate the radius of the base circle inscribing the triangle by multiplying the length of one of the sides by 0.577350269.

**STEP #2**

Using the CIRCLE command, draw the base circle at the location where the equilateral triangle is to be constructed (see Figure A–9).

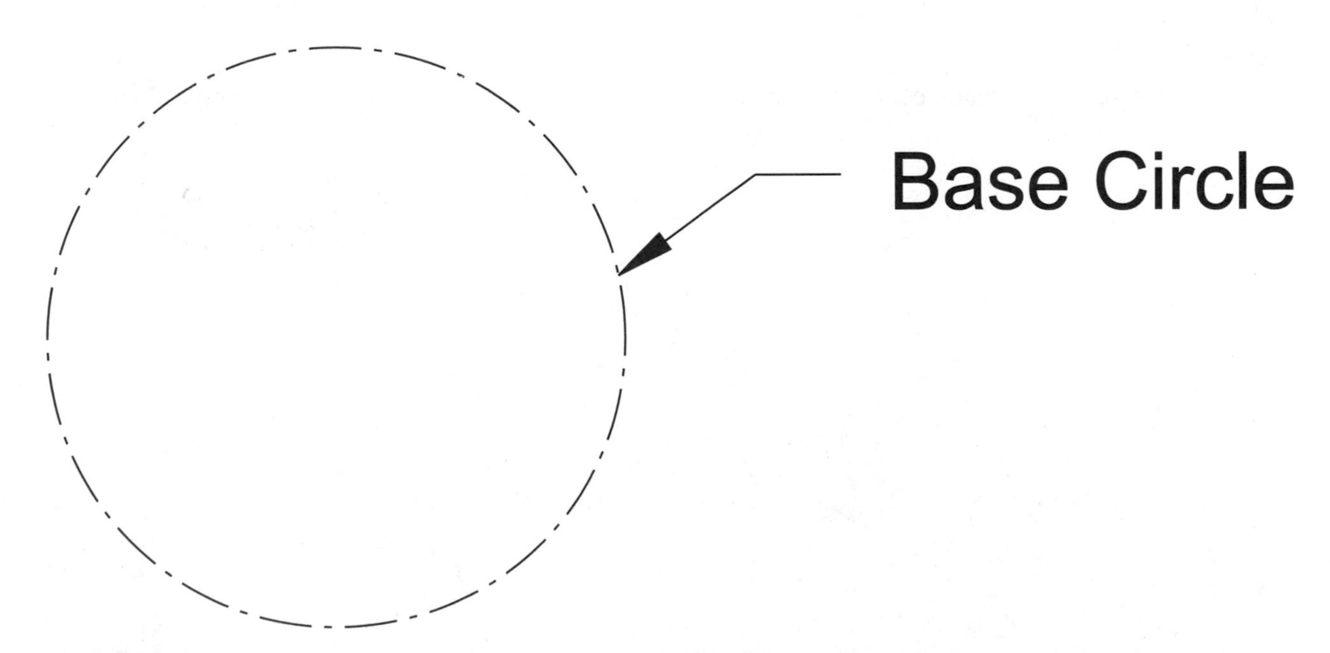


**Figure A–9** *Steps #1 and 2: Constructing an equilateral triangle using manual methods in AutoCAD.*

STEP #3

Using the DIVIDE command, the circle is split into three equal segments, placing nodes at each of the division points (see Figure A–10).

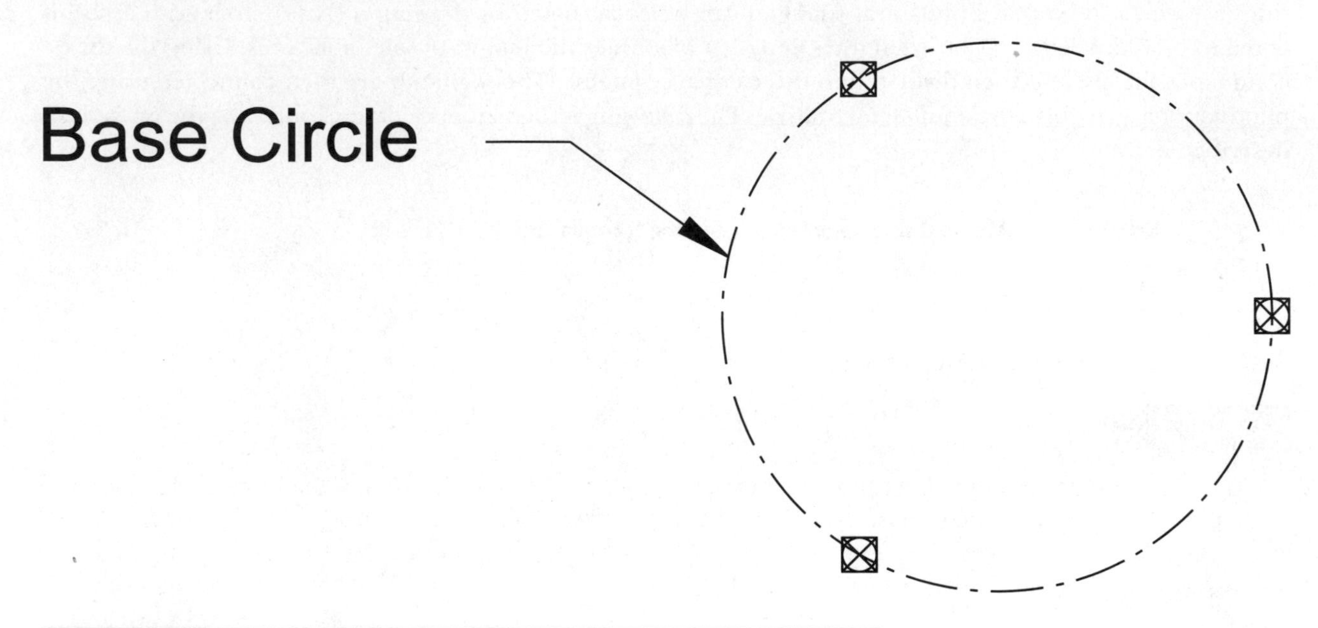


**Figure A–10** *Step #3: Splitting the circle into three equal segments.*

STEP #4

Using the PLINE command, connect the nodes produced in step #3 to form the equilateral triangle (see Figure A–11).

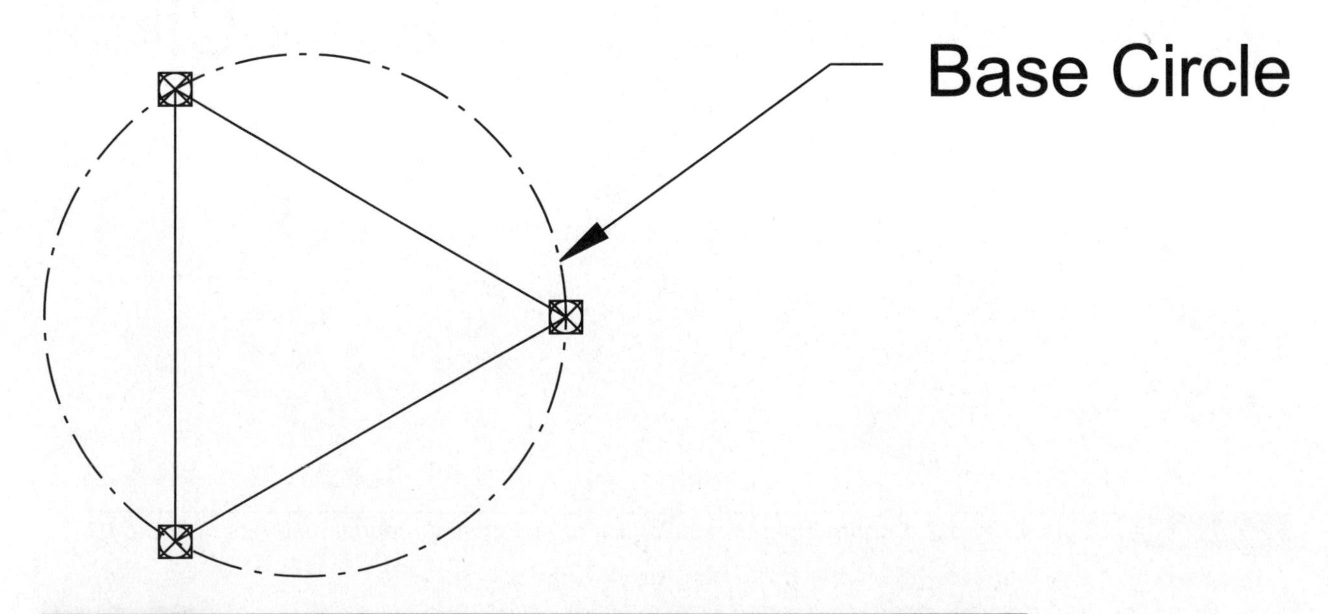


**Figure A–11** *Step #4: Connecting the nodes to complete the equilateral triangle.*

An equilateral triangle can also be constructed by first drawing a line equal in length to one of the sides of the triangle. Next, using the CIRCLE command, draw a circle whose center is positioned at one endpoint of the first line drawn and whose radius is equal in length to the side of the triangle. The same procedure is repeated, but this time the circle's center is positioned on the opposite endpoint of the base line. Finally, the triangle is constructed by drawing a line from the intersection of the two circles created to each of the endpoints of the base line, as shown in Figure A–15 (this is basically the same procedure used for the construction of a triangle with three sides given). The following is the sequence of steps outlining this procedure for the equilateral triangle.

**Line—Circle Method (Non-Isometric)**

**STEP #1**

Using the LINE command, draw a line equal in length to one of the sides of the equilateral triangle and at the location where the triangle is to be positioned (see Figure A–12).

**Figure A–12** *Step #1: Constructing an equilateral triangle using AutoCAD's CIRCLE command.*

**STEP #2**

Draw a circle whose radius is equal in length to one of the sides of the triangle and whose center is located at one endpoint of the base line constructed in step #1 ("A", see Figure A–13).

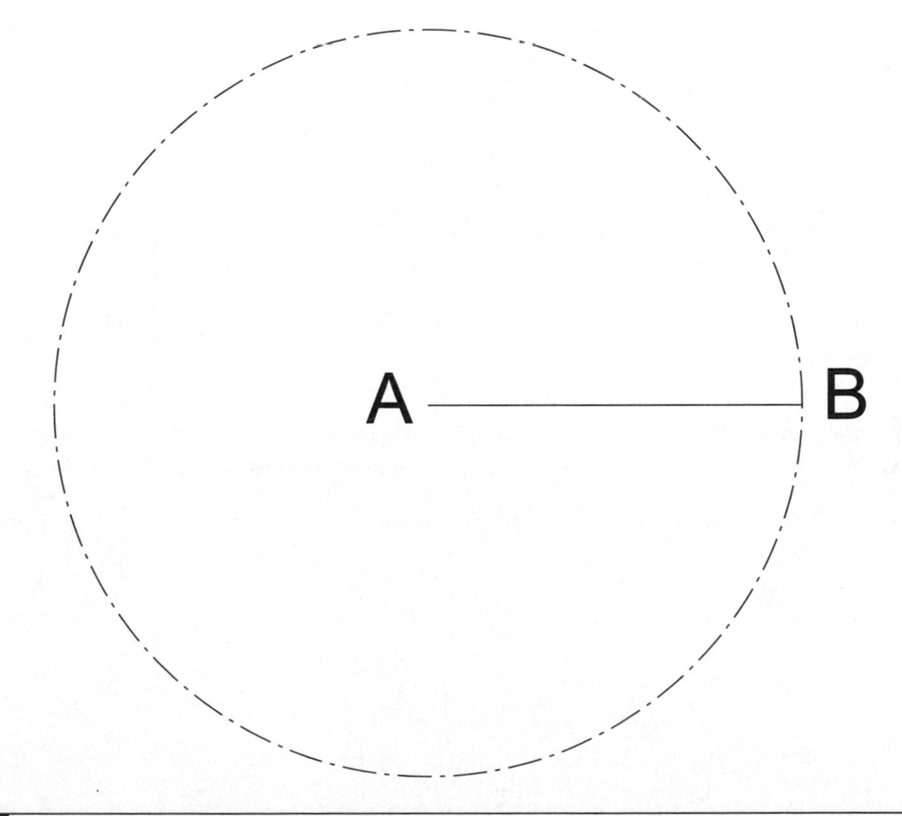


**Figure A–13** *Step #2: Constructing a circle whose radius is equal in length to one side of the traingle.*

## STEP #3

Draw a circle whose radius is equal in length to one of the sides of the triangle and whose center is located at the opposite endpoint of the base line constructed in step #1 ("B", see Figure A–14).

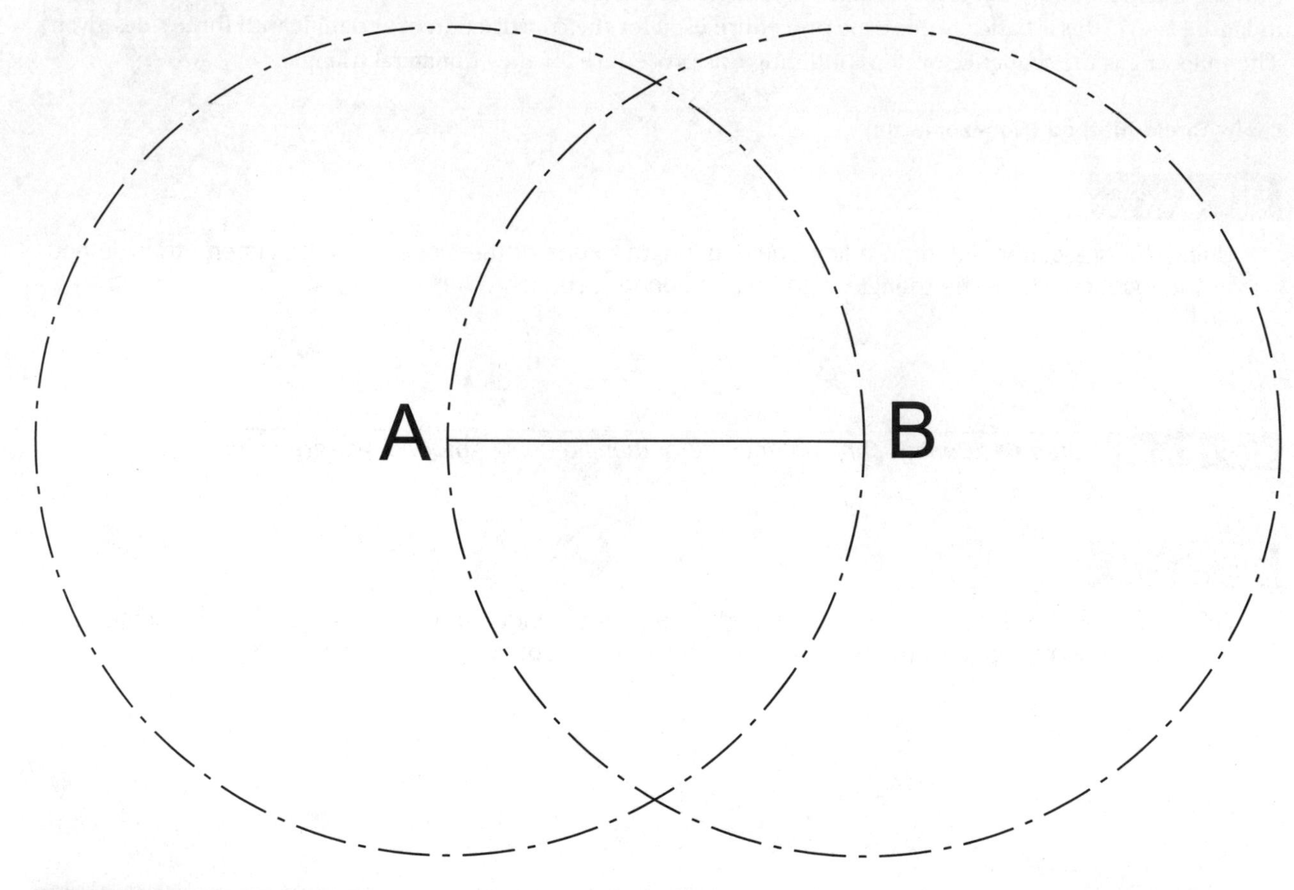


**Figure A-14** *Step #3: Constructing a circle equal in length to one side of the triangle and opposite from the one created in Step# 2.*

## STEP #4

Using the LINE command, draw a line from one of the endpoints of the first line drawn to the intersection of the two circles (from step #2 and #3) to the opposite endpoint (see Figure A–15).

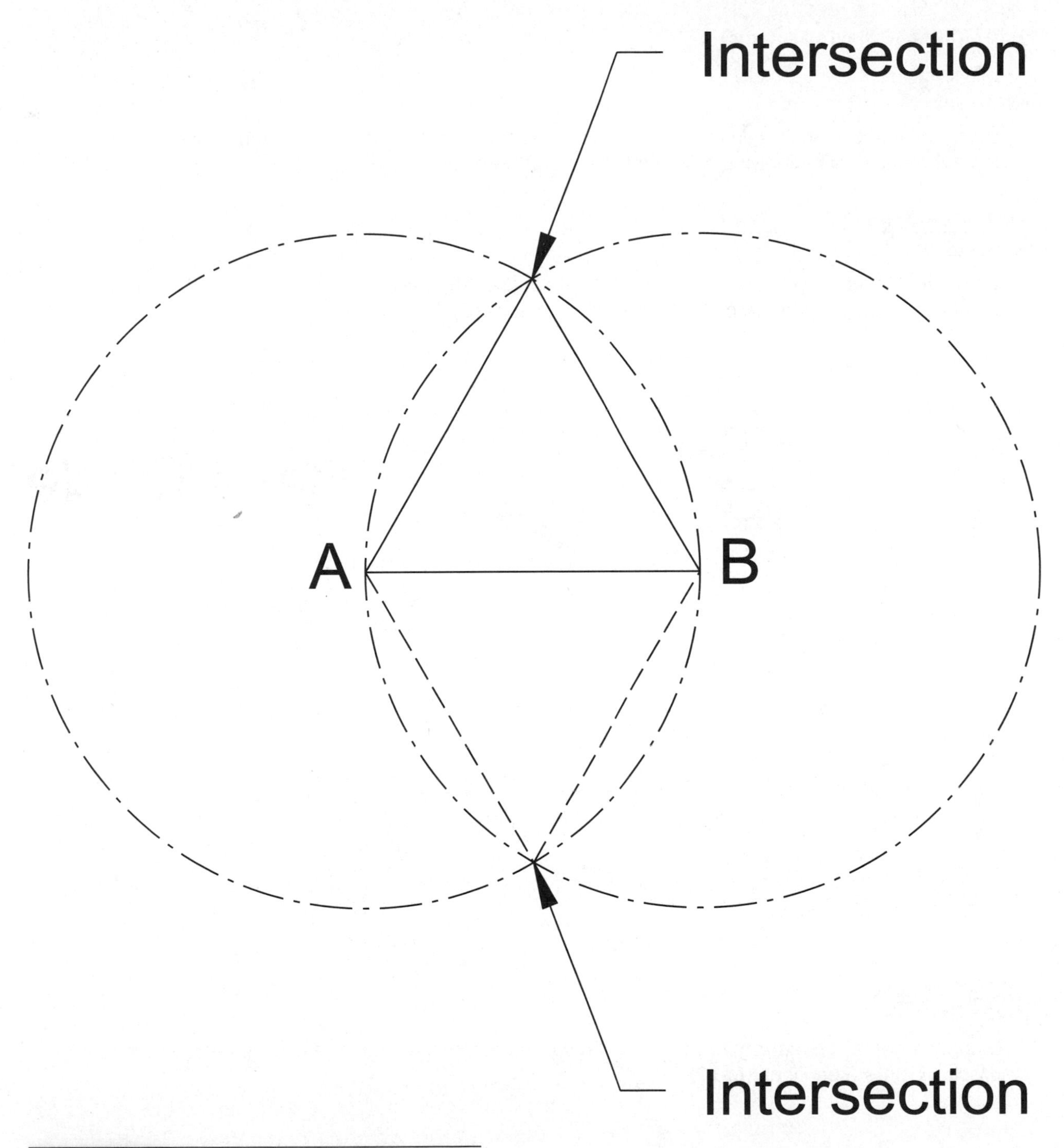


**Figure A–15** *Step #4: The completed figure.*

## Constructing an Isometric Equilateral Triangle

The main drawback of the POLYGON command is that it is not designed to construct isometric polygons. In order to draw an isometric equilateral triangle, the above procedure must be employed with one slight modification. Instead of using the CIRCLE command, the Isocircle option of the ELLIPSE command is used. The rest of the procedure is the same for either of the two methods previously mentioned (see Figure A–22).

**Circle—Divide Method (Isometric)**

**STEP #1**

Using the length of one side of the equilateral triangle, calculate the radius of the base circle inscribing the triangle by multiplying the length of one of the sides by 0.577350269.

**STEP #2**

Using the Isocircle option of the ELLIPSE command, draw the base circle at the location where the equilateral triangle is to be constructed (see Figure A–16).

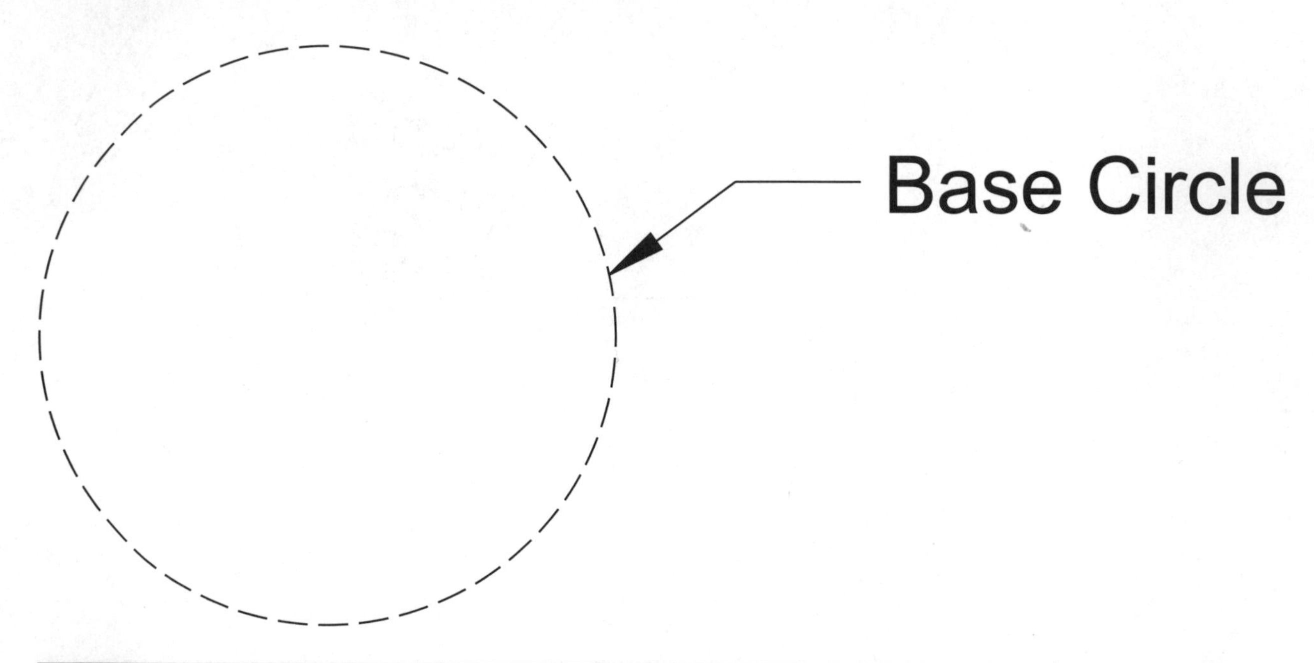


**Figure A–16** *Steps #1 and 2: Constructing an equilateral triangle using AutoCAD's* CIRCLE *and* DIVIDE *commands.*

**STEP #3**

Using the DIVIDE command, the circle is split into three equal segments, placing nodes at each of the division points (see Figure A–17).

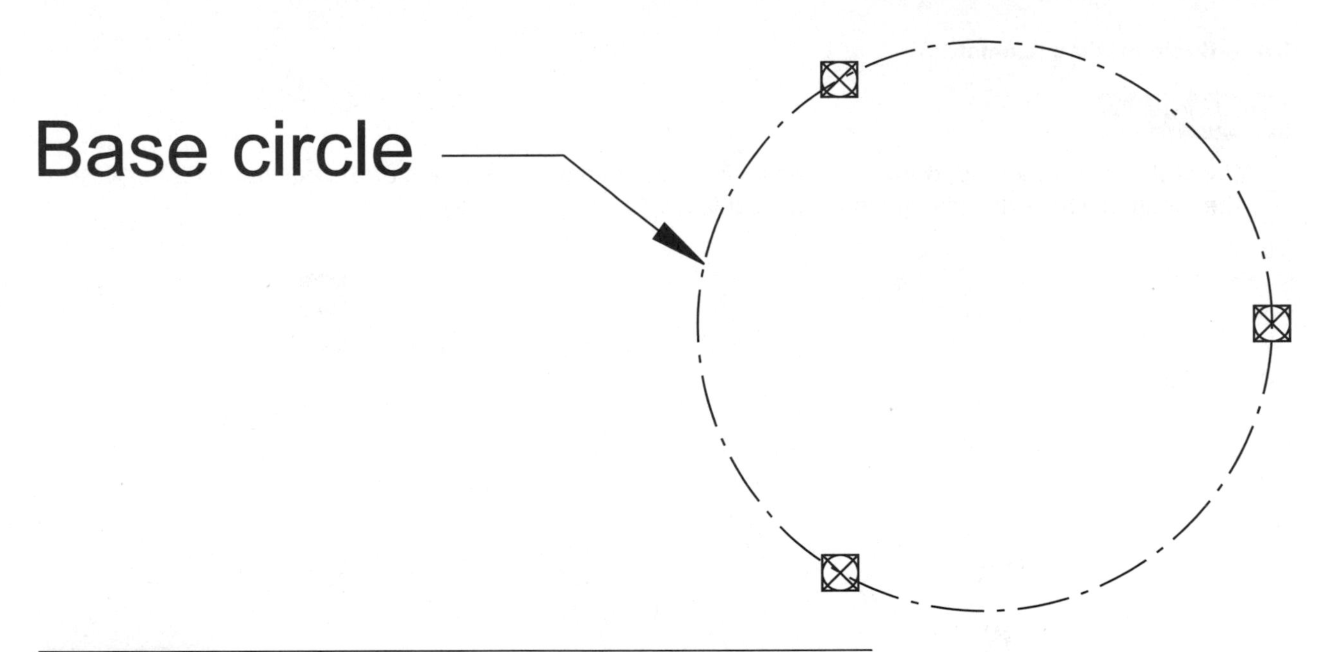


**Figure A–17** *Step #3: Dividing the circle into three equal segments.*

## STEP #4

Using the PLINE command, connect the nodes produced in step #3 to form the equilateral triangle (see Figure A–18).

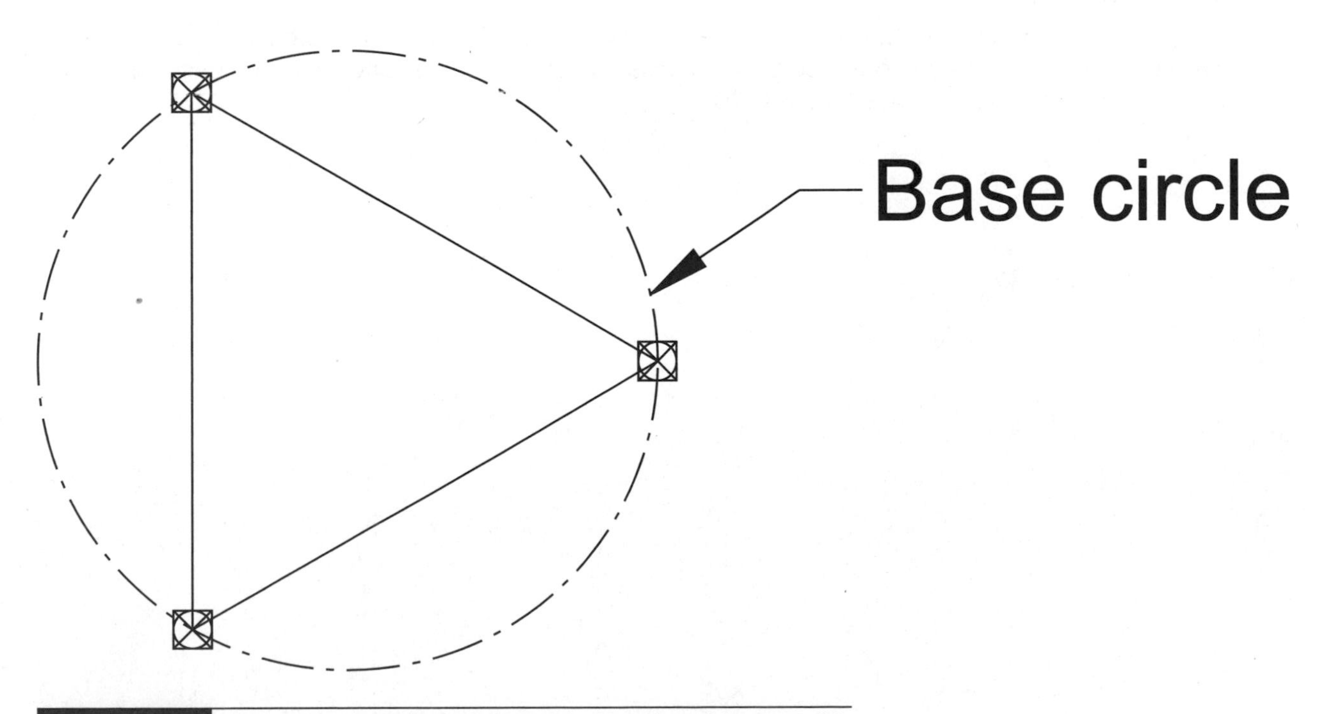


**Figure A–18** *Step #4: Connecting the nodes produced in Step #3.*

**Line—Circle Method (Isometric)**

**STEP #1**

Using the LINE command, draw a line equal in length to one of the sides of the equilateral triangle at the location where the triangle is to be positioned (see Figure A–19).

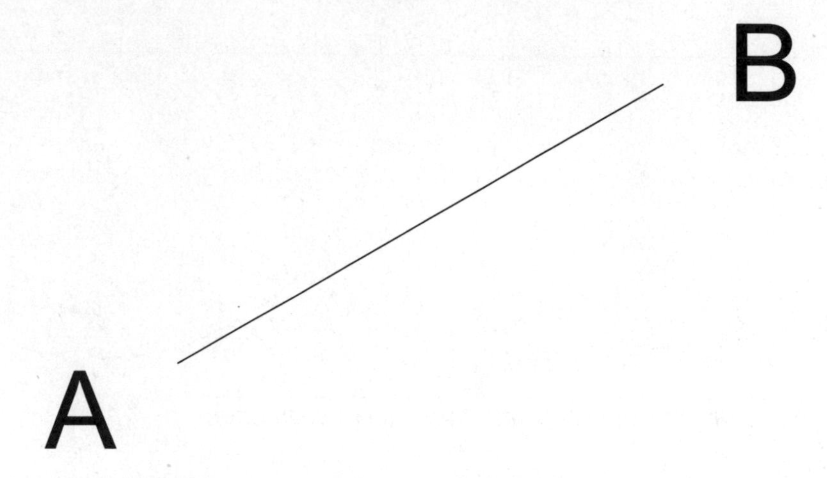


**Figure A–19** *Step #1: Constructing an equilateral triangle using AutoCAD's* LINE *and* CIRCLE *commands.*

**STEP #2**

Using the Isocircle option of the ELLIPSE command, draw a circle whose radius is equal in length to one of the sides of the triangle and whose center is located at one of the endpoints of the base line constructed in step #1 ("A", see Figure A–20).

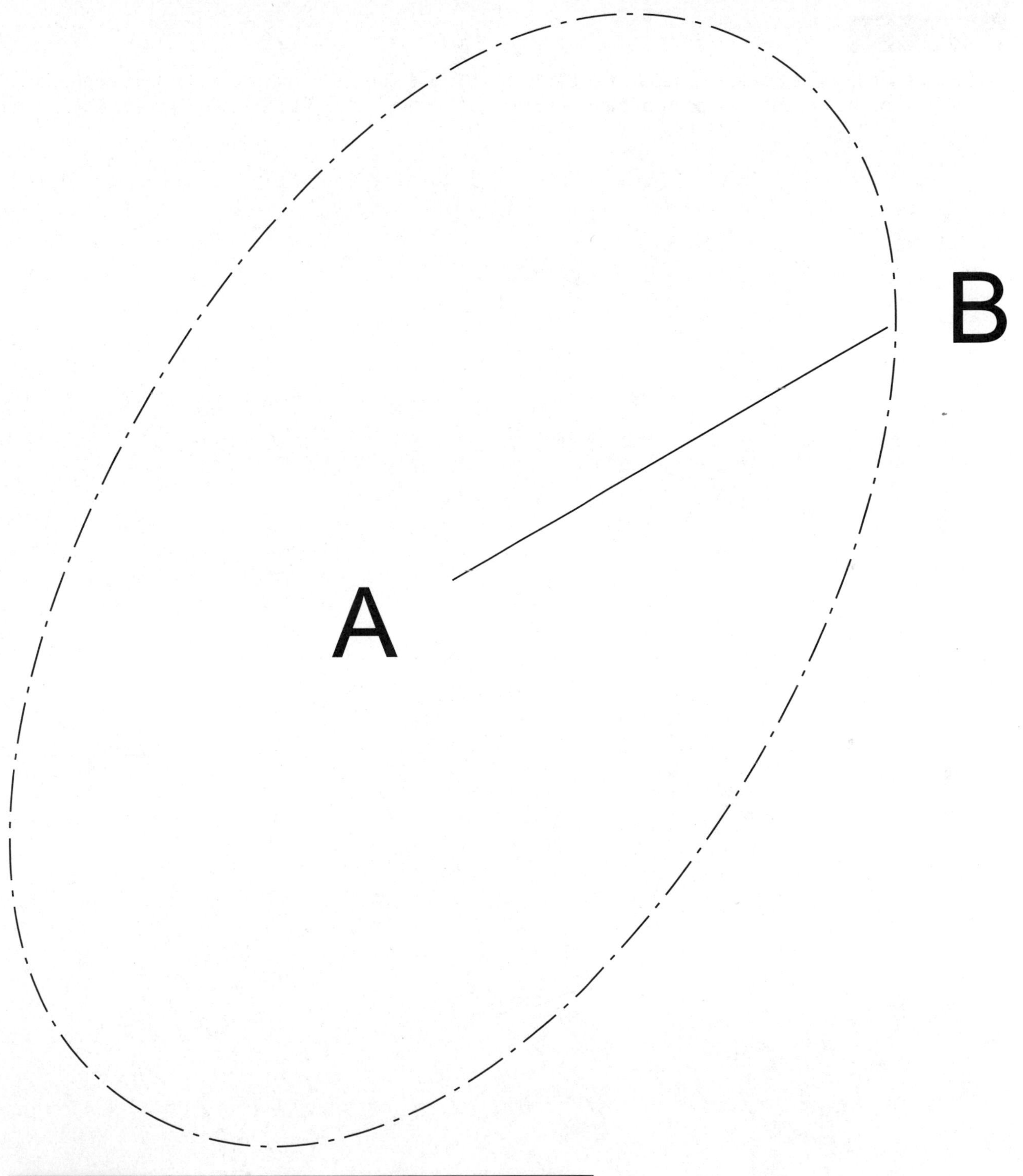


**Figure A-20** *Step #2: Using the Isocircle to create a circle.*

## STEP #3

Draw a circle whose radius is equal in length to one of the sides of the triangle and whose center is located at the opposite endpoint of the base line constructed in step #1 ("B", see Figure A–21).

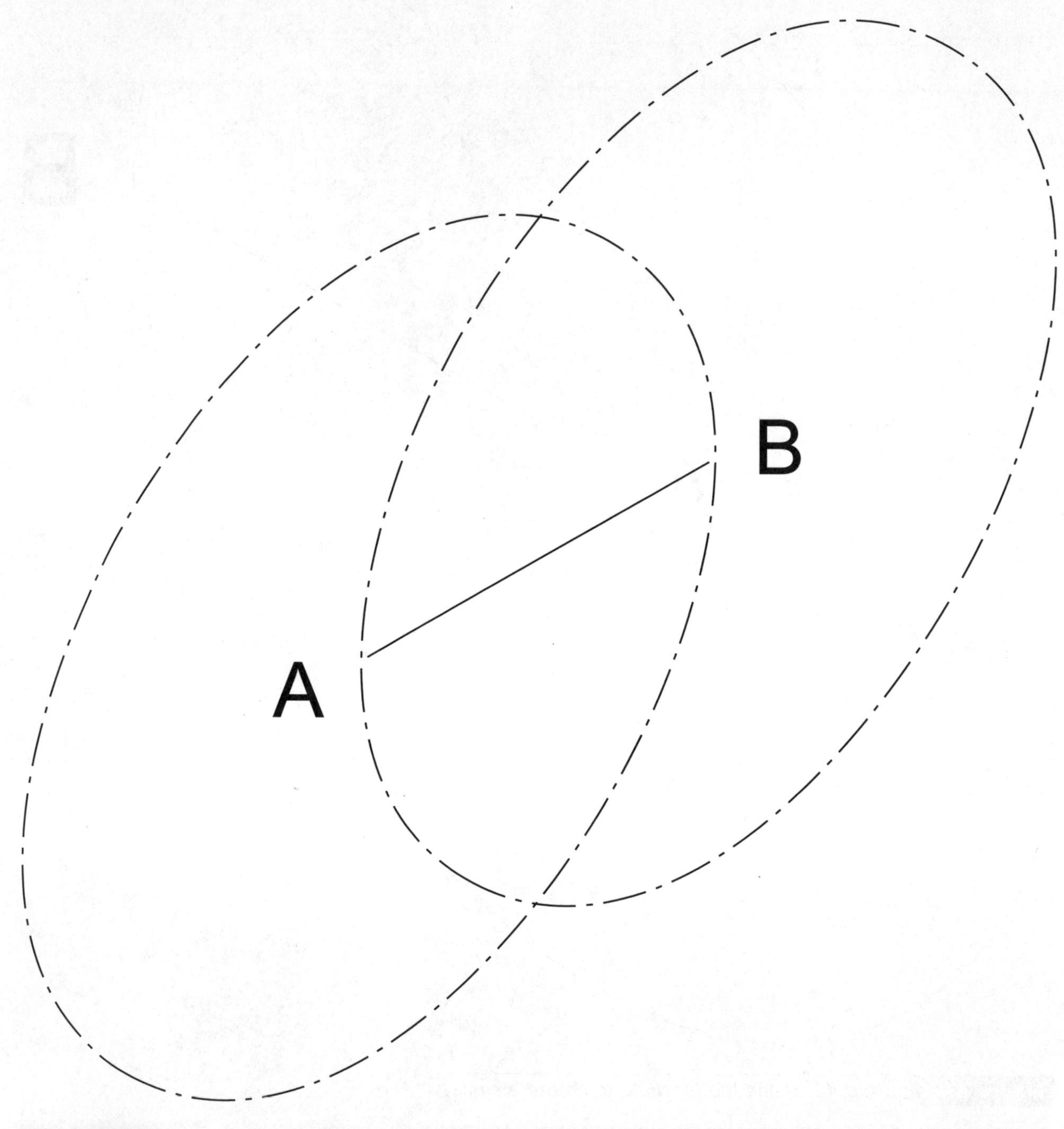


**Figure A–21** *Step #3: Constructing an equilateral triangle using AutoCAD's* LINE *and* CIRCLE *commands.*

## STEP #4

Using the LINE command, draw a line from one of the endpoints of the first line drawn to the intersection of the two circles (from step #2 and #3) to the opposite endpoint (see Figure A–22).

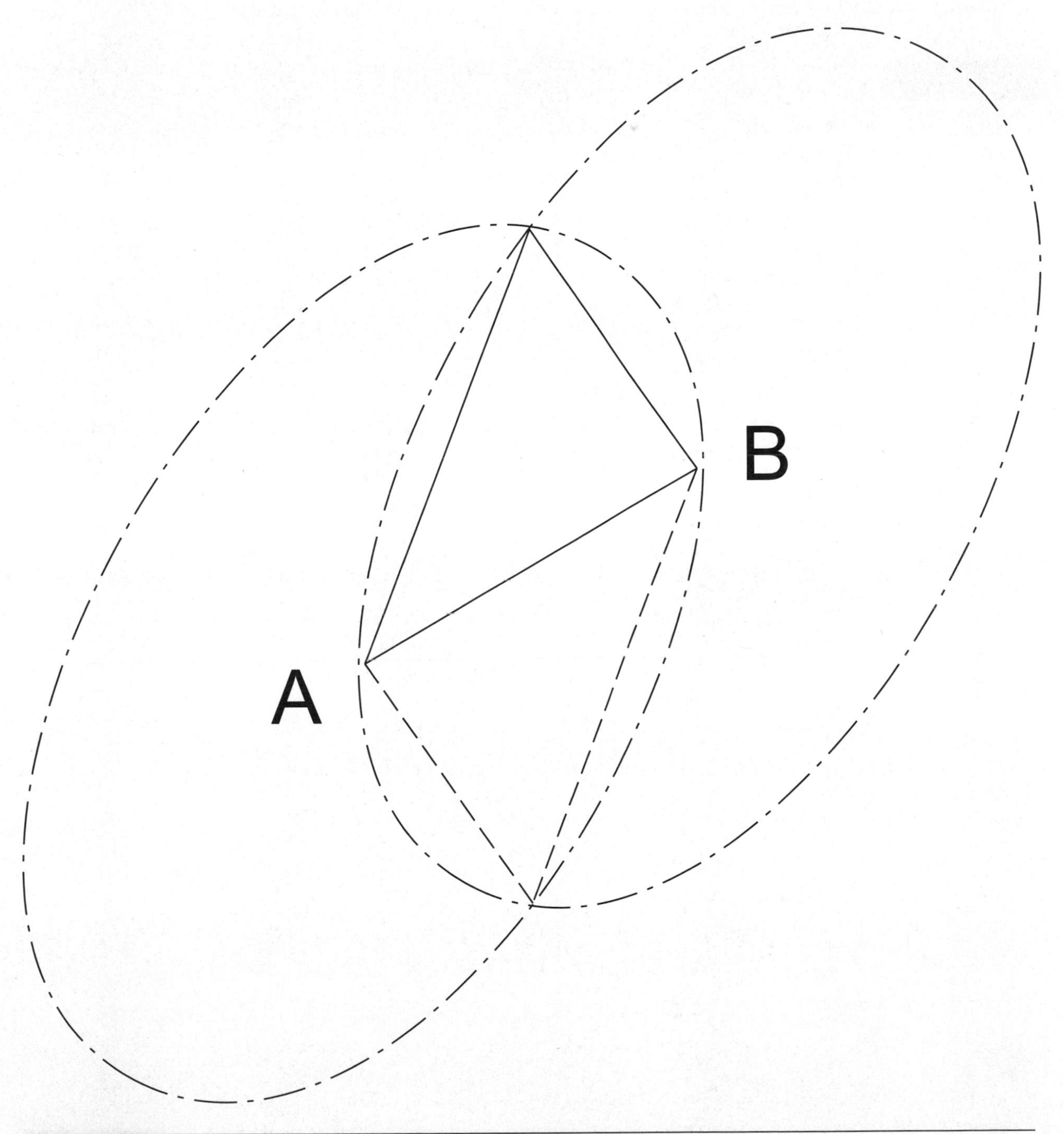


**Figure A–22** *Step #4: Constructing an equilateral triangle using AutoCAD's* LINE *and* CIRCLE *commands.*

## Constructing an Isometric Square

A square is a four-sided polygon with sides that are equal in length. It can be constructed by creating a circle whose radius is equal to 0.707106781 (Radius = $\{\sqrt{length^2 + length^2}\}/2$) times the length of one of the sides of the square. The circle is then divided into four equal segments and each segment is connected using the PLINE command. Listed below are the steps involved in constructing a square using AutoCAD.

### STEP #1

Using the CIRCLE command, construct a circle whose radius is equal to 0.707106781 times the length of one side of the square (see Figure A–23).

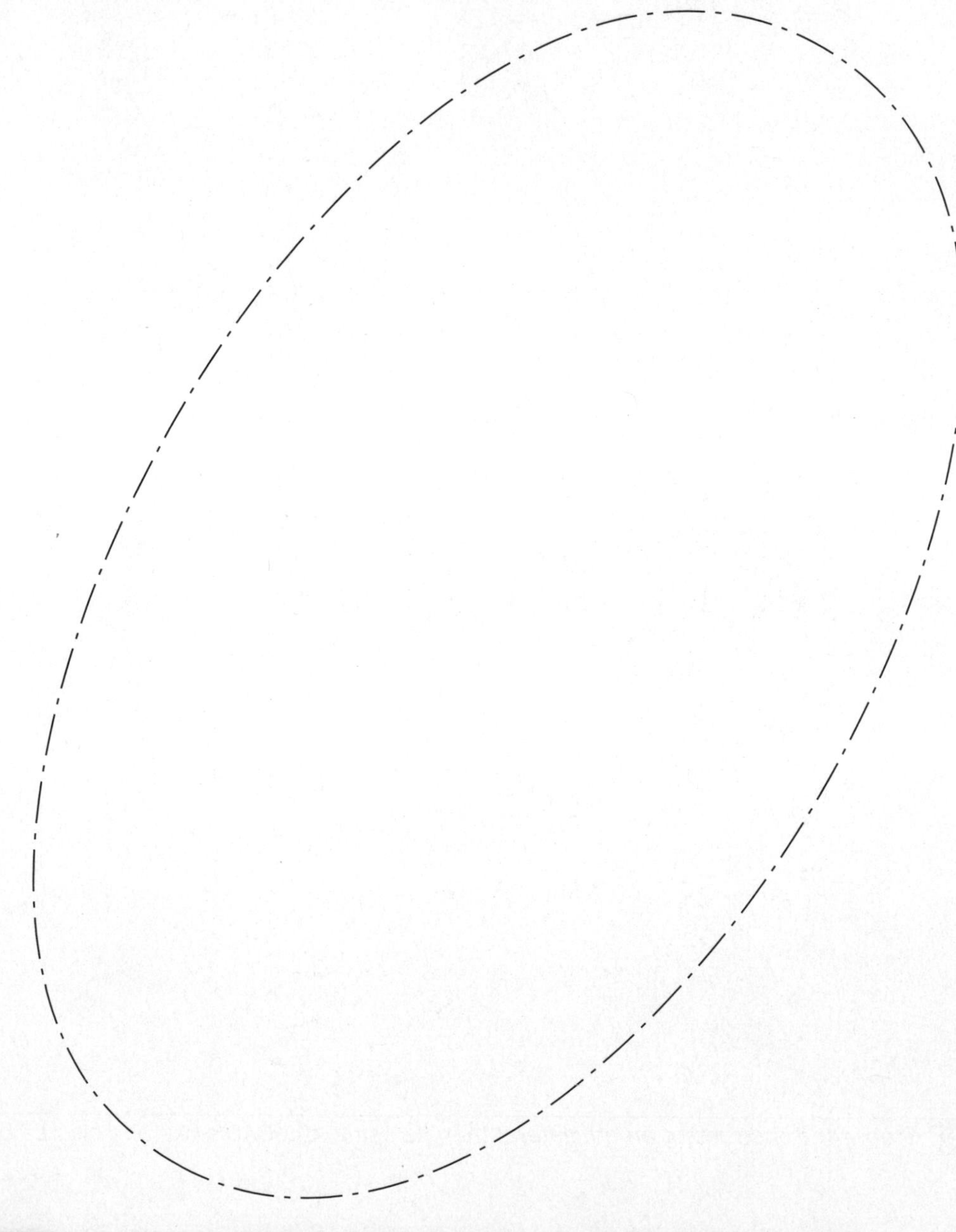

**Figure A–23** *Step #1: Constructing a square without using the POLYGON command.*

## STEP #2

Using the DIVIDE command, the circle is divided into four equal segments (see Figure A–24).

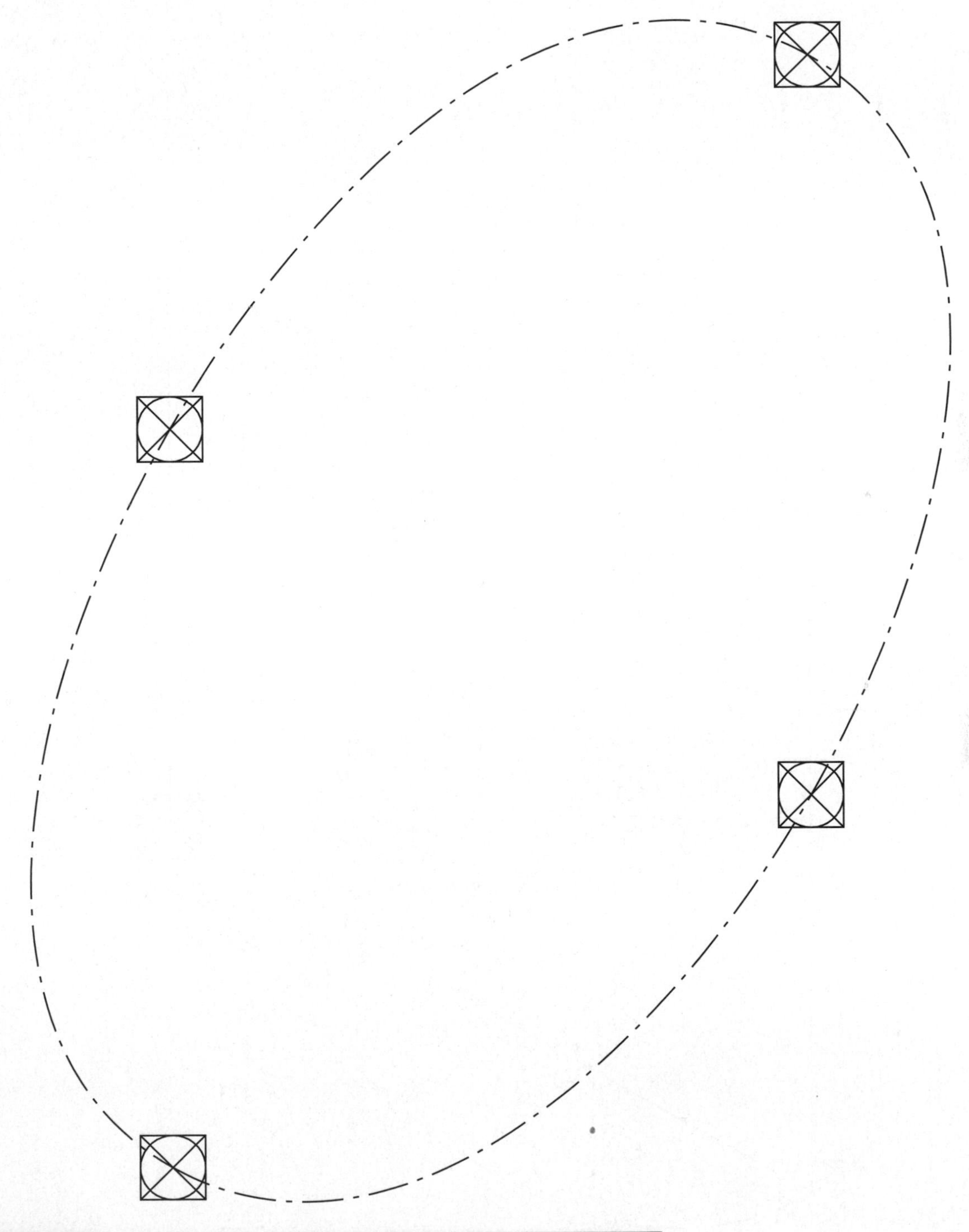

**Figure A–24** *Step #2: Dividing the circle into four equal segments.*

## STEP #3

Using the PLINE command, the node marking the division points on the circle are connected (see Figure A–25).

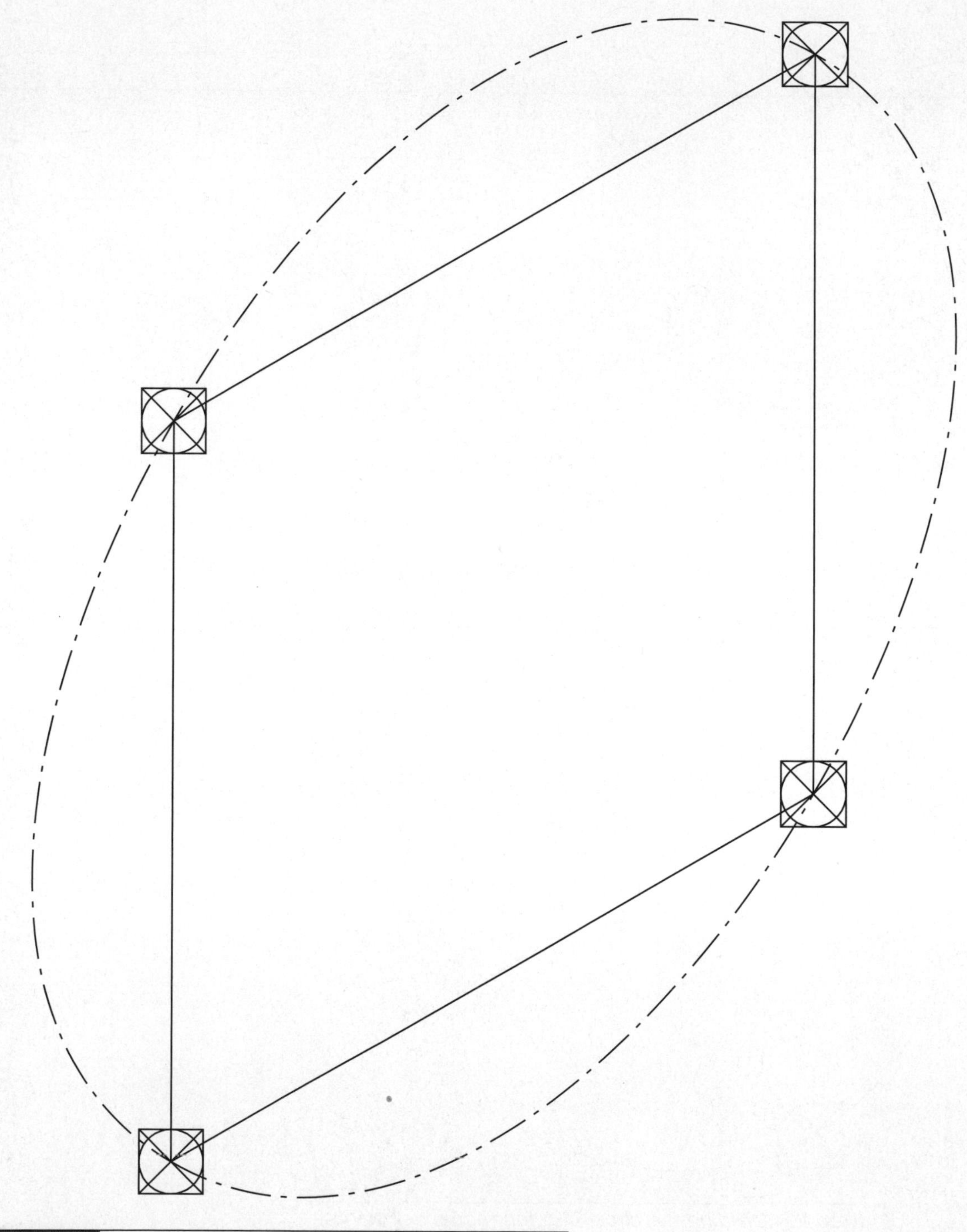

**Figure A–25** *Step #3: Connecting the nodes of the circle.*

**Note: To construct an isometric square, the procedure described above is employed with the exception that instead of using the CIRCLE command, the technician facilitates the Isocircle option of the ELLIPSE command.**

## Constructing A Hexagon using AutoCAD

A ***hexagon*** is a six-sided polygon with sides that are equal in length. When a hexagon is inscribed in a circle, the length of one of its sides is equal to the radius of the inscribed circle. Therefore, a hexagon can be constructed by drawing a circle whose radius is equal to the length of one of the sides of the hexagon and dividing the circumference of the circle into six equal segments. Listed below are the steps for constructing a hexagon.

### STEP #1

Using the CIRCLE command, construct a circle whose radius is equal in length to one of the sides of the hexagon (see Figure A–26).

**Figure A–26** *Step #1: Constructing a hexagon using AutoCAD.*

## STEP #2

Using the DIVIDE command, divide the circle into six equal segments (see Figure A–27).

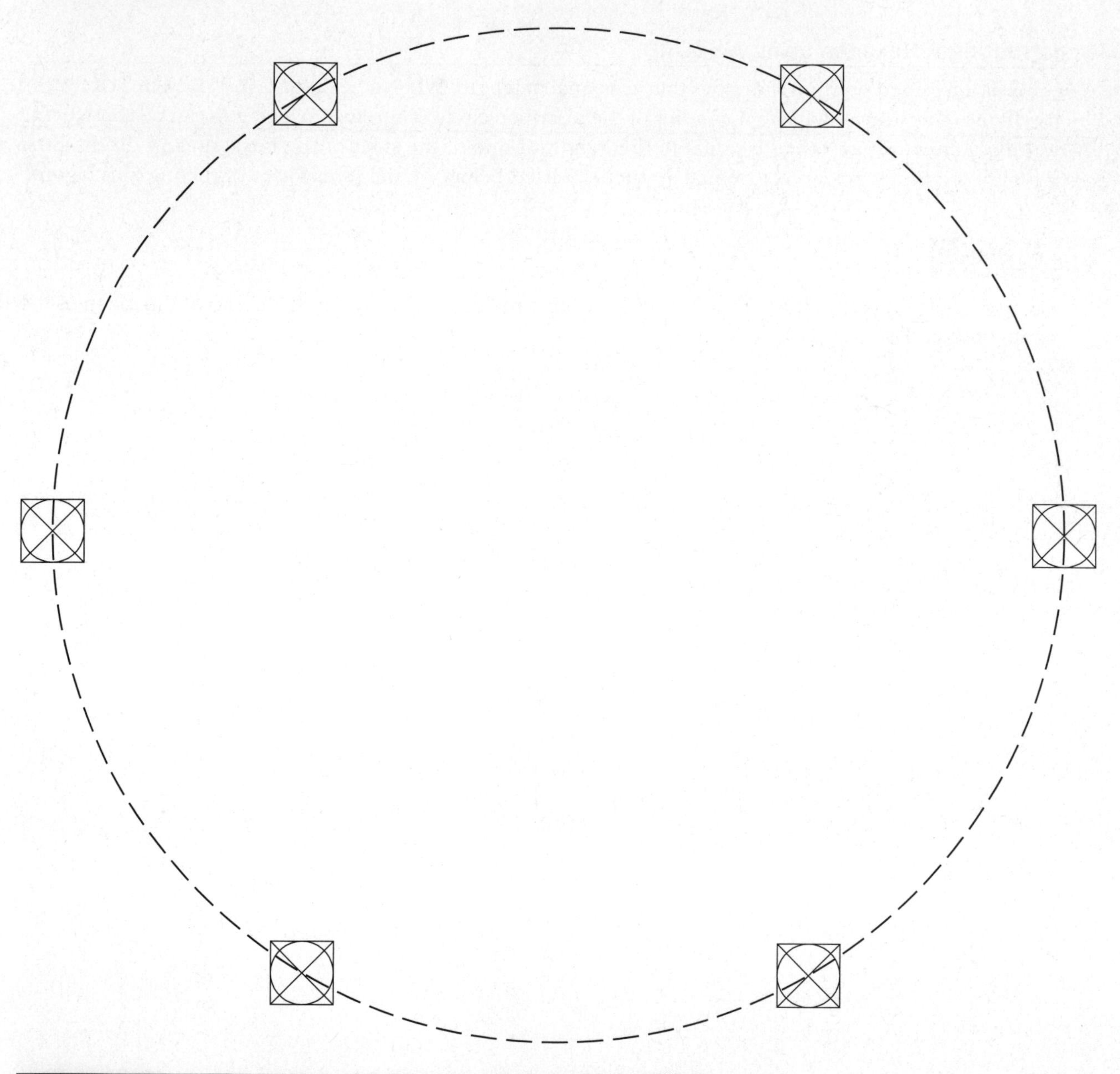

**Figure A–27** *Step #2: Dividing the circle into six equal segments.*

## STEP #3

Using the PLINE command, connect the nodes created by the DIVIDE command to form the hexagon (see Figure A–28).

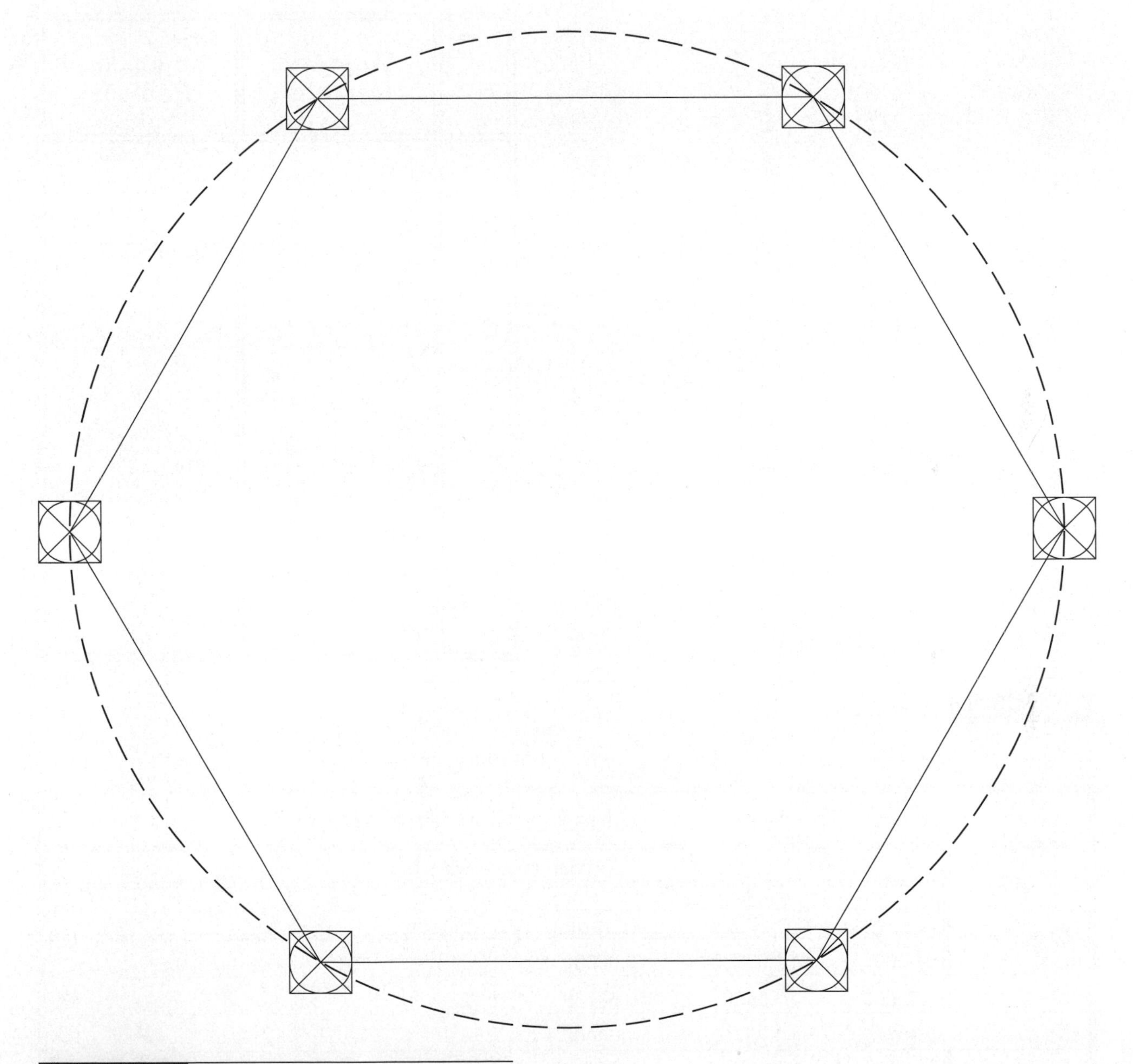

**Figure A–28** *Step #3: The completed figure.*

Again, the procedure for constructing an isometric hexagon is almost the same as for a standard hexagon. The only difference being that the Isocircle option of the ELLIPSE command is used in place of the CIRCLE command.

## Section 2: Trigonometry

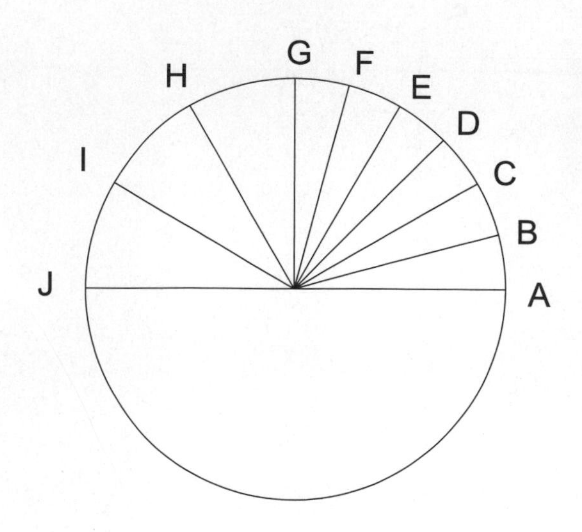


| Unit Circle | | |
|---|---|---|
| | Angle in Degrees | Angle in Radians |
| A | 0 | |
| B | 15 | $\pi/12$ |
| C | 30 | $\pi/6$ |
| D | 45 | $\pi/4$ |
| E | 60 | $\pi/3$ |
| F | 75 | $5\pi/12$ |
| G | 90 | $\pi/2$ |
| H | 120 | $2\pi/3$ |
| I | 150 | $5\pi/6$ |
| J | 180 | $\pi$ |

**Figure A–29** *Unit circle.*

| Trigonometric Values of Common Angles | | | | | | |
|---|---|---|---|---|---|---|
| Angle in Degrees | Trigonometric Functions | | | | | |
| | Decimal Value | | | | | |
| | Sin | Cos | Tan | Sin | Cos | Tan |
| 0 | 0.0000 | 1.00000 | 0.00000 | 0 | 1 | 0 |
| 30 | 0.50000 | 0.86602 | 0.57735 | 1/2 | $\sqrt{3}/2$ | $\sqrt{3}/3$ |
| 45 | 0.70710 | 0.70710 | 1.00000 | $\sqrt{2}/2$ | $\sqrt{2}/2$ | 1 |
| 60 | 0.86602 | 0.50000 | 1.73205 | $\sqrt{3}/2$ | 1/2 | $\sqrt{3}$ |
| 90 | 1.00000 | 0.00000 | Undefined | 1 | 0 | Undefined |
| 120 | 0.86602 | -0.50000 | -1.73205 | $\sqrt{3}/2$ | -1/2 | $-\sqrt{3}$ |
| 150 | 0.50000 | -0.86602 | -0.57735 | 1/2 | $-\sqrt{3}/2$ | $-\sqrt{3}/3$ |
| 180 | 0.00000 | -1.00000 | 0.00000 | -1 | 0 | Undefined |

**Figure A–30** *Trigonometric values of common angles.*

## Section 3: Properties for Common Geometric Shapes

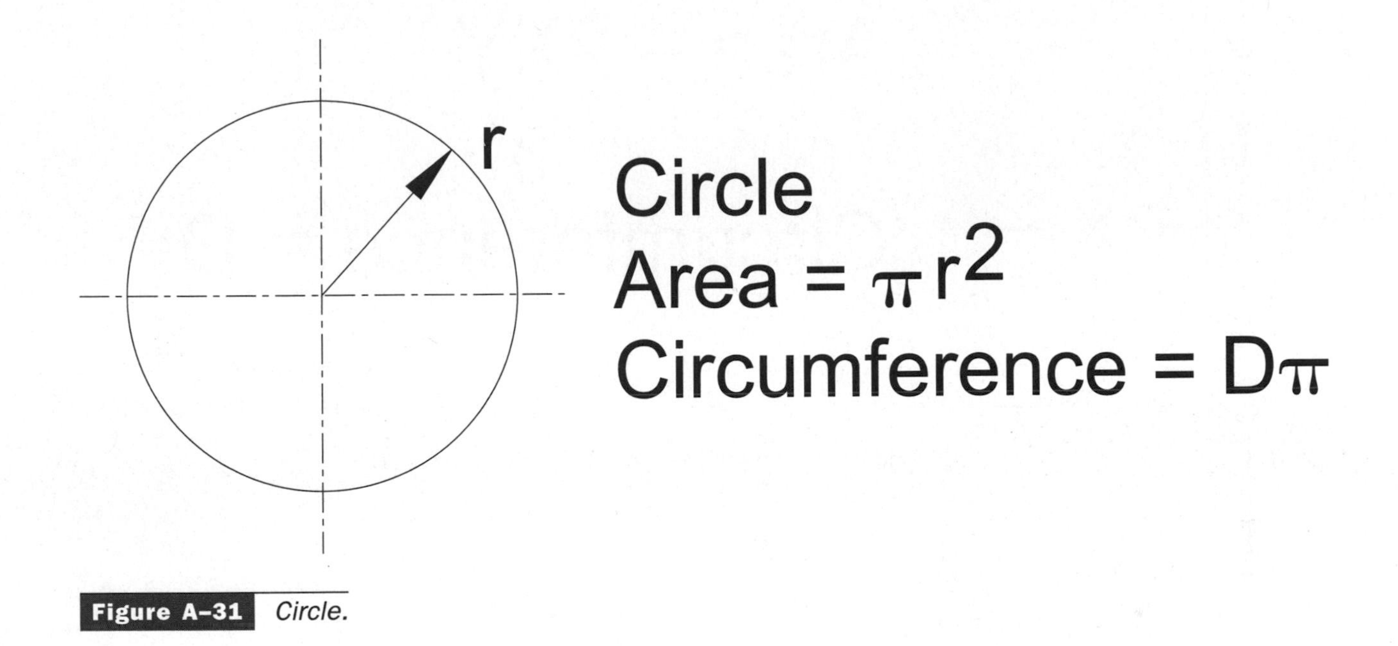


**Figure A-31** *Circle.*

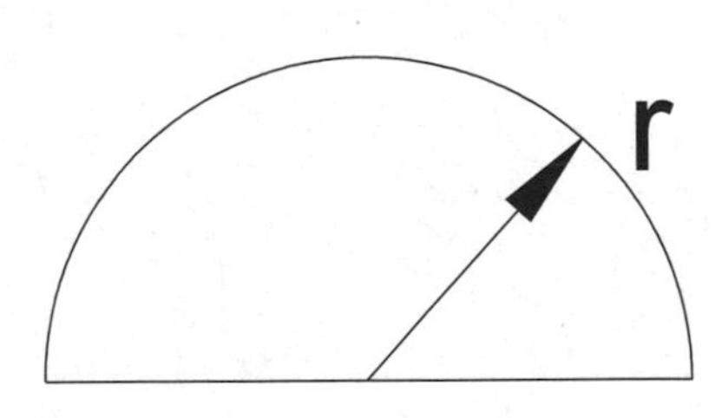


Semicircle

Area = $\frac{\pi r^2}{2}$

Circumference = $\frac{D\pi}{2}$

**Figure A-32** *Semicircle.*

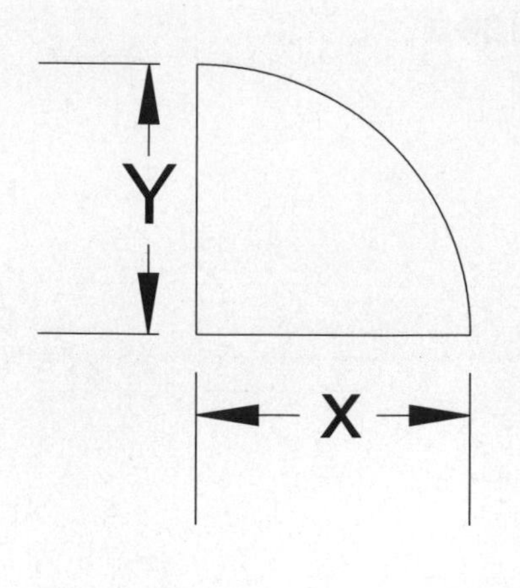


Semiparabolic

$$\text{Area} = \frac{2XY}{3}$$

$$\text{Circumference} = \frac{D\pi}{4}$$

**Figure A-33** *Semiparabolic.*

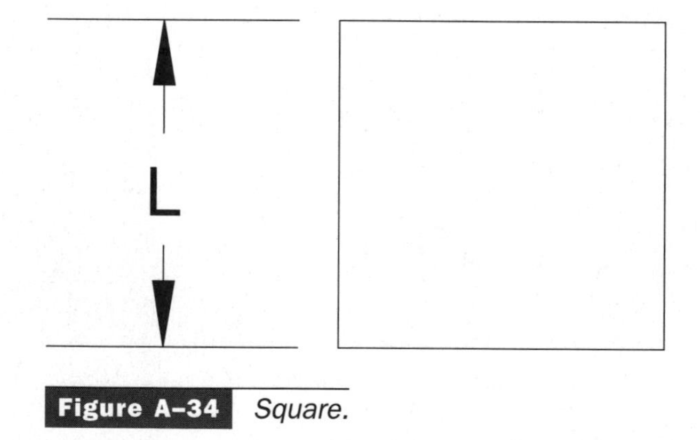


Square

Area = 2L

Perimeter = 4L

**Figure A-34** *Square.*

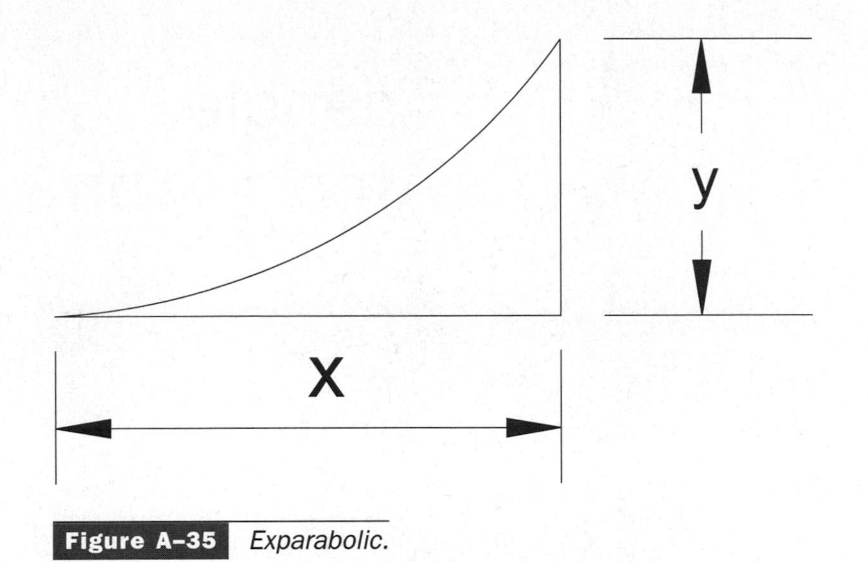


Exparabolic

$$\text{Area} = \frac{XY}{3}$$

**Figure A-35** *Exparabolic.*

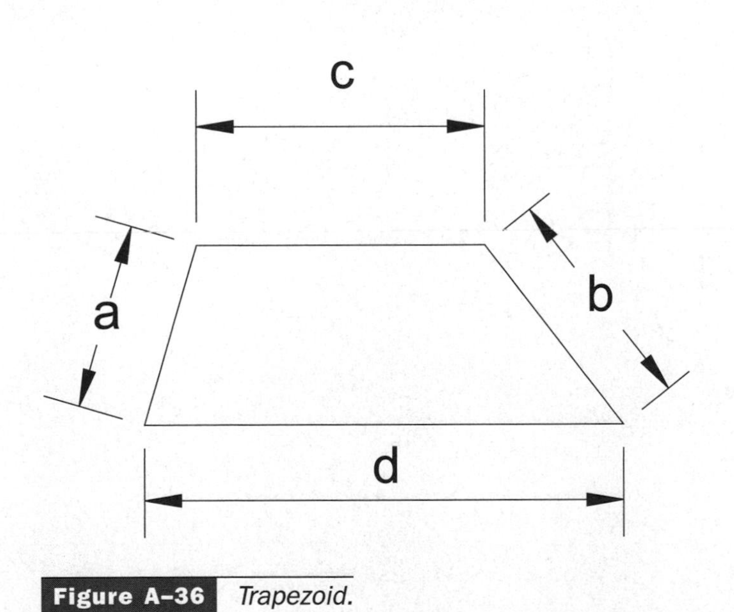


Trapezoid

$$\text{Area} = \frac{1}{2}h(a+b)$$

$$\text{Perimeter} = a+b+c+d$$

**Figure A-36** *Trapezoid.*

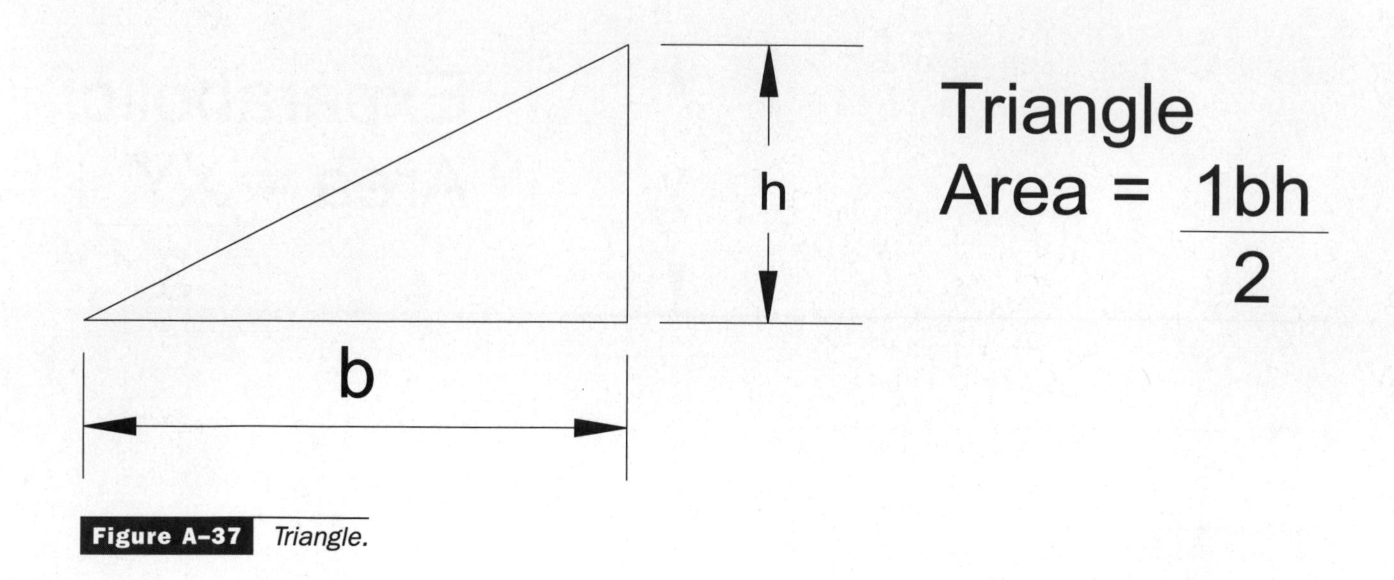


**Figure A-37** *Triangle.*

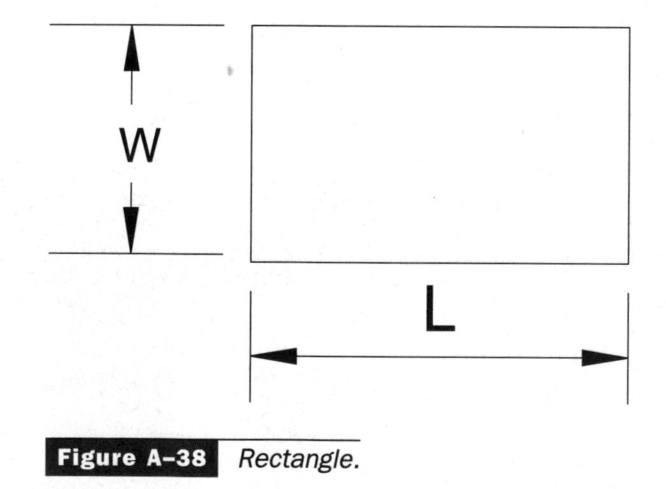


Rectangle

Area = LW

Perimeter = 2L + 2W

**Figure A-38** *Rectangle.*

## Section 4: Bend Allowance Tables

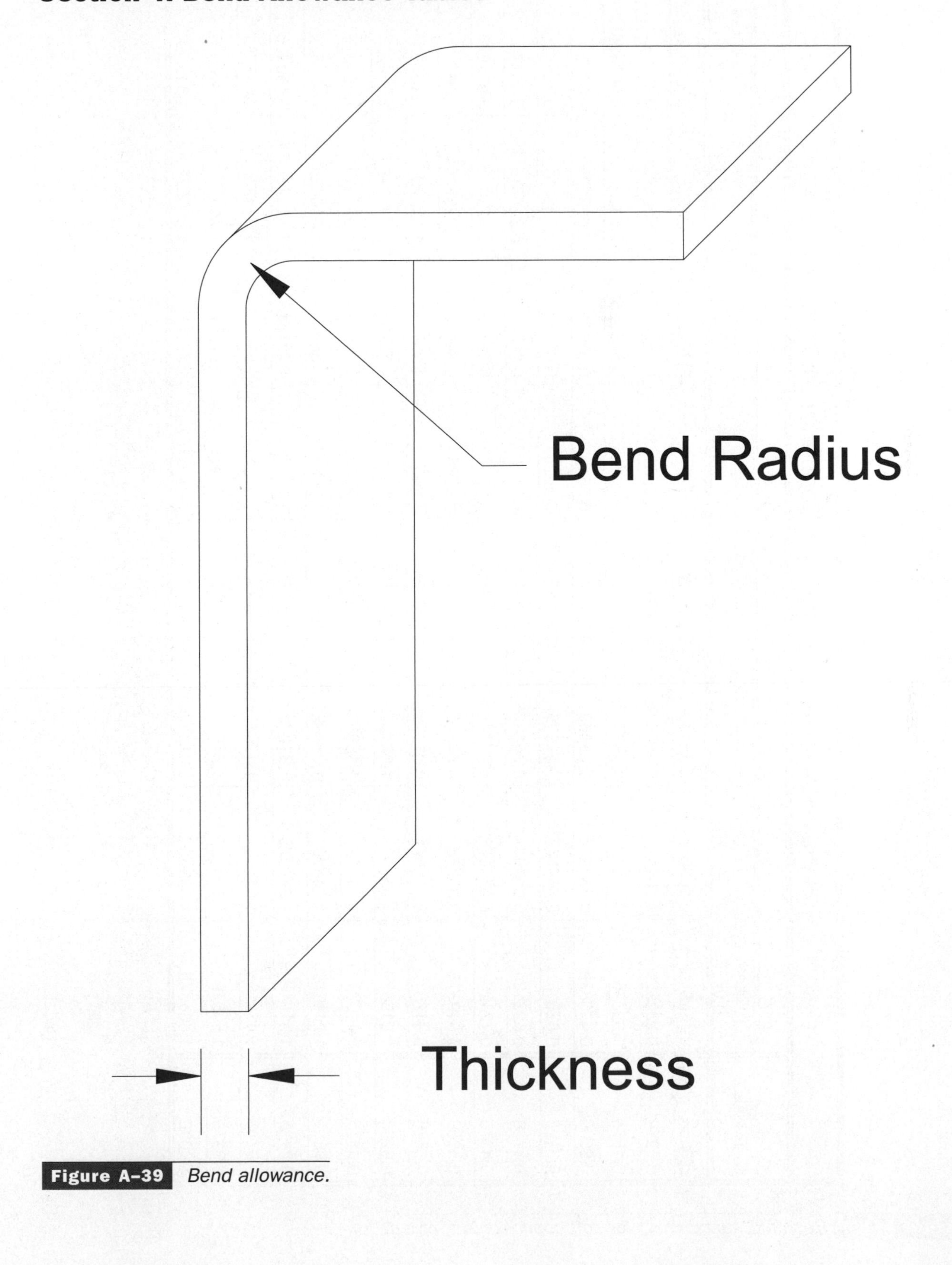


Figure A-39 *Bend allowance.*

## Bend Allowance Chart for Soft Copper or Soft Brass

| Bend Radius | Allowance | | | | | | | |
|---|---|---|---|---|---|---|---|---|
| | Thickness 0.0625 | Thickness 0.125 | Thickness 0.1875 | Thickness 0.25 | Thickness 0.3125 | Thickness 0.375 | Thickness 0.4375 | Thickness 0.50 |
| 0.0625 | 0.1325 | 0.166875 | 0.20125 | 0.235625 | 0.27 | 0.304375 | 0.33875 | 0.373125 |
| 0.1250 | 0.230625 | 0.265 | 0.299375 | 0.33375 | 0.368125 | 0.4025 | 0.436875 | 0.47125 |
| 0.1875 | 0.32875 | 0.363125 | 0.3975 | 0.431875 | 0.46625 | 0.500625 | 0.535 | 0.569375 |
| 0.2500 | 0.426875 | 0.46125 | 0.495625 | 0.53 | 0.564375 | 0.59875 | 0.633125 | 0.6675 |
| 0.3125 | 0.525 | 0.559375 | 0.59375 | 0.628125 | 0.6625 | 0.696875 | 0.73125 | 0.765625 |
| 0.3750 | 0.623125 | 0.6575 | 0.691875 | 0.72625 | 0.760625 | 0.795 | 0.829375 | 0.86375 |
| 0.4375 | 0.72125 | 0.755625 | 0.79 | 0.824375 | 0.85875 | 0.893125 | 0.9275 | 0.961875 |
| 0.5000 | 0.819375 | 0.85375 | 0.888125 | 0.9225 | 0.956875 | 0.99125 | 1.025625 | 1.06 |
| 0.5625 | 0.9175 | 0.951875 | 0.98625 | 1.020625 | 1.055 | 1.089375 | 1.12375 | 1.158125 |
| 0.6250 | 1.015625 | 1.05 | 1.084375 | 1.11875 | 1.153125 | 1.1875 | 1.221875 | 1.25625 |
| 0.6875 | 1.11375 | 1.148125 | 1.1825 | 1.216875 | 1.25125 | 1.285625 | 1.32 | 1.354375 |
| 0.7500 | 1.211875 | 1.24625 | 1.280625 | 1.315 | 1.349375 | 1.38375 | 1.418125 | 1.4525 |
| 0.8125 | 1.31 | 1.344375 | 1.37875 | 1.413125 | 1.4475 | 1.481875 | 1.51625 | 1.550625 |
| 0.8750 | 1.408125 | 1.4425 | 1.476875 | 1.51125 | 1.545625 | 1.58 | 1.614375 | 1.64875 |
| 0.9375 | 1.50625 | 1.540625 | 1.575 | 1.609375 | 1.64375 | 1.678125 | 1.7125 | 1.746875 |
| 1.0000 | 1.604375 | 1.63875 | 1.673125 | 1.7075 | 1.741875 | 1.77625 | 1.810625 | 1.845 |

**Figure A-40** *Bend allowance chart for soft copper or soft brass.*

**Bend Allowance Chart for Aluminum, Soft Steel, Medium Copper and Brass**

| Bend Radius | Allowance | | | | | | | |
|---|---|---|---|---|---|---|---|---|
| | Thickness 0.0625 | Thickness 0.125 | Thickness 0.1875 | Thickness 0.25 | Thickness 0.3125 | Thickness 0.375 | Thickness 0.4375 | Thickness 0.50 |
| 0.0625 | 0.138125 | 0.178125 | 0.218125 | 0.258125 | 0.298125 | 0.338125 | 0.378125 | 0.418125 |
| 0.1250 | 0.23625 | 0.27625 | 0.31625 | 0.35625 | 0.39625 | 0.43625 | 0.47625 | 0.51625 |
| 0.1875 | 0.334375 | 0.374375 | 0.414375 | 0.454375 | 0.494375 | 0.534375 | 0.574375 | 0.614375 |
| 0.2500 | 0.4325 | 0.4725 | 0.5125 | 0.5525 | 0.5925 | 0.6325 | 0.6725 | 0.7125 |
| 0.3125 | 0.530625 | 0.570625 | 0.610625 | 0.650625 | 0.690625 | 0.730625 | 0.770625 | 0.810625 |
| 0.3750 | 0.62875 | 0.66875 | 0.70875 | 0.74875 | 0.78875 | 0.82875 | 0.86875 | 0.90875 |
| 0.4375 | 0.726875 | 0.766875 | 0.806875 | 0.846875 | 0.886875 | 0.926875 | 0.966875 | 1.006875 |
| 0.5000 | 0.825 | 0.865 | 0.905 | 0.945 | 0.985 | 1.025 | 1.065 | 1.105 |
| 0.5625 | 0.923125 | 0.963125 | 1.003125 | 1.043125 | 1.083125 | 1.123125 | 1.163125 | 1.203125 |
| 0.6250 | 1.02125 | 1.06125 | 1.10125 | 1.14125 | 1.18125 | 1.22125 | 1.26125 | 1.30125 |
| 0.6875 | 1.119375 | 1.159375 | 1.199375 | 1.239375 | 1.279375 | 1.319375 | 1.359375 | 1.399375 |
| 0.7500 | 1.2175 | 1.2575 | 1.2975 | 1.3375 | 1.3775 | 1.4175 | 1.4575 | 1.4975 |
| 0.8125 | 1.315625 | 1.355625 | 1.395625 | 1.435625 | 1.475625 | 1.515625 | 1.555625 | 1.595625 |
| 0.8750 | 1.41375 | 1.45375 | 1.49375 | 1.53375 | 1.57375 | 1.61375 | 1.65375 | 1.69375 |
| 0.9375 | 1.511875 | 1.551875 | 1.591875 | 1.631875 | 1.671875 | 1.711875 | 1.751875 | 1.791875 |
| 1.0000 | 1.61 | 1.65 | 1.69 | 1.73 | 1.77 | 1.81 | 1.85 | 1.89 |

**Figure A-41** *Bend allowance chart for aluminum, soft steel, medium copper, and brass.*

**Bend Allowance Chart for Bronze, Hard Copper, Cold-Rolled and Steel**

| Bend Radius | Allowance | | | | | | | |
|---|---|---|---|---|---|---|---|---|
| | Thickness 0.0625 | Thickness 0.125 | Thickness 0.1875 | Thickness 0.25 | Thickness 0.3125 | Thickness 0.375 | Thickness 0.4375 | Thickness 0.50 |
| 0.0625 | 0.1425 | 0.186875 | 0.23125 | 0.275625 | 0.32 | 0.364375 | 0.40875 | 0.453125 |
| 0.1250 | 0.240625 | 0.285 | 0.329375 | 0.37375 | 0.418125 | 0.4625 | 0.506875 | 0.55125 |
| 0.1875 | 0.33875 | 0.383125 | 0.4275 | 0.471875 | 0.51625 | 0.560625 | 0.605 | 0.649375 |
| 0.2500 | 0.436875 | 0.48125 | 0.525625 | 0.57 | 0.614375 | 0.65875 | 0.703125 | 0.7475 |
| 0.3125 | 0.535 | 0.579375 | 0.62375 | 0.668125 | 0.7125 | 0.756875 | 0.80125 | 0.845625 |
| 0.3750 | 0.633125 | 0.6775 | 0.721875 | 0.76625 | 0.810625 | 0.855 | 0.899375 | 0.94375 |
| 0.4375 | 0.73125 | 0.775625 | 0.82 | 0.864375 | 0.90875 | 0.953125 | 0.9975 | 1.041875 |
| 0.5000 | 0.829375 | 0.87375 | 0.918125 | 0.9625 | 1.006875 | 1.05125 | 1.095625 | 1.14 |
| 0.5625 | 0.9275 | 0.971875 | 1.01625 | 1.060625 | 1.105 | 1.149375 | 1.19375 | 1.238125 |
| 0.6250 | 1.025625 | 1.07 | 1.114375 | 1.15875 | 1.203125 | 1.2475 | 1.291875 | 1.33625 |
| 0.6875 | 1.12375 | 1.168125 | 1.2125 | 1.256875 | 1.30125 | 1.345625 | 1.39 | 1.434375 |
| 0.7500 | 1.221875 | 1.26625 | 1.310625 | 1.355 | 1.399375 | 1.44375 | 1.488125 | 1.5325 |
| 0.8125 | 1.32 | 1.364375 | 1.40875 | 1.453125 | 1.4975 | 1.541875 | 1.58625 | 1.630625 |
| 0.8750 | 1.418125 | 1.4625 | 1.506875 | 1.55125 | 1.595625 | 1.64 | 1.684375 | 1.72875 |
| 0.9375 | 1.51625 | 1.560625 | 1.605 | 1.649375 | 1.69375 | 1.738125 | 1.7825 | 1.826875 |
| 1.0000 | 1.614375 | 1.65875 | 1.703125 | 1.7475 | 1.791875 | 1.83625 | 1.880625 | 1.925 |

**Figure A–42** *Bend allowance chart for bronze, hard copper, cold-roll, and steel.*

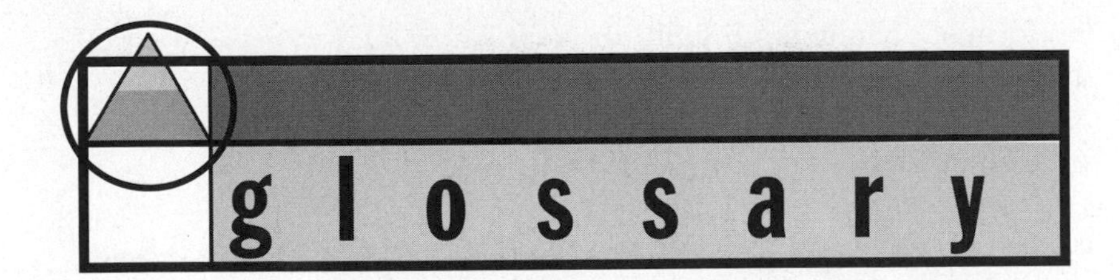


***Acceleration*** is defined as a change in velocity.

***Adjacent views*** are two orthographic views that share a common boundary line (see Figure 1–4).

***Apex*** is the point that is formed by connecting lines from the vertices of the base of a pyramid to a parallel plane.

***Aristotle*** reasoned that the natural state of an object was at rest, and would remain that way unless acted upon by another force or object.

***ASE*** is the link between the AutoCAD technician and the ***ASI*** libraries.

***ASI.ini*** is an ASCII text file that contains configuration information regarding the location and specific settings of the DBMS drivers, as well as the names of the databases to use and their locations.

***Auxiliary plane*** is the plane on which an auxiliary view is projected (see Figure 3–4).

***Auxiliary view*** is a drawing that uses orthographic techniques to project an object onto a plane other than one of the three principal planes. Its purpose is to show the true shape, size, or length of features that would otherwise be foreshortened in the principal views (see Figure 3–2).

***Axis of revolution*** is the axis in which a point is revolved (see Figure 5–3).

***Axonometric projections*** are drawings in which the object is rotated so that all three dimensions are seen at once (length, width, and depth) (see Figure 1–16).

***Azimuth*** is a term used to describe the angle measured clockwise from due north in the horizontal plane (see Figure 2–27).

***Bearing*** is a term used to describe the angle of a line measured from either due north or south in the horizontal plane (see Figure 2–26).

***Bend allowance*** is the amount of extra material needed to produce a particular bend.

***Cabinet oblique*** is a form of oblique projection that is similar to the cavalier oblique drawing because the line-of-sight is drawn at an angle of 45° from the X-axis, but in a cabinet oblique the protruding lines are drawn at a scale of one-half actual size (see Figure 1–23).

***Cavalier oblique*** is a form of oblique projection in which the projection plane is set parallel to the front of the object and the line-of-sight is drawn at a 45° angle from the X-axis. Also, the protruding lines (representing the depth) are drawn full scale (see Figure 1–22).

***Cell*** is the intersection of a row and a column where individual pieces of data reside.

***Change-of-position*** is a method of creating views of an object in which the line-of-sight is repositioned each time a new view is constructed (see Figure 5–1).

***Columns*** are identified with letters and are a vertical group of cells on a spreadsheet.

***Components*** in reference to vectors, are the pieces of data that define the positions of a line. Specifically these are the X-component, Y-component, and Z-component (see Figure 7–4).

***Concurrent noncoplanar vectors*** are a system of vectors that are concurrent, but reside on different planes.

***Concurrent vectors*** are a system of vectors that share a common point.

***Cone*** is a geometric shape consisting of a circular base and a point (located on a parallel plane) called an ***apex***; these are connected by line segments extending from the apex and tangent to the outside edge of the base.

***Coplanar vectors*** are a system of vectors that share the same plane (see Figure 6–24).

***Cosine function*** is defined as the ratio of the side adjacent to an acute angle divided by the hypotenuse, or cosine $A = b/c$ (see Figure 7–3).

***Cylinder*** is defined as two circular regions contained on parallel planes with the lateral surface formed by line segments connecting corresponding points on the circumference of the two circular regions (see Figure 6–12).

***Database*** is a collection of text and numerical data stored in a list created and managed by an application called a ***Database Manager*** or ***DBM***.

***Degree*** is equal to 1/360th of the circumference of a circle (see Figure 7–1).

| | |
|---|---|
| ***Descriptive geometry*** | is the branch of mathematics that precisely describes three-dimensional objects by projecting their essential reality onto a two-dimensional surface such as paper or a computer screen. |
| ***Development*** | is the pattern produced when a three-dimensional geometric shape is unfolded onto a flat surface (see Figure 6–2). |
| ***Diametric projections*** | are drawings that project an object so that only two axes form equal angles, instead of all three like in the isometric projection. |
| ***Dihedral angle*** | is the true angle between two intersecting planes (see Figure 4–57). |
| ***Dynamics*** | is a branch of science concerned with the study of force and its effects on rigid bodies not moving at a constant velocity. |
| ***Equilibrium*** | is a system of forces that is not accelerating. |
| ***Field*** | is a variable used to hold a record. |
| ***First-angle projection*** | is an orthographic projection produced by placing the object in the first quadrant (see Figure 1–10). This results in a drawing containing views that are arranged as follows: |

| | | | |
|---|---|---|---|
| | Bottom | | |
| Right | Front | Left | Rear |
| | Top | | |

| | |
|---|---|
| ***Flat pattern layout*** | is a pattern produced when a three-dimensional geometric shape is unfolded onto a flat surface (see Figure 6–2). |
| ***Flat patterns*** | are the patterns produced when a three-dimensional geometric shape is unfolded onto a flat surface (see Figure 6–2). |
| ***Folding lines*** | show where a plane begins and ends. They are used to transfer measurements from one view to the next (see Figure 1–13). |
| ***Force*** | is a vector quantity inclined to produce a change in the motion and/or the shape of the body on which it acts. |
| ***Foreshortened*** | is a line shown in a way in which the distance between the endpoints is less than it really is. The ***true angle*** is the angle which is formed by a line shown in true length and the edge view of one of the principal planes. |
| ***Free-body diagrams*** | are a diagram showing a force as emitting from a single point. |
| ***Frontal line*** | is a line that appears as true length when the observer's line-of-sight is set perpendicular to the frontal plane (see Figure 2–15). |

***Frontal plane*** is one of the principal planes used in descriptive geometry on which an object is projected (see Figure 4–7). A frontal plane will always be perpendicular to the horizontal and profile planes.

***Galileo*** concluded that bodies in motion had a tendency to stay in motion and bodies at rest tended to remain at rest unless acted on by another force. Galileo called this tendency ***inertia***.

***Gaspard Monge*** a young mathematician that solved three-dimensional spatial problems graphically, and by doing so, fathered the science of ***descriptive geometry***.

***General oblique*** is a form of oblique projection in which the line-of-sight is drawn at an angle other than 45° from the X-axis. Any angle can be used when creating this form of oblique, but the most common angles are 30° and 60°. Any scale can be used to create the protruding lines as long as it uses a scale greater than one-half and less than full (see Figure 1–24).

***Grade*** is the percentage of inclination between the line and the edge view of the horizontal plane. It is defined as the vertical rise of the line divided by its horizontal run, with the quotient multiplied by 100 (see Figure 2–37).

***Ground line*** or ***GL*** is the intersection of the ground plane with the picture plane.

***Ground plane*** or ***GP*** is the edge view of the ground.

***Header*** is a title used to describe the content of a column.

***Height*** is the distance from the apex of a pyramid perpendicular to the base.

***Height of a right cone*** is the distance from the center of the base to the apex.

***Horizontal line*** is a line that appears as true length when the observer's line-of-sight is set perpendicular to the horizontal plane (see Figure 2–16). Horizontal lines are also referred to as level lines.

***Horizontal plane*** is one of the principal planes used in descriptive geometry on which an object is projected (see Figure 4–8). A horizontal plane will always be perpendicular to the frontal and profile planes.

***Inclined line*** is a line that appears inclined in one of the principal planes and parallel to the other two projection planes.

***Inclined plane*** is a plane that is not classified as a normal plane and appears as a line (edge view) in one of the principal views while being distorted in the remaining two (see Figure 4–9).

| | |
|---|---|
| ***Isometric projections*** | are drawings that project an object so that the three principle axes are inclined equally 120° (see Figure 1–16). |
| ***Key*** | is when a group of columns are used in a database table to identify a specific row. |
| ***Line*** | is a non-curved entity that is void of any width, and stretches out to infinitely in both directions; therefore it is completely absent of a starting or ending point. |
| ***Line of intersection*** | is a line formed by the intersection of two planes. Points along this line will be common to both planes. |
| ***Line segment*** | is a portion of a line that is also void of any width, but is constrained by two end points. |
| ***Line-of-sight*** | is the angle from which an observer is viewing an object. |
| ***Locus*** | is used to represent all possible locations of the moving part (each possible point) (see Figure 2–3). |
| ***Miter line*** | is drawn at a 45° angle from the top view's bottom folding line upward to the side view's right folding line (see Figure 1–14). |
| ***Moment*** | is the tendency of a force to cause rotation about a point. |
| ***Multiview drawing*** | is a drawing containing two or more of the six primary views (see Figure 1–12). |
| ***Multiview perspective*** | is a perspective generated from a multiview drawing (see Figure 1–34). |
| ***Newton's First Law of Motion*** | states that *until acted upon by an unbalanced force, a body at rest will remain at rest, and a body in motion will remain in motion at a constant velocity and direction.* |
| ***Newton's Second Law of Motion*** | states that *when an unbalanced force is applied to an object, it causes the object to accelerate. This acceleration is directly proportional to the force applied and inversely proportional to the mass of the object. The object's acceleration will be in the same direction as the force.* |
| ***Newton's Third Law of Motion*** | states that *when an object exerts a force on a second object, the second object will exert a force on the first object that is equal in magnitude and opposite in direction.* |
| ***Noncoplanar vectors*** | are a system of vectors that do not share the same plane. |
| ***Normal plane*** | is a plane that is parallel to one of the three principal planes of projection (frontal, horizontal, profile) (see Figures 4—7, 4–8a and 4–8b). |

***Object lines*** are the contours that define the part (see Figure 1–6).

***Oblique circular cone*** is a cone that when a line is extended from the apex to the center of the base, that line is not perpendicular to the base (see Figure 6–24c).

***Oblique cylinder*** is a cylinder with lateral lines that are not perpendicular to the two circular regions (top and base) (see Figure 6–12b).

***Oblique line*** is a line that does not appear in true length in any of the principal views produced by the principal planes (see Figure 2–19).

***Oblique plane*** is a plane that does not appear in edge view in any of the three principal views.

***Oblique prism*** is a prism with lateral faces that are not rectangular regions (see Figure 6–9).

***Oblique projection*** is a form of drawing in which the projection plane is set parallel to the front surface of the object. However, the line-of-sight is at an angle to the projection plane, producing a view that reveals all three axes (length, width, and depth).

***Oblique regular pyramid*** is a pyramid with lateral regions that do not form isosceles triangles.

***One-point perspective*** is a projection in which all the visual rays converge into a single vanishing point (see Figure 1–31).

***Orthographic projection*** is a drawing containing two or more of the six primary views (see Figure 1–12).

***Parallel-line development technique*** is a technique that positions the object with one of its faces on the development plane.

***Parallelogram method*** is a technique that adds two concurrent vectors by arranging them into a parallelogram to find the resultant of a system of vectors (see Figure 7–5).

***Parametric solid models*** are a type of solid model that are dimension-driven (see Figure 1–37).

***Perspective projection*** is a type of projection in which the projection lines are not parallel (see Figure 1–31).

***Picture plane*** is an imaginary plane on which the object is projected (see Figure 1–31).

***Piercing point*** is the point in which a line intersects a plane (see Figure 2–27).

***Plane*** is a non-curved region in space that may or may not extend indefinitely in two directions. A plane can have length and breadth but will never contain a thickness (see Figure 4–3).

***Point*** is a theoretical exact position in space, which is defined by an X, Y and Z coordinate (see Figure 2–2).

***Polygon method*** is a technique that places vectors end to end, creating a polygon to find the resultant of a system of vectors (see Figure 7–8).

***Primary auxiliary view*** is produced by positioning the auxiliary plane parallel to the inclined surface of the object and perpendicular to one of the principal planes. Its constructions dependent upon two principal views.

***Prism*** is defined as a solid object that contains a base and top that form planes that are parallel to one another, and are identical in size and shape (see Figure 6–8).

***Profile line*** is a line that appears as true length when the observer's line-of-sight is set perpendicular to the profile plane (see Figure 2–17).

***Profile plane*** is one of the principal planes used in descriptive geometry, on which an object is projected (see Figure 4–8b). A profile plane will always be perpendicular to the horizontal and frontal planes.

***Projection plane*** is the current plane in which the projection is taking place (see Figure 1–6).

***Pyramid*** is defined as a polyhedron consisting of a polygonal region as the base and triangular regions for the lateral sides.

***Pythagorean theorem*** states that the square of the hypotenuse of a right triangle is equal to the sum of the square of the other two sides, or $R^2 = X^2 + Y^2$.

***Radial line technique*** involves laying out the true lengths of lines around a radius point for the purpose of creating a development of pyramids and cones (see Figure 6–21).

***Radian*** is an angle opposite to the arc length when that arc length is equal to the radius (see Figure 7–2).

***Record*** is the information entered into a database.

***Resultant*** is the product created by combining two or more components into a single vector.

***Revolution method*** is a technique used for creating views of an object in which the line-of-sight remains stationary and the object is revolved into a new position (see Figure 5–2).

*Revolution plane* is the plane containing a point being revolved (see Figure 5–3).

*Right circular cone* is a cone in which a line has been extended from the apex perpendicular to the center of the base (see Figure 6–24b).

*Right circular cylinder* is a cylinder with lateral lines that are perpendicular to the circular regions (top and base).

*Right prism* is a prism with lateral faces that are rectangular regions (see Figure 6–8).

*Right regular pyramid* is a pyramid with lateral faces that form isosceles triangles, and the base is a polygon.

*Right square prism* is a prism with lateral faces that are rectangular regions and the base and top form a square (see Figure 6–10).

*Rows* are identified by numbers and are a horizontal group of cells on a spreadsheet.

*Scalar* is a quantity in which a magnitude is defined. It is not a direction.

*Secondary auxiliary view* is an auxiliary view produced from a primary auxiliary view and a principal view. It is used to clarify features in a primary auxiliary view.

*Sine function* is defined as the ratio of the side opposite to an acute angle divided by the hypotenuse, or sine $A = a/c$ (see Figure 7–3).

*Sketch plane* is a plane used in mechanical desktop to construct the two-dimensional outline of a part's feature.

*Slant height (cone)* is the distance from the apex to a point on the outside edge of the circular base.

*Slant height (pyramid)* is the height along any of the faces of a pyramid.

*Slope* is the angle formed by the true length of a line and the edge view of the horizontal plane. Slope is always defined in degrees (see Figure 2–35).

*Solid models* are models that contain detailed information concerning the mass of the object.

*Space diagram* illustrates the physical relationship of the object and the physical location of the force.

*Static* is a system of forces that is not accelerating.

*Statics* is a branch of science that studies forces and their effects on rigid bodies at rest or at a constant velocity.

*Station point* or *SP* — is the location of the observer.

*Stretch-out line* — is defined as a line running the length of a prism or cylinder.

*Successive auxiliary view* — is an auxiliary view produced from either a primary auxiliary view and a secondary view or two secondary views.

*Surface* — is a bound region in space that is constrained by two dimensions only. A surface has length and breadth, but does not contain a thickness.

*Surface models* — are similar to wire frame models because they are constructed from three-dimensional entities such as lines, circles, arcs, etc., but differ because these models have a three-dimensional face applied to the object's surface (see Figure 1–36).

*Table* — is the way in which a list is presented.

*Tangent function* — is defined as the ratio of the side opposite divided by the adjacent, tan $A = a/b$ (see Figure 7–3).

*Third-angle projection* — is an orthographic projection produced by placing the object in the third quadrant (see Figure 1–7). This results in a drawing containing the views that are arranged as follows:

| | | Top | |
|---|---|---|---|
| Rear | Left | Front | Right |
| | | Bottom | |

*Triangulation method* — approximates the development by using a series of triangles to generate the development.

*Trigonometry* — is the branch of mathematics concerned with the relationship between the sides and angles of a triangle.

*Trimetric projections* — are drawings that project an object so that none of the three-principle axes form equal angles.

*True length* — is any view of a line that shows the exact distance between the endpoints.

*Truncated objects* — are objects that have been sliced by a plane.

*Two-point perspective* — is a projection of an object that is constructed from two vanishing points.

*Vanishing point* or *VP* — is the point at which all lines collapse into a single point (see Figure 1–32).

*Vector* — is a line that graphically represents magnitude and direction.

***Vector-polygon method*** is a technique for finding the resultant of vectors by arranging them into a three-dimensional polygon.

***Vertical line*** is a line that appears in true length in all views, and in which that line is shown as an elevation (see Figure 2–18).

***Visual rays*** are the lines that are extended from the object to the picture plane.

***Wire frame models*** are a series of three-dimensional points connected by entities such as lines, arcs, and circles (see Figure 1–35).

***Work plane*** is a plane used in mechanical desktop to define a parametric location for a sketch plane.

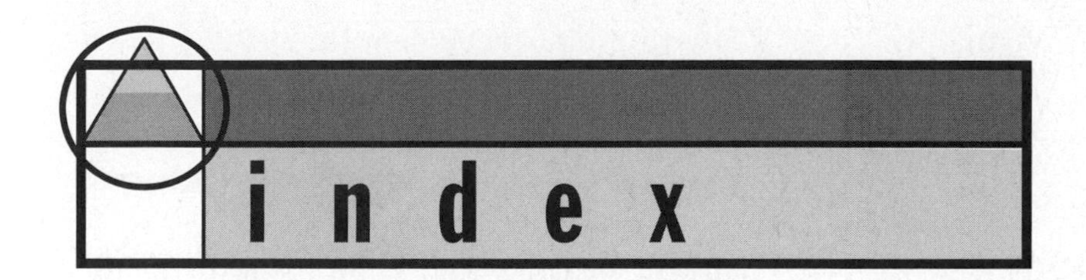

# A

# B



# C

# D

# E



# F

# G



# H



# I

# J



# K



# L

# M

# N



# O

# P

# Q



# R

# T

# Z

# License Agreement for Autodesk Press

## Thomson Learning™

### Educational Software/Data

You the customer, and Autodesk Press incur certain benefits, rights, and obligations to each other when you open this package and use the software/data it contains. BE SURE YOU READ THE LICENSE AGREEMENT CAREFULLY, SINCE BY USING THE SOFTWARE/DATA YOU INDICATE YOU HAVE READ, UNDERSTOOD, AND ACCEPTED THE TERMS OF THIS AGREEMENT.

### Your rights:

1. You enjoy a non-exclusive license to use the enclosed software/data on a single microcomputer that is not part of a network or multi-machine system in consideration for payment of the required license fee, (which may be included in the purchase price of an accompanying print component), or receipt of this software/data, and your acceptance of the terms and conditions of this agreement.
2. You own the media on which the software/data is recorded, but you acknowledge that you do not own the software/data recorded on them. You also acknowledge that the software/data is furnished "as is," and contains copyrighted and/or proprietary and confidential information of Autodesk Press or its licensors.
3. If you do not accept the terms of this license agreement you may return the media within 30 days. However, you may not use the software during this period.

### There are limitations on your rights:

1. You may not copy or print the software/data for any reason whatsoever, except to install it on a hard drive on a single microcomputer and to make one archival copy, unless copying or printing is expressly permitted in writing or statements recorded on the diskette(s).
2. You may not revise, translate, convert, disassemble or otherwise reverse engineer the software/data except that you may add to or rearrange any data recorded on the media as part of the normal use of the software/data.
3. You may not sell, license, lease, rent, loan, or otherwise distribute or network the software/data except that you may give the software/data to a student or and instructor for use at school or, temporarily at home.

Should you fail to abide by the Copyright Law of the United States as it applies to this software/data your license to use it will become invalid. You agree to erase or otherwise destroy the software/data immediately after receiving note of Autodesk Press' termination of this agreement for violation of its provisions.

Autodesk Press gives you a LIMITED WARRANTY covering the enclosed software/data. The LIMITED WARRANTY can be found in this product and/or the instructor's manual that accompanies it.

This license is the entire agreement between you and Autodesk Press interpreted and enforced under New York law.

### Limited Warranty

Autodesk Press warrants to the original licensee/ purchaser of this copy of microcomputer software/ data and the media on which it is recorded that the media will be free from defects in material and workmanship for ninety (90) days from the date of original purchase. All implied warranties are limited in duration to this ninety (90) day period. THEREAFTER, ANY IMPLIED WARRANTIES, INCLUDING IMPLIED WARRANTIES OF MERCHANTABILITY AND FITNESS FOR A PARTICULAR PURPOSE ARE EXCLUDED. THIS WARRANTY IS IN LIEU OF ALL OTHER WARRANTIES, WHETHER ORAL OR WRITTEN, EXPRESSED OR IMPLIED.

If you believe the media is defective, please return it during the ninety day period to the address shown below. A defective diskette will be replaced without charge provided that it has not been subjected to misuse or damage.

This warranty does not extend to the software or information recorded on the media. The software and information are provided "AS IS." Any statements made about the utility of the software or information are not to be considered as express or implied warranties. Autodesk Press will not be liable for incidental or consequential damages of any kind incurred by you, the consumer, or any other user.

Some states do not allow the exclusion or limitation of incidental or consequential damages, or limitations on the duration of implied warranties, so the above limitation or exclusion may not apply to you. This warranty gives you specific legal rights, and you may also have other rights which vary from state to state. Address all correspondence to:

Autodesk Press
3 Columbia Circle
P. O. Box 15015
Albany, NY 12212-5015